John W King

The China Pilot

The coast of China, Korea and Tartary

John W King

The China Pilot
The coast of China, Korea and Tartary

ISBN/EAN: 9783743393394

Printed in Europe, USA, Canada, Australia, Japan

Cover: Foto ©berggeist007 / pixelio.de

More available books at **www.hansebooks.com**

THE
CHINA PILOT.

THE COASTS OF CHINA, KOREA, AND TARTARY
THE SEA OF JAPAN,
GULFS OF TARTARY AND AMÚR,
AND SEA OF OKHOTSK;

AND THE

BABUYAN, BASHÍ, FORMOSA, MEIACO-SIMA, LU-CHU,
LADRONES, BONIN, JAPAN, SAGHALIN, AND
KURIL ISLANDS.

COMPILED FROM VARIOUS SOURCES

By JOHN W. KING, Master, R.N.

THIRD EDITION.

PUBLISHED BY ORDER OF THE LORDS COMMISSIONERS OF THE ADMIRALTY.

LONDON:
PRINTED FOR THE HYDROGRAPHIC OFFICE, ADMIRALTY;

AND SOLD BY

J. D. POTTER, *Agent for the Admiralty Charts,*
31 POULTRY, AND 11 KING STREET, TOWER HILL.
1861.

Price Five Shillings.

A FEW WORDS OF FREQUENT OCCURRENCE IN MAPS AND CHARTS.

Chinese.	English.	Chinese.	English.
Chah	Barrier.	Ma-tau	Jetty, port.
Chah-hwang-muh	Boom.	Miau	Temple.
Chau	District city, islet.		
Chin	Town.	Nan	South, southern.
Chuen	Channel.	Ni	Mud.
Chung-yuen	Mainland.	Nui	Inner.
Fau-fu	Buoy.	Pau-tai	Fort.
Fau-tau	Roadstead.	Peh	North.
Fu	Department city.	Peh	White.
		Po, Hu	Lake.
Gau	Harbour.	Pwang-sheh	Rocks.
		Sha	Sand.
Hai	Sea.	Shan	Hill, mountain.
Hai-kau	Bight, creek.	Shan-hu	Coral.
Hai-kioh	Cape.	Shan-ting	Mountain chain.
Hai-muu	Estuary.	Shan-tau	Bluff, cliff.
Hai-yau	Gulf.	Sha-sien	Shoal.
Hiah-kau	Strait.	Sha-tan	Bar.
Hiang-tsun	Village.	Sheh	Stone.
Hien, Chau	District city.	Sheh-tan	Reef.
Heh	Black.	Shui	Water.
Ho	River.	Si	West, western.
Ho-tun	Lighthouse.	Siau-ho	Rivulet.
Hu	Lake.	So	Town, village.
Hung	Red.		
Hwang	Yellow.	Tah	Pagoda.
		Tau	Island.
Kau	Mouth.	Tau-tu	Clay.
Kiang	River.	Ting, Ti-tau	Promontory.
Kiau	Bridge.	To-muh	Wooded.
King	Capital city.	Tsiau-pi	Cliff.
King-chi-chau	Peninsula.	Tsui-sha	Gravel.
Koh	Rocky peak, headland.	Tsui-wei	Rocky, stony.
		Tung	East, eastern.
Kwang-lau	Lighthouse.	Tutan	Ferry.
Kwan	Custom-house.		
Kuh	Valley.	Wan	Bay.
		Wi-moh ti	Isthmus.
		Wei	Outer.
Lau	Tower.	Wei	Military post.
Lin	Forest.		
Ling	Chain of hills.	Yen-tun	Beacon, buoy.

The vowels are to be sounded as in Spanish or Italian, or as in the following English words:—*a* and *i* as in ravine, *e* as in there, *o* as in go, *u* as in flute, except when followed by *ng*, as in the words Hung (Red), Tung (East), when *u* is to be sounded as in the English word Sung. The letter *h* following a vowel is intended to denote that it has rather a short sound, as Kuh (Valley) is sounded shorter than Fu (a second-class City).

ADVERTISEMENT.

THE Third Edition of the China Pilot contains Sailing Directions for the Coasts of China, Korea, and Tartary; for the Sea of Japan, Gulfs of Tartary and Amúr, and Sea of Okhotsh; and for the off-lying islands and dangers, comprised between the north coast of Luzon and Kamchatka, and long. 150° E. They have been compiled from the following sources:—

The Directions for Canton river, with the islands at entrance, are from the surveys of Lieut. D. Ross, of the Bombay Marine, in 1810, Capt. E. Belcher, C.B., R.N., in 1840, and Capt. W. T. Bate, R.N., in 1857–59; and the Si kiang or West River, from the survey of Lieut. C. J. Bullock, R.N., in 1859.

The coast from Hong Kong to the Yang-tse kiang, including that river and the Chusan archipelago, also Formosa, the Meiaco-sima, and Lu-chu islands, are from the surveys of Capts. Kellett and Collinson, C.B., R.N., in 1842–46. Also from Capt. Basil Hall, R.N., in 1816; Capt. F. W. Beechey, R.N., in 1827; George Johnson, Master, R.N., in 1840; Capt. E. Belcher, C.B., R.N., in 1845; Lieut. D. M. Gordon, R.N., in 1847; the remark books of H.M. ships; and the surveys of John Richards, Master, R.N., in 1854–55, and Comdr. J. Ward, R.N., in 1858–60.

The Yellow Sea, and Gulfs of Pe-chili and Liau-tung, are from the voyage of H.M. Ships *Alceste* and *Lyra*, and the explorations of Lieut. D. Ross in 1816; also from the surveys of Comdr. Ward, Lieut. Bullock, and assistants, and the remark books of H.M. ships up to the present date.

The north coast of the Yellow Sea, and the west and south coasts of Korea, are yet but imperfectly explored. The description is from the voyage of the *Alceste* and *Lyra* in 1816, the ship *Lord Amherst* in 1832, H.M. ships *Blonde* and *Pylades* in 1840, and the French frigate *Virginie* in 1856; Quelpart island and the Port Hamilton group were surveyed by Capt. Belcher in 1845.

The coasts of the Japan Sea, the Gulfs of Tartary and Amúr, and the Sea of Okhotsk, were but little known until visited by H.M. ships in 1854-57. Capt. W. R. Broughton, R.N., in 1797 sailed along the coasts of Tartary and Korea, and discovered the harbour of Chosan, which, in 1858, was surveyed by Comdr. Ward. From hence the description of the Korean coast to D'Anville gulf is from the voyage of the Russian frigate *Pallas* in 1854.

D'Anville gulf, Victoria bay, and the coasts of the Gulf of Tartary, are from the surveys and remark books of officers of H.M. ships in 1854-57; and the Gulf of Amúr from the Russian survey in 1849-54. Siau Wuhu and St. Vladimir bays were surveyed by Comdr. Ward in 1859.

The description of the off-lying islands and dangers are from various sources, but chiefly from Krusenstern in 1804; Basil Hall in 1816; Freycinet in 1819; Beechey in 1827; Siebold in 1828; Belcher and Collinson in 1845; Cecille in 1846; Gordon in 1847; the American Expedition to Japan in 1853-54; Richards in 1855; the remark books of H.M. ships, and documents in Hydrographic Office; Findlay's Pacific Directory; and the Nautical Magazine.

As this work embraces a large extent of coast and many islands and dangers which have been but imperfectly explored, it must necessarily be considered incomplete, and will furnish frequent occasions for revision and amendment. We give greater publicity to these requirements with the hope that Officers, both of the Royal and Mercantile Navy, will transmit to the Secretary of the Admiralty a notice of any errors or omissions they may discover, or any fresh information they may obtain, with a view to its improvement for the general benefit of the mariner.

I. W.

Hydrographic Office, Admiralty, London,
April 1861.

CONTENTS.

CHAPTER I.

MONSOONS, TYPHOONS, GALES, AND TIDES IN THE CHINA SEA, AND ON THE EAST COAST OF CHINA; AND GENERAL REMARKS ON MAKING PASSAGES.

	Page
South-west and North-east monsoons	1–3
Typhoons. Typhoon harbours	3–5
Gales. Currents in both monsoons. Tides	6–9
Passages on East coast of China	9–13

CHAPTER II.

APPROACHES TO CANTON RIVER, INCLUDING HONG KONG.—CHU KIANG OR CANTON RIVER, AND SI KIANG OR WEST RIVER.

Islands and anchorages in Great West channel to Canton river. Broadway river. Typa anchorage. Macao harbour and road. Cum-sing-mun harbour - - - - - - - - 14–21
Islands, channels, and anchorages, from Great Ladrone island to Hong Kong, and to Canton river - - - - - - 21–50
Hong Kong. Tytam bay and harbour - - - - - 33, 34
Lantao island. Cap-sing-mun passage. Urmston bay. Lintin island - 35–38
Directions from Singapore to Hong Kong and Canton river - - 40–50
Canton river. Directions through Boca Tigris and Bremer channel. Dangers in the river between Boca Tigris and Canton. Small bar. First and Second bars. Whampoa anchorage and docks. Canton; exports. Directions from Boca Tigris to Canton - - - 51–58
Si kiang or West river. Channels leading to the Si kiang from Canton river. Shao-king to Wu-chu fu. Tides, currents, and directions - 58–62

CHAPTER III.

EAST COAST OF CHINA.—HONG KONG TO AMOY.

Hong Kong to Breaker point. Tathong channel. Ninepin group. Port Shelter. Rocky harbour. Mirs bay. Tuni-ang group. Bias bay, Harlem bay. Sam-chau inlet. Pedro Blanco rock. Hong-haï bay. Tysami inlet. Chelang point. Hie-che-chin bay. Cupchi point. Tungao road. Breaker point. Tides - - - - 63–83
Breaker point to Amoy. Tong-lae point. Hai-mun bay and river. Hope

vi   CONTENTS.

                                                                                         Page

bay. Cape of Good Hope. River Han. Namao island. Lamock islands.
Challum bay. Shallow bay. Chauan bay. Owick bay. Jokako point.
Brothers rocks. Tongsang harbour. Rees pass. Hu-tau bay. Black
head. Red bay. Chapel island. Merope shoals. Tingtae bay. Tides.
Islands at entrance to Amoy. Amoy island and city. Amoy harbour - 83–104

CHAPTER IV.

EAST COAST OF CHINA.—AMOY TO THE WHITE DOG ISLANDS,
INCLUDING THE PESCADORES.

Amoy to White Dog islands. Quemoy island and bank. Leeo-lu bay.
Hui-i-tau bay. Chimmo bay. Chin-chu harbour. Pyramid point.
Port Matheson. Meichen sound. Pinghai bay. Ockseu islands.
Lam-yit island and channel. Hungwha sound and channel. Hai-tan
island and strait. Turnabout island. White Dog islands - - 105–122
Pescadores islands. Formosa banks 122–129

CHAPTER V.

EAST COAST OF CHINA.—WHITE DOG ISLANDS TO NIMROD SOUND.

River Min to Wan-chu river. River Min. Matsou island. Changchi
island. Alligator island. Ting-hae bay. Sam-sah inlet. Double
Peak island. Fuh-ning bay. Pih-scang islands. Lishan bay. Tae
islands. Pih-quan and Nam-quan harbours. Namki islands. Fong-
whang group; Bullock harbour. Wan-chu river - - - 130–145
Wanchu river to Nimrod sound. Lotsin bay. Kemong harbour.
Taow-pung island. Tai-chau islands, bay, and river. Barren bay.
Montagu island. San-mun bay. Sheipu road and harbour. Kweshan
islands. Harbour Rouse. Buffaloes Nose island. Nimrod sound - 146–157

CHAPTER VI.

EAST COAST OF CHINA.—NIMROD SOUND TO THE YANG-TSE KIANG,
INCLUDING THE CHUSAN ARCHIPELAGO.

Chusan archipelago. Luhwang island. Duffield, Gough, and Roberts
passes. Beak head, Vernon, and Sarah Galley channels. Chusan
island. Winds and weather. Ting-hai harbour. Directions to Ting-
hai harbour through Tower Hill, Melville, and Deer island channels.
Anchorage under Elephant island. Ching Keang harbour. Silver
island. Ta-outse harbour. Shaaon harbour. Directions from Shaaon
harbour through Kwei channel, and channel north of Lan-sew.
N.E. and East coasts of Chusan. Chinkeamun harbour. Lan-sew
bay. Video island. Fishermans group. Volcano and Saddle islands.
Parker, Rugged, and Gutzlaff islands. Tides - - - - 158–184
Kintang channel. Just-in-the-Way rock. Yung river. Ning-po fu
Tsie-kie. Yuyao; Directions from Ting-hai harbour to Yung river
through Kintang channel 184–187

River Yung to the Yang-tse kiang. Hang-chu bay. Seshan islands. Chapu bay. The Yang-tse kiang. Tides. Directions to the Yang-tse kiang east of Chusan archipelago. Directions for Wusung river - 188–201
Yang-tse kiang; Wusung to Hankau; Harvey point; Lang-shau crossing; Silver island. Nanking. Han-kau. Directions from Han-kau to Wusung - - - - - - - - 201–216

CHAPTER VII.

EAST COAST OF CHINA.—WHANG-HAÏ OR YELLOW SEA; GULFS OF PE-CHILI AND LIAU-TUNG; AND WEST AND SOUTH COASTS OF KOREA.

The Yellow sea. Whang ho, or Yellow river. Kyau-chu, or Glue city. Shan Tung. Weï-haï-weï or Oïe-haï-oïe harbour. Chi-fau or Yen-tai harbour. Teng-chau fu. Miau-tau islands and strait - - 217–231
Gulf of Pe-chili. South coast, from Miau-tau strait to the Peh-tang ho. Sha-lui-tien islands and banks. The Tien-tsin ho or Pei ho; tides. Winds and weather. Climate - - - - - 231–236
Gulf of Liau-tung. Port Adams. Hulu shan bay. Fu-chu bay. Kai-chu fu. Niuchwang fu. Great Wall of China. Ning-hai. Coast between the Great Wall and the Pei ho. The Peh-tang ho - 237–246
North coast of Yellow sea. Liau-tie-shan head. Encounter rock. Ta-lien-hwan bay. Blonde island - - - - - 246–250
West and south coasts of Korea. Chodo island. Joachim, Caroline, and Deception bays. Daniels island. Sir James Hall group. Marjoribanks harbour and Shoal gulf. Basil bay. Korean archipelago. South coast of Korea. Quelpart island. Port Hamilton - - 250–265

CHAPTER VIII.

PRATAS ISLAND AND REEF; NORTH COAST OF LUZON; AND BABUYAN, BASHÍ, FORMOSA, MEIACO-SIMA, AND LU-CHU ISLANDS.

Pratas island and reef; tides, anchorage, and caution - - - 266–268
North coast of Luzon. Port San Vicente. Directions for west coast of Luzon - - - - - - - - - 268–270
Babuyan islands. Musa bay. Port San Pio Quinto - - - 271–273
Batan or Bashi islands. San Domingo bay - - - - 274–277
Gadd Rock. Vele-Rete rocks. Botel-Tobago sima. Little Botel-Tobago sima - - - - - - - - - 278, 279
Formosa or Taï-wan island. West coast of Formosa. Lambay island. Leang-kiaou bay. Pong-li. Port Ta-kau-kon. Vuyloy shoal. Port Kok-si-kon. North coast of Formosa. Tam-siu, Ke-lung, and Coal harbours. East coast of Formosa. Kaleewan river. Sau-o bay. Chock-e-day village. Black rock bay - - - - - 279–294
Samasana island. Harp island. Pinnacle, Craig, and Agincourt islands. Hoa-pin-su, Pinnacle, and Ti-a-usu islands. Raleigh rock - - 294–296

viii   CONTENTS.

Page
Meiaco-sima group, Kumi island. Ku-kien-san, Pa-chung-san, and
  Taï-pin-san islands. Broughton bay. Port Haddington. Port Had-
  dington to Taï-pin-san   -   -   -   -   296–301
Lu-chu or Liu-kiu islands. Napha-kiang road. Directions from Hong
  Kong to Lu-chu islands. Directions into Napha-kiang road. Deep
  bay. Suco island. Port Onting or Melville. Shah bay. Barrow bay 301–309

CHAPTER IX.

ISLANDS SOUTH-EAST, EAST, AND NORTH OF THE LU-CHU GROUP; AND OFF THE SOUTH-EAST COAST OF NIPON.

Caution to be used in navigating between Formosa and the Mariana and
  Bonin islands -   -   -   -   -   -   -   310
Island south-east of Lu-chu. Borodino islands. Bishop rocks. Rasa
  and Kendrick islands. Douglas reef. Lindsay island. Mariana or
  Ladrones islands   -   -   -   -   -   -   310–320
Islands east of Lu-chu. Los Jardines, Sebastian Lobos, Forfana, Vol-
  cano, Mal abrigo, Arzobispo or Bonin islands -   -   - 320–324
Islands north of Lu-chu. Yori, Yeirabu, Tok, Iwo, Oho, and Kikai
  islands. Germantown reef. Sandon rock. Linschoten or Cecille
  archipelago. Powhattan reef. Trio rocks. Kuro sima. Ingersoll
  rocks   -   -   -   -   -   -   -   324–328
Islands off south-east coast of Nipon. Lots wife rock. Ponafin, Smith,
  Bayonnaise, Onanga and Fatsiziu islands. Broughton rock. Meac
  and Mecoura islands. Redfield rocks. Kozu, Sikini, Nee, Utoma,
  To, and Oho islands. Portsmouth breakers -   -   -   328–330
Winds and weather and typhoons between Formosa and the Bonin and
  Japan islands -   -   -   -   -   -   -   331, 332

CHAPTER X.

JAPAN AND KURIL ISLANDS, AND SOUTH-EAST COAST OF KAMCHATKA.

Japan islands; description; climate; winds -   -   -   333–365
Kuro-siwo or Japan stream. Van Dieman strait -   -   335, 336
South-east coast of Nipon. Ohosaka bay. Port Fiogo. Enora, Heda,
  Arari, and Tago bays. Simoda harbour. Yedo bay -   -   337–345
East coast of Nipon. Directions from Yedo bay to Tsugar strait   - 346
West coast of Nipon. Bittern rocks. Oga-sima peninsula. Tabu and
  Sado islands. Port Niegata. Yutsi sima. Cape Noto. Oki and
  Mino islands. Cape Louisa -   -   -   -   -   346–350
West coast of Kiusiu. Colnet, Obree, Wilson, Ykitsk, Harbour, and
  Firado islands. Meac-sima group. Pallas rocks. Udsi sima. Re-
  tribution rocks. Kosiki and Tsukurase islands. Nagasaki harbour - 351–358
Tsugar strait. Hakodadi harbour. Currents, tides, and directions - 358–364
East coast of Yezo. Volcano bay and Endermo harbour   -   - 365
Kuril islands   -   -   -   -   -   -   -   365–370
South-east coast of Kamchatka. Avatcha bay. Petropaulski harbour - 370–375

CHAPTER XI.

SEA OF JAPAN; GULF OF TARTARY; GULF AND RIVER AMÚR SAGHALIN ISLAND; LAPÉROUSE STRAIT; AND SEA OF OKHOTSK.

	Page
Sea of Japan. Winds and currents. Sentinel island. Tsus sima. Matu sima. Tako sima. Liancourt and Waywoda rocks	376–379
East coast of Korea. Chosan harbour. Unkofsky bay. Ping-hai harbour. Broughton bay. Yung-hing bay. Port Lazaref. Goshkevich bay. Tumen river	379–391
Coast of Tartary. D'Anville gulf. Capricieuse bay. Napoléon road. Port Louis. Gulf of Guerin. Port Bruce. Napoléon gulf. Hornet bay. Port Michael Seymour. St. Vladimir, Shelter, Sybille, Pique, and Bullock bays	392–400
Gulf of Tartary. Winds, fogs, and tides. Suffren bay. Fish river. Barracouta harbour. Castries and Jonquière bays	400–406
Gulf and River Amúr	406–408
Lapérouse strait. Aniwa bay. Tides and currents	409–412
East and north coasts of Saghalin island	412–416
Sea of Okhotsk. Salutation bay. Port Aian. Okhotsk harbour. Currents	417–420
Tide Table and Table of Positions	421–430

The South-west monsoon is strongest, and least liable to change, in June, July, and August, at which period there is at times much rain and cloudy weather all over the China Sea; in these months, and also in May, sudden hard squalls blow sometimes out of the Gulf of Siam,* as far as Pulo Condore and Pulo Sapata. When dense clouds are perceived to rise, indicating the approach of these squalls, sail ought to be reduced without delay.

From the Gulf of Siam to Cape Padaran, the South-west monsoon blows nearly parallel to the coast; and if close in, a light wind from the land is at times experienced in the night, succeeded by a short interval of calm on the following morning. The monsoon breeze then sets in, and generally continues brisk during the day. These land and sea breezes prevail most on the coast of Cochin China, from Cape Padaran northward to the Gulf of Tong King; for the sea wind dies away almost every evening on this coast during the South-west monsoon, and a land breeze comes off in the night, although not at a regular hour. This is followed by calms or light airs, which frequently continue until noon; the sea breeze then sets in from the south-eastward.

In March and April there are land and sea breezes on the coast of Luzon, with fine weather; but after the South-west monsoon sets in strong in June, and from that time until it abates in October, the weather is mostly cloudy; and the winds blowing from the sea upon that coast are generally accompanied with much rain.

On the south coast of China, the winds during the South-west monsoon prevail frequently at South and S.S.E. About Formosa, and between it and the coast, north-easterly winds often happen in July, August, and September.

The **NORTH-EAST MONSOON** usually begins in the northern part of the China Sea about the end of September or early in October; but in the southern part it seldom blows steadily till November; light southern or variable breezes prevailing the greater part of October. This monsoon generally sets in with a storm, which sometimes comes down without warning, and with a violence that has exposed several vessels to great danger: therefore when the monsoon is about to change, they should avoid anchoring in exposed positions, and weigh instantly if caught, as the swell rises with such rapidity as to cause a difficulty in getting the anchor. The first burst of the monsoon frequently lasts a week or 10 days. The weather in some years is settled and fine, during September

---

* For the winds and weather in the Gulf of Siam, see China Pilot, Appendix No. 1, page 20.

and October; but the equinox is a very precarious period, for within a few days of it storms are likely to happen.

In November, the North-east monsoon generally prevails; but it blows more steadily, and with greater strength, in December and January. The weather is frequently cloudy, with much rain and a turbulent sea, in these months; particularly about Pulo Sapata, and from thence to the entrance of Singapore strait; there are also considerable intervals of fine weather. On the coast of Palawan the winds are variable in October, November, and the early part of December, by which vessels pass along that coast either to the north-east or south-west, but the weather is often dark, rainy, and cloudy. On the coast of Luzon the winds are frequently variable during this monsoon, generally from the northward and north-eastward; but they veer to the north-westward and westward at times, and then blow strong, with cloudy weather and rain. In the Gulf of Tong King, in November, there are sometimes faint land breezes close to the coast; but the North-east monsoon prevails along the coast of Cochin China, as far southward as Cape Padaran, generally from September or the early part of October, to the beginning or middle of April.

In February the strength of the North-east monsoon abates. During this month and March it blows moderately, with steady weather all over the China Sea, inclining to land and sea breezes on the coast of Luzon. On the south coast of China, when the North-east monsoon prevails, the winds blow mostly from E.N.E. parallel to the shore; they veer, and blow off the land at times, and also from the south-eastward, but there are seldom any regular land or sea breezes on that coast.

Strong gales are frequently experienced on the eastern coast of China, in the North-east monsoon with a rising barometer; and a low barometer in the same season will generally be found to indicate a southerly wind. The barometer differs as much in some years as an inch during the two monsoons. The mean barometer from October 1847 to March 1848 was 30·33; while from April to September 1848 it was only 29·53. During the height of the monsoon the wind at night will be found frequently to draw off the land, affording an opportunity for a vessel to make a good board.

TYPHOONS.—These dangerous tempests occur in the northern part of the China Sea, near Formosa, the Bashí islands, and the north end of Luzon; also to the eastward of those islands, and between Formosa and the Japan archipelago. They do not extend into the Formosa channel, there being no record as yet of any having reached Amoy, but they are found to the northward of Formosa; one having visited the Chusan archipelago in 1843, much to the astonishment of its inhabitants, who stated that such a thing had not occurred for 50 years. Some violent storms have also visited Shanghai, but it is yet doubtful whether these are Cyclones or not. To the eastward of Formosa they extend as far as the Bonin islands; but the high mountain chain which runs nearly the whole length of the former island, and rises from 5,000 to 10,000 feet above the sea, must divert, not only their curve, but their direction. They usually blow with the greatest fury near the land: as the distance is increased to the southward from the coast of China, their violence generally abates, and they seldom reach beyond lat. 14° N., although a severe gale has been experienced at times two or three degrees farther to the southward.

These tempests are liable to happen in both monsoons; but they are usually less severe in the China Sea, if they occur in May, November, or December; although in the vicinity of Formosa and the Bashí islands there are sometimes furious gusts in November. From December to May they seldom or never happen; of late years, those that have been experienced in June and July were the most violent; many vessels have been dismasted and sustained other damage by them. The months of August, September, and October are also subject to these tempests. The September equinox is a very precarious period, particularly if the change or perigee of the moon coincide with the equinox: when this was the case, Typhoons happened several years at the equinox in September, on the coast of China, and many ships were dismasted on the 21st or 22nd of that month.

To be able to prognosticate the coming of these tempests would be very useful to navigators, but this cannot be done with certainty, for they frequently commence without giving much indication of their approach. The clouds having a red aspect is not a certain warning of the approach of a Typhoon; for, at the rising, but more particularly at the setting of the sun, the clouds, especially those opposite to it in settled weather, are sometimes tinged with a deep red colour by the reflected light. Neither is an irregular swell a good criterion to judge of their approach; for near the coast of China a cross swell frequently prevails during steady settled weather. A hazy atmosphere, preventing land from being seen at great distances, is no unfavourable sign on the coast of China; for this is generally its state in medium or settled weather.' A serene sky, with the horizon remarkably clear, should not be considered an indication of a continuance of favourable weather; for a series of fine weather and calms, favouring an increase of heat above the mean temperature, is likely to be succeeded by a Typhoon. When the horizon is very clear in some parts, and the summits of the hills or islands obscured by dense black clouds, there is some irregularity in the atmosphere, and stormy weather may be apprehended; but in reality, Typhoons are seldom preceded by any certain sign or indication. Marine barometers seem to afford the best means of anticipating their approach ; for, on the south coast of China, there is a greater fall of the mercury than might be expected within the tropics.

Many vessels have been driven from the entrance of Canton river to the Mandarins Cap, and even to the Taya islands near Hainan, during Typhoons ; for among the islands, and near this coast, these tempests generally commence between N.W. and North, then veer suddenly to N.E. and East, frequently blowing with inconceivable fury, and raising the sea in turbulent pyramids, which impinge violently against each other ; the current at such times runs strong to the westward. From eastward, the wind veers to the south-eastward and southward, and then becomes moderate.

In some years, no Typhoon happens on the south coast of China ; at other times, two or three of these storms have been experienced in one year ; but fortunately their fury is seldom of long continuance. There is, however, now considerably less danger in meeting these furious tempests, owing to the skill and research of Mr. W. C. Redfield, Colonel Sir William Reid, Mr. H. Piddington, and others who have collated a vast number of facts bearing upon the subject. It is found that their progress is, governed by a general law, and consequently the vortex can be avoided, and the vessel's safety assured by attention to a practical rule, which is this :*—Look to the wind's eye,—set its bearing by the compass,—and the 8th point to the *right* thereof, in North latitude will be the bearing of the centre of the storm. For example, suppose the vessel to be in lat. 18° N., the wind East, and the barometer and sky indicating a coming gale,—then, look at the compass, take the 8th point to the *right* of East, and South is the bearing of the brewing storm, *if it be* of a revolving type. In this case, the vessel will be on the northern part of the storm-field.

In the northern part of the China Sea, a low barometer for several days previous, an ugly threatening appearance, and heavy swell, will give sufficient warning, and, provided it be taken, will enable vessels to get sufficient sea room so as to avoid the centre of the storm, or to secure safe anchorage.

TYPHOON HARBOURS.—The following is a list of anchorages on the east coast of China where vessels will lie secure in a Typhoon or veering gale :—Tam-tu island ; Mirs bay ; Ty-sami inlet (for 12 feet draught) ; Namoa island (abreast Stewart's house) ; Tongsang harbour ; Amoy

---

* *See* Remarks on Revolving Storms, published by order of the Lords Commissioners of the Admiralty, 1853 ; also the Horn Book of Storms by H. Piddington, Esq., 1848.

harbour ; Quemoy island ; Pescadores islands (Makung harbour) ; Chinchew harbour (within the Boot sand); Hungwha sound ; southern entrance to Haetan strait ; Pih-quan harbour ; Bullock harbour ; Kelung harbour (Formosa) ; Chusan archipelago (Ting-hai outer and inner harbours, Chinkeamun and Chin Keang harbours, Fisher or Chang-pih island, and Ta-outse on the north-west side of Kintang).

GALES sometimes blow steadily from E.N.E. or N.E. several days at a time, in September or October, near the south coast of China. In the same months they are liable to happen on the west coast of Luzon. Here, they mostly commence at North or N.W., and veer to West, S.W., or South, blowing strong from all these directions, with heavy rain, and a cross turbulent sea ; but they seldom continue long.

Strong N.E. gales have been sometimes experienced on the coast of China during the South-west monsoon ; in one of these a vessel, after making the Great Ladrone July 16th, 1802, was driven by the 20th westward to the Mandarins Cap, with strong gales, hard squalls, and the current setting from 1 to 2 knots per hour westward. The north-east wind continued nine days, which obliged her to stand out to sea, and she did not arrive at Macao until the 26th.

In May, June, July, and August, severe gales are at times experienced in the north-western part of the China Sea, particularly between lat. 14° N. and Haïnan island, with the gulf of Tong King open. These gales generally begin at N.N.W. or N.W. and blow with violence out of the gulf, accompanied by dark weather and a deluge of rain : from N.W. they veer to West and S.W., still blowing strong, and abate as they veer more southerly. When these N.W. gales are blowing in the vicinity of Haïnan and the coast of Cochin China, strong S.W. or southerly gales generally prevail at the same time, in the middle of the China Sea.

CURRENTS in SOUTH-WEST MONSOON.—The currents in the China Sea are very changeable, their direction and velocity depending much upon local circumstances. Late in April, or early in May, they generally begin to set to the northward, in the south and middle parts of the sea, and continue to run in a north-easterly direction until September, while the South-west monsoon is strong ; but they are not constant in this monsoon, for at times, when the wind is moderate or light, the currents are liable to change and set in various directions. After the strength of the monsoon has abated, there is often little or no current in the open sea, running to the north-eastward ; but sometimes its direction is to the southward.

Along the coast of Cochin China, from Pulo Obi to Cape Padaran, the current sets mostly to the E.N.E., parallel to the shore, from April to the middle of October; and during the same period its direction is generally to the northward along the east coast of the Malay peninsula, from the entrance of Singapore strait to the Gulf of Siam. To the northward of Cape Padaran there is but little current in the South-west monsoon, near the Cochin-China coast; for, from thence to the Gulf of Tong King, a small drain is sometimes found setting to the northward, or at other times to the southward. When a gale happens to blow out of the latter gulf from the north-west and westward, the current at the same time sets generally to the south-west or southward, in the vicinity of the Paracel islands and reefs, or where these gales are experienced; and this current running obliquely, or contrary to the wind, a turbulent and high sea is thereby produced.

On the south coast of China the current is much governed by the wind; when strong S.W. winds prevail, it runs along shore to the eastward, but seldom strong. Near, and amongst the islands, westward of Macao, there is generally a westerly current, occasioned by the freshes from Canton river, which set in that direction; frequently sweeping along the islands from Macao to St. John between W.S.W. and W.N.W., about 1 or 2 knots per hour. This westerly current is, however, not always constant in the South-west monsoon, for it slacks at times; then a weak tide may sometimes be experienced running eastward.

On the coasts of Luzon and Palawan, the current generally sets northward in the South-west monsoon, but frequently there is no current, and near these coasts it seldom runs strong. Near the Bashí islands, it sometimes sets to the eastward when strong westerly winds prevail; but generally strong to the northward, or between N.N.W. and N.E.

The strength of the current on the eastern coast of China increases with the freshness and duration of the monsoon, varying from one to as much as 3 and even 4 knots per hour; and 'this requires to be especially guarded against when hove-to off a port or running for one in thick weather. Thus a number of vessels in the South-west monsoon have run into Hu-i-tau bay instead of Amoy; and again in the North-east monsoon have picked themselves up off Red bay instead of Chapel island. The current will slack a little at particular times of tide, but during the survey of this coast in 1843, it was seldom found to run to the south in the southerly monsoon, or to the north in the other. At the Pescadores islands, in the month of August, a current was sometimes experienced of 4 knots per hour, running to the north, whilst with the ebb it slackened for two or three hours, but seldom ceased entirely.

CURRENTS in NORTH-EAST MONSOON.—The current in the China Sea, during the North-east monsoon, generally, runs to the south-westward, with a velocity depending on the strength of the wind. When the force of the monsoon is abated, or during moderate and light breezes, there is often little or no current.

In the western parts of the China Sea, along the coasts of Cochin China and the Malay peninsula, the current generally begins to run to the southward about the middle of October (sometimes sooner on the former coast); and continues until April. During the month of March its direction is constantly to the southward about Pulo Aor, with light easterly winds and calms at times. On the coast of Cochin China, and adjacent to Hainan island, a current varying from South to S.W., commences sometimes about the middle of September; near the land, from lat. 15° N. to 11° or $11\frac{1}{2}$° N., it increases in strength; but its rate decreases in proportion as it flows southward. During the prevalence of the North-east monsoon, from about lat. 14° N. to Cape Padaran, the current near the coast frequently runs 40 or 50, and sometimes 60 miles to the southward in 24 hours; the rate, however, is variable, and it is only in the limits above mentioned that it is occasionally so strong, for its strength abates at Cape Padaran, and runs with less velocity to the S.W., towards the entrance of the Gulf of Siam.

On the south coast of China, the current during the North-east monsoon runs almost constantly to the W.S.W., nearly parallel to the land; and sometimes with inconceivable rapidity, when a Typhoon or a storm happens. At the distance of 20 or 30 leagues from the coast, it seldom runs so strong as near it; and in 30 or 40 fathoms soundings there is much less current than in shoal water, near the shore and amongst the islands. The westerly current sometimes slacks, and, contiguous to the land, is succeeded by a kind of tide.

Between the island of Formosa and the China coast the current runs to the southward during the North-east monsoon. When strong N.E. winds prevail, its direction is generally to the S.W. or southward, between the south end of Formosa and the north end of Luzon; but here, in light variable winds, it often sets to the northward. On the west coast of Luzon the current is changeable, sometimes setting southward along the coast, at other times northward. On the coast of Palawan it is also mutable, governed by the prevailing winds, but seldom runs strong in any direction, unless impelled by severe gales. To the eastward of Formosa, about Boteltobago sima, it frequently runs strong to the northward and north-eastward, so early as the 1st of March; and although changeable at times, it sets mostly in that direction during the South-west monsoon; and in the opposite direction during the North-east monsoon.

TIDES.—The tidal wave strikes upon the eastern coast of China, from Hong Kong to the Yang-tse kiang, nearly at the same period; it being high water on full and change days in the neighbourhood of the Lema islands, at about 8h. 30m., and at the outer islands of the Chusan archipelago it is an hour later. The rise and fall, however, increases considerably to the northward; probably owing to the obstruction which the wave receives from the Philippine islands; and in some instances the diurnal inequality is great. The establishment for Hong Kong and other places along the coast is given in the body and at the end of this work, and it will be perceived that to the eastward of Hong Kong, and as far as Breaker point, the tides are irregular and weak; the current occasioned by the monsoon overcoming them.

After passing Breaker point, the coast trends more northerly, and the flood stream will be found useful to vessels bound to the northward. The rise and fall increases, passing from 7 feet at Namoa island to 12 feet at Tongsang, and 20 feet at Amoy. Between Amoy and the river Min, the rise of the tide varies from 16 to 18 feet at the springs, and the flood enters on the north as well as on the south side of Hai-tan strait.

To the northward of the Min, the flood sets more determinately to the north; it seldom, however (unless off headlands or in narrow channels), overcomes the current caused by the monsoon, but has the effect of slackening it.

Throughout the Chusan archipelago and the estuaries to the north, great care and attention to the tides is necessary; particular instructions for this purpose will be found in the body of the work; and it only remains here to caution the navigator that, as his vessel approaches the coast to the northward of Chusan the tides increase in rapidity, and unless precaution is taken, she will be set among the small islets of this rugged archipelago.

GENERAL REMARKS ON MAKING PASSAGES ON EASTERN COAST OF CHINA.

PASSAGE EAST of FORMOSA.—A vessel bound from Hong Kong to Ning-po, or Shanghai, or even to Fu-chau fu, during the North-east monsoon, should be in good condition for contending with rough weather and for carrying sail. The best plan appears to be, to work along the coast as far as Breaker point, and then stretch across to the south end of Formosa, and work up eastward of that island. By remaining in with the coast of China, she will have the advantage of the land wind at night, of smoother water, and the ebb-tide out of the deep bays, which will generally be under her lee on the starboard tack, and in the event of its blowing too hard to make weigh, there are numerous convenient anchorages. It will be prudent to keep within 10 miles of the coast, to avoid being swept to the southward whilst standing off the land; but as this cannot be done at night without risk, a vessel should, if possible, anchor in the evening, and weigh in the middle watch, when the wind, generally coming more off the land, will enable her to make a good board on the off shore tack. By passing eastward of Formosa, also a heavy short sea in the Formosa channel will be avoided, as well as the constant set to the southward during this season.

After rounding the south end of Formosa, off which there is generally a troublesome sea, a vessel should make short tacks, if requisite, to keep within the influence of the Kuro-siwo or Japan stream (page 335), which has sometimes been found running to the northward at the rate of 30 or 40 miles per day.

There are no harbours on the east coast of Formosa, except Sú-au bay, towards the north end of the island, and deep water will be found close to the land. The mountains rise almost immediately from the sea; their sides in some places are cultivated, and a good many houses will be seen. H.M.S. *Plover* anchored on an uneven bottom in Black Rock bay, the vessel swinging from 13 to 22 fathoms, and rode out a gale from the S.W.; but it is by no means to be recommended.

Having weathered the north end of Formosa, it will be still advisable to keep to the eastward, and not approach the continent until the parallel of lat. $30\frac{1}{2}°$ N. is gained. Should, however, a vessel be driven to the westward, she may always calculate on smooth water, and be able to tide it through the southern part of the Chusan archipelago; and if disabled and in want of spars, she can remain at the southern side of Duffield Pass, and supply herself from the Fu-chau wood junks.

AMOY to RIVER MIN.—If bound to Amoy and the ports between that place and the river Min, a vessel will generally find a difficulty in getting round Breaker point; for the tide here is of no use, and all there is to assist is the likelihood that the wind will draw off the land after midnight, when, by being inshore, a good board can be made, and possibly the Cape of Good Hope reached. Hai-mun bay cannot be recommended, but still it would be better to anchor there than to be carried round the point. In this case, should West hill be obscured, run in under the point, lower a boat, and let her find the sunken rock (page 84), and then come in with good weigh to windward of Parkyns rock,—if drawing less than 13 feet,—and shoot up round the boat into Fort bay.

Having reached the Cape of Good Hope, the flood will assist a vessel to round it, and the ebb out of the Hau river will be a weather tide; in the latter case, and not intending to go inside Namoa island, endeavour to get along the south side of the island, where there is an eddy tide, and anchor in South bay, should the weather prove too bad to proceed on the flood: both tides will be found strong off Three Chimney point, and the same may be said of Jokako point, round which vessels should take the first of the flood on the port tack.

Farther northward about Rees island, the flood tide in strong winds causes an uneasy sea which will distress a vessel much. Red and Ting-tae bays will be found good stopping places; and the latter should be preferred, though at the loss of 2 or 3 miles, to anchoring in an exposed position in the entrance to Amoy harbour; as when the N.E. winds freshen off here on the flood, they generally bring a mist in with them, which makes it difficult to find the entrance, and at the same time a vessel will have trouble to get out of the harbour against the tide.

To the northward of Amoy are Leeo-lu, and Hu-i-tau bays, both of which afford good shelter. Chimmo bay is not so good; but with a long scope of good ground tackle vessels may ride in it. The current in the monsoon overcomes the tide here; and advantage must be taken of every slant of wind, bearing in mind that it is likely to draw off the land in the middle watch, and in the event of anchoring for shelter this is the time to start, should the wind moderate: by waiting for daylight vessels lose their offing; and will have to make an off-shore board at a loss. The fogs are at times thick, but the lead is not a bad guide, as the soundings generally change from sand to mud as the shore is approached. There is also fair anchorage under Pyramid point, but not so good as that under the South Yit; and if the vessel is looking up North or anything east of it, the ebb out of Meichen sound will be of assistance.

From the Lam-yit islands or the south end of Hai-tan strait to the White Dogs is beyond doubt the most difficult part of the passage. With steamers the strait will afford the best route; but sailing vessels should decidedly keep outside, and stretch over to the north-west coast of Formosa (page 287), where they are likely to get a slant of wind, and the advantage of a weather tide; and as this portion of the coast has been surveyed, by attention to the soundings no vessel can come to any harm.

RIVER MIN to CHUSAN ARCHIPELAGO.—North of the river Min the ebb is generally a weather tide,—unless the wind is far to the north, —and out of the river, Ting-hai and Sam-sah bays, vessels will get a good lift; with the flood, the indraught into the latter will be sensibly felt as far out as Larne islet, and increases to 2 and 3 knots as the main is closed. As a general rule, tack for the inshore tide, when the moon is on the meridian.

Tung-ying island will be found a snug anchorage, and here the coast should be forsaken (unless in a vessel under 12 feet draught), and the deep water to the eastward kept in. The tide will afford but little assistance until the vessel arrives at the Chusan archipelago; the flood causes an uneasy sea in the shallow water, while the ebb has too much southing in it, unless the wind is eastward of E.N.E.; but Nam-ki and Pih-ki-shan islands will afford good shelter.

On reaching the Chusan archipelago, take the Beak Head channel unless the tide is nearly done, in which case there is Harbour Rouse and the south side of Luhwang island as anchorages under the lee; and as the first of the ebb runs to the northward through the Foto channels, the tide through may be saved, and anchorage gained on the Ketau shore. From hence, if bound to Ting-hai harbour, contrive to arrive at the west end of Tower Hill island about slack water; otherwise in light winds the vessel is liable to be carried on to Just-in-the-Way, and even through the Blackwall channel.

In working through the north part of the Chusan archipelago, as the set of the ebb and flood trends nearly East and West, advantage can always be taken of the tide, and vessels may count on feeling the influence of the ebb within an hour of the moon's meridian passage. When in the vicinity of Gutzlaff island, the first of the flood takes a direction to the southward of West, running into Hang-chu bay.

The eddy tide, generally speaking, will carry vessels clear of the large islands; but when they are approaching detached rocks, great attention is required to prevent being set in amongst them.

PASSAGES in S.W. MONSOON.—There will not be the same difficulty in getting to the southward against the southerly monsoon, as there is in going to the northward against the other, as it is not so permanent in its direction, and land and sea breezes prevail; the current will be found running strong to the northward in the Formosa channel, but vessels are not liable to the same detention which they often experience in the northerly monsoon. Care, however, must be taken not to overshoot the port. Fogs prevail in the early part of the season, and render the navigation at times as harassing as it is in the North-east monsoon; they, however, generally lift in the vicinity of the land, and a ship's length from where the bowsprit can hardly be seen will carry her into sunshine.

The chief difficulty to overcome in making the passage between the Gulf of Pe-chili and Hong Kong during the southerly monsoon is the strong easterly or north-easterly current. After passing the parallel of the Yang-tse kiang, it will be advisable to keep near the China coast; for although a vessel may lie up South or S. by E. on the starboard tack, it should be remembered that she is making little better than a S.E. course in consequence of the easterly set. A stretch to the north-westward, though apparently a loss of ground, will ultimately prove useful.

H.M.S. *Pique* in making this passage in July and August, 1838, was not favoured when close in shore by any land and sea breezes, nor had the least slant, but generally lost the wind. A weather tide was occasionally felt when near the shore in the Formosa channel.

Although the constant adverse current makes this a tedious passage against the monsoon, there is nothing to prevent a vessel of moderate sailing qualities making the passage at this season. The *Pique* had seldom more than single-reefed topsails, and the sea was generally smooth; she made the passage from the Gulf of Pe-chili to Hong Kong in 31 days.

CHAPTER II.

APPROACHES TO CANTON RIVER, INCLUDING HONG KONG.—THE CHU KIANG OR CANTON RIVER, AND THE SI KIANG OR WEST RIVER.

VARIATION 0° 30′ EAST IN 1861.

As vessels bound to Canton river from the southward in the South-west monsoon endeavour to make Great Ladrone island bearing about North and then proceed towards the river by the Great West channel; a description will first be given of the islands and anchorages on the western side of this channel, from San-chau island to Cum-sing-mun harbour, and then returning to the Ladrone islands the mariner will be taken through the different passages to the eastward of these islands to Hong Kong, and to the entrance of the river.*

SAN-CHAU, which forms the western side of the entrance to the Hueng river or Broadway, is the next large island to the north-eastward of Tylou island, and its south-east point bears W. by N. $15\frac{1}{4}$ miles from the Little Ladrone. The space between San-chau and Tylou is shoal, with some islets and rocks adjoining the north-east end of the latter. The depths decrease gradually off San-chau, but it is not so bold to approach as the islands to the westward, for soundings of 3 to 4 fathoms extend a considerable distance from it; there is a conical islet and some rocks nearly touching its east point, with 3 fathoms close to them.

MONTANHA, or Wung Cum island, forming the eastern side of the entrance to the Broadway, is a large high island to the N.E. of San-chau, and close to it on the north-east side is Ko-ho island. These two islands form the south side of the Typa anchorage; and the Great West channel is bounded by them on the west, and by Potoe and the other islands adjacent on the east.

HUENG or BROADWAY RIVER, the entrance to which has sufficient depth to admit a vessel of moderate draught a considerable way up, may be found useful to such as intend to make a long stay near Macao, or to those who have parted from their anchors, and draw too much water to attempt the Typa anchorage.

---

* See Chart, East Coast of China, Sheet I., No. 2,212; scale, $d = 14$ inches.

HUENG OR BROADWAY RIVER.

The Water islands are two small islets lying close off the south end of Montanha; and N.W. ¾ N. a mile from them lies Inside islet, having a small inlet, called Lark bay, between it and Morgan point (608 feet above the sea), the west extreme of Montanha. These islands are on the east side of the Broadway entrance, and Coffin island, bearing S.W. by W. ¼ W. distant 4 miles from the Water islands, is on the western side. At 5 miles in a S. ¾ E. direction from Montanha peak and 2½ miles from the Water islands is a shoal patch of 12 feet.

TIDES.—It is high water, full and change, in the entrance of the Broadway at 11h. and springs rise 7½ feet. The neaps are very irregular, there being then only one flood and one ebb, of any considerable strength, during the 24 hours. The direction of the flood outside is governed principally by the winds: with strong easterly winds it comes from E.S.E.; and when south-westerly winds prevail, from South. The ebb runs generally to the S.W. Inside the river the tides take the direction of the channel.

DIRECTIONS.—The best time to enter the Broadway is with the first of the flood, and if a vessel at anchor in Macao road be obliged to run for it with a N.E. or East wind, about three-quarters ebb will be the best time to leave the road, that she may meet the first of the flood at the entrance, where it flows sooner than in the road. Having rounded the east point of Ko-ho island, about 1½ miles distant in 4½ fathoms, steer at any convenient distance round Apomi point, the high south-east extreme of Montanha, which has 3 fathoms near it, deepening gradually to the eastward towards Potoe island.

When abreast the point, the Water islands will be seen in one with each other, near the western extreme of a bay with a sandy beach. As there are not more than 2 fathoms in this bay at low tide, it should be avoided by steering to pass about half or three-quarters of a mile to the southward of these islands, in 2¾ or 3 fathoms, then haul round the western island, preserving the same depth and distance. Do not exceed the distance of one mile to the westward of this island, for beyond that the water shoals fast to 2¼ fathoms, towards the San-chau shore. From abreast the islands about a N.N.W. ½ W. course, giving a berth of three-quarters of a mile to Inside islet, will lead up to abreast the west point of Montanha, in 3 and 3¼ fathoms at low water, off which a vessel may anchor and be sheltered till the termination of the gale.

From the above point the water shoals gradually towards Ross island on the west side the channel; there is generally a line of fishing stakes extending westward from the point, with passages among them for vessels. Mong-chau, or Ballast island, bears N.N.W., distant

2½ miles, from the west point of Montanha, and between them there are two passages leading to Macao, but both so shoal at low water as only to afford a passage for boats.

N.W. ¾ N. about 1¼ miles from the west point of Montanha, and fronting the first of the above passages, there is a rock which shows at low water about the size of a small boat. The channel is about a cable's length to the westward of this rock; for W. ¼ S. about a mile from it there is another rock, which also shows at low water, and shoal banks bound the channel on both sides. From the West point of Montanha to Ballast island the water is shoal, the edge of the bank leaving only a narrow passage on the east side the eastern rock, with 1½ fathoms in it at low water. Pak-tang, a small island with a sharp hummock on its north-east end, lies on the western bank, W. ½ N. distant 3 miles from Ballast island: the bank, composed of mud, has only 6 feet on it, and extends 1½ miles from Pak-tang towards Ballast island, and commencing at the western rock, trends to the N.N.W. the whole length of the channel, contracting it to about the breadth of from half a mile to a mile, with 2¾ and 3 fathoms in it.

If intending to proceed farther up than the west point of Montanha, steer N.N.W. towards the rock fronting the first passage to Macao; the soundings will be about 3 fathoms at low water, and the rock should be passed within a cable's length on the west side, to avoid a shoal patch of 2 fathoms in mid-channel. When abreast the rock, steer N.N.W. ½ W. 1½ miles, and the vessel will then be abreast Ballast island, in 2¾ fathoms water. This is a safe and convenient anchorage, about 6 miles to the S.W. by W. of Macao, and the boats are kept in sight when passing to or fro from that place. Fresh water may be obtained in a small bay to the northward, under Beacon hill, which is 690 feet high, and has a remarkable stone on its summit.

The channel for vessels, between Ballast island and the bluff point to the northward, becomes narrow. If intending to proceed higher up a N.N.W. ½ W. course will lead about a mile above the bluff point, in 3 and 3¼ fathoms, and this point ought to be passed at about three-quarters of a mile. If drawing more than 14 feet, wait here for the last of the flood, to pass the Tang rocks, lying a little to the northward, and off which there are only 3½ or 4 fathoms at high water.

From the bluff point, steer N.N.W. ¾ W. to pass a long half mile westward of the Tang, and when abreast them, steer about N.W. ¾ N., or directly for the entrance of the river, keeping about half a mile off Nam-ye-kok point, which forms the east side of entrance; it has a pagoda on it, and is well covered with trees. Here, the depths begin to increase, and in steering to pass not more than a quarter of a mile off

Moto fort, to avoid a rock lying in mid-channel, the soundings will be 4 and 5 fathoms. About 4 or 5 miles above this fort, the Broadway separates into two branches: the easternmost, called Hong-shan river, communicates with Canton, and by it the trade is mostly carried on between this city and Macao; and the western branch leads to the Si kiang or West River. The wide opening eastward of Nam-ye-kok point, called the Flats, has a boat passage through it leading to Macao.

If the wind does not admit sailing directly into the entrance of the Broadway, there is room for short tacks between the Water islands and the rocky islets off San-chau, taking care of the latter shore, which is shoal. Farther in, the channel contracts a little, but the tides are of sufficient strength to back and fill past the rocks that lie opposite the passage to the Typa, or where the channel may seem rather narrow for working.

The following extracts have been selected from Sir J. J. Gordon Bremer's dispatch, dated March 1841, to point out the track of the H.C. steam-vessel *Nemesis*, when forcing a passage through the Broadway from Macao to Whampoa:—

" At 3 a.m. on the 13th the *Nemesis*, with the boats of the *Samarang* in tow, weighed from Macao road, and proceeded over the flats between Patera and Macarira islands to the Broadway river. At 8 a.m. came in sight of Macao fort. On reaching Hok-tau point the river is divided into two channels; that to the right takes a sudden sharp turn and becomes very contracted in its breadth; Tei-yat-kok, a field battery recently constructed of 14 guns, was seen strongly posted on a rising ground on the left bank of the river (surrounded by overflowed paddy fields), which enfiladed the whole line of the reach leading to it. On entering the reach in which they were, they observed on the right bank of the river a new battery, scarcely finished, with ten embrasures, but without guns, and Hoc-kang fort close to it, well built of granite, surrounded by a wet ditch, and mounting 14 guns and 6 gingalls. Abreast of these (which they flanked) the river was strongly staked across.

The *Nemesis* having got through the centre passage of the stakes, which was just wide enough to admit of her passing, arrived at 4 p.m. off the large provincial town of Heang-shan. The dense population thickly crowded the banks, boats, junks, housetops, the large pagoda, and surrounding hills; both sides of the river were packed with trading craft of the country, the centre of the river, which is very narrow, having merely sufficient space to allow the steamer's paddle-boxes to pass clear of the junks moored to its banks. At 6 p.m. the steamer passed on into a narrow shallow channel, scarcely more than the breadth of a canal, where she anchored head and stern for the night. At daylight on the 14th, weighed and proceeded up the river in the steamer's draught of water, and not broader than her own length, grounding occasionally on both sides; at 7h. 50m. arrived at the large village of Honghau, with a fort of the same name at the upper part, which flanked a strong and broad line of stakes 20 feet wide, completely across the river, filled up in the centre by large sunken junks laden with stones. After the *Nemesis* had made good her passage through the stakes, which was effected after 4 hours' incessant labour, she arrived at 4 p.m. off a military station, where she anchored for the night.

At day-light on the 15th, the steamer continued her course upwards, and at 7h. 30m. arrived off the large village of Zam-chau. On moving up to Tegnell, a large town on the left bank of the river, three forts were passed, all dismantled and abandoned, and on proceeding up to Whampoa three more dismantled forts were observed. At 4 p.m. the *Nemesis* came to in that anchorage, having in conjunction with the boats destroyed five forts, one battery, two military stations, and nine war junks."

TYPA ANCHORAGE.—The eastern entrance to this anchorage is formed between two high islands, that on the south side named Ko-ho or Apomec, and that on the north side named Typa or Kaikong. Ko-ho is separated from the north-east point of Montanha by a narrow gut with 24 feet water in it, decreasing to 9 or 10 feet farther in towards the Typa. The anchorage is between the west end of Typa island and the east end of Macarira island, and affords secure shelter in from $3\frac{1}{2}$ to 4 fathoms water. H.M. ships *Herald* and *Modeste* refitted here during the operations in China in 1841.*

TIDES.—In the Typa anchorage, and in Macao harbour, it is high water, full and change, at 10h. 0m. The springs rise about 7 feet; in the Typa they run $1\frac{1}{2}$ and 2 knots per hour, when not influenced by the winds. The ebb runs out of the mouth of the Typa, but it sets across the entrance when outside the points.

DIRECTIONS.—Vessels entering or leaving the Typa should endeavour to weigh at half-flood. In entering steer for the north extreme of Ko-ho island, and pass it pretty close, the deepest water being on this side the entrance; continue to steer along until the summit of Sylock island is in line with the north extreme of Ko-ho. Keep this latter mark on, or the north point of Sylock just in sight, bearing about E. $\frac{3}{4}$ S., leads in the deepest water; and when the east end of the middle hill on Typa island opens westward of a rocky mount forming the south-west point of

---

* See Plan of Macao, No. 1,290; scale, $m = 3$ inches.

the same island, haul gradually to the northward, and anchor near the west point of Typa, with the south point of Sylock open of the north extreme of Ko-ho.

Here the depth is $3\frac{1}{4}$ to 4 fathoms at low tide, and vessels are sheltered from all winds by the high lands around; the deepest water is near the west point of Typa, for the bay abreast, at the east end of Macarira, is shoal. The watering cove is at the head of this latter bay, and from the north point a reef of rocks projects nearly a quarter of a mile to the eastward; a vessel ought not to go so far northward as to approach this reef. In the middle and eastern parts of the Typa the depths are only 14 and 15 feet at low tide, in the fair channel leading to the anchorage, but no injury can be received by grounding, the bottom being remarkably soft.

MACAO HARBOUR.— Macao, a Portuguese settlement in China, stands on a small peninsula projecting from the south-east end of the island of Macao, on the south-west side of the entrance of Canton river. The peninsula is nearly 2 miles long, and less than a mile wide at its broadest part, and is connected with the island by a low, narrow, sandy isthmus, across which extends a barrier wall to exclude foreigners from the interior of the island. The town is built on the declivities round the harbour, the shore beneath being embanked, so as to form a marine parade, backed by a terrace of white houses.

The mean temperature at Macao, from observations taken for a number of years by the late Mr. Beale, is as follows:— Jan. 51°; Feb. $51\frac{1}{2}°$; March $65\frac{1}{2}°$; April $72\frac{1}{2}°$; May $75\frac{1}{2}°$; June $81\frac{1}{2}°$; July 86°; Aug. $84\frac{1}{2}°$; Sept. $81\frac{1}{2}°$; Oct. 73°; Nov. $64\frac{1}{2}°$; and Dec. $57\frac{1}{2}°$.

The harbour is formed between the peninsula and the large island of Patera to the westward. Its entrance is narrow but the depths are 26 feet at low water close to fort San Iago, which is built on the east point; and from thence the soundings are 21 and 20 feet along the eastern shore to the town.

PILOTS.—The river pilots are procured at Macao, and each receives a chop from the residing mandarin, to deliver to the officer stationed at the Boca Tigris, describing the force of the ship, and to what nation she belongs.

MACAO ROAD is shoal, the depth being generally from 3 fathoms at low water springs on the west side, to $4\frac{1}{2}$ or 5 fathoms close over to Samcock and the other islands that bound the east side; there is however said to be much less water in it of late years, but as the bottom is soft loam or loose mud there is no danger of a vessel striking on her anchors, for they immediately bury in it.

Vessels of large draught usually anchor in deep water near the islands, with Macao bearing between W. by N. and W.N.W., distant 6 or 7 miles, which renders the communication with that place difficult and dangerous in blowing weather. With Ko-ho point S. by W. ¼ W., and Macao W.N.W., distant 4 or 5 miles, a large vessel may anchor in 4½ fathoms at low water, and be more conveniently situated for procuring a pilot. If drawing under 18 feet she can anchor with Macao on the same bearing, about 1½ miles off the entrance of the Typa, into which she may run if a gale is approaching.

Small vessels may anchor in the South-west monsoon in the entrance of the Typa, nearest to the south point, in about 3 fathoms at low water, and a little within Ka-o islet, on the north side of Ko-ho point. In the North-east monsoon they can anchor close to the northern shore, abreast a sandy beach, between the Cau-chau or Nine islands and Macao, in 3 or 3¼ fathoms ; here, they will generally have smooth water and an easy communication with the shore.

DIRECTIONS. — The usual route to Macao harbour is through the Typa anchorage, there being 13 feet at low tide in the fair track between the Typa and the harbour ; but only 12 and 11 feet in the large space between Typa island and Macao. The channel trends in a direct line from the Typa to the harbour, and to avoid the sunken rock, named Pedra-mea, lying about a quarter of a mile eastward of the north-east point of Macarira, keep the north-east point of Montanha open eastward of Macarira ; or, in passing it keep rather more than mid-channel towards Typa island.

From thence, steer direct for the entrance of the harbour, there being no other danger except the Pedra Areeka rock, on the east side of the channel, from which the south point of the outermost of the two high Ma-lo-chau islets, to the south-west of the entrance, bears W. by S. ½ S., distant 1¼ miles, and the point of fort San Iago N.W. ¼ N. about half a mile. The north-east point of Montanha in line with the east point of Macarira leads clear to the westward of the Pedra Areeka, and a vessel will not be too near it if she does not go eastward of a line drawn from the west point of Typa island to fort San Iago point. This point should be rounded pretty close in entering and the eastern shore kept aboard to the anchorage abreast the town, where a disabled ship may be hove down and repaired.

CAU-CHAU, or Nine islands, are a group of islets near the eastern shore of Macao island, about 4 miles to the north-east of the town. They lie close together, and the depth is 3 fathoms at about half a mile to the eastward of the outermost islet, which bears N.E. by E. from Senhora de Penhos church at Macao; S.W. about three-quarters of a mile from this islet there is a rock always above water.

CUM-SING-MUN HARBOUR.—From Macao the eastern shore of Macao Island trends to the N.N.E. about 11 miles to Bluff head, where it turns abruptly to the westward and forms a deep bight called Cum-sing-mun harbour. This harbour is safe for small vessels, and it would be a desirable haven for vessels of large draught to run for from the anchorage off Lintin, at the approach of a Typhoon, were it not for the extensive flat outside being too shoal, the depths being only 2 to 3 fathoms to the distance of 2 miles outside the entrance; but they increase quickly to 7 and 8 fathoms when within half a mile of Bluff head, which is the proper side to steer for in coming from the eastward, and also to keep nearest to when running into the harbour.

The entrance, which is about half a mile wide, and is formed between the south part of Kee-ow island and Bluff head, bears W. by S. 10 miles from Lintin peak, and is 8 miles to the northward of the Cau-chau islands. Between the head and the small islet and sunken rocks, near the opposite shore, the depths are irregular, from 14 to 6 fathoms; but inside, about half a mile West, or W. by S. from the small islet, the bottom is soft, affording safe anchorage in 6, 5, or 4 fathoms, taking care, however, to avoid the shoal patches shewn in the chart.*

GREAT LADRONE, (Man-san of the Chinese,) being the outermost island directly fronting the entrance to Canton river, is generally used as a landfall by vessels bound there from the southward during the South-west monsoon; and with the Little Ladrone adjoining to the westward, and Potoe to the north-westward, bounds the east side of the Great West channel, leading to the river.

This steep bold island may be easily known by its north-west part forming a round mount or dome, (1,465 feet high,) which, being more elevated than the other parts, can be seen, in clear weather, about 27 miles from a vessel's deck, and 40 miles from the masthead; none of the other islands have a similar appearance, although most of them are high. The island is about 2 miles in diameter, with a rocky aspect close to the sea, but it is safe to approach, the depths near it being 14 or 15 fathoms; on the south-west part there is a small inlet, named Pumice Stone bay, where fishing boats take shelter in the North-east monsoon.

LITTLE LADRONE, (Pocking-han of the Chinese,) is of a convex

---

* *See* Plan of Cum-sing-mun Harbour, No. 1,253; scale, *m* = 3 inches.

sloping form, not so much elevated as the Great Ladrone and separated from its west side by a narrow channel carrying a depth of 9 to 18 fathoms, but too confined for a vessel unless in a case of necessity. Near the west side of the island the depth is about 10 fathoms, decreasing gradually to 7 fathoms about half a mile to the southward of Potoc ; there are 12 fathoms near its south point, and 14 and 15 fathoms near the south and south-east sides of the Great Ladrone.

A small rocky islet lies close to the north-east part of the Little Ladrone, and North nearly three-quarters of a mile from this islet there is a Black rock covered at high tide with 10 fathoms close around : it will be prudent therefore, in passing this locality at high water when the rock is covered, to keep about mid-channel between the Little Ladrone and Tong-ho island, which lies 2½ miles to the northward. This is the only danger near the Little Ladrone, excepting a high rock close to its northwest side, having a depth near it of 9 and 10 fathoms.

POTOE, or Passage island, bearing N.N.W. ¼ W. 5½ miles from the south-west end of the Little Ladrone, is a flat sloping rock, visible about 9 miles, with 5 to 6 fathoms near it all around ; but it ought not to be approached too close in light winds, as the eddies occasioned by the freshes out of the river may render a vessel ungovernable, and probably drift her towards it, or Wong-mou, the adjacent island. The channel between it and the south-east point of Montanha is about 5 miles wide, and safe.

WONG-MOU and LIUNGNIB ISLANDS.—Wong-mou, lying 1½ miles to the E.N.E. of Potoc, is 1¼ miles long, in a north and south direction, and has a peaked hill on its northern part ; at nearly half a mile from its west side there are some rocks above water. Liungnib, lying a mile to the eastward of Wong-mou, has a round islet off its south end.

At about three-quarters of a mile to the N.W. from the north end of Liungnib lie two rocks, which cover at springs, and break in blowing weather ; therefore, in passing the north end of this island, keep at least a mile from it.

PAK-LEAK ISLAND, called also Putoy, lies N.E. by N. nearly 1½ miles from the Great Ladrone, and on its north-east part stands a remarkable cone hill, 855 feet high, which is visible from Macao. The island is of irregular shape, and on the southern side the hills are much covered by black rocks. On its eastern side and fronting Hoa-ock islet there is a cove where fishing boats find shelter ; on its northern side are some small bays in which fresh water may be procured ; and near its north-east point there is a rocky islet, on which the fishermen have erected a hut and a fishing stage. A rock, awash, lies close off its south extreme.

CLIO ROCK, on which H.M.S. *Clio* struck, 12th December 1841, lies about 2 cables from the west side of Pak-leak, with the north-west extreme of the island bearing N. by W. distant 4 cables' lengths.

TONG-HO ISLAND, bearing N. by E. ½ E. about 2½ miles from the Little Ladrone, is 1¾ miles long, east and west, and of moderate and unequal height. There is a small cove on the north-east part of this island, into which the ship *Boddam*, drawing 21½ feet water, was taken by her pilot on the approach of a Typhoon. It appears this vessel, after being disabled during one of these tempests by the loss of her masts and rudder, and having fixed temporary ones, was proceeding towards the river, when the pilot, perceiving another Typhoon coming on, ran her into this cove, where she remained in safety during a violent storm.

Boddam cove is about 2 cables wide, with 24 feet water in the entrance, 17 and 18 feet well inside, at low water springs, and the bottom all soft mud. Here a vessel may lie at anchor, or if she has none, be run into the mud without risk. On each side the land is steep from the water's edge, terminating in a valley at the head of the cove, where there is a sandy beach and plantain trees. Being the chief rendezvous of the fishing boats in bad weather, or a place of refuge from the pirates, it is protected by a fort on the north-west point of the entrance. The rocks along the north-west side of the cove have 12 feet, mud, within 3 or 4 yards of them.

Supplies.—Good water may be obtained at Boddam cove, also beef, fish, poultry, and some fruit.

DIRECTIONS.—This cove will not be readily distinguished until within about 2 miles of the north-east part of Tong-ho. In steering for the entrance, take care to give a berth to a sunken rock, lying about 1½ cables, to the north-eastward of the fort point ; when the head of the cove bears S.W. by W., the vessel will be to the south-east of the rock.

Having brought the cove fairly open on the above bearing, steer for the point on the south-east side of the entrance, and pass it within the distance of half a cable ; for the north-west point, where the fort is built, is bordered by rocks. At about 2 or 3 cables to the south-east of the entrance there is also a reef of rocks, which extends between 1 and 2 cables from the south-east part of the island ; these are mostly all in sight at high water, and easily avoided by steering from the offing directly for the south-east point of entrance. There is a sunken rock off the north-west part of the cove, but when passing this part of the island it will be avoided by keeping about three-quarters of a mile off shore. The flood sets N.W. outside the entrance, and the ebb S.E. They both run pretty strong, but there is scarcely a drain in the cove.

CHUK-WAN ISLANDS.—These two islands lie about E. by N. 1½ miles from Pak-leak, and the larger island, the eastern one, has a high rocky islet, named Sharp island, lying off its south-east point, and a small bay on its north side. There are 14 fathoms water between Hoa-ock and the western island, and 11 and 12 fathoms to the northward of the group.

RALEIGH ROCK, on which H.M.S. *Raleigh* struck on the 14th April 1857, is a small pinnacle, upon which a moderate sea breaks at low water springs, with 9 and 10 fathoms close to. Its position is lat. 22° 2′ N., long. 113° 47′ E., nearly in mid-channel between Pak-leak island and the South White rock, distant 2½ miles from the latter. When on the rock the gap in the centre of the South White rock is in line with the right extreme of a small wedge-shaped island off the eastern side of Lafsami island bearing N.E. b. N.; the highest part of Ai-chau island E. ⅓ N.; and the peak of the great Ladrone is over the western slope of Pak-leak S.S.W. ½ W.

NORTH and SOUTH WHITE ROCKS are two high white rocks half a mile apart, lying North about 3½ miles from the western or small Chuk-wan island. From the southern rock the north-east point of the eastern Chuk-wan bears S.S.W. ¼ W., distant 4½ miles; the peak or highest part of Ty-lo W. by N. ½ N., nearly 6 miles; the north point of Liungnib W. b. S. ⅛ S., 6 miles; the southern part of eastern Chi-chau N.E. b. E. ⅔ E., 5¼ miles; and the western Ai-chau island S.E. by E. ½ E., distant 6 miles. About a mile to the south-east of the southern rock, there is a small black rock, visible only at low springs, having 9 fathoms water close around. Between the two high rocks, but a little more westerly, there is a smaller rock above water.

CAUTION.—The White rocks may be seen in fine weather in time to avoid them, and the depth is about 9 fathoms near their eastern side, 8 fathoms on the western and northern sides, and 9 fathoms in the channel, between them and Chuk-wan; but since the loss of the *Raleigh* by striking on the Raleigh rock, it will be prudent not to use this channel until it has been more accurately examined.

AI-CHAU ISLANDS.—These two islands lie N.E. by E. ½ E., 4 miles from the eastern Chuk-wan, and the eastern or larger island is separated from the smaller one on its west side by a very narrow channel with 4 and 5 fathoms in it at low water. The depth on the southern side of these islands is 14 fathoms, on the north and east sides 12 and 13 fathoms, and on the west side 8 or 9 fathoms.

HILL ISLET, lying N.E. distant 1½ miles from the northern part of the eastern Ai-chau, has 11 and 12 fathoms water at a short distance from the rocks around it.

SAMOUN GROUP.—The Samoun or Three gates, are a group of three small islands lying 2½ miles to the eastward of Ai-chau, and extend about 3¼ miles in a N.W. and S.E. direction, with narrow passages between them. Near the north-west part of the north-west island, called Hak-chau, there are two peaked islets, and on the northern side of the group, between the eastern and middle islands, there is another high rocky islet, named Gauze, with a bed of rocks lying to the southward of it; the south end of the eastern island is the highest part of the group and forms a round mount. There is a small harbour on the south-west side of the largest island, which would afford shelter to two or three vessels against a N.E. gale. The anchorage is in 6 to 10 fathoms, muddy bottom.

These islands lie to the north-west of the Taitami channel, and a vessel may pass either to the southward of them, or between them and Lingting in from 12 to 15 fathoms water.

LINGTING ISLAND, bearing W. ¾ N. distant 15 miles from the North-east head of the Lema islands, is of rugged appearance, about 1¾ miles long, east and west, and rises to a peak near its centre. There are two rocks, one awash and the other above water, lying to the eastward of the north point of the island, and bearing N. by E. and S. by W. of each other; the outer one, awash, lies E.N.E. distant nearly a mile from the north point, and the other S. by W. about half a mile from the outer one, with depths near them of 13 fathoms, but foul ground between.

The Needle rocks, on which H.M.S. *Doris* struck in 1813, are two heads lying within a few yards of each other, about 1½ cables' lengths to the south-west of the low rocky north-west extreme of Lingting, and are so sharp that it is difficult to keep the lead fixed on their points; at low springs they have about 6 feet water on them, at which time, with a swell, they may probably show either breakers or a rippling. From the outer rock the south-west extreme of the Lema islands is just shut in behind the south-west point of Lingting, and the highest part of Lamma island is a very little way over the low north-west point. A vessel will avoid them when passing round the north-west end of Lingting by not approaching it within half a mile, and by keeping the south-west extreme of the Lema islands a little open to the south-west of Lingting.

The depths close to the north point of Lingting are 18 or 19 fathoms, decreasing to 14 and 15 about a mile distant; to the southward and westward of the island, there are 10, 11, and 12 fathoms over a soft bottom.

CAUTION.—When passing to the northward of Lingting in the night, it will be prudent to keep 1 or 2 miles off, on account of the two rocks lying off its north-east side.

TY-LO ISLAND, the southern of the range of small islands bounding the east side of Macao road, is high near the western part, sloping a little to the eastward. It lies N. ½ E. from the north end of Liungnib, from which it is separated by a good channel 2¾ miles wide, but in using it take care to avoid the rocks off the northern point of the latter. Ty-lock, lying about half a mile northward of Ty-lo, is a small rocky islet, having a large rock on its summit.

SAM-COCK ISLAND, the largest of the above range, lies 1¼ miles in a N.N.E. direction from Ty-lock, and is of moderate height, rugged in appearance, and in the form of a pyramid. Between this island and Ty-lock there is a small islet, named Sy-lock, and two rocks above water; but the channels between these are so narrow, that they should not be attempted on account of the strong eddies, which render vessels frequently unmanageable. In passing between Sam-cock and Chung-chau-si, the next island to the northward, keep in mid-channel or nearest to the latter, in 6 or 7 fathoms water, as there are only 3¼ fathoms at a quarter of a mile from the north point of Sam-cock, 3¼ fathoms about a quarter of a mile off the west point, and only 3 fathoms the same distance off its eastern point.

Water.—On the northern part of Sam-cock there is a small bay or cove for boats, and the island affords fresh water.

CHUNG-CHAU-SI, or West Water island, the northernmost of this range, lies N.N.E. about 1¼ miles from Sam-cock, and there are 7 fathoms water near it to the eastward, and 5 and 6 fathoms to the northward and westward. The depths are 5 or 5½ fathoms along the western side of this range, from Ty-lo to Chung-chau-si, and 7 fathoms on the eastern side; the ebb stream runs strong from the northward along the western side, and the flood in eddies from the south-eastward.

FOUR-FEET ROCK.—This small dangerous needle rock, with only 4 feet on it and 10 fathoms close around, lies E.S.E. 3 miles from Chung-chau-si, and from it the summit of Ty-lo bears S.W. by W., the centre of Sam-cock W. ½ S., and the small islet lying off the north-west end of Chung-chau N.N.E. ¼ E. When Chuck-tu-aan island (which bears S.E. by S. 3 miles from Chung-chau-si) and the small islet off the north-west end of Chung-chau are on the same bearing, about N.N.E. ½ E. and S.S.W. ½ W., the rock will be between the two, but nearest the former; therefore if a vessel has occasion to enter Macao road by this channel, and keeps about three-quarters of a mile off Lafsami and the south side of Chung-chau, she will pass in mid channel, and have 10 or 12 fathoms water decreasing to 7 fathoms as she nears Chung-chau-si.

CHUNG-CHAU, or Water island, which with the islands to the southward of it bound the south-west side of the Lantao channel, lies about S.W. by W. 2½ miles from the south-west point of Lantao, is high, and near its north point there is a peaked hill. It is 1½ miles long, N.W. and S.E., and has not any hidden dangers near its northern side. The soundings in the channel between it and the south point of Lantao will be irregular owing to the strong eddies generally prevailing hereabout; the depths being 7 fathoms near the point of Lantao, 18 or 20 fathoms in mid-channel, and 28 or 30 fathoms close over to Chung-chau. There is a cove for boats on the north side of the island, and a short distance to the westward of its northern point there is a round and high islet, with a large rock on its summit; round this islet to the northward and westward the depth is 15 fathoms.

NAU-TAU-MUN, or Bullocks Head Gate, the next island to the south-east, is small but high, and is separated from Chung-chau by a narrow channel through which H.M.S. *Doris* ran, and found shoal water near Chung-chau. The depths near the north side of Nau-tau-mun are 15, 16, and 17 fathoms, rather irregular; but to the southward, in the bay, there are only 3, 4, and 5 fathoms.

LAFSAMI ISLAND, lying to the southward of Nau-tau-mun and separated from it by a narrow channel, is larger than either Chung-chau or Nau-tau-mun, and is inhabited on the south-western side, where fresh water is to be had in a small bay. The depth on the north side of the island in the Lantao channel is very irregular, from 17 to 25 fathoms in overfalls, about a quarter of a mile off, and on the south side 10 and 11 fathoms. This island from some views forms a peak; and at a short distance to the eastward of its south point there is a rocky islet, on which the fishermen have huts, and a winch for heaving up their nets.

CHI-CHAU ISLANDS.—Chi-chau is the largest of two islands lying 2¼ miles to the E.S.E. of Lafsami, and forms the south side of the east entrance of the Lantao channel. This island is high, of round appearance, inhabited on the west side, and separated by a narrow channel from the smaller island, which is lower, and lies on its western side; a sunken rock lies off its north-east point, and a patch of 4 fathoms about a quarter of a mile from its north point. There is a safe channel 1½ miles wide, and carrying 9 and 10 fathoms, between the west point of the smaller island and the rocky islet lying off the eastern side of Lafsami; it may be taken by a vessel bound up the river when she enters the islands from the south-east betwen Chuk-wan and Ai-chau.

SOKO ISLANDS.—A-chau, the southern of the two Soko islands, lies S.E. ¾ E. distant nearly 4 miles from the south point of Lantao, and forms the north side of the eastern entrance of the Lantao channel. The south point of A-chau is high, and rises very steep, having 7 fathoms water close to; the depths between it and Chi-chau are 11 or 12 fathoms in mid-channel, 13 fathoms nearly over to Chi-chau, deepening suddenly to 25 or 30 fathoms in a hole or swatch close to Chi-chau.

The other island lies a short distance to the northward of A-chau, and is about a mile long, east and west, and very narrow in the middle. A sand spit extends nearly West upwards of 1¼ miles from the west side of this island, and on the west extreme of the spit there are 2¾ fathoms at low water, decreasing quickly to 2 and 1¼ fathoms towards the island.

A rocky islet and two rocks above water lie between the two Soko islands, nearest to the south-west point of the northern one; there is also a high rocky islet lying nearly a mile to the eastward of A-chau, and which may be passed at half a mile to the southward in 7 fathoms water, but the ground is foul between it and A-chau.

Water.—Fresh water may be procured at a little sandy beach on the northern side of A-chau.

KYPONG ISLANDS are the southernmost group of the archipelago fronting the entrance to Canton river. Pak-tsim, the largest and north-eastern island, bearing E. by S. 16 miles from the Great Ladrone, has near its western extreme, two high remarkable peaks, called the Asses Ears, which make it easily known, as they rise from the same base almost perpendicularly from the sea to the height of 980 feet and sloping suddenly down on the north-east side, are united to a piece of moderately elevated land, which terminates that part of the island.

Tsi-mi-wan, the next island to the south-west, is of considerable size, and separated from the south-west point of Pak-tsim by a channel about half a mile wide. A range of islets extends 4½ miles in a south-westerly direction from Tsi-mi-wan; the outermost islet (90 feet high) called Gap rock, but Man-mi-chau by the Chinese, has a small gap in it, and is the south-westernmost islet of this group. Between the south end of

Tsi-mi-wan and Peaked rock (180 feet high), the easternmost islet of the range, there is a passage 1¼ miles wide, with 18 fathoms least water in it. A Rugged rock, 50 feet high, lies about 1½ miles N.W. ½ W. from the south end of Tsi-mi-wan. The passage between Nut island and the islet nearest to it to the southward, is about half a mile wide, and carries a depth of from 10 to 26 fathoms. There is also, between Gap rock and the other islets to the eastward, an opening a mile wide, with from 16 to 18 fathoms water in it, and safe to pass through with a steady wind.

Kwei-tau, or Tortoise Head, lying about three-quarters of a mile from the east point of Pak-tsim, is a white rocky islet, having other rocks between it and the point, neither of which ought to be approached. Gay-une is another islet, rather more than a mile to the northward of the north end of Pak-tsim: there is a passage between it and the latter, which, however, ought not to be attempted unless from necessity; for there is said to exist some straggling rocks on which the sea breaks at times. The following danger requires the greatest care to avoid when vessels are passing through the Tai-ta-mi channel between the westernmost of the Lema islands and these rocks.

CAMBRIDGE ROCK, on which a vessel of this name struck, August 30, 1820, when running through the channel between Pak-tsim and the Lema islands, is of a spiral form with only 17 feet water on it, and sometimes breaks. It lies N. by W. ¼ W. 2¼ miles from Kwei-tau, and N.N.E. ½ E. 1¾ miles from the north point of Pak-tsim; and from it the highest part of Chi-chau island is in line with Hill islet bearing N.W., and the south-east side of Gay-une islet is on with the north-west extreme of Rugged rock, S.W. ¼ W. There are 4 and 5 fathoms on the rocks which surround the spiral rock, from thence the depths increase to 23 fathoms in the Tai-ta-mi channel, between the rock and the Lema islands. This channel is 2¾ miles wide, and safe by borrowing towards the latter islands in passing through.

LEMA ISLANDS consist of three large and one small island, extending in an E.N.E. and W.S.W. direction 12½ miles. The easternmost and largest island, named Tamkan, is 6 miles long and a mile broad, of moderate height and undulating, and separated from Ye-chau, the middle island, by the narrow Yat-moun channel.

The Yat-moun channel by Capt. Bate's survey of 1850, is free of danger and carries a depth of from 12 to 19 fathoms, but by the following extract* from the log of the ship *Cordelia*, it would appear there

---

* Horsburgh's Directory, Vol. II., seventh edition, page 398.

is a sunken rock in mid-channel and that this passage should not be attempted unless from necessity :—

"November 14th, 1834 : the current and swell setting the *Cordelia* bodily on the land, and having the Yat-moun channel open, steered for it, keeping near the south-west end of Tam-kan to prevent the vessel from being carried by the current on a small rocky islet lying off the north-east point of Ye-chau ; afterwards endeavoured to steer in mid-channel, but the eddy current swept the vessel into the surf that rebounded from the point of Tam-kan, when at the same time a sunken rock appeared about mid-channel, upon which the vessel must have been lost by following the track intended. Although blowing strong outside, the sails flapped to the mast as we entered the channel, which ought not to be adopted unless in a case of extreme necessity, and then the shore of Tam-kan should be kept close aboard to avoid the rock."

Ye-chau is the middle and highest of the Lema islands, and from most positions it appears flat on the top. Close to its north-east part lies a small rocky islet, visible when the Yat-moun channel is open.

Poun-tin, the third or southern of the large islands, is separated from Ye-chau by the narrow Ye-chau channel, having from 19 to 30 fathoms water in it. This island (1,210 feet high) forms more in a peak than either of the other two, and has a point projecting to the westward with a hummock on it, named E-chau head. To the southward of this head lies Tai-ta-mi, a small but high island, with a narrow channel between it and the head. Tai-ta-mi forms the north-east boundary of the Tai-ta-mi channel, which has Cambridge rock, Pak-tsim island, and the Kwei-tau bounding its western side.

DIRECTIONS.—The Lema islands on their southern side are all steep and rocky, not affording even a single bay for a boat to shelter in, and the soundings are 22 or 23 fathoms about 1½ miles from their coast ; on their northern sides the depth is generally 15 or 16 fathoms close to the shore. Vessels in the North-east monsoon should endeavour to pass between the north end of Tam-kan and Putoy, which lies 6 miles to the northward, and its north end when viewed from the E.N.E. forms a small peaked hummock.

Notwithstanding the Lema islands appear barren, there are a few men residing on them, preparing charcoal from small quantities of brushwood found between the rocks, which they send to Macao for sale. Fresh water may be obtained along the north side of Tam-kan at several places and close to the westward of its north-east point, in a little cove, called Joss House bay, there is a Chinese place of worship, and about this part the Compradore's boats await vessels after the end of August, when the easterly winds set in. The Yat-moun and Ye-chau channels should not be used unless in a case of emergency, or when the wind blows directly through, as they are narrow, with deep water, and have generally a strong current sweeping through them. Yat-moun is the widest, and of moderate depth, but if the *Cordelia* rock is in existence, it is very dangerous.

LAMMA ISLAND lies off the south-west side of Hong Kong, and its south-west point bears N.W. by W. ½ W. 13 miles from the north-east extreme of the Lema islands, and N.E. 5¼ miles from the north point of Lingting. The island is of rocky appearance, about 4 miles long, north and south, and 2 miles wide, but is narrowed near the middle by a deep cove on its east side, and a long bay on its west side, so that between them the island is not more than a quarter of a mile across. The north end of the island is about a mile distant from the south-west part of Hong Kong.*

From the north point of the long bay, on the west side of the island, the shore trends N. ½ E. a mile to another point, off which are some sunken rocks† lying half a mile from the shore. The south-east point of the island is remarkable from its being a small round hummock, of bright green appearance on the top, and rocky near the water's edge; this part of the island, as far as the eastern point, is rocky close to the shore, with 13 or 14 fathoms water half a mile off.

The cove on the east side of the island is to the northward of its eastern point, and is about 1¼ miles deep and two-thirds of a mile wide. It carries a depth of 8 to 3½ fathoms, and a vessel may anchor in 6 or 7 fathoms water, over rocky bottom, about half a mile in from the entrance, and ride in security, being land-locked. A small islet, named George island, 234 feet high, lies close to the northward of the north point of the cove.

LAMMA CHANNELS.—The East Lamma channel, between Lamma island and Hong Kong, is about a mile wide, and carries a general depth of from 17 to 23 fathoms; but a vessel will find a good and sheltered anchorage between George island and the north point of Lamma in 7 or 8 fathoms. There appear to be no dangers in this channel, but a rock‡ is said to lie off the south-east point of Mas-kong or Round island, which is on the Hong Kong shore, fronting Deep water or Heong-kong bay. About a mile to the north-east of the north point of Lamma island, and near the western point of a deep cove, named Aberdeen or Shekpywan harbour, on the Hong Kong shore, there is a cascade of good water conveniently obtained.

---

* *See* Chart of Hong Kong, No. 1,466 ; scale, *m* = 2·4 inches.
† Horsburgh's Directory, Vol. II., seventh edition, page 394.
‡ This rock is doubtful ; it is not shown in Capt. Belcher's survey of 1841.

On the western side of Lamma, between it and the islands lying off the east side of Lantao, the depth is generally 5 fathoms on a mud bottom; when bound through the East Lamma channel from the southward, the soundings will decrease rapidly to 7 and 6 fathoms after rounding the north point of Lamma. About a third of a mile N.N.E. from the north point, is a rocky patch of 8 fathoms, surrounded by depths of 14 to 21 fathoms.

CHUNG ISLAND lies near the south-east side of Lantao, and N. ¼ W. 5 miles from Lingting. Its north and south parts are high, but it is narrowed near the middle, which is low, by two bays, one on the east, the other on the west side of the island. A vessel of moderate draught will find good shelter, during an easterly gale, in the western bay in 3¼ fathoms. There is no danger in passing the south end of the island, there being 7 and 8 fathoms close to, and 5 and 6 fathoms near the western part; but East, about 3 cables from the eastern part of the island, there is a small rock, which dries at low water, and has 6 and 7 fathoms close around it.

To the northward of Chung, and at a short distance from Lantao, there are several small islands and rocks above water; but the channels between them and the Lantao shore are narrow, shoal, and unfit for large vessels.

Water.—Fresh water can be procured at the bay on the western side of Chung.

PU-TOY, lying off the south end of Hong Kong and N.N.W. ¼ W. 6 miles from the North-east head of the Lema islands, is the southern island of a group which bounds the northern side of the Lema channel. The island is of moderate height, the appearance in general barren, there being only a small quantity of brushwood in the valleys. On its western side there is a cove for boats, with a small rocky islet. The depth of water between Pu-toy and the north-east end of the Lema islands is 18 and 19 fathoms.

LO-CHAU, or Beaufort island, lying to the northward of Pu-toy, and separated from it by a narrow channel, is high, flattened at the top, and very steep all around; about its north-western brow there is a small peak, with a few large and remarkable rocks on it. Off its south-west side, at the distance of half a mile, are some large rocks above water, having no hidden dangers near them.

SUN-KONG, bearing East about 1½ miles from Lo-chau, is a small but high island, rising in a peak, 466 feet high, towards the centre; near the

CHUNG ISLAND.—HONG KONG ISLAND.

north-western part of the island there are some rocks considerably above water.

WAG-LAN, bearing N. ½ W. 7 miles from the North-east head of the Lema islands, and East about three-quarters of a mile from Sun-kong, is a small barren rocky islet, the easternmost of this group, having 16 and 17 fathoms water at a small distance round it to the eastward.

HONG KONG ISLAND, about 9 miles long, N.W. by W. and S.E. by E., 2 to 5½ miles broad, and with an area of about 29 square miles, lies between Lamma island and the main, from which it is separated by a narrow channel a quarter of a mile wide, named Lyemun pass. The appearance of the island is somewhat picturesque, but on the whole it is generally barren and unprepossessing. It consists for the most part of rocky ranges, the highest summit of which, named Victoria peak, is at the north-west part of the island, and rises to the height of 1,825 feet above the sea level. It was first ceded to Great Britain by the treaty of Canton in January 1841, and again by the treaty of Nanking in August 1842. The British settlement of Victoria is on its north side, nearly abreast the peninsula of the mainland which forms the west side of Kaulun bay. Water abounds everywhere, and each valley of the least pretension sends its stream to the cultivated grounds near the shore, where a portion is retained for irrigation, and the remainder is permitted to find its way to the sea.

The population of Hong Kong in 1841 was only 5,000; but in 1858 it amounted to 75,503, of whom only 1,462 were white, and the remainder coloured. In 1858 the total number of vessels entered in ballast and with cargoes was 1,007, amounting to 716,476 tons. The native boats frequenting the island import sugar, alum, sulphur, rice, nut-oil, dye-barks, provisions, &c.; taking in exchange opium, manufactured goods, saltpetre, and stones quarried in the island.

HONG KONG ROAD.—The shores of Hong Kong are indented by numerous bays, of which the most considerable are on the south-east side of the island. There is good anchorage throughout the entire channel between the island and the main, except in the Lyemun Pass, where the water is deep; but the best anchorage is in Hong Kong road, in front of the settlement, where the depth is from 5 to 9 fathoms over good holding ground. During the Typhoon months the anchorage in the northern part of the road is considered preferable, in consequence of the shelter afforded by Kaulun or Kowloon peninsula to the north-east, the point from which the wind blows hardest. The inner anchorage in Victoria bay is in 6 and 7 fathoms water, about half a mile off shore, abreast the Ordnance jetty, where a vessel will be sheltered from the eastward by Kellett island and the rocks off Matheson point, and be out of the strength of the tide. There is a patent slip at this latter point.

The directions for approaching Hong Kong road from the westward are given in page 41.

TIDES.—It is high water, full and change, in Hong Kong road at 10h. 15m., and the springs rise about 4¾ feet. The tides around the island are irregular, flowing and ebbing without any apparent change of direction at the surface, and sometimes there appears to be only one tide in 24 hours.

TYTAM BAY and HARBOUR.—There are several small bays on the southern shore of Hong Kong, all of which are safe for small vessels; but at the south-east part of the island is a deep inlet, named Tytam bay, which is 2½ miles deep, 1¼ miles wide at entrance, free from danger, and carries a depth of 10 to 6 fathoms. Tytam head, the western point of entrance, is a high bluff, with 13 and 14 fathoms water near it; from thence the western shore of the bay trends about N. by E. three-quarters of a mile to a small sandy bay, with a rocky islet fronting the beach. About half a mile to the northward of the islet the land forms a round projecting point, to the northward of which is a large bay, with a sandy beach, in which is Tytam village.

Tylong head, or Cape D'Aguilar, off which are two green islets, forms the eastern point of entrance to Tytam bay, and from thence the eastern shore of the bay bends round to the northward for 2 miles, and terminates in a small inlet, called Tytam harbour, carrying 4 to 6 fathoms, but its head, to the north-west, is shoal and rocky. This bay would be useful to a vessel, in the event of her being near Wag-lan at the close of the day, with the probability of a dark and tempestuous night, for by running in she will at any rate be snug, even if a Typhoon should happen during the night.

Water.—At the head of Tytam harbour there is a rivulet of fresh water, which, however, cannot be procured without inconvenience when the tide is low. Water may be obtained at Tytam village on the western shore of the bay.

TIDES.—There is very little tide in Tytam bay, and, like all the places hereabouts, it is difficult to fix the time of high water, owing to the variety of channels, and the wind greatly influencing the tidal streams; but the rise and fall is about 7 or 8 feet at springs, and about 3 or 4 feet at neaps. The ebb sets through between Lo-chau and Hong Kong to the eastward.

DIRECTIONS.—If bound to Tytam bay from the eastward the route may either be taken to the northward of Wag-lan, Sun-kong, and Lo-chau islands through the Shingshimún Pass, or to the southward of these islands through the Lema channel; then round the Castle rock to the westward of Lo-chau. But the passage to the northward is preferable, for after opening the bay a vessel may haul to the northward into any convenient berth; whereas, by taking the southern route, if the wind be northerly, she will have to turn in.

If Shingshimún Pass be taken, give Wag-lau and Sun-kong a berth of about half or three-quarters of a mile, and steer for the Pass, which is formed by the high island of Lo-chau to the southward, and by the two green islets off Tylong head to the northward: in this track a vessel will carry 17 and 16 fathoms water from Wag-lan, and, by keeping in mid-channel, will have 27 and 30 fathoms deepening as Lo-chau is neared, and shoaling to 14 or 16 fathoms near the islets. The depths will shoal fast to 10 or 11 fathoms when about 1 or 1½ miles to the westward of the two islets. From thence steer for the anchorage off Tytam village, on the western shore, in 6½ fathoms. In this position a vessel will be well sheltered from all winds, except those from South, which cannot affect her much, as the islands and rocks contiguous to the entrance prevent much swell from rolling in.

LANTAO, or Ty-ho, the large high island lying to the westward of Hong Kong, is 14 miles long, N.E. by E. and S.W. by W., and its greatest breadth is 5½ miles. About the centre of the island the land is very high, making in peaks, the highest and westernmost of which rises 3,050 feet above the sea level.

West Coast.—Close to the western shore of Lantao, at 1¼ miles from the south point, there is a peaked hill, which at high water is insulated. From this hill to the point a mud flat extends about a third of a mile off shore, with only 2 fathoms water on it; therefore, in passing this part of the island do not decrease the depth under 7 fathoms, as the soundings will shoal fast from 17 to 7 fathoms near the edge of the flat.

About a mile to the N.N.W. of the peaked hill, and three-quarters of a mile from the nearest shore, there is a rock above water, having near it a depth of 15 fathoms, and between it and the shore 7 fathoms, decreasing quickly towards the latter. N.E. by N. 1¼ miles from this rock is a bluff point, and to the eastward of the latter a bay, in which is the village of Ty-ho, where there is a creek or rivulet into which a boat may go at high tide. To the southward of the village are two bays, both of which are shoal, but fresh water may be procured in them.*

---

* *See* Plan of Tong-ku or Urmston Bay, with north-west side of Lantao Island, No. 1,222; scale, $m = 1$ inch.

North Coast.—On the northern side of Lantao there are two projecting points, three-quarters of a mile apart, between which is the bay and village Sah-lo-wung; and directly fronting the eastern point of the bay and about a quarter of a mile distant, is a small islet, having a rock awash a short distance to the northward of it. Between this islet and Saw-chau, the depth is too small for a vessel of large draught at low water; towards Saw-chau is the deepest water, $3\frac{3}{4}$ and $4\frac{1}{4}$ fathoms, shoaling near the Lantao shore to 3 and $2\frac{3}{4}$ fathoms, on a soft mud bottom.

Immediately eastward of the small islet off Sah-lo-wung is another bay formed by Chu-lu-cock island extending north and south $1\frac{3}{4}$ miles; in this bay is Tung-chung village. Red point, the north-east extreme of Chu-lu-cock, has a remarkable rocky appearance, and is frequented by a company of stone-cutters, who cut the granite rocks into slabs for building. The south point of this island is so near the Lantao shore, that in passing it is difficult to distinguish it to be an island. In Tung-chung bay the water is shoal, being only 2 and $2\frac{1}{2}$ fathoms; and there is but little water on the eastern side of Chu-lu-cock; from thence the northern shore of Lantao is not inhabited.

About $1\frac{1}{4}$ miles E.N.E. of Red point lies a small green island, and three-quarters of a mile farther in the same direction another small island, which are the Brothers of Dalrymple, or Motoe of the Chinese. A rock above water, lies about half a mile to the southward of the East Brother, and about a mile off the Lantao shore.

The depths near the Brothers are 7 and 8 fathoms, shoaling from the eastern islet towards the northern shore into 4 or 5 fathoms, and making the channel narrow; a small reef borders the western side of the West Brother. According to Captain H. Smith, of H.M.S. *Druid,* there is a good channel with 8 or 10 fathoms water between the East Brother and the large rock to the southward of it; the rock is high above water, and bold on all sides. From the East Brother the north-east point of Lantao bears E. by N. 4 miles.

South Coast.—The southern coast of Lantao forms two large bays with shoal water in them. The larger and eastern bay, to the north-eastward of the Soko islands, has in it a small islet and some rocks above water, and a populous village at its head. The depth is 2 fathoms within the rocks, and 4 and 5 fathoms in the entrance of the bay, where there appears to be good anchorage. The western bay is less capacious than the other, and carries a depth of 2 to 5 fathoms.

Off the east entrance point of the eastern of the above bays, and separated from it by a narrow channel, is a high green island named Patung, and close to its west side are some rocks above water. A small vessel will find good anchorage to the westward of these rocks, anchoring with them bearing about S. by E., three-quarters of a mile distant, in 5 fathoms water. Fresh water may be procured at the sandy beaches on the south shore of Lantao. In the channel formed between Lantao and Patung there are 7 fathoms water; the ebb tide here runs to the eastward.

CAP-SING-MUN PASSAGE, or Throat Gates, formed between the north point of Lantao and the main, is separated into two channels by Mah-wan island. The channel on the west side of Mah-wan, being extremely narrow with dangerous eddies, ought never to be used by any vessel, especially as the channel to the northward of that island, is wider, with good anchorage, a regular tide, and the advantage in the North-east monsoon of being to windward.

DIRECTIONS.—In proceeding through the Cap-sing-mun passage from the westward, keep close over to the mainland to avoid a reef, extending a third of a mile from the north-east point of Mah-wan; then keep in mid-channel between that island and Chung-hue island, which lies a mile to the eastward, and after rounding the south end of the latter, if bound for Hong Kong road, steer for the west end of Won-chu-chau or Stone-cutters island.

SAW-CHAU, lying 3 miles to the northward of Lantao, and S.E. 4½ miles from Lintin, is a small narrow island nearly a mile long, with a sharp hummock on its north end. To the northward of Saw-chau about one mile distant, there is another island, higher and more rocky in its appearance, named Tong-ku; and S.W. by S. from the south point of Tong-ku, and W. by N. from the north point of Saw-chau there are two rocks above water, about a mile distant from each island: the western rock is named White rock from its white appearance. The depths near the eastern sides of Saw-chau and Tong-ku are 5 to 9 fathoms; to the westward of Tong-ku, 4½ fathoms; and near the rocks 5 fathoms, at low water.*

URMSTON BAY or Tong-ku harbour, bounded by the islands Tong-ku and Saw-chau to the west, and Castle Peak land to the east, is a safe anchorage, and tolerably sheltered from all winds. The best berth is in about 8 or 9 fathoms, with the peak of Tong-ku just open of the south end of Lintin, and nearer to the main land than to Tong-ku. This safe bay or harbour was named Urmston by the captains of the fleet who anchored there in August and September, 1823, at the recommendation of Sir James Brabazon Urmston, President of the Company's factory at Canton during the discussion with the Chinese, relative

---

* *See* Chart of Canton River, Sheet 1, No. 1,782; scale, $m = 1\frac{1}{2}$ inches.

to the affair of the *Topaze* frigate in 1821-2, at Lintin; the anchorage was found secure, with smooth water when it blew a gale from eastward. Fresh water was procured in abundance.

DIRECTIONS.—The proper channel into Urmston bay for vessels of large draught is to the northward of Tong-ku, and has a depth of 7 and 8 fathoms; and the approach to it from southward, is between the east side of Lintin South sand and the islands of Saw-chau and Tong-ku. The passage to the southward of Saw-chau has only 3¼ fathoms in it, and between Saw-chau and Tong-ku, the depth is only 2½ fathoms.

The channel between White rock and the east side of Lintin spit is about 2 miles wide, with 7 and 8 fathoms, decreasing towards the spit to 5 fathoms. If working to the northward, do not stand so far west as to shoal to 5 fathoms, or to bring the east side of Lintin to the northward of N. by W. With the south end of Saw-chau bearing E.N.E. and Lintin peak North, a vessel will be on the southern edge of the spit in 4¾ or 5 fathoms, sand and mud.

LINTIN ISLAND, lying to the W.N.W. of Urmston bay, is about 7 miles in circumference, and its summit terminates in a high conical peak, which bears N.E. 14 miles from the outermost of the Cau-chau or Nine islands. A narrow spit of sand extends about 4½ miles to the southward from the south end of the island, having 3½ fathoms water on its outer part, but only 9 feet within 2¾ miles of the island, and rather less in some places. The spit is steep-to on the west side, with 10 fathoms near it, 7 fathoms touching its verge, then 3 fathoms; and the water suddenly deepens from 3 fathoms on the east side to 7 or 8 fathoms. When the island is approached within 5 miles, a vessel of large draught should not when standing eastward towards this spit bring the west end of the island to the westward of N. ¾ W. or should tack immediately after deepening to 9 or 10 fathoms; but in the night she ought not to deepen to above 7 or at most 8 fathoms.

A narrow sandbank also extends 13½ miles in a N.N.W. direction from the north side of Lintin, and on its northern part is a narrow ridge called Lintin bar, the southern end of which, in 2½ fathoms, bears W. by N. ¾ N. from Fan-si-ak islet, and N.N.W. ¼ W. about 6¾ miles from Lintin peak; the least water on the bar is 12 feet, and its northern end in 2¾ fathoms lies N.W. by N. 10½ miles from Fan-si-ak, with Sam-pan-chau just open westward of the west extreme of Anung-hoy island.

The anchorage off Lintin is in 10 or 12 fathoms, about 1½ miles from the sandy beach on its south-west side; under 10 fathoms the water shoals quickly to the shore.

LINTIN ISLAND—LANKEET ISLAND.

Water.—Fresh water may be obtained at the eastern extremity of the beach on the south side of Lintin, and at times a few bullocks and vegetables may be procured from the inhabitants of the village.

Tides.—It is high water, full and change, at the anchorage off Lintin island at noon, and the rise at springs is 7 or 8 feet. The streams run nearly North and South, and the ebb in the freshes sometimes sets at the rate of $5\frac{1}{2}$ or 6 miles per hour. In the North-east monsoon the neaps are very irregular, sometimes only one flood perceptible during 24 hours with a small rise when the other flood should prevail.

Fan-si-ak Islet.—Two rocky islets, the easternmost of which is the largest and called Fan-si-ak, and the other White rock, lie nearly North $4\frac{3}{4}$ miles from the peak of Lintin. When these islets are in one bearing E. by S. $\frac{3}{4}$ S., the southern extremity of Lintin bar is on the same bearing. The east side of the channel between the south extremity of the bar and Lintin is bounded by mud-banks, with irregular soundings on them of $2\frac{1}{2}$ and $3\frac{1}{2}$ fathoms at low water.

Lankeet Flat, or bar, extending from the north end of Lintin bar, across the channel to the shoal mud-bank on the west side, and N.W. towards Lankeet island, consists of sand and mud, with hard bottom in some places. The depths on it are 3 and $3\frac{1}{4}$ fathoms at low tide, and $4\frac{1}{2}$ to $4\frac{3}{4}$ fathoms at high water springs; a vessel drawing more than 20 feet ought not to pass over it until about half flood. Close to the northward of this flat there are generally some fishing stakes, and boats fastened to them, and others between Lintin and Lankeet; care should be taken not to run over the boats, which generally show lights in the night.

Lankeet Island, bearing N.N.W. $\frac{1}{2}$ W. 19 miles from Lintin peak is formed of two hills, sloping into a low point at the west end, where there is a well of fresh water by a small temple close to some trees; but the island is not inhabited. A spit extends S.S.E. $\frac{3}{4}$ E. $2\frac{1}{2}$ miles from its south-west point, with only 2 and $2\frac{1}{4}$ fathoms over it at low water. Between this spit or flat and a long narrow sand to the westward there is a channel leading close past the west point of Lankeet, to the western part of Ty-cock-tau island.

To proceed up this channel, keep a large white patch on Ty-cock-tau in line with the outermost of the rocks projecting off the west end of Lankeet; with this mark on, a vessel will have $4\frac{3}{4}$ or 5 fathoms at high water, about 4 miles from Lankeet; and will carry the same depth till nearly abreast the west end of the island, where she will have about 6 fathoms in Lankeet road. This is a convenient place for a vessel to moor when circumstances require her stores or sick to be landed. All the space between Lankeet and Ty-cock-tau is shoal, having in many places only one fathom at low water.

SAM-PAN-CHAU, or Boat islet, bearing N. by E. ½ E. about 1½ miles from the east end of Lankeet, is small, of middling height, resembling a boat turned bottom upwards. An extensive rocky bank, partly above water, projects N.W. from it, and joins the shoal bank extending from Lankeet to Ty-cock-tau. There are regular depths of 7, 8, and 9 fathoms to the eastward of Sam-pan-chau. This islet is the best guide for crossing over Lankeet flat, between the northern part of Lintin bar and Lankeet.

DIRECTIONS from SINGAPORE to HONG KONG and CANTON RIVER.—When June approaches, and the South-west monsoon is set regularly in, the track from Singapore to China by the Outer passage to the eastward of Pulo Sapata and over the Macclesfield bank is preferable, the winds being more steady in the open sea than near the coast. About full and change of the moon and as early as April, a westerly breeze will sometimes be found blowing out of the Gulf of Siam to carry a vessel to the Macclesfield bank, and afterwards easterly winds to run her to Hong Hong.*

This passage becomes precarious if a sailing vessel is not up with Pulo Sapata early in October;† for near this island, about the middle of that month, strong southerly currents begin to prevail with light northerly winds, variable airs, and calms, by which many vessels have been delayed for several days, and have made no progress to the northward. Fresh southerly winds have been met with, even so late as 1st of November, but these instances are rare.

Sailing vessels leaving Singapore for China in February, March, and part of April, may expect a tedious beating passage. In March, April, or May, they can proceed by the Inner passage along the coast of Cochin China, which is generally the most expeditious route in these months.

Some vessels proceeding by the Outer passage have carried strong S.W. and southerly winds, when others inside the shoals have at the same time experienced N.W. and westerly gales blowing out of the Gulf of Tong King, with dark weather and rain, and have been in danger of being driven among the shoals; the Inner passage ought, however, to be chosen in the strength of the South-west monsoon if the vessel is a dull sailer, for the sea will be smooth, and being near the land she may reach an

---

* *See* Charts:—China Sea, Sheets 1 & 3, Nos. 2,658, 2,660; scales, $m = 0.05$ of an inch.

† Directions from Singapore to Hong Kong in N.E. monsoon, will be found in page 83 of China Pilot, Appendix No. 2.

anchorage if required. The gales out of the gulf are not frequent, and the land may be kept in sight nearly all the time.

In taking the Inner passage, steer from Pulo Aor along the coast to the Redang islands, thence across the Gulf of Siam, and along the coasts of Cambodia and Cochin China, keeping the latter aboard to Cape Touron. From thence steer for the south-west part of Haïnan, coasting along this island and passing between it and the Taya islands; then cross over to make the coast of China about Tieu-pak, or Hailing island. The islands from thence to Hong Kong may be coasted along at discretion, or shelter may be found amongst them on emergency. If the vessel takes this route before the middle of March or the 1st of April, the passage will be tedious unless she is a good sailer.

TO HONG KONG.—If bound to Hong Kong in the strength of the South-west monsoon, with the wind steady between S.E. and S.W., endeavour to make the Great Ladrone island bearing about North, then steer between it and the Kypong islands, and between Lingting and the Lema islands, for the west Lamma channel. After the middle of August, when easterly winds are likely to prevail several days together, as they are more or less at all seasons, it will be necessary to make the North-east head of the Lema islands, and proceed in by the Lema channel, towards the west Lamma channel. The east Lamma channel is also safe in both monsoons, for although the water is deep, if the wind falls light it is safe to anchor in, and there is little or no tide.*

Hong Kong road is generally approached by sailing vessels from the westward, on which side it is protected by Green island and Kellett bank, which extends nearly 1½ miles northward from the latter island, and carries a depth of 3½ fathoms. The road is sometimes approached from the eastward through the Lyemun Pass during the North-east monsoon, but the winds are generally baffling under the high land.

When abreast of Green island, if the vessel be of heavy draught, keep the peak of Lamma island (Mount Senhouse, 1,140 feet high) open westward of Green island S. ¾ E. until the Devil's peak (on the mainland near Lyemun Pass) is in line with the White rock on the south point of Won-chu-chau or Stone-cutters island, when a S.E. by E. course will lead northward of Kellett bank, and direct for the anchorage.

Vessels of proper draught can proceed over Kellett bank, or through the 4 fathoms channel between Green island and the south part of the bank, by passing about 1½ cables to the northward of the island, and then steering for the road.

---

* See Chart of Hong Kong, No. 1,466; scale, $m = 2$ inches.

The narrow channel may be taken between Green island and Hong Kong, if the wind is fair, and blows *right through*. Many sailing vessels have run through this channel, amongst which were H.M. ships *Modeste*, *Wellesley*, *Cornwallis*, and *Vernon*. It has depths of 10 to 12 fathoms in the middle, shoaling to 8, 6, and 4½ fathoms after passing the small islet eastward of Green island.

To CANTON RIVER through LEMA, LANTAO, and TAI-TA-MI CHANNELS.— The Lema channel, formed by the Lema islands on the south, and by the Pu-toy group on the north, is about 6 miles wide, and safe to navigate, with regular depths of 17 to 19 fathoms, and soft bottom. This channel should, if possible, be always adopted by sailing vessels bound to Hong Kong or Canton river in the North-east monsoon, to effect which they ought to make the North-east head of the Lema islands, bearing to the westward. If the weather be thick, and the wind blows strong at East or S.E., it may be prudent to heave to, when land cannot be discerned above 3 or 4 miles.* The depths are 19 to 21 fathoms, close to the head, and about 18 fathoms at the entrance of the channel. If the weather will not permit a vessel to enter the channel, she should not shoal under 25 or 26 fathoms: in these depths she will drift clear outside all the islands.

If, however, a vessel should happen to be near the entrance of the Lema channel in the evening, and from the falling of the mercury in the barometer, or by the appearance of the weather, a Typhoon is expected, she should run immediately for Tytam bay, or for Tathong channel, or the east Lamma channel, as may be most convenient ; in either of which she will be secured from the tempest, if an anchorage is gained before night.

During S.W. or westerly winds, it will sometimes be found difficult to enter the Lema channel from the eastward, by turning through, as there is generally a set from West to East, occasioned by the ebb coming from the westward out of the numerous channels, and the flood coming from the S.W. ; with a strong S.W. wind the velocity of the stream is about 1½ knots per hour to the eastward, only slacking a little when it ought to change its direction. Pu-toy island may be approached with safety to a quarter of a mile, and the whole north side of the Lema islands to half a mile.

---

* The *Nautilus* of Calcutta, September, 1802, made Pedro Blanco, and after running to the westward, hove to for the night, keeping in from 18 to 14 fathoms. A strong easterly gale had prevailed in the night, which increased, with thick weather, at daylight, when they found themselves close to the east side of one of the rocky islands northward of the Lema channel, on which the vessel struck, and soon went to pieces.

Through LANTAO CHANNEL.—From about a mile to the southward of Pu-toy, a West course for 19 miles will lead to the entrance of the Lantao channel, passing northward of Lingting and southward of Lamma, the depth decreasing from 17 fathoms off Pu-toy, to 12 and 13 fathoms after passing Lingting, and to 7 and 8 fathoms as the channel is approached; there are 12 fathoms in mid-channel in the entrance, decreasing to 7 or 8 fathoms towards A-chau. Lingting, which is of considerable height, and terminates at the summit in a conical peak, may be passed on either side as the wind requires. If passing to the southward give a wide berth to the sunken rocks off its north-west point; and to those off the north-east point if passing to the northward; but the channel to the northward of this island is preferable, for in daylight it has no hidden danger, and a vessel may work from side to side. In the night it will be prudent not to come nearer the north side than $1\frac{1}{2}$ miles to avoid the two small rocks (page 26), off its north-east point.

Chi-chau, when seen from the eastward, has a remarkable appearance, and is a good guide; it makes like a high, round, detached island, with distant rugged land to the westward of it, which is Lafsami and Chung-chau. Having entered the Lantao channel, the course through is N.W. by W., and the depth will be variable, not under 8 or 9 fathoms, or above 25 fathoms; this inequality may be owing to the ebb tide running in strong eddies, particularly in July or August, when its velocity is sometimes $4\frac{1}{2}$ knots per hour on the springs. With a light wind, at times, it is very difficult to manage a vessel hereabout; on some occasions two or three boats, assisted by the sails, have been baffled in their attempts to tow the vessel's head round. After passing between Chi-chau and A-chau, the water will deepen from 10 to 17 fathoms in mid-channel near the islands which front the south-west point of Lantao, and there are 7 fathoms close to the point. Having rounded the point at a moderate distance, steer to the northward for Lintin, or to the westward for Macao road, as circumstances require; in the latter case the depth will gradually decrease to $5\frac{1}{2}$ or 5 fathoms.

In turning through the Lantao channel, when standing northward do not decrease the depth under 7 fathoms, in a vessel of large draught, nor pass the line of bearing between the south points of Lantao and A-chau. There is a good channel, one mile wide, between the northern Soko island and the Lantao shore, which may be taken by a vessel when blowing fresh from the northward. In this case, after passing the south point of Patung, a small rocky islet will be seen in the bay on the southern shore of Lantao; steer West through this channel, until this islet is shut in behind the western point of the bay, when keep towards the south point of Lantao, and the depth will be 4½ fathoms, muddy bottom, between the point of the sandy spit and the Lantao shore.

From the small islet off the north side of Chung-chau, Lintin bears N. ⅓ E. 13 miles; the sand spit extending off the south side of Lintin is on the latter bearing, therefore a vessel will clear it if this islet is kept S. by E. until Lintin peak bears N. by E., then steer for the west point of Lintin. In a dark night, a N.N.W. or N.W. by N. course should be steered from the middle of the Lantao channel until the depths shoal to 6 fathoms, then steer North, keeping a good look out for the fishing stakes; on this latter course, if the soundings deepen above 7 fathoms, keep a little westerly until the vessel is near or above Lintin, where she can anchor. By not deepening above 7 fathoms, she will not be too near Lintin sand spit, there being 9 and 10 fathoms close to it. The ebb tide, from the west end of Lintin to the eastward, sets South; but over on the western shore its direction is S.E.

Through TAI-TA-MI CHANNEL.—If proceeding towards Canton river through the Tai-ta-mi channel, between the Lema and Kypong islands, after clearing the Cambridge rock (page 29), steer to the northward for Lingting, pass between it and the Samoun group, and then proceed through the Lantao channel; or pass between the Samoun group and Ai-chau, and then steer for the Lantao channel or to the N.W. direct for Lafsami. Having approached Lafsami keep within a mile of its western side, and of the south part of Chung-chau, to avoid the 4-feet needle rock; after passing Chung-chau, steer for Lintin, or for Macao road.

If the channel be taken between the Great Ladrone and Gap rock, or the narrow passage between the latter and the Kypong islands, steer to northward, and proceed along the west sides of Ai-chau and Lafsami. Or if bound into Macao road, there is a more direct passage about a mile wide, with 13 fathoms water, between Pak-leak and Chuk-wan, then on the north side of Tong-ho and Liungnib, and to the southward of Ty-lo, which track lies nearly in a direct line towards the road. Although this channel is safe in the day-time great care must be taken to avoid the Raleigh rock (page 24). There is a safe passage between the Great Ladrone and Pak-leak, with 14 to 9 fathoms water in it, but recollect the sunken rock lying half a mile from the north side of the Little Ladrone, and also the Clio rock; a vessel taking this route should pass to the southward of Potoe.

To CANTON RIVER through the GREAT WEST CHANNEL.—This channel, on the west side of the Ladrone islands, is generally used by vessels bound to Canton during the strength of the South-west monsoon, and to do this they endeavour to fall in with the Great Ladrone bearing about North or N. by E.; but late in the season when the winds incline to the eastward, or at any other time when they are expected to come from the northward or eastward, it will be prudent to make the North-east head of the Lema islands, and proceed towards the river by the Lema and Lantao channels, page 42. Here the risk of being horsed to the westward by the freshes setting out of the Great West channel is avoided, and a northerly wind will lead to an anchorage in the river. When Typhoons happen on the coast, they generally commence in a moderate gale from the northward, which is a leading wind for these channels, and as the wind commonly veers to the eastward before it blows hard, a vessel with the first of the gale may get well up the river above Lintin, where these storms blow with less violence than outside among the islands.

As the approach to Canton estuary is probably more safe than that of any other large river in the world, there being no sandbanks at the entrance, and the channels amongst the islands outside being mostly all free from hidden danger, a stranger should not hesitate to push through the nearest convenient channel without a pilot, if the weather is tolerably clear. But the streams must be attended to, as they set in different directions amongst the islands to the south-eastward according to the prevailing winds; a strong easterly wind generally producing a westerly current or tide, which abates in strength when the ebb should be setting to the south-east. If an outside pilot can be obtained at a moderate rate he may be useful, to run the vessel into some cove or place of shelter, if a storm should be approaching, or if she be in a disabled state. Macao road should not be used if there is an appearance of bad weather, but run well up the river above Lintin.

About South 30 miles from the Great Ladrone, the depths increase to 27 or 28 fathoms; about 60 miles from it, to 42 and 44 fathoms; and soundings extend on the same meridian to about lat. 20° N.; from hence they continue westward towards Hainan head; but converge towards the land, with deeper water eastward of the meridian of the Ladrone islands. A vessel falling in with the land in thick weather may easily distinguish whether it is that of the islands to the eastward of the Great Ladrone; for the Kypong and Lema islands have soundings of 23 and 24 fathoms close to; whereas the islands between the Great Ladrone and St. John to the westward have only 10 and 11 fathoms at a considerable distance outside. These are also large and of regular appearance, resembling a coast more than islands; but those to the eastward, are detached, high, and uneven, excepting Tam-kan, the largest of the Lema islands, which is long and of an undulating form.

The freshes out of Canton river set almost constantly from the south end of Montanha, along the shores of the islands to the westward, at the rate of 1 to 2 knots an hour, particularly with strong easterly winds; and although at times there seems to be on the surface a flood tide setting eastward, or into the entrance of the river, the freshes underneath continue to run westward, by which sailing vessels are rendered ungovernable, even in fresh winds. Many vessels from this cause, after getting near Montanha, or between it and Potoe island, have been drifted nearly to St. John island whilst making every endeavour, with moderate winds, to keep their heads to the eastward. Steering, therefore, for the Great West channel, never borrow near San-chau, or the other islands to the westward, unless it is blowing strong from the S.W. to avoid being drifted to the westward. The freshes abate at times, and then weak tides set to the eastward; but as these are not of long duration, a vessel should keep on the eastern side of the channel in deep water towards the Ladrone islands and Potoe, and anchor instantly if she finds the current is drifting her westward.

In the strength of the South-west monsoon, (as before stated in page 45,) endeavour, if the wind be steady between S.E. and S.W., to make the Great Ladrone, bearing about North, and never fall in with the islands to the westward; this is the more necessary after the middle of August, when easterly winds are likely to prevail several days together, as they are, more or less, at all seasons. If a vessel falls to leeward about St. John, in September or October, she will generally make a tedious passage to Macao if she keeps close along the islands, where the current or freshes setting westward will oblige her frequently to anchor; as these freshes prevail only in shoal water, near the islands, the best plan to adopt is, to stand well off the land, and take every advantage of the favourable shifts of wind, to get to the eastward.

Having arrived abreast of Chung-chau-si, or if the vessel has anchored in Macao road, with a leading wind she may weigh with the ebb, if she can haul over north-eastward for Lintin; for the tide will then act upon her port bow, and keep her off the western shore: whereas, with an easterly wind, the flood is likely to force the vessel into shoal water near that shore. With a fair wind, steer about N.N.E. $\frac{1}{2}$ E. from Macao road for Lintin; if it be night from $4\frac{1}{2}$ to 5 fathoms are good soundings; for at low water springs, greater depths ought not to be expected, until several miles north-eastward of the road.

In turning up with the northerly wind and flood tide, tack from the west side of the channel in about 4 fathoms, according to the vessel's draught, the lead being a safe guide along the western shore, where the bottom generally consists of mud. The islands eastward of Macao road may be safely approached, having 5 fathoms near them, and when past Chung-chau-si, the depths will increase to 9 and 10 fathoms on the east side the channel, towards Lantao. Working from hence to Lintin in the night, stand to $4\frac{1}{2}$ fathoms in the west part of the channel, and do not deepen above 7 or $7\frac{1}{2}$ fathoms to the eastward. Here, the tides become stronger as the vessel proceeds upwards.

In Macao road, and between it and Lintin, the tides are frequently irregular, setting in a different direction at the surface to what they do underneath, by which vessels are rendered ungovernable in light winds. The ebb is stronger, and continues longer than the flood; the freshes often running out below, when a flood tide at the surface is setting into the river.

From LINTIN to the BOCA TIGRIS.—When within 7 miles of Lintin steer for its west point bearing about N. $\frac{1}{2}$ E., and when abreast the point run to the northward in soundings from 5 to $6\frac{1}{2}$ fathoms: with a westerly wind, borrow on the west side of the channel; if it is easterly keep in 6 to $6\frac{1}{2}$ fathoms with the flood tide. It will be safe to proceed 9 or 12 miles above Lintin, even in the night, with a working wind, the lead being a certain guide, by tacking from the west side of the channel in $4\frac{1}{2}$ fathoms, and from the east side in $6\frac{1}{2}$ fathoms; but when about 6 or 7 miles northward of Lintin, tack in $5\frac{1}{2}$ fathoms from the east side of the channel, for the deepest water is near the edge of Lintin sand, and if a vessel begins to shoal on its verge to 5 fathoms, she will not have room to tack.*

Lantao is frequently obscured by clouds or haze, but when its summit is visible the west peak of that island affords a good mark for running up this channel in the day. Steering N. by W. or N. by W. $\frac{1}{2}$ W. from the west end of Lintin, draw gradually the high west peak of Lantao on with the west end of Lintin, and continue to bring it more easterly until it is on with Lintin peak, or a little open eastward of that peak, and keep it so, until the vessel is more than half way from Lintin towards Lankeet. Then, if the wind is contrary, Lantao west peak may be brought nearly to the east end of Lintin, in tacking from the east side of the channel, and well westward of Lintin peak when tacking from the west side; but on a nearer approach to Lankeet, the west peak of Lantao must not be brought westward of Lintin peak.

When within 5 miles of Lankeet, the west peak of Lantao must not be brought more westerly than touching the east end of Lintin, when in the west side of the channel; and to a considerable way open with the same,

---

* *See* Charts, Sheets 1 and 2 of Canton River, Nos. 1,782, 1,741; scales, $m = 1\frac{1}{2}$ and 3 inches.

when on the eastern side. Here the depths decrease, and there is only about a fathom more water in the east side than in the west side of the channel. A narrow mud bank, with 2¼ and 2 fathoms at low tide, bounds the west side of the channel in this part, and extends in a N.N.W. direction about 4 miles, terminating nearly 1¼ miles to the S.W. of Lankeet. There is a channel of 4½ and 5 fathoms westward of this mud bank, into which or upon the bank, the pilots sometimes get vessels in the night; but with those of large draught they are more inclined to borrow close over to the eastward, whereby they have frequently grounded upon Lintin bar; it will therefore be prudent, when the pilot appears confused or uncertain of his position, to anchor before the vessel shoals her water.

From a position about half a mile off the west end of Lintin, a N. b. W. ¾ W. course would lead fair through the channel, close on the east side of Sam-pan-chau, were the tides to run in that direction; but from Lintin they set N.N.W. and S.S.E. nearly as far as the north end of Lintin bar, and from thence to Sam-pan-chau about N.W. by W. and S.E. by E.

Steering to the northward, with the west peak of Lantao open a little eastward of Lintin peak, page 47, or keeping between 4½ and 5½ fathoms if the weather is cloudy, Lankeet island will be seen making like a saddle, and shortly afterwards two small islets or rocks will appear close to its eastern end. These rocks will be nearly on with the middle of the opening of the Boca Tigris when first seen, and should not be brought more easterly; nor in working ought they to be brought to touch the point of Tiger island, which forms the west side of the opening, until within 4½ miles of Lankeet; being then northward of Lintin bar, a vessel may edge over to the eastward. There is no good cross mark to know when clear of the bar; but a pagoda on the western shore bearing S.W. ⅔ W., will lead northward of its extremity. From the northern end of the bar Sam-pan-chau is a little open with Anung-hoy point N.N.W. ½ W., and the little hill on the east end of Lankeet N.W. ⅓ N., distant about 4¾ miles.

Shortly after the rocks off the east end of Lankeet are on with the middle of the opening of Boca Tigris, or rather more westerly, Sam-pan-chau will be recognized when within 6 or 7 miles of Lankeet, and will then appear under the land, a little eastward of the high round summit of Anung-hoy, a high, round hill, sloping to a point on the west side, and forming the eastern boundary of the Boca Tigris. Anung-hoy peak in line with Sam-pan-chau hummock, N. by W. ¾ W., leads westward of Lintin bar, and eastward of Lankeet spit. With a working wind keep Sam-pan-chau between the eastern shoulder of Anung-hoy hill and the west point of the same; but that islet must not be opened westward of Anung-hoy point until clear of the north end of Lintin bar.

With an easterly wind, to prevent being set by the tide towards Lankeet, keep on the east side of the channel, with Sam-pan-chau shut in a little eastward of Anung-hoy point, or nearly on with it. When within 4 miles of Lankeet a vessel may stand well to the eastward in working, opening Sam-pan-chau considerably westward of the point, being then to the northward of the extremity of Lintin bar; do not, however, stand so far over as to bring Anung-hoy point to touch Chuen-pee, but tack before they come on, for farther eastward the water is shoal. After opening Sam-pan-chau with Anung-hoy point, which with a westerly wind need not be done until abreast Lankeet, steer then direct for the land of Anung-hoy, giving Sam-pan-chau a berth to the westward of half a mile or more at discretion, in 9 or 8 fathoms; the depths from thence will be 9, 8, and 7 fathoms to the entrance of the Bocca Tigris, increasing to 13 and 15 fathoms abreast South Wantong.

If in a vessel of small draught, a cast of 3½ or 4 fathoms hard ground be got before Lankeet is seen, in a clear night, she may be certain of its being on Lintin sand, and will deepen fast on hauling westward into the channel.

To CANTON RIVER through FAN-SI-AK CHANNEL.—It would be imprudent to attempt the channel on the east side of Lintin in a vessel drawing 23 feet water, being very narrow just above and about Tree island, with a considerable swell in it when blowing strong from the northward. The southern part of the channel between the White rock and the east side of Lintin south spit is about 2 miles wide, with 7 and 8 fathoms, decreasing towards the spit to 5 fathoms. In working to the northward, do not stand so far west as to shoal to 5 fathoms, or to bring the east side of Lintin to bear North of N. by W. When northward of Tong-ku, if the vessel is of 20 or 21 feet draught keep the eastern shore aboard, avoiding the spits of shoal water at the points of the islands until off the north end of Mah-chau, the shoal off the south end of which will be avoided by not shutting Tree island in with Mah-chau, or by not bringing the highest peak of Mah-chau westward of N. ¼ W., when the White rock is in one with the north end of Fan-si-ak, which is the mark for the south end of Mah-chau spit.

From hence to Tree island, when standing towards Lintin bar or Fan-si-ak bank, keep the lead going, and tack in 4 fathoms or less, according to the vessel's draught; but the lead will be the best guide, as the bank is much curved in shape. Standing to the eastward, do not bring the north or highest peak of Mah-chau to the westward of South, and when the south point of Sui-chan bears N.E. ¼ N. do not bring the tree on Tree island to the westward of N. by W. ¼ W., to avoid the shoal spit of 2 or

3 fathoms, which extends S.S.E. from that island nearly a mile. When thus far, endeavour to pass between Tree island and the fishing stakes No. 1 (in the chart) placed near it; this island is safe to approach close to the rocks, but on the Channel banks, on the western side of these stakes, the water shoals suddenly to $2\frac{1}{4}$ fathoms, irregular soundings, sand and mud.

Being close to the west end of Tree island, do not bring the tree to bear more to the southward than S.E. $\frac{1}{2}$ E., this bearing being close on the edge of the shore bank. Standing to the westward, do not bring the White rock off Fan-si-ak to the eastward of the saddle on the east end of Lintin, or the east end of the fishing stakes No. 3 (in the chart) to the northward of N.W. by N., the lead not being a sufficient guide for the Channel banks. If the fishing stakes be not removed, they appear to be a preferable guide to the landmark, being always discernible, but either may be used in clear weather. When within half a mile of the stakes No. 3, the passage becomes wider, extending from the shore bank to Lintin bar, with 4, $4\frac{1}{4}$, and $4\frac{1}{2}$ fathoms in it at low water, shoaling gradually on either side, so as to render the lead a guide in tacking, the bottom being very soft mud.

If close to Tree island with a leading wind, steer direct for the centre of the fishing stakes No. 3, and pass them on either side, as circumstances require.

There is another range of fishing stakes (numbered 4), bearing S.W. $\frac{1}{2}$ W. of No. 3, which will, when near them and bearing South, warn a vessel of her proximity to Lintin bar.

If the vessel is under 20 feet draught a wider range may be taken, but she ought, if possible, to follow the above directions, and at any rate pass between Tree island and the fishing stakes No. 1, or close to their western end, and avoid the Channel banks. If drawing 17 feet or under she may pass up or down any part of the channel, keeping to the eastward of Fan-si-ak well over towards Mah-chau, avoiding the shoal spits which project from the ends of the islands.

TIDES.—It is high water, full and change, in the Fan-si-ak channel at 1h. 0m., but the rise is irregular, especially on the neaps, the rise and fall being then only $2\frac{1}{2}$ to 3 feet, and from 6 to $8\frac{1}{2}$ feet on the springs; velocity from 3 to 4 knots, and from 2 to $2\frac{1}{2}$ knots on the neaps.

A vessel proceeding up with a working wind should weigh instantly the tide slackens sufficiently for her to make any progress, in whatever part of the channel she may have anchored. The passage between Lintin and Fan-si-ak should not be attempted in vessels of large draught, having only $2\frac{1}{2}$ to $2\frac{3}{4}$ fathoms in it at low water.

CANTON RIVER.

The entrance to Canton river, (Chu kiang of the Chinese,) is formed between Chuen-pee and Ty-cock-tau, or perhaps more strictly between Ty-cock-tau and Anung-hoy. It is divided into two channels by the Wan-tong islands; that to the eastward of the islands is generally used by vessels of large draught, and is named Boca Tigris (Tigers mouth entrance) and Fumun by the pilots; that to the westward of the islands is called Bremer channel.*

CHUEN-PEE POINT, the south extreme of Chuen-pee island, is close to a small peak called Chuen-pee hill, and N.N.E. ¼ E. 1¾ miles from Sam-pan-chau. On the north-west point of Chuen-pee there is a small watch-turret, with a fort under it on the north-west side; and midway between this point and Chuen-pee point there is a ledge, named Pratt rock, lying a quarter of a mile off shore, with from 6 to 9 fathoms close outside.

Anchorage.—A small sandy beach will be seen on either side of Chuen-pee point, and fresh water may be obtained in the bay on the southern side; but vessels of large draught cannot anchor near it, the soundings being shoal on a sandy flat, extending from the point to the eastward and south-eastward. The anchorage is in 6 to 7 fathoms, about a third of a mile from the beach on the northern side of the point.

TIDES.—At this anchorage it is high water, full and change, at about 2h. 0m., and springs rise 7 to 8½ feet.

ANSON BAY, between Chuen-pee and Anung-hoy islands, is very shoal; from the depth of 5 or 6 fathoms the soundings suddenly decrease to 2 fathoms within a line joining the north-west point of Chuen-pee and Anung-hoy point, affording only a harbour for boats in Junk creek.

BOWER POINT, the south-east extreme of the island of Ty-cock-tau, forms the western point of entrance to Canton river. From this point to Sam-pan-chau, the west side of Chuen-pee channel is bordered by a shoal flat over which boats can only pass to East and West Ow-chau, the two small islets lying southward of the point.

ANUNG-HOY POINT (called Namshan by the pilots), the south-west point of Anung-hoy island, bears N. ½ E. 2⅓ miles from Bower point, and

---

* *See* Charts, Sheets 3, 4, and 5 of Canton River, Nos. 1,740, 1,742, 1,739; scales, $m = 3$ inches.

N.W. ½ N. 3 miles from Chuen-pee fort, and with Keshen point half a mile to the north-west, forms the eastern side of the Boca Tigris. The principal fortifications for defending the strait are built on this face of Anung-hoy, and Anung-hoy peak rises immediately behind them to the height of 1,500 feet.

WANTONG ISLANDS.—North and South Wantong are two small islands lying nearly in mid-channel abreast Anung-hoy point, and form the western side of the Boca Tigris. They bear N.N.W. and S.S.E. from each other, distant a third of a mile apart, and are surrounded by a bank which extends 1½ miles in a S.E. by S. direction from the southernmost island, at which distance the depth is only 4 fathoms.

DIRECTIONS.—From abreast Sam-pan-chau with a leading wind a N.W. by N. course for 4½ miles will lead to the entrance of the Boca Tigris; but with a turning wind be careful when standing towards Chuen-pee not to borrow too close to Pratt rock. When standing westward towards the shoal flat extending south-eastward of South Wantong, tack before the eastern extreme of Tiger island touches the eastern part of the fort in North Wantong.

The Boca Tigris has deep water and uneven bottom, and is much contracted by the Chain rock, above water, lying E.S.E. a quarter of a mile from the east point of North Wantong; and although the passage between it and Anung-hoy point is too narrow for working a large vessel, she can always back and fill through with the tide. The tide runs strong through in eddies, and vessels generally keep nearest the eastern shore in passing. If detained here by the Chinese authorities, the best position to anchor is in 7 or 8 fathoms, abreast of and about a quarter of a mile northward of the fort and turret on North Wantong, taking care, however, to avoid the Wantong rock, lying North nearly a cable's length from its eastern point.

The Bremer channel, to the westward of the Wantong islands, carries a depth of 10 to 5½ fathoms, and was frequently taken by Her Majesty's ships during the operations in Canton river in 1841. If intending to use this channel, the first village seen to the northward of Bower point open south of the first bluff point above Ty-cocktau fort (this latter point has the appearance of an island), will lead to the southward of the south extreme of the shoal flat off South Wantong, and the east extreme of Tiger island just open to the westward of the west end of South Wantong will lead along its western edge in 5 fathoms. When abreast North Wantong, about 1½ cables' lengths from its west point, steer about North to avoid the shoal flat on the western shore.

DUFF ROCK.—This dangerous pointed rock, with only 18 feet over it, lies about N.N.W. ⅜ W. nearly a mile from the eastern end of North Wantong. It has 7 to 9 fathoms around it at low water, and from it the small round hummock on the western part of South Wantong is seen over the western slope of North Wantong, between the small redoubt with a tree on it and the point; and the high land of Goefou island is touching the western brow of Tiger island. A vessel will pass eastward of this rock by not bringing Sam-pan-chau to touch the east end of North Wantong, until she has approached Tiger island so near as not to see the high land of Goefou to the westward of it.

TIGER ISLAND.—N.W. ½ N. 1¾ miles from the east end of North Wantong is the Tigers Claw, the south-east extreme of a remarkable high island, called by the Chinese Ty-fu and by Europeans Tiger island, the summit of which appears cleft. A shoal extends to the south-east from the Claw, and at the distance of a quarter of a mile the depth is only 3¼ fathoms. There is a fort on the north-east side of the island.

BATE ROCK, with only 14 feet on it, was discovered by the late Captain W. T. Bate, R.N., in 1857. It lies 2 cables to the northward of the north extreme of Tiger island, with the fort on that island bearing S.E. ¼ S. ; the highest part of the island (eastern summit) S. ⅔ E. ; and the north-west extreme of the island nearly in line with a small granite boulder on the summit of the hill on the western shore of the river S.W. ⅞ W. It is steep-to, having 10 fathoms water, mud, close to the eastward, and 7 fathoms between it and the island. To pass outside or to the north-east of the rock, keep the east extreme of North Wantong open of Tiger island fort.

TOWLING FLAT.—About a third of a mile to the eastward of the fort on Tiger island, is a projecting point of the Towling flat. Both this flat and sand are much increased, and have extended considerably to the westward since the survey of this river in 1840. The sand is now (in 1860) an island covered with vegetation, and never wholly under water even at the highest tides. The old mark for the western edge of the flat in 3 fathoms, was Tomb point (the next point north-west of Chuen-pee point) in line with Keshen point S.E. ⅔ S.; but Tomb point must now be kept well open in passing the flat.

Vessels turning to windward from the Boca Tigris towards Tiger island may stand to the eastward and shut in the high land of Chuen-pee with Anung-hoy until abreast the south-east point of Tiger island. If of large draught they had better back and fill, between the island and the flat, as the tides are strong.

The SMALL BAR is the name given, in the chart of this river by Lieutenant D. Ross, I.N., 1815, to a small 2½ fathoms bank of hard ground lying nearly in mid-channel, about 4¾ miles northward of Tiger island. Since that period the bank appears to have grown up and extended at least half a mile further south. When surveyed by Captain Bate, R.N., in 1857, it was a mile in extent, north and south, and near its centre, west two-thirds of a mile from Blake point, a patch was found with only 10 feet on it; the depths on the other parts of the bank were 2 to 3 fathoms.

H.M.S. *Calcutta* grounded and remained 13½ hours on this bank when proceeding up Canton river, 8th November 1856. When upon it Geofou rock, off the east end of Geofou island, bore South 4 miles; Saw-shee hill E. ½ S., and Second Bar pagoda N.W. by N.

SECOND BAR.—The channel for vessels of large draught becomes very narrow abreast a large inlet on the eastern shore of the river, called Second Bar creek, the entrance to which bears S.E. by E. ¼ E. from the Second Bar pagoda. The southern part of the Second bar begins at a third of a mile from the shore abreast a small creek immediately southward of Second Bar creek; from thence the bar extends 2¼ miles in a N. by W. ½ W. direction, and carries 9 to 16 feet at low tide.

FIRST BAR, at nearly 7 miles above the Second bar, is formed between a shoal bank of sand bordering the south side of First Bar island, and a shoal spit projecting from the eastern point of the low Flat islands. The least depth on it in 1857 was 20 feet.

BRUNSWICK PATCHES, on one of which the ship *Brunswick* struck in 1798, and the *Wyndham* was totally lost in 1815, lie about two-thirds of a mile above the west end of First Bar island, on the northern shore of the river. The rock on which the *Brunswick* struck is described to be about half a cable long, in a N.E. by E. and S.W. by W. direction, with irregular depths of 8 to 18 feet on it at low water. When upon it the Second Bar pagoda bore S. by E. ¾ E.; South Chop-house S.S.E. ¼ E.; Whampoa pagoda W. ½ N.; west extreme of First Bar island S.E. ¼ E.; and a large house inland N. ¼ W.; and with this house bearing from N. ¼ W. to N. by E. a vessel will be in the line of the rock. There are channels between the patches, and to the southward of them, but in the absence of leading marks the narrow 4-fathoms channel to the northward of the patches, close along the north shore, is the most generally used.

WHAMPOA ANCHORAGE is in American reach, between the north sides of Danes and French islands, and the eastern part of Whampoa island, on which the town of Whampoa is built. It is a safe anchorage with a moderate tide, and from 5 to 6 fathoms, soft mud bottom; there is, however, scarcely room for two large ships to moor abreast, which occasions the lower part of the shipping to lie in English reach, abreast the entrance of Junk passage.

TIDES.—At Whampoa, it was high water, full and change, in the month of March at 1h. 40m.; in April, at 1h. 15m.; and in May and June, at 0h. 30m.; and the rise at springs was 7 or 8 feet. In March the day and night tides rose to the same level. From April to October the day tides were the higher; and from November to February the lower. In May and June, spring tides rose 4 feet, and the neaps 2 feet higher than in March.

DOCKS.—H.M.S. *Actæon*, April 1858, was taken into Mr. Cooper's stone dock on the northern part of Danes island, drawing $15\frac{1}{4}$ feet forward, and $15\frac{1}{2}$ feet aft. The dock is 350 feet in length, and 280 feet on the blocks. The gate 75 feet wide at top, 71 feet at bottom. Vessels of not more than 17 feet draught can be taken into it at very high tides. There are numerous mud docks for vessels of smaller draught.

CANTON.—The city of Canton, the capital of the province of Quang-tung, stands on the north bank of the river, about 30 miles above the Bocca Tigris, 70 miles from Macao, and 75 miles from Hong Kong. It is (in 1858) surrounded by a strong wall between 6 and 7 miles in circumference, the foundation of which is of sandstone, the upper part brick. The wall is 25 to 40 feet high, and 20 feet thick, having an esplanade on the inside, and pathways leading to the ramparts on three sides.

The city consists of three divisions separated from each other by thick walls running east and west; the north or larger division is called the old city, and is occupied by the Tartar population; the south division is the new city, but the suburbs are more extensive than either of the walled divisions. The houses are built so near the wall, on both sides, that it is hardly visible when walking round it except on the north side. There are 12 outer gates, each defended by a two-storied house, which commands the wall on either side. There are 4 gates in the partition wall, and two water gates, through which boats pass from east to west across the new city.

A ditch once encompassed the wall, but it is now dry on the north side; on the other three sides, and within the city, it, and most of the canals, are filled by the tide, and empty at low water. A five-storied pagoda stands on a wall, to the north-east of the great north gate, and is loop-holed for musketry, and would hold 500 men; it commands the whole city, and also the two forts to the north-west of it on the outside. Gough fort is about 500 yards outside the city north wall, and is slightly commanded by the fort inside. The population of the city and suburbs is estimated at about one million.

The principal article of exportation is tea; the other exports consist chiefly in porcelain, raw and wrought silks, nankeen cloths, camphor, alum, quicksilver, turmeric, &c. &c.; and the imports from England are woollen and cotton manufactures, clocks and watches, hardware, iron in bolts and bars, lead, tin plates, Indian produce, spices, miscellaneous articles, &c. In the year 1844 the total export of tea was 72,567,111 lbs., of which 15,825,000 lbs. went to America; 2,338,867 lbs. to Holland; 101,956 lbs. to France; 17,109 lbs. to Belgium; 449,643 lbs. to the Hanse towns; 470,887 lbs. to India; 557,765 lbs. to Australia; 28,001 lbs. to the East India islands; 80,864 lbs. to the Cape of Good Hope; 43,675 lbs. to Peru; 2,488 lbs. to Mexico; 6,000 lbs. to Brazil; 130,800 lbs. to Nova Scotia; and 334,423 lbs. to unknown destinations. The total shipment of tea to Great Britain in 1845-6, was 57,622,803 lbs.; in 1846-7, 53,448,339 lbs.; but a considerable portion of the latter shipment, perhaps 10,000,000 lbs., was from Shang-hai. In 1848, 176 British vessels of 73,975 tons, with goods to the value of 8,653,033 dollars, left, and 171 of 72,315 tons, with goods to the value of 6,534,597 dollars, entered the port of Canton.

TIDES.—It was high water, full and change, in the river off Canton at 2h. 40m. in March, and at 1h. 40m. in May and June; the springs rose about 5½ feet. During the North-east monsoon the tides in the river rise 2 to 3 feet higher in the night than in the day; but in the Southwest monsoon the day tides are the highest.

DIRECTIONS.—Having entered Canton river, by the Boca Tigris, be careful when approaching the Duff rock not to bring Sam-pan-chau to touch the east end of North Wantong, until the high land of Geefou island is shut in with the western part of Tiger island. In passing through the channel between the latter island and Towling flat, observe that Tomb point, on Chuen-pee island, kept well open of Anung-hoy north fort, will lead westward of the western edge of the flat; and that the eastern end of North Wantong kept open of the fort on Tiger island will lead north-east of Bate rock. With a working wind, a vessel of large draught had better back and fill through this channel, as the tides in it are strong.

After passing Tiger island keep the watch tower on Chuen-pee fort open of Anung-hoy north fort, until Bower point, the east extreme of Ty-cock-tan, is in line with the eastern side of Tiger island; then steer up the river with this latter mark on and it will lead in the deepest part of the channel, but nearest to Towling island, in 7 or 8 fathoms water.

This mark will not answer much farther than to bring the remarkable high part of Geefou island on with the highest land to the westward, or bearing S.W., but keep more eastward, and open Bower point again. From thence steer to the northward, pass on either side of the Small bar, and attend to the soundings on the chart.

There are two Fairway marks for crossing the Second bar, one, the west point of Tiger island in line with Grassy Tongue bearing S. by E. easterly, leading through the western channel, and the other, Wantong tower, in one with the grassy edge of the land at Amherst point about S. by E. ½ E., leading through the eastern channel, which is the route generally taken; but the services of a pilot are here indispensable to a vessel of 20 feet draught, without the channel be previously buoyed, for the knolls or shoal patches being formed of sand and gravel mixed with mud, are subject to alter in position by the freshes of the river and the spring tides, which also render the navigable channel changeable. A pilot can be obtained from amongst the fishermen on the spot, who then buoy the channel with their sampans, but sufficient time should be given them to sound with their bamboos and to take their stations properly, or else a vessel is likely to take the ground. Vessels often ground and lie in a dangerous state for a tide; and this often proceeds from two or three pushing over together, as there is no time to be lost after the water has risen sufficiently for a vessel drawing 23 or 24 feet to pass over.

Vessels of large draught proceeding up the river from an anchorage below the Second bar in the North-east monsoon, or with a weather tide, should be under weigh by the last quarter flood, to save the tide across the bar; for the channel between the knolls being very narrow, they must back and fill through; if of moderate draught they may weigh much earlier. The difficulty in crossing the bar is in ascertaining correctly the shoal patches on either side the channel, and it will be best to place the boats on them at the first of the flood. When the Second Bar pagoda bears W. by S., the bar is crossed, and the bottom will be soft and loose, unlike that on the bar, which is in parts hard and stony.

After passing the Second bar, keep between a third and half a mile from the eastern shore until First Bar island is approached, when the river begins to be contracted and requires great caution. When Whampoa pagoda is observed just on with the northernmost clump or hill on Danes island, haul out more into the middle of the river to avoid the shoal ground off the south side of First Bar island.

As no safe marks can be given for leading towards the First bar, between First Bar island and the easternmost of the Flat islands, it will be prudent for a stranger, without a pilot, to buoy the south-east extreme of the spit extending off the eastern Flat island, and also the Brunswick patches. The best route appears to be, when the South Chop-house on the southern shore of the river, bears S.S.W., to haul over to First Bar island to avoid the spit, and then steer in about N.W. ½ N., passing about a cable's length along the western face of the island. When Whampoa pagoda is seen clear to the northward of the Flat islands, steer for the northern shore, which must be skirted at about half a cable's length, passing through the narrow 4-fathoms channel northward of the Brunswick patches.

As the northern patch is approached, or when the large house inland (page 54) bears about N. by W., be careful in preserving the distance of half a cable from the shore, and when the house bears eastward of N. by E. the danger will be passed. From thence steer towards Whampoa through Cambridge reach, borrowing towards the northern shore. Entering English reach the southern or Danes island shore is generally preferred, to avoid the shoal flat off Junk and Watson islands, taking care to give a berth to the cluster of rocks, covered at half flood, near Jardine point, the east point of entrance to French river. The anchorage off Whampoa is in from 5 to 6 fathoms, over a soft mud bottom; but there is scarcely room for two large ships to moor abreast, which occasions the lower part of the shipping, when there are many arrivals, to be moored in English reach.

Several of the other passages leading to Canton were used by H.M. ships during the operations against that city in 1841. The *Blenheim*, drawing 23 feet, proceeded through Blenheim passage and anchored in Brown reach; the *Nimrod*, of 15 feet draught, also used this passage, and entered Macao Fort passage between 49th and 26th points; the *Modeste*, of 13½ feet draught, through American reach and Elliot passage, passing southward of Narrow island; the *Alligator*, of about 15 feet draught, through Junk creek and Whampoa channel as far as Kuper island; and the *Sulphur* of about the same draught anchored to the westward of the city in Sulphur reach.

SI KIANG OR WEST RIVER.

The following description of the leading features of the Si kiang or West river, and other creeks and passages explored by the gun boat squadron under the command of Capt. J. J. M'Cleverty, C.B., is by Lieut. C. J. Bullock, H.M.S. *Dove*, February 1859.

The expedition, consisting of the *Haughty, Forester, Staunch, Starling, Watchful, Clown, Kestrel, Woodcock,* and *Janus* gun boats, left the anchorage of Canton on the 16th February, and proceeding down the river entered the Sai-wan passage by the Si-chi-tau channel (Hills passage), which passes directly east of the Second Bar hills.*

SAI-WAN PASSAGE was easily navigated until abreast the town of Sai-wan, where the stream divides into two channels round a large middle ground of hard sand (there was also said to be rock) named Sai-wan bank, on which several of the vessels grounded in attempting to cross the river, both going and returning.

Sai-wan Bank will be avoided by keeping the south shore of the river aboard in approaching Sai-wan, and hauling close round the point where the river suddenly winds east of the bank; it may also be crossed near the island west of Sai-wan in 7 feet at low water.

TAM-CHAU PASSAGE.—The rocky mound, 50 feet high, opposite Sai-wan is an excellent guide into the Tam-chau passage, the second or third turning to the southward being taken. The squadron proceeded by the first, which was found to lead into the Wilder passage, passing westward of Lan-keet; contrary to information locally received it turned round the Tam-chau hills, and joined the Tam-chau passage. It is in some parts very narrow, but affords a good channel for gun boats. Its communications towards Tiger island and Tam-chau were not examined; the latter was said to be impassable for gun boats.

TAI-LUNG CHANNEL.—The squadron returned and entered the Tam-chau passage turning into the Tai-lung channel, which runs out of it westward.

Tai-lung Rocks.—No obstruction was found in the Tai-lung channel till within 3 miles of the village of Yun-kai-tau, Tai-lung Hill pagoda bearing N.E. by E. At this position the river splits round a flat island, a mile in length. The north passage is almost barred by a bed of rocks, some of which are dry, with a channel of 7 feet only at low water. The south passage is the best, but it is very narrow at its east end, and made still more so by a rock—dry at low water—lying nearly in the centre of the passage, and on which the *Woodcock* struck, when leading the squadron through; it lies nearer to the south shore. The passage had been sounded the previous day, without the rock being discovered.

---

* *See* Charts: Canton river, Sheet 3, No. 1,740, scale $m = 3$ inches; and the Si kiang or West river, Sheets 1, 2, and 3, Nos. 2,733, 2,734, 2,735; scales, $m = 0.72$ of an inch.

Yung-kai-tau Rock, awash at half tide, lies about 3 or 4 cables westward of the village of Yung-kai-tau, above the west Shun-tak branch; to avoid it, keep the north shore aboard. The *Clown* struck on this rock on her passage down.

Forester Rock, which probably sometimes dries, lies in mid-channel, with 5 to 7 fathoms on either side of it, at the eastern end of the limeburning village (name not known), 3 miles from the West river. The best guide for its position is, the commencement of some rocky hills on the north bank abreast which it lies; keep this shore aboard. The *Forester* struck on this rock on her return voyage.

On the 19th of February the squadron entered the Si kiang or West river, also called the Blue river from the clearness of its waters, and on the 20th reached the San-shui junction, anchoring the same evening at Shao-king. Here they broke new ground.

The direct channel from San-shui to Fat-shan and Canton is said not to be navigable by vessels drawing above 4 feet water; there was, however, no means of verifying this report. About 2 miles below or to the eastward of San-shui, on this branch, there is a flourishing commercial town called Sai-nam, which probably might be easily reached by small river steamers. The rise and fall of tide here appeared to be 5 or 6 feet.

SHAO-KING, a walled city of the second class, with an extensive suburb lying to the westward, is situate on the left or north bank of the river, 20 miles above the San-shui junction, and 80 miles from the sea at Macao, in latitude, by observation, 23° 03′ N., and longitude, by rough computation, 113° 03′ E. The river is here 6 cables to 7½ cables broad, with a depth of from 5 to 6 fathoms on the north shore.

TIDES.—The tidal influence was felt at Shao-king, there being a rise and fall at this season of 3 feet or more: but the stream, though completely checked, never turned.

GREAT PASS.—From 3 to 6 miles below or to the eastward of Shaoking is a fine pass, the river flowing amongst mountain ranges, which rise to the height of 2,000 or 3,000 feet.

This pass is 3½ miles in length, and nearly straight, and in its narrowest part from 200 to 300 yards wide; the water is deep, but its depth was not ascertained.

KWANG-LI.[*]—A village so called stands on the north shore, midway between the San-shui junction and Shao-king.

---

[*] Li, village; Chau or Hien, district town; Fu, department city; King, capital city.

FIRST BAR.—Between this village and the Great pass the river is three-quarters of a mile wide, and becomes shallower. A passage was found of only 2 fathoms close to the north shore, and it is probable there is no better channel. Here, therefore, at 75 miles from the sea, the river ceases to be navigable at this season for vessels drawing over 12 feet.

Below Kwang-li the river splits round a richly cultivated island. The South channel is full of shoals; and a long spit runs out from either end of the island.

SHAO-KING to WU-CHU FU.—On the 22nd February, the *Watchful*, *Janus*, and *Woodcock* started for Wu-chu fu. There the river winds through a continuously hilly country of sandstone and granite, chiefly in northerly and westerly directions. The hills, varying from 100 to 1,500 feet in height, are in general densely wooded, and many highly cultivated. Near Shao-king, limestone hills appear in rugged and picturesque groups; one crops out on the river of a most picturesque form, and is called by the Chinese Kai-yik-kwan, or the Cock's Comb, which it strongly resembles. A group also lies 2 miles north of Shao-king, to which they give the name of the Seven Stars, after the constellation of the Great Bear.

At 50 miles above Shao-king, and on the left or north shore of the river, a single mass of granite (in the form of a thumb) rises perpendicularly some 300 feet out of a range of hills of 1,500 to 1,800 feet elevation. Its local name is Kum-kwoh-shek, but it is also called Fa-pew, or the flowery tablet, and it is the most remarkable object in the river. After passing this, the navigation becomes dangerous, and the river bed studded with rocks.

The district city Wu-chu or Ng-chu is 75 miles above Shao-king. Its latitude by observation is 23° 28′ N. (22 miles north of Canton), and its longitude, approximately, 112° 14′ E. The breadth of the river here is about 3 cables between the sandbanks, and nearly a mile from shore to shore, but it is with difficulty navigable by junks higher up at this season; the first rapids being (by report) about 12 miles above Wu-chu.

Wu-chu fu stands at the confluence of the stream on which is Kwei-ling, the capital of Kwang-si. This communication was open, though the intermediate country was in the hands of the rebels. It had the appearance, observed from the heights, of being easily navigable by gun boats.

TIDES and CURRENT.—The level of the Si kiang at this season (February) at Wu-chu, was from 25 to 30 feet below the river banks; probably 25 feet below the summer level in July and August.

The velocity of the stream never exceeded $2\frac{1}{2}$ or 3 knots.

There was a rise and fall of 18 inches at spring tides. The stream never turned, but was checked during the flood.

BOUNDARY.—The boundary of the provinces Quang-tung* and Kwang-si is 4 miles below Wu-chu fu.

DIRECTIONS.—The Si kiang is moderately deep from Shao-king to Wu-chu, having generally from 3 to 5 fathoms water in the channels for nearly 60 miles above Shao-king; in a gorge in Yuet-shing reach a depth of 29 fathoms was obtained.

At 15 miles below Wu-chu is a 12 feet bar. After this the navigation becomes more and more difficult as Wu-chu is approached. For the last 4 miles it is most intricate, and barely 7 feet was found between the rocky ledges. The *Woodcock* grounded twice on rocks, and the *Watchful* on a bank of hard sand a few miles below the city.

There are numerous villages along the river banks and a military station at every 3 miles. The walled towns of Fong-chuen, and the walled city of Tak-hien (of the second class), are only worthy of notice. They stand on the left bank, the former 12 miles, the latter 55 miles, below Wu-chu.

The longitude of Wu-chu must at present remain uncertain, as gloomy, overcast skies prevailed throughout the cruize. The distances were determined by the mean of the runs up and down, as shown by Massey's patent log. Some idea may be given of the rate at which these explorations may be conducted by the following particulars:—The 75 miles of river from Shao-king to Wu-chu was accomplished in $3\frac{1}{2}$ days of 12 hours a day; this gives an average of $21\frac{1}{2}$ miles a day, or $1\frac{3}{4}$ miles an hour. The average speed of the gun boats (40 H.P.) being about 4 knots, 92 miles of running were required to complete the 75, showing an adverse current of 17 knots. The return passage was made in 20 hours, the log showing 61 miles, *i.e.*, 45 miles per day, or $3\frac{3}{4}$ miles per hour.

---

* Tung signifies East, and Si, West.

CHAPTER III.

EAST COAST OF CHINA—HONG KONG TO AMOY.

VARIATION, from 0° 30′ East to 0° 15′ West in 1861.

TATHONG CHANNEL is formed between the west side of Tamtu island and the east side of Hong Kong, and close to the latter, about 1¼ miles northward of Tylong head, lie two small rocky islets: between these islets and Tathong point is the Tathong rock, above water.

Vessels having run out from Hong Kong road through the Lyemun pass, and wishing for anchorage, either for the night or in consequence of bad weather, will find a good berth in the bay on the northern side of Tamtu island in 6 fathoms; but it must be borne in mind that the water shoals to 2¾ fathoms at 3 cables' lengths from the Joss house on the north side of the bay.*

TAMTU or Tunglung island, 820 feet above the sea and 3 miles in circumference, is separated from the mainland by a channel called the Fotaumun pass, which is only 1¼ cables wide between the rocks which lie off both points in the channel, and carries a depth of 3 fathoms. A sunken rock lies S.E. ⅓ E. distant 4 cables from the north point of Tamtu; when on it, the west end of Steep island (the first small islet to the north-eastward) just shows clear of a remarkable headland named Yih bluff, bearing N.N.E. ½ E.

The south point of Tamtu forms a low peninsula, and to the southward of its west point there is a flat islet or rock lying a cable's length from the shore, with reefs inside it. Upon the first point outside the Fotaumun pass stands a ruined fort.

STEEP and TRIO ISLETS.—Steep islet is 1¼ miles to the northward of the eastern entrance of the Fotaumun pass, and 4 cables from the shore; at 1½ miles farther north lie the Trio islets. There is an indentation in the coast, with 8 fathoms water, between Trio and Steep, but it is exposed to easterly winds and swell.

---

* *See* Charts :—Hong Kong, No. 1,466, scale $m = 2$ inches; Sheet 2, East Coast of China, No. 1,962, scale $d = 14\frac{1}{2}$ inches; and Mirs Bay, No. 1,964, scale $m = 0.8$ of an inch.

NINEPIN GROUP lies 3 miles eastward of the Fotaumun pass; the two largest islets bear North and South of each other, and the channel between them is 2 cables wide. The southern face of the South Ninepin is a precipitous cliff, 330 feet high; off its south-west side there is a smaller islet, and towards its northern point the land becomes lower, with a peaked rock in the offing. The surface of the North Ninepin is nearly of the same elevation, with the exception of a cleft near its northern end; an islet lies off its south-west extreme.

Ninepin Rock, or East Ninepin, 222 feet high, lies nearly a mile eastward of the North Ninepin, and assumes the appearance that its name indicates only when seen in a N.W. or S.E. direction; otherwise the name is liable to mislead. Close to its north-west side is a smaller islet, and there are detached rocks upon its north-east and west sides.

One-foot Rock, lying S. ¾ W., not quite 7 cables from the Ninepin, has only a foot over it at low water. When on this rock the south end of the South Ninepin bears W. ¼ S., and is in line with the shoulder of the hill northward of the highest part of Tamtu; and the right extreme of the rock on the north side of the North Ninepin is in line with the summit of Shelter island in Shelter bay, N.W. ⅓ W. The south end of the South Ninepin on with Fotaumun pass, W. ¾ N., will lead south of it.

North Rock, lying N.W. ¼ N. distant 9 cables from the Ninepin, is nearly awash; a reef, which breaks at low water, lies a short cable's length to the south-east of it.

TIDES.—At the Ninepin group it is high water, full and change, at 10h. 0m., and the rise is 5 feet. The channel between the group and Steep islet is nearly 2 miles wide, and carries a depth of 14 to 16 fathoms; at full and change in May 1845, the flood ran to the S.S.E., and the ebb to the S.S.W., the former at the rate of 0·3 of a knot and the latter half a knot.

PORT SHELTER.—To the northward of the Ninepin group the mainland forms a deep bay, containing Port Shelter and Rocky harbour. Port Shelter, the western of the two, runs back to the northward 5½ miles, and its head is separated from the south-west portion of Mirs bay by an isthmus 1½ miles wide, overlooked by the Hunchback hills, 2,315 feet above the sea, which with Sharp peak, 1,540 feet high, on the west side of the entrance to Mirs bay, form conspicuous marks by which this portion of the coast may be recognized.

When steering for Port Shelter, pass eastward of Trio and Table islets on account of some rocks which extend 3 cables from the point to the westward of them. Nearly a mile northward of Table islet is the southern

CHAP. III.] NINEPIN GROUP.—PORT SHELTER.—ROCKY HARBOUR. 65

point of Jin island, with a peaked rock lying 2 cables to the southward of it; and E. ¾ N. rather more than a cable's length from the peaked rock, there is a rock awash at high water.

Shelter island, 1¼ miles to the north-west of Table islet, should likewise be left to the westward when steering for Port Shelter, as the ground is foul between it and the main. Good anchorage will be found on the north-west side of Shelter island, in 8 fathoms, but give the north point of the island a berth of a cable's length, and avoid the Nine-feet patch, which lies 6 cables to the northward in the centre of the bay; the marks for which are:—Table island on with the north end of North Ninepin, bearing S.E. ¾ S.; the opening between Keui and Jin islands nearly East; and Shelter island S. by W. Southerly, one cable's length from the west point of Shelter island, is a rock awash at low water; and there is a patch of 2¾ fathoms lying half a mile to the westward of it.

Sharp island lies North 1½ miles from Shelter, with fair anchorage on its eastern side, but exposed to southerly winds; and from which, passing north of Keui island, there is a junk or boat passage leading into Rocky harbour.

ROCKY HARBOUR is formed by Keui and Jin islands on the west, and by High, Basalt, and Bluff islands to the east and south-east. The southern entrance between Bluff and Jin islands is a mile wide; the rock awash at high water off the latter, has been mentioned above. On the east side of Jin, at 2 cables from the shore, is Bay islet, which is low and flat.

Three-feet Patch.—Midway between Bay islet and the north end of Bluff island lies a rocky patch with only 3 feet on it, from which the west point of Bluff island is in one with the summit of North Ninepin, S. ¼ E., and the southern summit of Bay islet bears W.N.W. The North Ninepin and Bluff islands touching, leads westward of it; and the west end of the rock lying off the south-west end of North Ninepin, in one with the west point of Bluff island, leads eastward; also, a vessel will be northward of it when the Pyramid rock opens clear of the north-east extreme of Bluff island, S.E. by E. ¾ E.

Three-fathoms Patch lies 6 cables to the northward of the Three-feet patch, with the summit of Bay islet bearing W.S.W., Pyramid rock S.E. ¼ S., and Green islet, the small islet on the eastern shore, E. ¾ N., and distant 3 cables' lengths.

ANCHORAGE will be found in the North-east monsoon on the eastern side of Rocky harbour, in the neighbourhood of a small cove northward of Green islet, where there is a mandarin station and a village. Inside the cove the depth is 6 fathoms, but the space is confined, owing to sunken rocks. In the South-west monsoon vessels will be better sheltered by anchoring to the North-west of Bay islet.

BASALT ISLAND lies 4 cables to the south-east of Bluff island, and the depth between them is 5 fathoms at low water. The former is 8 cables long, north and south, and rises to the height of 572 feet above the sea; the southern faces of both it and Bluff island are very precipitous.

TOWN ISLAND lies half a mile to the northward of Basalt island. The channel between them is 4 cables wide, but it should not be used without a leading wind or in a handy vessel, as the chow-chow water,* or whirling eddies, might lead them into difficulty. It is also obstructed by islets and a rock awash at high water, and to the eastward of the Three-feet patch at the entrance of Rocky harbour, the ground is foul with some casts of 3 fathoms.

HIGH ISLAND, $7\frac{1}{2}$ miles in circumference and 910 feet above the sea, is separated from Town island by a channel which carries a depth of $3\frac{1}{4}$ fathoms, but in some places it is barely a cable wide. At $1\frac{1}{2}$ cables eastward of the latter is Hole island, so called from its being perforated. To the northward of these islands there are two low islets. The channel between High island and the main has not more than a foot in some places at low water.

FUNG BAY.—Conic isle lies N.N.E. $2\frac{1}{4}$ miles from Hole island, at not quite a cable's length from the shore, and immediately westward of it is a small bay $3\frac{1}{2}$ cables wide and three-quarters of a mile deep, which might be used in the North-east monsoon. Fung bay, the next inlet to the northward, is $1\frac{3}{4}$ miles wide, and has two islets and a rock in the middle of it; but it is too much exposed to the eastward to be of any use to the navigator. Sharp peak, noticed in page 64, overlooks this bay, and bears from the Ninepin N. $\frac{1}{4}$ E., nearly 10 miles.

MIRS BAY is a deep inlet, 15 miles to the north-east of Hong Kong, and its entrance, between Fung head and Mirs point, is $5\frac{1}{2}$ miles wide; its extent northerly is 11 miles, and in an east and west direction 18 miles. Gau-tau, a rocky islet 90 feet high, lies about 2 miles within the entrance, and S.W. by W. about half a mile from it is a rocky ledge, part of which is always uncovered. South Gau island, 96 feet

---

* Chow-chow water is a term applied to those ripplings occasioned by the meeting of adverse currents, the agitation of which is frequently so violent as to render a vessel unmanageable when within their influence.

high, is 1½ miles to the S.W. by W. of this ledge, and half a mile off shore.

The hills near Mirs point rise to the height of 1,200 feet, and just off its southern extremity lies a small islet, named Griffin rock, and east of it some rocks, at a cable's length from the beach. The first point to the westward of the islet is perforated.

GRASS ISLAND.—The point 1¼ miles N. by W. of Fung head has two islets off it, and from thence the western coast of Mirs bay trends suddenly to the westward, then northerly 1½ miles, where there is an opening 3 cables wide leading into Long harbour, bearing West from Gau-tau; the navigable channel however has only 2 fathoms water in it, and is barely a cable wide, with shoal water extending from both shores.

On the north side of the opening lies Grass island, which is 1¼ miles long, north and south, three-quarters of a mile wide, and 420 feet high; and at 3½ cables eastward of this island is a large black rock, named North Gau, with a reef, awash at high water, lying N.W. ½ N. 4 cables from it.

PORT ISLAND, nearly 2 miles in circumference and 420 feet high, lies nearly 6 cables northward of Grass island, and its north-east point, which is narrow, projects 3 cables from the body of the island.

Water.—There is a convenient watering-place on the northern side of Port island.

LONG HARBOUR, the entrance to which lies a mile S.S.W. of Port island, is 3½ miles deep, and at its entrance is 6 cables wide. Both shores are steep-to, with the exception of the south-west end of Grass island, where there is a cove with a rock off its north point; and at about a cable's length to the northward of this rock and half a cable from the shore is a rocky patch of 3¼ fathoms; some rocks also, which show at low tide, extend nearly a cable's length from high water mark at the south-west end of the island. To the southward of Grass island, the harbour widens to 1¼ miles, and then gradually decreases towards its southern extremity, where it is separated into two coves; the depth is 4 fathoms at a mile from the head of the harbour.

JONES COVE, the next inlet westward of Long harbour, is a mile deep N.N.E. and S.S.W., and 3 cables wide; but it, as well as Long harbour, is open to a considerable swell from the N.N.E.

On the western side of the cove there are three islets, and at 2 cables to the northward of the largest, Flat islet, are two rocks, awash at high water, from which the summit of Port island bears N.E. ¾ E., and the north point of Grass island E. ⅔ N.

TOLO CHANNEL, leading into Tolo harbour, is the next deep inlet westward of Long harbour, on the western shore of Mirs bay. The entrance, between Port island and Bluff head, is nearly 1¼ miles wide, and from thence the channel trends S.W. by W. 7 miles to White head, forming a Sound not less than 7 cables wide, with steep shores, and carrying a depth varying from 6 to 14 fathoms. There is a small cove on the northern shore of the channel, at 2 miles within Bluff head.

Within the channel at 3¾ miles from Bluff head is Knob reef, and a flat reef at 2 cables' lengths farther to the S.W.; and 2¼ miles farther lies Bush reef, north of which, 3½ cables, is Harbour island. The main land to the southward is nearly a mile distant from this latter reef, but the 3-fathoms' line extends 4 cables from the shore on this side of the channel. Although there is a navigable channel on either side of these reefs, the one northward of them is preferred, being the wider, and having 7 to 10 fathoms water. Abreast of Knob reef, on the northern shore, there is a large cove.

Tide Cove.—At White head (which is a peninsula with the Hunchback hills, 2,315 feet high, with very precipitous face, rising immediately behind it), the Tolo channel separates into three arms, the south-western of which, named Tide cove, extends 3½ miles beyond it, and the water shoals gradually from 5 fathoms to the bottom of the cove, from whence there is a footpath to Kowloon village in Hong Kong harbour, the distance across from water to water being 2¼ miles, and the greatest elevation to surmount 920 feet. In the middle of the cove, at 2 miles from White Head, is a reef which covers at high water, and from which a remarkable waterfall on the western shore bears S.W. by W. ½ W.

Tolo Harbour, the north-west arm, also extends 3½ miles from White head, and has in its entrance Centre isle, and to the northward some smaller islets, with anchorage between them and the main.

Plover Cove, the north-east arm, would in all probability be found the most eligible place to ride out a Typhoon; it runs back 2¼ miles to the eastward beyond Harbour island, and carries a depth of 6 to 4 fathoms.

ROUND, CROOKED, CRESCENT, and DOUBLE ISLANDS.—N.W.b.N. 2¾ miles from Port island, is Round islet, the easternmost of an extensive group lying in the north-west part of Mirs bay; the largest of the group are Double, Crescent, and Crooked islands. Double island, the southernmost, lies N.W. 6 cables from Bluff head, and the channel which separates its south-west point from the main is only large enough for boats. The passage between it and Crescent island is a cable wide, with 4 to 7 fathoms in it; and between Crescent and Crooked islands, the narrowest part of the channel is 2 cables wide, with a depth of 10 and 12 fathoms.

The east end of Crooked island is a remarkable peaked head, and between it and the mainland, to the northward, the depths are 9 to 4 fathoms, muddy bottom. A good harbour, Crooked Harbour, will be found on the west side of Crooked island; and a very secure basin, named Double Haven, is formed to the southward, by Crescent and Double island, the entrance into which on the north side is 3 cables wide: within it the depth is 7 fathoms. On the north-west side of Crooked island is a large village.

PENG-CHAU ISLAND, 3 miles in circumference and 148 feet high, is in the north-east corner of Mirs bay, and bears N. $\frac{1}{4}$ E. $4\frac{1}{2}$ miles from Gau-tau. The geological formation of this island is totally different from the adjacent land, being alluvial, shale stones forming its beaches. The distance between it and the main land to the eastward is rather more than a mile, forming a convenient harbour sheltered from all winds. E.N.E. from this island is the remarkable peak of East Cone, 750 feet high, overlooking Typung bay, the distance across being $1\frac{1}{4}$ miles, and the land but little elevated. The village of Namoh stands on the isthmus; and in the bay to the south-west of it there is a peaked rock and a sunken reef.

ANCHORAGE.—The water gradually shoals to the westward of Crooked island, and this part of Mirs bay affords good anchorage. The northern portion of the shores of the bay are steep-to. There is anchorage in the North-east monsoon all along the eastern shore of the bay to the southward of Peng-chau; but the number of fishing platforms on stakes in 8 and 9 fathoms water render the navigation awkward in the dark. There is anchorage in south-west winds to the westward of the South Gau, in 8 or 9 fathoms.

TIDES.—It is high water, full and change, in Tide cove, on the western shore of Mirs bay, at 10h. 0m., and springs rise about $6\frac{1}{2}$ feet; but during the neaps the water remains nearly at the same level.

Off Mirs point in April, 2 days after the change of the moon, the ebb made to the E. b. N., the greatest velocity being 0·3 of a knot per hour. With the flood, there is a great indraft into Mirs bay and Rocky harbour, which must be guarded against in shaping a course from Tuni-ang island to pass outside the Ninepin.

On full and change in May, the flood inside the Ninepin rock ran to the S.E., and the ebb to the S.W., the former at the rate of 0·3 of a knot, the latter half a knot per hour. In March, the moon being 19 days old, the ebb ran to the S.W. 2 knots, 9 miles in the whole tide.

The COAST from Mirs point trends N.E. b. E. 8 miles to Teyih point, and between the points there are two sandy bays, off the westernmost of which, and at 4 cables from the shore, lies Coast islet, having 4 fathoms inside it.

DIRECTIONS.—As the ebb stream runs to the southward along the western shore at the entrance of Mirs bay, a vessel working to windward with a S.W. wind will get to the westward speedily by keeping near it, passing between the Ninepin group and Tamtu; but as soon as the Lema channel opens out she will meet with a strong set to the eastward.

During the month of August and part of September, if a vessel is to eastward of the Lema islands, she will find it difficult to proceed along shore to the westward if the wind is from that quarter; she ought therefore either to stand off to the southward for two or three days, if near the full and change of the moon, when bad weather may be apprehended, or anchor in Mirs or Harlem bay for an easterly wind, which in these months usually happens every few days, close in with the coast.

TUNI-ANG GROUP, lying 6 miles eastward of Mirs point, fronting the peninsula which separates Mirs and Bias bays, consists of eight islets, including Single island and Acong rock. The largest islet, the northernmost, is 5 miles in circumference, and the summit rises like a cone to the height of 960 feet; off its western end are two islets; the nearest, Net island, is sugar-loaf shaped, and at low tide there is but a foot water between it and Tuni-ang.

Peak rock, lying a quarter of a mile westward of Net island, with a depth of 4 and 5 fathoms between, appears like two islets with a shingle beach connecting them. N.W. $\frac{1}{4}$ W., 4 cables from Peak rock, lies a ledge of rocks, the northern edge of which is always visible; and between them is a reef which breaks at low water.

Immediately to the southward of Tuni-ang island lie three islets, called by the Chinese Samun (or three passages), which form a harbour sheltered from all winds, except those between W. N. W., round westerly, and S.W. b. S. The southern islet is 3 miles in circumference, and distant $1\frac{1}{4}$ miles from Tuni-ang; the channel between it and Cone island to the northward is not quite 2 cables wide, with a depth in it of 9 and 10 fathoms. The passage between Cone and Tuni-ang is the same breadth, but crooked, and carries only $2\frac{1}{2}$ fathoms at low water. The channel between Samun and Single island is $1\frac{3}{4}$ miles wide; the latter island is even-topped, and 200 feet high.

The Acong is a remarkable pyramid rock lying 6 cables to the N.E. of Single island, with a depth of 15 fathoms between them. There is a rock with 16 feet upon it at low water, lying N.N.E. ¾ E. about a mile from Acong, on which bearing it is on with the south-east point of Single island. When on this rock, which is so steep all around that there was great difficulty in finding it, Cone island bore N.W. by W. ¾ W., and was in one with a remarkable gap in Tuni-ang.

ANCHORAGE.—In the North-east monsoon the trading junks anchor in 9 fathoms to the southward of Net island, and abreast a fort on Tuni-ang; but the ground is foul within 2 cables' lengths of the fort point. The best anchorage is off the south-west point of Cone island in 7 and 8 fathoms water. During the prevalence of south-westerly winds there is anchorage, in 9 and 10 fathoms, abreast a bay on the north-east side of Tuni-ang.

MIDDLE ROCKS.—N.E. ⅓ E. from the summit of Tuni-ang lie the Middle rocks, which are just awash at high water. From them Acong rock bears S. ¾ W.; Bate island, off the east point of Bias bay, N.N.E.; and Lokaup island N.W. b. N. 4 miles. At 3 cables to the south-west of these rocks is a reef which breaks only at low water; the marks for it are, the east end of Cake islet (on the east side of Lokaup) in line with the Pillars, bearing N. b. W. ¾ W.

The channel between Tuni-ang island and Teyih point, the west point of Bias bay, is 1½ miles wide; both shores are steep-to, with the exception of the reef already mentioned, lying off Peak rock near the north-west point of Tuni-ang, and a rocky ledge extending south-westerly from the first point east of a remarkable white rock on the north shore. The hills on this side attain an elevation of 2,600 and 2,800 feet.

BIAS BAY is a capacious and deep inlet, similar to Mirs bay. It has a chain of islands fronting its western shore, which is indented by two large bays, at the head of the principal of which is Typung harbour.

Bias point, the eastern point of entrance, is fronted by rocks to the extent of nearly a mile. The channel between them and the land is unsafe, but the passage between these rocks and the rock lying S.E. of Bate island may be used, being 8 cables wide, with a depth inside of 4½ and 5 fathoms.

Bate island is 8 cables long, north and south, and half a mile wide, and besides the rock which lies 3 cables south-east of it, there is another rock, awash at high water, lying N.N.E. 6 cables from its north end, and from which the south point of Lokaup island bears S.W. by W. ½ W., and the rock south-east of Bate island S. by E.

LOKAUP ISLAND, off the south end of which there are some pyramidal rocks, bears N. b E. 6 miles from Tuni-ang, and the channel between it and the west point of Bias bay is 3 miles wide, with a depth of 9 fathoms. The island is about 2 miles long, and nearly separated in two places; the highest part, 330 feet above the sea, being near the south end. There is anchorage on either side of it, according to the prevailing winds. There are six islets around this island, three on the west, two on the north, and one on the east side. The north islet of the group, named the Pillars, is remarkable from its two square pillars; there is a reef off the west end of the small island south of the Pillars.

TYPUNG HARBOUR, so named from the walled town of Typung on its northern shore, is on the west side of Bias bay, and although contracted is capable of affording good shelter for moderate-sized vessels, except with easterly winds, when the anchorage under Lokaup island should be preferred. The entrance is 6 miles W. $\frac{1}{2}$ S. from the north end of Lokaup, and on the northern side there is a smooth conical hill, off which a reef commences, extending half a mile from the shore; the southern side, which is steep-to, must therefore be kept aboard. Vessels drawing more than 15 feet should not proceed farther westerly than the third point on the south side, as the bottom of the bay is shoal.

MIDDLE GROUP.—About a mile northward of the Pillars is Middle group, consisting of six islets. Green island, 254 feet high, the southernmost, has an islet off its west end; and at three-quarters of a mile to the northward is Reef islet, to the S.E. of which is a reef that breaks at low water; the centre of this reef bears N. by E. $\frac{3}{4}$ E. from Green island, and S.S.E. $\frac{3}{4}$ E. from the summit of Reef islet. There is also another rock awash at low water, lying North 3 cables from reef islet; when on it, summit of Red islet bears E. $\frac{2}{3}$ S. There is a third rock, N. $\frac{1}{3}$ W. $1\frac{1}{4}$ miles from Reef islet, and N.W. $\frac{1}{2}$ N. from Red islet.

HARBOUR GROUP, consisting of nine islets, lies in the middle of Bias bay, not quite a mile to the northward of Middle group. The southernmost are two small islets named the Twins, to the N.E. of which, at 2 cables' lengths, is Shoal island, having rocky ground extending north-westerly 2 cables from it, on some parts of which there are only 3 feet water. Shoal island is separated from Narrow island by a channel $3\frac{1}{2}$ cables wide; should it be used, the shore of the latter must be kept aboard to avoid the shoal just mentioned. Narrow island is three-quarters of a mile long, north and south, and 2 cables wide. Round island lies rather more than 2 cables to the northward of Narrow island, with a depth of 5 and 6 fathoms between them; to the northward of it at 2 cables lies a flat rock nearly awash. N. by W. 6 cables from Round island is the North Cone, a conical rock surrounded by reefs; vessels wishing to anchor to the westward of Narrow island will find this the best channel to enter by. N.N.W. 2½ miles from Narrow island is Low island.

At a quarter of a mile to the westward of the Twins is Tree-a-top islet, and westward of it, at half a mile, is a Sugar Loaf shaped island, having between them a good channel to enter inside the group. To the westward of Sugar Loaf is Big island, off the north face of which is a small islet, and further north a flat rock, with a reef, which shows only at low water; when upon this reef the highest part of Narrow island bears S.E. by E., and Nobby reef N.E. by E. To the N.W. of Big island is Sand patch, a low rock surrounded by sand; between it and the island there are 3½ fathoms water. On the south side of Big island there is also a rock awash at high water.

The passage to the westward, between Big island and the main is three-quarters of a mile wide, but a reef lies nearly in mid-channel and only shows at half tide; it bears W. by S. ¼ S. from Sugar Loaf, and N.W. by N. from Green island.

DUMBELL BAY, the next inlet northward of Typung harbour on the west side of Bias 'bay, runs back westerly 6 miles from Big island, and carries a general depth of about 3 fathoms.

TRIPLE ISLAND.—From Bias point the eastern coast of Bias bay trends northerly 9½ miles; the first islet on this shore is Triple, lying 2¼ miles northward of Bate island.

There is anchorage in the North-east monsoon between Bate and Triple islands. The channel between Triple and the main is 6 cables wide, with a depth of 3 fathoms; at a cable's length from the eastern shore of the island is a small rock which is never covered.

TSANG-CHAU ISLAND is a low flat islet with a smaller one S.E. of it, lying 6½ miles northward of Triple island. The passage between it and the main land is a mile wide, with a depth of 2 fathoms; but rocks extend from the shores on each side of the channel.

FAN-LO-KONG HARBOUR.—To the northward of Tsang-chau the eastern coast of Bias bay bends round to the eastward, forming the harbour of Fan-lo-kong, the entrance to which is 1½ miles wide, with a depth in mid-channel of 4 fathoms. At 4 miles to the north-east of Tsang-chau the soundings decrease to 3 fathoms, and shoal water extends 2 miles farther to the head of the harbour. The village of Fan-lo-kong is on the northern shore. This will probably be found the best anchorage in Bias bay in a Typhoon.

PAGODA ISLAND bears from Tsang-chau N.W. by W. ¾ W. 4 miles, the soundings varying from 4¾ to 2¼ fathoms between them; the water shoals towards Pagoda, which lies 3 cables from the northern shore of Bias bay, with a depth of only 9 feet inside of it; to the W.S.W. of the island are some rocks.

MENDOZA ISLAND, 480 feet high, and 2¾ miles in circumference, bears S.E. by E. ¼ E. 7½ miles from Bate island, and a vessel will find shelter from a S.W. wind on its northern side. On its western side there is a small islet separated from it by a channel a cable wide, and carrying 9 feet water. Tsincoe island, 167 feet high, lies 6 cables to the northward of Mendoza, the depth between them being 11 fathoms; near its centre there is a remarkable cleft.

FOKAI POINT, bearing N.E. by E. 3¾ miles from Mendoza, is the south extremity of a high promontory, connected to the main by a low sandy isthmus; the land near the point is high, and has the appearance of an island when viewed from eastward or westward. On the summit of the Fokai hills is an artificial mound 670 feet above the sea, and on the hill over the south-west point stands a large fort. On the east side of the isthmus lie three rocky islets; and E. by N. 8 cables from the northernmost islet, is a reef showing at low water, from which the east extreme of Fokai point bears S. by W. ½ W., and the Pauk Piah rock E.S.E.

HARLEM BAY, formed to the westward of the Fokai promontory and northward of Mendoza island, affords secure anchorage in the Northeast monsoon; but it cannot be considered safe during a Typhoon, when the winds are liable to shift suddenly to different points of the compass. A good berth will be found to the northward of Hebe islet in any convenient depth of water. This islet is flat-topped, and 70 feet high, and a ledge of rocks, which covers at high water, extends 3 cables north-eastward of it.*

The distance between Mendoza and the west extreme of Fokai point is 2¾ miles, and between the two, at 6 cables from the latter, and 10 or 12 feet above the sea, is Middle rock, which may be passed on either side. On the western foot of the Fokai hills stands a fort, and a tall chimney on the hill over it: to the northward of the fort is a creek, which extends northerly along the sandy isthmus, and into which junks run at high

---

* See Plan of Harlem Bay, scale, m = half an inch, surveyed by Lieut. D. Ross, I.N., 1812.

water. S.W. by W. 3 cables from Hebe islet is a rocky patch, of $3\frac{1}{2}$ fathoms water, bearing North from Middle rock, and N.W. $\frac{1}{3}$ N. from the west extreme of Fokai point.

TIDES.—It is high water, full and change, at Tuni-ang island at 8h. 0m.; at Tsang-chau island in Bias bay at 8h. 30m.; and at Hebe islet in Harlem bay (two days before full moon) at 10h. 0m. In the month of April the current in this neighbourhood set constantly to the westward, increasing its velocity upon the flood, but its rate did not exceed a knot.

DIRECTIONS.—When bound to Bias bay from the eastward, after passing about a mile westward of Mendoza island, steer N.W. by W. for the opening between Lokaup and Bate islands, carrying a depth of 13 to 10 fathoms over muddy bottom. If there is a turning wind, when standing westward do not bring Bate island to the eastward of N. $\frac{1}{4}$ E., nor Acong rock to the southward of S. by W. $\frac{1}{2}$ W., until Tsincoe island bears to the southward of East, to avoid the Middle rocks. From thence either proceed up the bay to an anchorage in 5 fathoms water, about $1\frac{1}{2}$ miles from the eastern shore, 3 miles northward of Triple island, or to the southward of Lokaup to an anchorage in the bay or in the harbour of Typung. There are several populous villages on the eastern shore where no doubt refreshments could be obtained.

If bound to Harlem bay, round Fokai point in 13 fathoms about half a mile off, and either haul up between the shore and the Middle rock, or pass between the rock and Tsincoe island. If the wind be easterly, it will perhaps be better for a vessel of moderate draught to adopt the former channel, as she will fetch the anchorage without tacking, taking care, however, to avoid the $3\frac{1}{2}$ fathoms patch to the south-west of Hebe islet; but a large ship should pass westward of the Middle rock, although she should have to tack, as she will then be far enough from the high land to avoid the variable flaws of wind, and the disagreeable consequences that might arise from being baffled in a narrow channel.

SAM-CHAU INLET.—From Fokai point the coast trends N.E. by N. 12 miles to Ross head, and at the distance of 9 miles is Coast islet, lying 4 cables from the shore. Shoal water, over rocky bottom, extends 6 cables to the southward of this islet, and here, close to a flat rocky head, there is an opening a cable wide into the extensive inlet of Samchau, the channel, carrying 5 and 6 fathoms, being close to a narrow cliff on the southern shore; but in strong easterly winds the sea breaks across it. The entrance bears W. by N. $\frac{1}{2}$ N. from Si-ting islet, and E. $\frac{1}{2}$ N. from Harlem peak, which, rising 2,070 feet above the sea, forms a conspicuous landmark. S.S.W. $\frac{1}{4}$ W. nearly $2\frac{1}{2}$ miles from Coast islet lies a sunken rock, from which Si-ting bears East 6 miles, and Harlem peak N.W. ½ W.

Commander P. Cracoft, of H.M.S. *Reynard*, who visited this inlet in chase of pirates, says, "The mouth of the inlet is very little wider than the breadth of a ship; there is also an inner bar with an equally narrow passage; and across both these bars the tide runs with a velocity of 5 knots. The depth in the channel varies from 6 to 8 fathoms, and deepens to 10 fathoms above the upper bar, where there is ample room for a vessel to swing; but such is the intricacy of the navigation that a personal examination should be made, and the state of the tide carefully ascertained, before attempting the entrance."

PEDRO BLANCO ROCK, in lat. 22° 18½′ N., long. 115° 7′ E., bears S. by E. ⅞ E. from Pauk Piah rock; S.S.W. ½ W. from Ty-sami mound; E. ¼ N. 42 miles from the south extreme of the South Ninepin; S.E. ¾ E. 19½ miles from Mendoza island; and S.W. by W. ¼ W. 83 miles from Flat reef, Breaker point. When bearing North it appears as two rocks; the summit is of a white colour. It is bold to approach, having 20 fathoms close to the southward, and 18 fathoms to the northward, decreasing gradually to 13 fathoms in the neighbourhood of the Pauk Piah.

PAUK PIAH and WHALE ROCKS.—The Pauk Piah is a flat rock, 4 feet above high water, from which the summit of the Fokai hills bears W. ½ N. 7 miles.

S. by W. 2½ miles from the Pauk Piah lie the two Whale rocks, upon which the sea sometimes breaks. They rise abruptly from the depth of 12 fathoms, and when on them, the west extreme of Fokai point is on with the summit of Bate island, and bears W. by N. ¼ N., the summit of Fokai N.W. by W. ½ W. 7 miles, and the summit of Mendoza West a little northerly.

TUNG-TING and SI-TING are two rocky islets about 50 feet above the sea, lying S.E. ½ S. and N.W. ½ N. from each other, distant 1½ miles apart; there are sunken and detached rocks lying around them both, and the depth of water in their vicinity is 9 fathoms. From Si-ting the summit of Fokai point bears S.W. by W. ¾ W. 11 miles, and the Pauk Piah S.S.W. ¼ W. 6¾ miles.

N.W. by W. 1½ miles from Si-ting lies a rocky patch upon which the sea sometimes breaks. Hat islet bears from it N.E. ¼ E., and Harlem peak W. ⅔ N.; Mace point, open North of Hat islet, bearing about N.E. ¾ E., will lead to the northward. There is also the Single rock which breaks only at low water or when there is a heavy sea, and from

CH. III.] PEDRO BLANCO ROCK.—HONG HAI BAY.—INSIDE ISLAND. 77

which Si-ting bears S.W. by W. $\frac{3}{4}$ W.; Tung-ting S.W. by S.; Hat islet N. by E., and Harlem peak W. $\frac{1}{4}$ N.

HONG-HAI BAY, about 15 miles to the north-east of Fokai point, is extensive, but in the upper part the water shoals to 3 and 4 fathoms, and it is open to S.W. and South winds. There are several islands in the bay, the largest of which, Hong-haï, is in the middle of it.

Vessels are recommended not to pass to the westward of Tung-ting and Si-tung, nor into the north-west part of Hong-haï bay, as they will experience a heavier sea there than outside.

HONG-HAI ISLAND, bearing N.E. $\frac{1}{2}$ E. 8 miles from Si-ting, is half a mile long, east and west, 3 cables wide, and will afford shelter on its northern side from southerly winds. S. by E. $\frac{3}{4}$ E. from its summit, which is 240 feet high, there are two rocks, visible at low water; they lie 3 cables from the shore with the south-west point of Hong-haï bearing N.W. by W. $\frac{3}{4}$ W., and in line with the south end of Inside island, and the east point of Hong-haï N. by W. westerly, and in line with the highest part of Mace point.

HAT ISLET is a peaked rock lying $2\frac{3}{4}$ miles westward of Hong-haï. It is called by the Chinese Ke-sin-she (a fowl's heart), which it more resembles than a hat; there are detached rocks about it.

SHOAL BAY is formed at the head of Hong-haï bay, 3 miles N.N.E. of Hong-haï island. Its entrance is 2 miles wide, and, within the heads, the depth is less than 3 fathoms. There is an inlet, with only 6 feet over the bar at low water, which communicates in its north-east part with Hie-chechin bay; by report it is navigable for small boats only.

At three-quarters of a mile eastward of Club point, the east point of Shoal bay, there is a rocky ledge, part of which is always above water.

INSIDE ISLAND, 5 miles to the N.W. of Hong-haï, is 460 feet high, a mile long north and south, and but little more than a cable wide. At 3 cables from its south-west end are some detached rocks; and in the bays east and west of it no greater depth than $2\frac{1}{2}$ fathoms will be found at low tide. There is usually a long ground swell here, rendering it advisable for vessels not to stand farther into the bay than Hong-haï island.

West, 3 miles from Inside island, is the embouchure of a large stream, but with only 6 feet over the bar at low water.

TY-SAMI INLET, the entrance to which bears E. ¾ N. 9 miles from Hong-haï island, has a channel leading into it half a mile wide, and carrying 2½ fathoms at low water. The northern shore of the entrance is shoal-to, and rather more than half a mile from the beach are some rocks, which show at low tide, and from which Ty-sami mound bears S.S.E., and the low conical hill at the back of the town E. by N. ½ N.*

The southern edge of the channel is bordered by a sandbank, which commences under Ty-sami mound, and extends 1¼ miles from the shore, until its north end bears West from Entrance head, where it shoals suddenly, and has but 3 feet on its edge. The north end of the sandy spit under Entrance hill (the hill on the south side of the entrance), in line with the conical hill at the back of the town bearing E. ¾ N., will lead into the inlet on the south side of the channel. Ty-sami mound is an artificial cone on the highest part of the hills over the south-east point of Hong-haï bay; its elevation is 970 feet above the sea.

TIDES.—In Hong-haï bay it is high water, full and change, at 10h. 0m. and the rise is 6½ feet.

GOAT ISLAND lying S.E. 3 miles from Tsiech point, the eastern outer entrance-point of Hong-haï bay, is the southernmost and largest of a numerous group, amongst which there are no navigable channels. S.W. ¼ W. from its summit, and S.S.E. ½ E. from Ty-sami mound, lies a dangerous rock, which shows only when the tide is low and the wind high. At rather more than a mile inland from the beach to the northward of Goat island, is the walled town of Tsieching.

ANCHORAGE.—There is good anchorage in the N.E. monsoon on the north-west side of Goat island, which, with the group of islets to the northward of it, shelters well from the heavy sea. This roadstead is much used by the opium vessels, which approach as close to the shelving beach as the depth of water will allow.

REEF ISLANDS lie S.E. by E. 3 miles from Goat island; and E. ¼ N. 1⅓ miles from the latter, and N.W. ¾ W. 2 miles from the north end of the former, there is a rock on which the sea breaks at low water. The southern island of the group is the largest, and reefs extend a cable's length in a southerly direction from its east end.

Vessels may pass between the Reef islands and some rocks awash, lying 1¾ miles to the northward, the depth being 7 and 8 fathoms; but it must be borne in mind that the shoal water extends rather more than 2 cables

---

* See Plan of Ty-sami Inlet enlarged, scale 1 inch to a mile, on Sheet 2, East Coast of China.

CHAP. III.]   TY-SAMI INLET.—HIE-CHE-CHIN BAY.   79

to the northward of the islands; the north end of the danger bearing W. ½ S. from Chelang point. Vessels should not pass between the rocks awash and the coast.

CHELANG POINT, bearing E. by N. ¼ N. 5 miles from the Reef islands, is very remarkable, of moderate height, composed of red sand, with many ragged rocks scattered over it. The point has two islets and a reef off it, and the depth is 13 fathoms within a mile of the outer islet, which is 80 feet high.

There is a fort on the western extremity of this headland, and to the northward of the fort a small bay, which will afford shelter in the North-east monsoon; but a sunken rock, having only a foot water over it, lies N.W. by W. 5½ cables from the fort, and from it the summit of Chelang point bears S.E. ⅔ E., and is in line with the southern rock off the fort point, and Flat rock bears S.W. ⅓ W. Flat rock lies W. by N. 1¾ miles from Chelang point, and there is a small sunken rock lying N.W. from it, and West from the fort.

KIN-YU or Kemsue is a rocky islet, half a mile long, in a N.E. and S.W. direction, lying N.E. ⅓ N. 3¾ miles from Chelang point, and under its highest or north-east part there is a high rock. Its shores are bold-to, but the islet is too small to afford shelter. The channel between it and Che-chin point is 1½ miles wide, and carries a depth of 7 and 8 fathoms; but off the point is a large white rock surrounded by reefs.

HIE-CHE-CHIN BAY, formed between Paukshao point on the west, and by Tongmi point on the east, carries a depth of 7 to 5½ fathoms at the entrance, and 3 or 3½ fathoms within a mile of its head, over soft muddy bottom. It will afford shelter from westerly and northerly winds, and from the North-east monsoon, but it is quite exposed to the southward and south-east. At the head of the bay the land is low, and there is a sandy beach; the eastern side of the bay is high and mountainous. The village of Kinsiang stands in the north-east bight of the bay, immediately under Round hill; to the northward of Kinsiang point there are not more than 3 fathoms at low tide. Two rivers empty themselves at the head of the bay, with bars of less than 9 feet water, and the sea usually breaks all across them; the western river communicates with Hong-hai bay, and affords a passage for boats and even small junks.

Near Tongmi point, which bears E.N.E. 14 miles from Chelang point, there is a remarkable conical hill 455 feet high, named Chino peak, which, with the islets of Tung-ki and Si-ki, render this side of the bay easy to recognize. The peak bears N.W. ¼ N. 2⅓ miles from Tung-ki, which is about 18 feet above the sea, and has some detached rocks lying on its eastern side, and three rocks awash at low water, half a cable's length from its north-west side; the channel between it and the main is a mile wide, and carries a depth of 9 to 12 fathoms. Si-ki islet, 80 feet high, rises abruptly and is cleft at the summit; from it Tung-ki bears E.N.E. 3¼ miles, and Chino peak N.N.E. ¼ E. Between the two islets the soundings are 11 and 12 fathoms. A mile North of Tung-ki, and East three-quarters of a mile from Tongmi point, is a cluster of rocks nearly level with the water's edge.

PAUKSHAO BAY, on the western side of Hie-che-chin bay, affords good shelter, unless the wind comes to the eastward of South, there being a depth of five fathoms with Paukshao point bearing westward of South. Paukshao point is of moderate height, with numerous rocks scattered over its surface. The other point to the westward has a high battery on it; and between this latter point and the high land to the northward there is an opening into a deep harbour, the entrance to which is nearly barred by rocks, and the harbour too shoal for vessels drawing over 6 feet. There is said to be a sunken rock lying N.E. about half a cable's length from Paukshao point.

CHINO BAY is on the eastern side of Hi-che-chin bay, to the northward of Chino peak, and on its shore there is a fort and small village, abreast of which the water is shoal, the 2 fathoms line of soundings being half a mile from the shore. West from Chino peak the Chino reef extends 4 cables from the shore, the outer rock of which does not show at high water unless there is a considerable swell; when upon it Tung-ki bears S.E. ½ E., Si-ki S. by W. ¾ W., and the East White stone, in the northern part of the bay, is in line with Round hill, bearing N.N.W. ¾ W.*

A dangerous coral rock, with only 7 feet water on it, on which the *Sarah Lucy* struck, lies 8½ cables to the south-east of the Yellow Stone. It has 4½ fathoms, mud, close to, and from it the Yellow Stone bears N.W. ¼ W.; small rocky isle at mouth of the creek leading to Kieshi-wei N. by E. ½ E.; and the extreme of Chino point S.S.E. easterly. When proceeding to the anchorage, keep the East White Stone open to the westward of the Yellow Stone.

The best anchorage is in a depth of 3¾ fathoms farther to the northward about East of the Yellow Stone, which is the southernmost of all the rocks, with the exception of the Sarah Lucy, in the north-east part of the bay. The walled town of Keishi-wei, bearing E. by N. 3 miles from the Yellow Stone, will be seen over the low land from this

---

* *See* Chart:—East Coast of China, Sheet 3, No. 1,963; scale $d = 14\frac{1}{2}$ inches; with enlarged Plan of Chino Bay on it, scale m = one inch.

anchorage; there is a creek leading up to it which will admit junks at low water.

Between the Yellow Stone and the rocks, three quarters of a mile N.N.W. of it, there is a channel carrying 4½ fathoms; but vessels are recommended not to approach that part of the bay northward of the Yellow Stone, as there are several sunken rocks, one of which bears N.W. by W. ½ W. 1$\frac{4}{10}$ miles from the Yellow Stone, on which bearing it is in line with the northern end of Chino bay hills; from it the East White Stone bears N.E. by E. ⅜ E. and the West White Stone, N.W. ¼ W. As this rock lies to the south-westward of all those above water, care must be taken to avoid it in working up the bay; the East and West White Stones will be known by their being the largest of the group.

Vessels drawing less than 18 feet may stand into the bay to the northward of the West White Stone, where the depth is 3½ to 2¼ fathoms, the water shoaling gradually towards the beach.

HUTUNG POINT.—From Tongmi point the coast trends in an E. by N. ½ N. direction about 15½ miles to Cupchi point. At the distance of 4½ miles is Black Rock point, with black rocks off it, and a square white rock on its south-west side; N.W. 1½ cables from the white rock is a sunken rock. This bay is not deep enough to afford shelter.

About half way between the above points is the mouth of the river Hutung, which falls into the sea westward of Hutung point, but it has only 6 feet water over the bar. There is a fort on its south bank, and close to the fort a remarkable dome-shaped building, apparently intended for a fire beacon; this is a good mark in hazy weather, being so easily recognized, indeed there is nothing resembling it on this part of the coast.* S.S.E. 1½ miles from the fort is a small islet, surrounded by reefs and detached rocks, one of which, to the eastward, is of a curious shape, from which it has obtained the name of Figure rock.

At 3 miles eastward of Hutung point the hills come down to the beach, and on one of their peaks is a conspicuous knob. At a mile from the beach lies a flat rock with sunken dangers between it and the shore; there is also a rock awash to the S.E. of it.

CUPCHI POINT, 210 feet above the sea, has a rugged summit, and near its south end there is a dilapidated fort.† South 1½ miles from the point is Turtle rock, 14 feet above the sea, and inside of it two islets, and four patches of rock. The junks pass between Turtle rock

---

* *See* view on chart.

† *See* enlarged Plan of Cupchi Point, scale, m = 1 inch, on Sheet 3, East Coast of China.

and the rock next to the northward, though sunken rocks lie westward of both, and much discoloured water, which, however, is a help to detect them.

Between the islets and the point the channel is 2 cables wide, but the bottom is rocky and uneven, and a rock on which the steamer *Five Brothers* was wrecked, 28th Feb. 1859, lies 60 fathoms South of the point. The least water on this rock is 12 feet, and as many sunken dangers are in its vicinity, it would be imprudent for a stranger to attempt the passage. A ledge of rocks extends 2 cables from the point westward of the fort, its outer end breaking at low water.

A remarkable little black conical hill, named the Black mount, rises 230 feet above the sea from a red sand down, at $4\frac{1}{2}$ miles to the north-east of Cupchi point, and half a mile from the beach. Reefs extend from the shore half a mile along this part of the coast.

ANCHORAGE.—There is good anchorage during the North-east monsoon to the southward of the Shag rock, which lies N.N.W. of Cupchi point, at half a mile off shore, and is 3 feet above high water; it has $2\frac{1}{2}$ fathoms around it, except on its S.S.E. side, where there is a projecting reef. On the main, abreast this rock, is a fort standing on the left bank of a river leading to the walled town of Kiahtsz. The town is $1\frac{3}{4}$ miles from the fort, having to the southward of it a pagoda two stories high. On the bar of the river there are 9 feet at low water,* but the channel over it is both crooked and narrow. Nearer the entrance there is a second fort, and on the sandy point opposite a martello tower.

TUNGAO ROAD.—The village of Tungao stands in a bight of the coast N.E. by E., 15 miles from Cupchi point, the intervening shore being low and sandy. On the bar of the river, west of the village, the sea breaks heavily at low water, and outside the bar the water shoals suddenly, so that vessels approaching the anchorage in Tungao road should not bring the fort at the village to bear eastward of N.E. $\frac{1}{2}$ N., when within $1\frac{1}{2}$ miles of it; this will be found a good roadstead in the North-east monsoon. There are two pagodas in the neighbourhood, one on the low land at the east side of the river's mouth; the other on the hills 2 miles to the northward.

S.E. by E. $2\frac{1}{2}$ miles from Tungao is a white rock, which forms a good

---

* Capt. P. Cracroft, who visited this locality in H.M.S. *Niger*, Feb. 1859, found only 6 feet on the bar at low water. On the 3rd, the day before new moon, it was low water on the bar at 10 a.m., and high water at about 4 p.m., with a rise of about 5 feet. There can be no doubt, however, that the time of high water on this part of the coast varies with the force of the monsoon.

mark by which this part of the coast may be recognized; half way between the rock and the village is a creek with a fort upon the hills eastward of it. The land near the coast is low, with several fishing villages in the sandy bays; the boats belonging to which are numerous, and being of different shape and smaller than those of Hai-mun and Cupchi, will enable a vessel to identify her position in a fog.

BREAKER POINT, lying 9 miles eastward of Tungao road and E. by N. ⅜ N. 23 miles from Cupchi point, may be known by a black dome-shaped hill rising 280 feet from a red sand drift on the point, from whence the hills trend northward and westward, dipping suddenly at their extremity. At the south extreme of the point is a remarkable rocking stone, and off the south-east and south-west points of land on each side of the stone are two small islets; a fort stands on the point within. Detached reefs lie off the shore, which should not be approached within half a mile.

At 2 miles westward of Breaker point is a small islet having a Flat rock, part of which is always uncovered to the S.E. of it. W. ½ S., distant 8 cables from this rock, and South from the islet, lies a sunken rock on which the sea seldom breaks. The bay westward of Breaker point, and which is fronted by Flat rock and the islet, cannot be recommended as a place of shelter, being full of rocks.

TIDES.—It is high water, full and change, at Kin-siang point in Hie-che-chin bay, at 7h. 0m.; at the Shag rock, north-west of Cupchi point, at 8h. 0m.; and in Hai-mun bay and the Cape of Good Hope at 9h. 0m.; and the rise is 6 or 7 feet.

In Tungao road it was high water, full and change, in January, at 3h. 0m. At 5 miles eastward of the road the ebb ran to the westward a knot per hour on the 12th day of the moon, and no flood tide was perceptible during that month. There is a tide race with the flood off the Cape of Good Hope.

From observations on the tidal streams, from January to May, between Breaker point and Hong Kong, the ebb runs to the eastward, but, generally speaking, very little tide was experienced. To the eastward of Breaker point, however, the flood sets to the eastward, which is its direction throughout the north-east coast of China; the times of high water, full and change, from Hong Kong to the Yang-tse kiang, not deviating more than one to three hours before the moon's transit, unless obstructed by local causes, with the exception of the vicinity of Breaker point, where it was high water at 3h. 30m. p.m. at the full moon, January 1845.

TONG-LAE POINT bears N.E. by N. 5 miles from Breaker point, and about a mile westward of it is the entrance to a creek leading to the walled town of Tong-lac. On the eastern side near the entrance is a fort, under which indifferent shelter might be found in the North-east monsoon by a vessel of not more than 12 feet draught, but she would be in an awkward position should the wind come to the southward of East. Sunken rocks abound along this portion of the coast, one of which lies 6 cables from the land, with the fort bearing N.W. b. N., and Rocky point N. by E. ¾ E.

Rocky point is low and bears N.E. 1¾ miles from Tong-lac point; from thence the coast trends northerly for 4½ miles, joining a headland with reefs extending a quarter of a mile to the south-eastward.

HAI-MUN BAY is between the above headland and Hai-mun point, N.E. ½ E. 7 miles from it, and carries a general depth of 6 and 7 fathoms. The highest part of the hills at the back of this point forms two peaks, on the highest of which is an artificial mound 590 feet above the sea. There are three pagodas on the land to the northward of the bay, two of which are on the hills, and can be seen in clear weather from Namoa island, the other is on the low land.

Parkyns rock, on which the sea breaks at low tide, lies 9 cables to the southward of Hai-mun point, and from it the artificial mound bears N. ¾ E.; Cape of Good Hope, N.E. by E. ¾ E.; and Rocky Head point, on the eastern side of entrance of Hai-mun river, N.W. Rocky Head point in line with the west peak of Pagoda range, bearing N.W. ¼ N., will lead close to the south-west side of the rock; there is a passage between it and the main.*

A rocky ledge with 2¼ fathoms on its southern end extends 6 cables from the fort point. The above mark leading close to the Parkyns rock points to its southern edge, with the fort bearing N.E. There is also a rock awash at low water lying W. ¾ N. half a mile from the fort, the mark for which is the west peak of Pagoda range in line with a large stone in the centre of the first sandy beach eastward of Rocky Head point bearing N.W. ¼ N. The channel between this rock and the western shore of the bay is 4 cables wide.

There is another rock showing at low water, the bearings from which are the south extreme of Hai-mun point E. b. S., Rocky Head point N.E. ½ N., and the west peak of Pagoda range N.W. b. N.

HAI-MUN RIVER has 10 feet on its bar at low water. The town is built on the left bank, one mile from the entrance, and north of the town the river turns to the westward. The land being low to the northward, a canal communication with the Shantau estuary will most likely be found.

---

* See enlarged Plan of Hai-mun Point, scale, $m = 1$ inch, on Sheet 3, East Coast of China.

HOPE BAY is formed between Hai-mun point and the south extreme of the Cape of Good Hope, which bears E. by N. ¾ N. distant about 9 miles. The detached rocks lying along the coast for 3½ miles to the N.E. of Hai-mun point, renders it advisable for a vessel not to close this part of the shore nearer than half a mile, until beyond that distance, when the sandy beach is steep-to.

There is a secure anchorage in the North-east monsoon on the southern side of the cape, to the north-west of Tide point; the smoothest water will be found in the outer little sandy bay near a fort and a large tree. Sunken rocks extend a cable's length from the fort point; otherwise this sandy bay is clear and the lead will be the best guide.*

On the western side of this sandy bay is a remarkable peaked rock; and 1½ miles to the N.W. of this rock is the entrance to a creek which makes the Cape of Good Hope an island, and communicates with the Shantau estuary and the river Han. The creek has 7 feet water over the bar, which is barely a cable across, and is defended by a fort. Reefs extend south-westerly 3 cables from the latter fort to a rock awash at high water, rendering the straight channel impassable to the large fishing boats at low water; at this time of tide they leave the rock to the eastward and pass between it and two islets off the fort. In the event of a vessel being wrecked on the coast, and the crew wishing to reach the depôt vessels at Namoa island, this would be the best route, as few boats could live in the tide race off the cape.

CAPE of GOOD HOPE is the north-east extreme of a hilly peninsula, the highest part, 480 feet above the sea, appearing like a dome. The eastern face of the peninsula is steep-to, and has three points projecting from it; the northernmost is the cape, the middle one is named Ma-urh point, and the third Tide point from the tide race which sets round it. On the north side of the peninsula is Green islet, having a patch of rocks between it and the land.

At 3 miles northward of the cape and half a mile from the shore is Bill islet, 56 feet high; and S. b. E. 4 cables from Bill islet is Squat rock, with a reef lying 2 cables westward of it, and showing at low tide. Rocks extend from the points on the main abreast these two islets, and in the narrow channel there are 3 fathoms water. Abreast of Bill islet the coast trends N.W. b. N. 3 miles to Sugarloaf island, which has a reef extending a cable from its east point.

RIVER HAN.—From Sugarloaf island the coast trends westward, forming the south side of the entrance of the River Han, the bar of which

---

* See enlarged Plan of Cape of Good Hope, on Sheet 3, East Coast of China; also Entrance of the River Han, No. 2,789; scale, $m = 0.75$ of an inch.

has 2½ fathoms on it at low tide. N.W. b. N. three-quarters of a mile from Sugarloaf is Double island, from which the town of Swatau or Shantau, upon the north bank of the river, bears W. by N. ½ N. 4¼ miles. At half a mile to the south-east of the town the depth is 8 fathoms, and at low tide the water in the rainy season is fresh. Swatau is the port of Chinhae, which is distant about 2 miles to the north-east. The country in the vicinity is highly cultivated ; sugar-cane and tobacco grow luxuriantly.

*The tides inside Double island are very irregular ; in ordinary weather the rise and fall is from 8 to 9 feet. Any vessel from 13 to 14 feet draught, with a fair wind, could in fine weather with no ground swell on, enter the port at any time of tide with safety.

DIRECTIONS.—The Cape of Good Hope may be rounded closely if necessary when bound to the river Han, steering for the Pagoda hill, which makes like an island to the northward, until Bill islet and Squat rock are plainly seen. Having passed about 2 cables eastward of Bill islet, bring the middle of the islet in line with the extreme of the Cape, and it will lead in mid-channel and towards Double island. When the high point opposite Shantau opens northward of Double island, a vessel may stand towards the latter, close to the fishing stakes, but its south-east side should not be approached within 3 cables' lengths. Double island will not show double until it bears N.W., and Sugarloaf island is likely to mislead a stranger as it has not the slightest resemblance to its name. There are establishments on Double island belonging to English merchants, and receiving ships for opium lie here.

The channel between Double island and the mainland to the northward is half a mile wide, the mud drying that distance from the shore, which is low. Joachim bank is an extension of this mud flat in a south-easterly direction, its southern edge in 2 fathoms bearing E. b. S. 2½ miles from Double island. A good guide to lead a vessel of 14 feet draught to the eastward of this bank is, to keep Brig island open of the east end of Fort island, bearing N.E. ½ N. ; but great care must be taken, for in bad weather the sea breaks heavily, and in light winds the flood sets strongly over it.†

FORT ISLAND, with a fort on the table-land at its western end, lies E.N.E. nearly 2 miles from a pagoda (260 feet above the level of the sea), which bears N. ½ E. 10½ miles from the Cape of Good Hope: the pagoda stands on an isolated hill near the coast, and the land in its vicinity is so low, that the hill when first seen appears like an island. The channel between Fort island and the main is shoal.

---

* See Plan of Namoa Island, No. 1,957 ; scale, $m = 0.7$ of an inch.
† Nautical Magazine, page 335, year 1859.

BRIG ISLAND, so called from a rock at its southern extremity which appears like a brig when seen in an east or west direction, lies N.E. ¾ E. 4 miles from Fort island, the depths varying from 2½ to 4 fathoms, the most water being towards Brig island.

NAMOA ISLAND, 12 miles long, east and west, and 5½ miles wide at its broadest eastern part, is separated from the main by a channel about 3½ miles wide with depths varying from 3 to 6 fathoms. The peaks of this island (of which there are three) form the most prominent land-marks in the neighbourhood, and rise to the height of 1,700 and 1,900 feet above the sea. Notwithstanding its barrenness, the island is exceedingly populous, the fisheries affording a livelihood to the greater portion of the inhabitants. Clipper point, its western extreme, is fronted to the southward by knolls of sand which shift, and which render local knowledge necessary when steering for the anchorage on the west side of Namoa. The eastern entrance, between the north point of Namoa and fort head, is much wider and has a general depth of 7 fathoms.

KNOLLS off WEST END of NAMOA.—S.E. b. E. 4½ miles from Pagoda hill, with the west end of Namoa island in line with Breaker island bearing N.E. ¾ N., there was formerly a shoal with only 11 feet water on it; in August 1844, there were several knolls, none of which however had less than 13 feet. The following are their bearings:—

The west point of Namoa island in line with Breaker island N.E. b. N. is the mark for three of the knolls. From the westernmost knoll, with 13 feet on it, Pagoda hill bears N.W. b. W.; from another with 17 feet it bears W.N.W.; and from a third with 18 feet it bears West. With Pagoda hill W. b. N., and the west point of Namoa N.N.E. there is a knoll with only 14 feet on it. All these are sand, and will probably be found to shift, owing to the freshes from the river Han.

BAYLIS BAY and CLIPPER ROAD.—Baylis bay is the first inlet on the west side of Namoa island, to the northward of Clipper point, and there is a fort on the ridge to the westward of it, and an outwork on the beach. There are three knolls off this bay, bearing from the fort as follows: the first, W. b. N. rather less than a cable's length from the fort point, with only 5 feet water over it; the second, N.W. ¼ N. a cable's length from the point, with 9 feet on it; and the third, N.W. ¾ N., a quarter of a mile from the same point, with 11 feet on it. From the latter the summit of Brig island bears N.W. ½ N., and the summit of Fort island W. by S. ⅓ S.

During the North-east monsoon the opium vessels anchor off this bay, remaining here from October unt May. In the other monsoon they lie 1½ miles to the north-eastward in Clipper road, abreast Steward's house, as the swell setting round Clipper point renders that anchorage inconvenient. With the exception of a few narrow passages of about 90 or 100 feet wide, the channel inside Namoa island is staked across; but vessels soon shoot through them.

From Baylis bay a bank commences and borders the north-west coast of Namoa for 2½ miles; its greatest distance from the shore is 4 cables, which is abreast Steward's house; the lead gives no warning, and there are only 9 feet upon its edge.

Supplies.—These two anchorages must be considered more as safe roadsteads than harbours, as from the velocity of the tide and the fetch of the sea, laden boats would frequently have much difficulty in passing to and fro. Water may be procured with great facility and there is no difficulty in obtaining fresh provisions.

FOLKSTONE ROCK, with only 5 feet on it, lies with the south extreme of the Brig rock in line with the north-west head of Fort island, bearing S.W. by W. ½ W.; Coffin island (the largest of a cluster of islets 3 miles north of Brig island) N.W.; and the flagstaff at Steward's house is in line with a whitewash rock at the back of it, S. by E.

The south extreme of Brig rock, just open of the north-west extreme of Fort island, will lead south of the Folkstone, and also of the shoal which extends nearly all the way from Brig island to Breaker island; the latter (a peaked rock with several others around it, which must not be approached nearer than 2 cables upon its western side) bears from Brig island N.E. by E. ⅜ E. To the eastward of Breaker island, shoal water extends a great distance from the northern shore; its south edge in 3 fathoms bears East 3 miles from the island.

SHOAL BAY.—From Opium point the north coast of Namoa trends to the south-east, forming a deep indentation, named Shoal bay, in which there are two islets and several rocks; the land at the bottom of this bay is low, and only a mile across to the southern side of the island.

NANGAOU BAY, the next bay eastward of Shoal bay, has at its head a walled town, the residence of the magistrate of the district. Vessels drawing less than 18 feet may stand into this bay until Pagoda island bears E. by N.; but during the North-east monsoon there is a considerable swell in it, and the entrance of Challum bay, on the opposite shore, will be found a more eligible anchorage, and vessels will be in a better position to avail themselves of the land wind, which usually draws to the northward in the morning.

The North point of Namoa has a double peak over it, and forms the eastern boundary of Nangaou bay: rocks extend 3 cables from its north-eastern face.

PAGODA and SOUTH BAYS.—From Clipper point the southern coast of Namoa trends nearly East 5 miles, where there is a small bay with a pagoda upon its eastern point. This portion of the island corresponds with Shoal bay on the northern shore.

South bay is 4 miles eastward of Pagoda bay, and affords good shelter in the North-east monsoon; rocks extend $1\frac{3}{4}$ cables southerly from its eastern point. Vessels drawing 18 feet may run into this bay until the extreme of the point bears S.E. About half a mile to the south-east of the point is a low flat island, called Crab islet, and in the channel between it and Namoa the ground is foul. At about $1\frac{1}{4}$ miles eastward of South bay is a bold bluff (the southernmost point of Namoa), with three tall chimneys on it.

TIDES.—It is high water, full and change, in Clipper road, Namoa island, at 11h. 15m., and springs rise about 7 feet. The streams on the north side of the island run parallel to it at the rate of 1 to 3 knots. The flood comes in on the north as well as on the south side of the island, but the streams in the neighbourhood of Nangaou bay are not so strong as at the western end of the island, where they run 4 knots at the springs, the ebb coming from the eastward.

LAMOCK ISLANDS are a group of four islets, with two patches of rocks extending altogether in a N.E. and S.W. direction $7\frac{1}{2}$ miles. From their south extreme, the west end of Namoa bears N.W. $\frac{1}{2}$ W. 22 miles; from their north end, the east point of Namoa bears N.W. $13\frac{1}{2}$ miles, and the south-eastern Brother N.E. by E. $25\frac{1}{4}$ miles. At the south-west end of the group are two square rocks, 15 feet above high water, and named Boat rocks; they are about the size of boats, and have several reefs between them.

White rock, lying N.E. $1\frac{1}{2}$ miles from the Boat rocks, is sufficiently large to afford protection to boats. The distance between the White rock and High Lamock island is 3 miles, with a safe channel between, the depth varying from 8 to 14 fathoms. High Lamock, 250 feet above the sea, is thickly covered with brushwood. The channel between it and East Lamock is $1\frac{1}{2}$ miles across, and in about the middle of the channel there is a rock, with a reef, which shows at low tide, extending southerly $1\frac{1}{2}$ cables' lengths from it. The three northern islets lie close together; the northern one, which has a pyramid on it, is without vegetation.

Mr. Anderson, master of the ship *Sir E. Ryan*, states, that he saw a rock (when in command of the *Times* schooner) to the N.E. of the Lamock islands. He described it as "a rock awash 3 miles from the North rocks, with all the Lamocks in one;" H.M.S. *Plover*, however, searched for it without success.

TIDES.—In the month of April, with High Lamock island bearing E. ½ S. 17 miles, the tides made as follows: 1st hour of ebb S.W. ¼ W. 1½ knots; 2nd hour, S.W. by S. 1½ knots; 3rd hour, S.W. by S. 1½ knots, and the 4th hour, S.S.W. half a knot. Flood, 1st hour, N.E. by E. 1 knot; 2nd hour, E.N.E. 1½ knots; 3rd hour, E.N.E. 1½ knots; 4th hour, E.N.E. 1½ knots; and the 5th hour, E.S.E. half a knot. And in September, with High Lamock bearing E. by N. 4 miles:—The ebb, 1st hour, S. by E. half a knot; 2nd hour, S. by W. 1 knot; 3rd hour, S.S.W. 1 knot; and the 4th hour, S.S.W. 1½ knots. The flood ran to the N.E. the whole tide, the total amount being 10½ knots.

Thus in passing inside the Lamock islands, attention to the tide as well as to the vessel's course is necessary.

DOME ISLAND and LAMON ROCKS.—Between the Lamock islands and Namoa are four islets; the northernmost of which is the highest, and from its appearance is called Dome island. The two southern islets, lying east and west of each other, are named Ruff rock and Oeste rock; to the southward of the Ruff are the Dot and Sul rocks. A reef extends a third of a mile to the southward of the Sul; the east end of the Oeste in line with the east end of Plat island bearing N.W., leads to the southward. Plat island is flat topped, and is lower than the Ruff or the Oeste.

SINTA ROCK, with only 2 feet water upon it, lies S.E. ⅜ S., nearly 5 miles from Dome islet, with the south-west extreme of Ruff rock in line with the summit of Plat island, bearing W.N.W. the East point of Namoa island N. by W.; and the highest part of High Lamock island E. by S. ¾ S.

YENG ROCK, awash at low water, lies N. ¾ E. 5 miles from the Sinta, with the north end of Crab islet in one with the south-west point of Namoa, bearing W. by N.; Dome islet W. by S. ¼ S.; High Lamock S.E. ⅜ S., and the east end of Namoa N.N.W. ⅜ W. The North point of Namoa seen clear of the East point, leads to the north-eastward.

HALF-TIDE REEF.—There is another patch of rocks which show at half tide, between Dome islet and Namoa, bearing from the former, from N. by E. to N.N.E. ½ E. distant one mile. They lie rather more than a mile from the Namoa shore, S.E. by S. from Three Chimney bluff.

CHELSIEU ROCKS, 20 feet high, are a cluster of four rocks, bearing East nearly 7 miles from the North point of Namoa island.

The DIOYU is a reef just awash at high water, lying N.W. by N. 3½ miles from the Chelsieu; but should high tides and smooth water prevent its being seen, Pagoda island in Nangaou bay, in line with Saddle peak, on Namoa, bearing S.W. by W. ⅜ W., will lead northward of it.

CHALLUM BAY is fronted by the northern side of Namoa island, and, as before stated in page 88, its entrance affords better shelter during the North-east monsoon than Nangaou bay.

To enter this bay, pass within a mile westward of a barren rock, named Middle islet, bearing N.E. by E. 5⅓ miles from Breaker island. The anchorage is between Entrance island and Middle islet, in 6 to 3 fathoms; the bay north of Entrance island is shoal. Should a vessel pass eastward of Middle islet, it must be within half a mile, as there is shoal water (11 feet) extending 9 cables from Fort head. When running in, steer for the East point of Entrance island, and beware of the starboard shore, as the water shoals suddenly on that side, and there is a sand bank which shows at low water, lying nearly half a mile to the southward of the west end of Challum island. Under Fort head is a rock nearly covered at high tide, and also one in the bay between it and Difficult point; otherwise the coast line here is steep-to.

DIFFICULT ISLET, 110 feet above the sea, lies nearly 3 miles to the E.N.E. of Fort head, the east point of entrance to Challum bay: and on the highest part of the hills over Difficult point (the point west of the islet) is a square fort.

TERNATE ROCK, with only one foot water on it, lies E. by N. 1⅛ miles from the summit of Difficult islet, which will then be in line with the third and last sandy hill on the northern part of the range extending from Fort head. Pagoda island, in Nangaou bay, in one with Namoa peak S.S.W. ¾ W., will lead close to the eastward of it.

SHALLOW BAY.—Six and a half miles E.N.E. of Difficult point is the entrance to a shallow bay with a pagoda on an island within it; the boundary line of the two provinces, Fu-kyen and Quang-tung, passes through this bay.

CHAUAN BAY.—The entrance to this bay is 10 miles to the N.E. ¼ E. from the north extreme of Namoa island, and its western point, which is the eastern point of Shallow bay, has a small islet off its south extreme.

The walled town of Tongyung stands on a peninsula on the western shore of the entrance abreast Pagoda island, and although the channel between it and the island is 3 cables wide, it is not a good one to enter by, as rocks extend from both shores. Tongsang basin runs back N.N.W. 11 miles from the Middle islands, and there is said to be a river at its head; 3 fathoms water were obtained at the highest point reached, but the channel was very narrow. There is a boat passage on the western side of the basin leading into Chauan bay, the entrance to which bears West from Fall peak. In the north-west portion of the basin is a range of high mountains, remarkable for their rugged appearance.

CAUTION.—When running into Tongsang harbour, sail should be reduced in time, if the wind is fresh outside, for violent squalls will be experienced under Thunder head.

When proceeding eastward, the coast on the eastern side of Thunder head must not be approached within a cable's length, as there are some rocks along it; the south face of the head, however, is steep-to.

TIDES.—It is high water, full and change, in Tongsang harbour at 11h. 30m.; springs rise 12 feet.

REES ROCK, which covers at high-water springs, lies S.E. by E. $\frac{3}{4}$ E. $1\frac{3}{4}$ miles from Fall peak, with the chimneys on Chimney island, forming the eastern side of Rees pass, bearing N.E. by N.; a rock, on which the sea breaks at low water, lies a cable's length to the eastward of it. Pass islands bear N. $\frac{3}{4}$ E. $1\frac{1}{3}$ miles from Rees rock, and the ground between them and the rock being foul with shoal water, should not be approached. Junks use a channel 2 cables wide between the Pass islands and the main.*

REES PASS is formed between Pass and Rees islands, and on its eastern side, W. by S. from the chimneys and 3 cables from the shore of Chimney island, is a shoal with only $2\frac{1}{4}$ fathoms on it. The Rees islands are barren, and only inhabited by a few fishermen.

H.M.S. *Plover* rode out a heavy gale veering from N.E. to E. by N., in this Pass; she anchored in 6 fathoms, at 2 cables westward of the black rock at the southern end of the sandy bay under the chimneys. There is also anchorage under South-east island in 6 fathoms, with its south point bearing East. It is said that in the northerly monsoon a vessel will not gain anything by going through the Pass, for on clearing

---

* See enlarged Plan of Rees Pass, scale, $m = 2$ inches, on Sheet 4, East Coast of China.

the north end of Chimney island as much swell will be experienced as will be found eastward of the group.

The Simplicia Wreck rocks lie 6 cables to the N.E. of South-east island, and at their eastern end are several rugged rocks, on the outermost of which the ship *Simplicia* went to pieces on the 8th of October, 1844, having struck upon a reef which shows at low water, and lies a cable to the north-east of the outer rock.

CAUTION.—In the neighbourhood of the Rees islands the sea rises rapidly after the commencement of a breeze, and over tops, leading to the supposition that there must be some change in the soundings.

DANSBORG ISLAND, lying 2 miles to the N.E. of the Simplicia Wreck rocks, is 6 cables long, N.E. and S.W., and a quarter of a mile wide; there are three peaks on it of nearly equal heights. Skead islet is 1¼ miles to the W.N.W. of this island, and between them, at the distance of 4 cables from the islet, there is another small islet with a reef extending from its west point; a reef also projects from the east point of Skead islet.

CHING ROCK, which covers at half tide, lies N. by W. ¾ W. 1$\frac{4}{10}$ miles from Skead islet, with the north-east peak of Dansborg bearing S.E. ½ E.; the chimneys on Chimney island S.W. ½ W.; the Awoota rock, W. by S. ⅔ S.; and Black head, N. by E. The Ching is the highest head of a reef, of some extent, the north-eastern rocks of which break only at low water, and extend 2 cables eastward of its highest part. The eastern Simplicia open east of Skead islet, leads to the eastward of the reef.

GOO ROCK is 2 miles to the S.W. by W. ⅓ W. of the Ching, and shows at the last quarter ebb; when on it the chimneys on Chimney island bear S.W. ¼ S., the Awoota rock, W. ¾ S., the summit of the Simplicia Wreck rocks, S.E. ⅞ S., and Skead islet, E. ¾ S. The Awoota lies close to the main, N.W. ½ W. 2¼ miles from the chimneys on Chimney island.

HU-TAU-SHAN, or BLACK HEAD, 5½ miles to the northward of Dansborg island, comprises five separate hills, the southern of which, Black head, is the most remarkable. On the northern of the five hills is a walled town.

The Hu-tau-shan river, which disembogues on the western side of the head, has deep water inside the entrance, but it is not available for navigation without being buoyed, as the channels, besides being narrow and intricate, are liable to continual change.

The walled town of Tongyung stands on a peninsula on the western shore of the entrance abreast Pagoda island, and although the channel between it and the island is 3 cables wide, it is not a good one to enter by, as rocks extend from both shores. Tongsang basin runs back N.N.W. 11 miles from the Middle islands, and there is said to be a river at its head; 3 fathoms water were obtained at the highest point reached, but the channel was very narrow. There is a boat passage on the western side of the basin leading into Chauan bay, the entrance to which bears West from Fall peak. In the north-west portion of the basin is a range of high mountains, remarkable for their rugged appearance.

CAUTION.—When running into Tongsang harbour, sail should be reduced in time, if the wind is fresh outside, for violent squalls will be experienced under Thunder head.

When proceeding eastward, the coast on the eastern side of Thunder head must not be approached within a cable's length, as there are some rocks along it; the south face of the head, however, is steep-to.

TIDES.—It is high water, full and change, in Tongsang harbour at 11h. 30m.; springs rise 12 feet.

REES ROCK, which covers at high-water springs, lies S.E. by E. ¾ E. 1¾ miles from Fall peak, with the chimneys on Chimney island, forming the eastern side of Rees pass, bearing N.E. by N.; a rock, on which the sea breaks at low water, lies a cable's length to the eastward of it. Pass islands bear N. ¾ E. 1½ miles from Rees rock, and the ground between them and the rock being foul with shoal water, should not be approached. Junks use a channel 2 cables wide between the Pass islands and the main.*

REES PASS is formed between Pass and Rees islands, and on its eastern side, W. by S. from the chimneys and 3 cables from the shore of Chimney island, is a shoal with only 2¼ fathoms on it. The Rees islands are barren, and only inhabited by a few fishermen.

H.M.S. *Plover* rode out a heavy gale veering from N.E. to E. by N., in this Pass; she anchored in 6 fathoms, at 2 cables westward of the black rock at the southern end of the sandy bay under the chimneys. There is also anchorage under South-east island in 6 fathoms, with its south point bearing East. It is said that in the northerly monsoon a vessel will not gain anything by going through the Pass, for on clearing

---

* *See* enlarged Plan of Rees Pass, scale, $m = 2$ inches, on Sheet 4, East Coast of China.

CHAP. III.]   REES PASS.—HUTAU BAY.—BLACK HEAD.   95

the north end of Chimney island as much swell will be experienced as will be found eastward of the group.

The Simplicia Wreck rocks lie 6 cables to the N.E. of South-east island, and at their eastern end are several rugged rocks, on the outermost of which the ship *Simplicia* went to pieces on the 8th of October, 1844, having struck upon a reef which shows at low water, and lies a cable to the north-east of the outer rock.

CAUTION.—In the neighbourhood of the Rees islands the sea rises rapidly after the commencement of a breeze, and over tops, leading to the supposition that there must be some change in the soundings.

DANSBORG ISLAND, lying 2 miles to the N.E. of the Simplicia Wreck rocks, is 6 cables long, N.E. and S.W., and a quarter of a mile wide; there are three peaks on it of nearly equal heights. Skead islet is 1⅓ miles to the W.N.W. of this island, and between them, at the distance of 4 cables from the islet, there is another small islet with a reef extending from its west point; a reef also projects from the east point of Skead islet.

CHING ROCK, which covers at half tide, lies N. by W. ¾ W. 1 4/10 miles from Skead islet, with the north-east peak of Dansborg bearing S.E. ½ E.; the chimneys on Chimney island S.W. ½ W.; the Awoota rock, W. by S. ⅔ S.; and Black head, N. by E. The Ching is the highest head of a reef, of some extent, the north-eastern rocks of which break only at low water, and extend 2 cables eastward of its highest part. The eastern Simplicia open east of Skead islet, leads to the eastward of the reef.

GOO ROCK is 2 miles to the S.W. by W. ¼ W. of the Ching, and shows at the last quarter ebb; when on it the chimneys on Chimney island bear S.W. ½ S., the Awoota rock, W. ¾ S., the summit of the Simplicia Wreck rocks, S.E. ⅞ S., and Skead islet, E. ¼ S. The Awoota lies close to the main, N.W. ½ W. 2¼ miles from the chimneys on Chimney island.

HU-TAU-SHAN, or BLACK HEAD, 5½ miles to the northward of Dansborg island, comprises five separate hills, the southern of which, Black head, is the most remarkable. On the northern of the five hills is a walled town.

The Hu-tau-shan river, which disembogues on the western side of the head, has deep water inside the entrance, but it is not available for navigation without being buoyed, as the channels, besides being narrow and intricate, are liable to continual change.

A spit extends 3 miles in a south-westerly direction from Black head, and some parts of it are dry at low water; its eastern edge bears W.S.W. from the head.

ANCHORAGE.—Vessels might ride out a strong breeze under Black head in 4 fathoms water at the distance of 3 cables from the shore, particularly if the wind holds to the northward; should, however, a gale come on, or the wind draw round to the eastward, the sooner this anchorage is quitted the better. Under these circumstances refuge will then be found by running through Rees pass and anchoring close under Chimney island, or in Tongsang harbour.

The COAST from Black head to Red bay, 10 miles to the north-east, with the exception of one hill and two hillocks, is a sandy plain. At the distance of 6 cables eastward of Tagau point is Hut islet and some rocks, a portion of which are always uncovered. Spire islet, lying 2 miles to the north-east of Hut islet, has a remarkable square column on it, and two low flat rocks to the westward. N. by E. one mile from Spire is Cleft rock, surrounded by reefs, which render it dangerous to be approached within the distance of 3 cables. Abreast Cleft islet is Crab point (one of the few places where the natives showed a disposition to attack the *Plover* during the survey of the whole coast). Knob rock, which is steep-to, bears S.E. by E. $\frac{1}{4}$ E. $3\frac{1}{4}$ miles from Spire islet, and East $6\frac{1}{2}$ miles from Black head.

RED BAY will be found a fair roadstead in the North-east monsoon, and its position may be readily known by the two high Black rocks off its east point, as well as by the low red sand hills at the back of it. A reef, having 3 fathoms close to, extends north-westerly from the southern of the two rocks, leaving a passage for small boats between it and the main at low tide. N.W. by W. 7 cables from the southern Black rock is a reef which covers at high water; the anchorage lies between them, and the water will shoal gradually after passing the Black rocks. At the head of the bay there is a village and a creek, the entrance to which is dry at low water.*

DIRECTIONS.—In working up to Red bay from the southward, take care to avoid the Shun reef, lying East 6 cables from a low hill on the shore, and 3 miles to the south-westward of the anchorage, which may be done by tacking when the Black rocks are in one with the point off which they lie. When upon this reef the eastern Black rock bears E. $\frac{1}{2}$ E.

---

* *See* enlarged Plan of Red Bay, scale, m 1 inch, on Sheet 4, East Coast of China.

In navigating this portion of the coast during the North-east monsoon, the wind will be found to hang to the northward from 2h. a.m. to 10h. a.m. and in the eastern quarter the remaining period. Deeply laden vessels will find it more advantageous to seek shelter in one of the harbours or roadsteads above mentioned during a strong north-east wind, than to keep underweigh, as ground can seldom be gained in consequence of the depth of water.

The COAST from Cork point, the north-east point of Red bay, trends N.E. ½ N. 18½ miles to Chin-ha point, and is steep-to, with the exception of a reef off Cork point, and a sand spit with some rocks on it extending in a southerly direction from House hill (a low hill with a house on its summit) which bears N.W. by W. ½ W. from Lamtia island. An inlet runs a long way back, to the westward of House hill, but it is shallow.

LAMTIA and NOTCH ISLANDS.—Lamtia island, bearing N.E. 9 miles from Cork point, is of basaltic formation, and its southern side rises abruptly from the sea: a reef extends 7 cables to the N.W. from it. Notch island, of similar formation, and with a rock awash off its south-east point, lies N. b. W. 3 miles from Lamtia island.

CHAPEL ISLAND, in lat. 24° 10′ 18″ N., long. 118° 13½′ E., lies N.E. ¾ N. 47½ miles from the South East Brother, and S.S.E. 11¼ miles from the Chauchat rocks at the entrance of Amoy harbour. Its height is about 200 feet, the whole of the island being nearly of the same elevation, and it is perforated at its southern end; there are also two remarkable mounds like chimneys on either end of the island. When this island bore South, and about midway between it and the entrance to Amoy harbour, Captain Ross, of the Indian Navy, passed over a sand-bank with 6 fathoms water on it, but no less could be found.

*MEROPE SHOALS are between Chapel island and the coast. The South Merope has only 5 feet on its shoalest part, at its southern end, from which Chapel island bears N.E. by E. ¼ E. 7¾ miles, and Lamtia island N.W. b. W. 5 miles; from thence it extends, with depths of 3 and 4 fathoms, nearly 5 miles in a N.E. direction. There are 2½ fathoms about 1¾ miles to the westward of its shoalest part, and probably shoal water extends to the southward, as its limits in that direction are not defined; it is said not to break except in heavy weather, or at very low tide.

When approaching this shoal from the southward do not bring the high land of Cork point to the southward of W. by S., while Chapel island is to the eastward of N.E. Chin-ha point N. by W., or Nantai Wúshan pagoda N.N.W. will lead eastward of the shoalest part.

The North Merope is 8¼ miles W. b. N. of Chapel island, and its eastern edge bears N.E. from Lamtia island. This shoal is formed by a number of Pinnacle rocks which show at half ebb, and have deep water between them.

TINGTAE BAY is 4 miles to the northward of the North Merope shoal, and affords shelter for small vessels in the North-east monsoon. Its vicinity may be easily known by a flat table head with three chimneys on it, forming the eastern point of the bay, and the ruins of a walled town on the hill above it. The pagoda of Nantai Wúshan, 1,720 feet above the sea, stands on the hills immediately at the back of this bay. The coast here continues in a north-easterly direction 3 miles farther to Chin-ha point, when it takes a sudden to the N.W., forming Amoy harbour.

TIDES.—It is high water, full and change, at the entrance of Chauan bay at 11h. 0m., and springs rise 6½ feet; and at the beach under Fall peak in Tongsang harbour, and at Chimney island in Rees pass, at 11h. 30m., and the rise is 12 feet. The flood tide enters Tongsang harbour at the rate of three-quarters of a knot per hour.

Off Jokako point, 4 days before the change of the moon, the ebb ran 4½ knots in one tide; the two first hours being from the N.E. by E., and the last four from the N.E.

In Rees pass on October 25th, with a gale from the N.E., the ebb tide ran from the N. by E. in all 12¾ miles; there was no perceptible flood. Also in October, the Awoota rock bearing S.W. ½ W. 3 miles, the 1st hour of flood ran W.S.W. half a knot; 2d hour, S.W. half a knot; 3d hour, W.S.W. 1 knot; 4th hour, W.S.W. 1 knot; 5th hour, W.S.W. half a knot; and 6th hour, S.W. by W. half a knot. The first hour of the ebb ran N.E. half a knot; 2d hour, N.N.E. half a knot; 3d hour N.N.E. 1 knot; 4th and 5th hours, N.N.E. a quarter of a knot.

Again, with Fall peak bearing W. by N. 7 miles; the 1st two hours flood ran S.S.E. 1½ knots; and the 3d S.E. by E. 1 knot. The 1st hour of ebb ran N.E. by N. 1 knot; 2d hour, 2 knots; 3d hour, N.N.E. 1½ knots; 4th hour, N. by E. 1½ knots; and the 5th hour, North 1 knot. Another observation in Rees pass, the moon's age being 11 days, gave the set of the ebb from the N.N.E., last hour North 1¼ knots; and the flood from the S.W. at the rate of half a knot per hour; wind N.E., force 7.

At Red bay in October, the moon being 19 days old, the rise and fall was 11 feet, and the ebb ran W. by N. and W.N.W., the whole amount of tide 2½ knots. Flood, 1st hour, W.S.W. 1 knot: then E.S.E. for the remainder of the tide, whole amount 1½ knots. Again with the moon's age 9 days: ebb ran North, and then N.W., 1 knot per hour, and the flood E.N.E. half a knot.

With Lamtia islet, bearing W.S.W. 7 miles; 6 days after the change in December, the ebb made from the E.N.E., then N.E., and for the last three hours N.W.; total amount in the tide, 3¾ knots. The flood came from the S.S.W., then South; total amount, 4½ knots.

Under Wu-seu island, at the entrance of Amoy, on the 4th day of the moon, the ebb came from the N.W. at the rate of 1¼ knots: and the flood from the S.S.E. at half a knot. Between Wu-seu and the main the tides are more rapid, and vessels should not attempt to pass.

WU-SEU ISLAND, 300 feet high, is on the western side of the entrance to Amoy outer harbour, and on its summit are three chimneys, which are intended for alarm signals. The island is 1¼ miles long, north and south, and near the middle only 2 cables broad. Its north-east and south-east faces are steep cliffs, and there are three sandy bays on its western side, and one on its eastern; in the northernmost bay on the western side is a large village and the ruins of an ancient fort.

The opium vessels used to lie between this island and the island of Wu-an to the westward of it, but the anchorage was found too confined, and not so convenient of access as that under Tae-tan island; it will be prudent not to pass westward of Wu-seu, as the channels inside are only partially surveyed.*

A rock, which is sometimes covered, lies between Wu-seu and Chin-ha point, with that point bearing S. ¼ W., and Nantai Wúshan pagoda W.N.W.

The CHAUCHAT are a patch of three flat rocks nearly awash at high tide, lying about half a mile eastward of Wú-seu. When on them, the three chimneys on Wu-seu are in one with Nantai Wúshan pagoda W. ⅜ S.; and by keeping Tae-pan point open northward of Tsing-seu N.W. by W., it will lead outside or eastward of them, should high tides and smooth water prevent their being seen. There is a channel, half a mile wide, between these rocks and Wu-seu, but in consequence of the chow-chow water there, it will be better to keep to the eastward.

The CHIN-TSEAO are two patches, covered at high water, lying N.W. by N. half a mile from the north end of Wu-seu; between them and the main are several islets and half-tide rocks.

---

* See Plan of Amoy Harbour, No. 1,767; scale, m = 1 inch.

TSING-SEU is a table-topped island lying three quarters of a mile to the north-west of the Chin-tseao; it rises precipitously from the sea, and forts are built upon its summit, which is 250 feet above high water. The entrance to Amoy outer harbour is between this island and Chih-seu, 8 cables to the north-east.

At the distance of 6 cables to the north-west of the Chin-tseao and 2 cables to the southward of Tsing-seu, lies a rock, which only shows at very low tides. To the westward of Tsing-seu are many sunken rocks, on one of which the ship *Blundell* struck in 1850. Vessels should therefore not use the channel between this latter island and the main.

TAE-PAN SHOAL.—The western side of Amoy outer harbour between Tsing-seu and Tae-pan point, which has an islet off it and lies 4 miles to the N.W. $\frac{3}{4}$ W., is shoal and has several reefs in it; but they will be avoided when standing westward by keeping the pagoda on Ki-seu island open north-east of Tae-pan point: to avoid the shoals on the north-east side of the harbour, do not bring the north end of Seao-tan to the southward of S.E. by E.

CHIN-SEU is a small islet, 60 feet high, lying N.E. $\frac{1}{4}$ E. 8 cables from Tsing-seu. Rocks extend in a southerly and an easterly direction half a cable's length from this islet, and it is connected to two other small islets, named Hwangkwa and Tao-sao, by a rocky bed which blocks the passage. Foul ground extends N.W. 4 cables from Tao-sao, and terminates in a reef which bears North half a mile from Chih-seu, and W. $\frac{1}{2}$ S. from the north extreme of Seao-tan. A vessel will keep north of this reef, by having the channel open between Seao-tan and Tae-tan.

SEAO-TAN ISLAND, $6\frac{1}{2}$ cables long, east and west, and 200 feet high, has three chimneys on it, and a sandy bay on its northern side. It lies to the E.N.E. of Hwangkwa, and the channel between, which is 3 cables wide, is frequently used; but as foul ground extends to the southward of both islands, and shoal ground runs off 2 cables to the N.N.W. of the west point of Seao-tan, a heavy or unhandy vessel had better use the channel between Tsing-seu and Chih-seu. There is a signal station on Seao-tan which communicates with Amoy.

TAE-TAN ISLAND, the highest of this group, and lying to the north-east of Seao-tan, is about 8 cables long in a N.N.W. and S.S.E. direction, with a low sandy isthmus in the centre: its east end is the highest (300 feet above the sea), and has a small circular watch-house and three chimneys on it; its western end rises to a conical peak, on which is a small circular fort. The channel between this island and Scao-tan is 2 cables wide, but as vessels are likely to have baffling winds, it would not be prudent for a stranger to use it.

It is said that since the survey of this locality in 1843 the soundings on the bank westward of Tae-tan have much decreased, and that a vessel drawing more than 12 feet must wait for high water to run through this channel, as where the depth of $3\frac{1}{2}$ and $4\frac{1}{4}$ fathoms are marked in the chart at 3 cables northward of Scao-tan, there are now only $2\frac{1}{4}$ fathoms. The rocks to the northward of Tae-tan also extend much farther out, and two separate ones are visible at low water springs.* Between Tae-tan and Amoy the channel is shoal under 2 fathoms; but, as before noticed, the foul ground on the north-eastern side of Amoy outer harbour will be avoided by not bringing the north end of Scao-tan to the southward of S.E. by E.

AMOY ISLAND, about 22 miles in circumference, occupies the northern portion of the great bight between Chin-ha and Hu-i-tau points, in the eastern portion of which is the island of Quemoy and Hu-i-tau bay. The city of Amoy stands on the south-west part of the island, abreast the small island of Kulangseu, which affords protection to the inner harbour. The traffic is considerable. In 1847, 117 vessels entered the harbour, with an aggregate burthen of 16,494 tons; and the value of the imports, by British ships, during the same year, was 179,758*l.*, by foreign vessels 75,976*l.*; of the exports in British ships 7,139*l.*, in foreign vessels 8,568*l.* The principal exports are, crockery ware, umbrellas, tea, sugar, sugar-candy, paper, tobacco, camphor, and grass-cloth. Population in 1847, 250,000.

The south point of Amoy is sandy, with several black rocks extending 2 cables from the shore. On the slope of the hill which forms the point is a circular battery; W. by S. 6 cables is a second battery, and between the two at 3 cables from the shore, a half tide rock, which will be avoided by a vessel of light draught by tacking before a cliff point with a battery and three chimneys on it, comes in one with a sandy point with a large stone (named Cornwallis Stone) at its south extreme, three quarters of a mile to the north-west. From the cliff point the 3 fathoms line of soundings extends 2 cables from the shore, otherwise the shore to the westward, which is a continuous sandy beach, is steep-to, and the lead a good guide.

From Cornwallis Stone the shore trends rather more to the northward for a quarter of a mile, where there is a creek dry at low water, and

---

* Captain Luard, H.M.S. *Serpent*, and Henry McAusland, Master of H.M.S. *Reynard*, 1850.

at the back of the creek an extensive suburb, and an isolated hill, the summit of which is a large mass of granite. N.W. ½ W. three quarters of a mile from the Stone are several rocks which cover at half tide, the outermost lying 1¼ cables from the shore; on the point from whence they extend is a mass of granite. The city commences at this point and stands very little above the sea level; the houses extend down close to the beach, on which the trading junks lie aground. The ridge of hills upon this face of the island do not rise above 600 feet. They are abrupt and barren, with numerous large boulders of granite; a square upright mass of which, on the highest part of the western extreme of the ridge, rises 528 feet above the sea, which is about the average height of the chain.

KULANGSEU ISLAND lies off the south-west shore of Amoy, and the channel between is 675 yards wide. The island is nearly 3 miles in circumference, and there are two distinct ridges upon it, which might be separately defended, the highest point being 280 feet above the sea. The island is principally of granite, and fresh water from wells is plentiful.

This island has detached rocks lying off nearly all the points; there are several that cover at high water off the north point, one of which, with only 1½ feet on it at low tide, lies N.N.W. ½ W., nearly 4 cables from the point, with the north extreme of Watson island bearing N.W. by W. ¾ W., and Nantai Wúshan pagoda South; and foul ground extends at least half a mile from the south-east point.

COKER ROCK, with only 4 feet on it, lies W. by N. ¼ N. not quite 6 cables from the Cornwallis Stone, and nearly 2 cables from the beach. From the rock, Ki-seu pagoda is just open southward of the rocks lying off the south extreme of Kulangseu, and a *white beacon* erected on the east side of Kulangseu is in line with the western edge of a *whitewashed* rock. There is another head with 10 feet on it lying about 16 yards eastward of the 4 feet; the rock open West of the beacon leads to the westward of it, and open twice its breadth to the East, leads to the eastward.*

A pinnacle rock, with only 12 feet on it and 4 and 5 fathoms close around it at low water, lies 3 cables to the northward of the Coker, with the flagstaff N.E. ½ N., and the *white* rock off Kulangseu W. by S. ½ S.

---

* A large stone, with a staff erected on it, has been placed on the Coker rock to mark its position, but it seldom remains more than a few days.—*T. R. Collingwood, Master of H.M.S. Comus*, 1857.

AMOY HARBOUR.

ANCHORAGE.—The outer harbour of Amoy has extensive anchorage in 7 to 16 fathoms, good holding ground, and unless vessels are badly found it is not probable that any gale could hurt them. The usual anchorage is to the southward of Cornwallis Stone; a good berth for a large ship is in 8 fathoms with the white beacon on the east side of Kulangseu open South of the whitewashed rock N.W. by N.; and the first entrance of the sally-port in the fort opening North-west of Cornwallis Stone N.E. by E. There is also good and safe anchorage in 7 to 17 fathoms in the channel on the west side of Kulangseu.

TIDES.—It is high water, full and change, in Amoy inner harbour at 12h. 0m. The rise and fall of the tide from one day's observation, on the full moon in September, was $14\frac{1}{4}$ feet; at this period, however, the night tides exceeded the day by 2 feet. The change in the depth in all probability three days after full and change exceeds 16 feet, which would be of infinite importance to vessels requiring repairs; particularly as sites for docks and ample materials for making them are to be found upon the island of Kulangseu, as well as in other parts of the harbour.

DIRECTIONS.—When bound to Amoy from the southward, after rounding the Lamock islands and the Brothers, steer about N.E. b. N. for Chapel island, keeping between 10 and 12 miles off the coast to avoid the South Merope shoal. The Nantai Wúshan pagoda is a good land mark by which the entrance of Amoy may be recognized when in the neighbourhood of Chapel island, which may be passed close to on either side, and from thence a N. b. W. $\frac{1}{2}$ W. course will lead towards the entrance of the harbour. As the Chauchat rocks are approached keep Tae-pan point open North of Tsing-seu to pass to the eastward of them, and from thence steer between Tsing-seu and Chih-seu into Amoy outer harbour.

In approaching the harbour from the eastward give Dodd island a berth of a mile, and after passing Leeo-lu head, which is steep-to, be careful not to shut in the island with the head until Ki-seu island opens South of Tae-tan island, W. by N. $\frac{3}{8}$ N. to clear Quemoy spit. From thence steer for Tsing-seu, keeping Tae-pan point open North of Tsing-seu to avoid the Chauchat rocks.

The channel into the inner harbour, between Kulangseu and Amoy, is so narrow, and sunken rocks lie off both its shores, that a stranger should not attempt it without a pilot. In steering in keep the white beacon open eastward of the whitewashed rock on the eastern side of Kulangseu, to pass eastward of Coker rock. The best anchorage is between the small islet off the city point and Hauseu island. The inner harbour, however, may be reached without difficulty by passing through the channel westward of Kulangseu, taking care to give Druid head, the south-west point of the island, a berth of at least a cable's length, and recollecting that shoal water extends half a mile from the main land on the opposite shore. After passing Druid head keep well over towards Watson and Hauseu islands, and do not steer to the eastward until clear of the spit extending from the north point of Kulangseu.

The channel around the island of Amoy is so narrow and winding that directions would be useless; the chart is the best guide. The bay of Sungseu, on the north side of which the city of that name is built, runs back 7 miles to the westward from Kulangseu; it is, however, shoal, and only navigable for small craft.

CHAPTER IV.

EAST COAST OF CHINA—AMOY TO THE WHITE DOG ISLANDS, INCLUDING THE PESCADORES.

VARIATION 0° 15' to 1° 0' West in 1861.

QUEMOY ISLAND is separated from Amoy by a channel from 5 to 7 miles wide, in the middle of which is Little Quemoy island. Between Tae-tan island and Little Quemoy the channel is deep, but it is narrowed by reefs.

The channel between Little Quemoy and Quemoy is half a mile wide. To enter, bring the north-east point of Little Quemoy on a N. ½ E. bearing, and steer for it until the pagoda on Quemoy bears West, then haul to the N.E. by N. for a mile and anchor in about 9 fathoms secure from all winds. Vessels drawing less than 15 feet may borrow over on the Little Quemoy shore.

QUEMOY BANK extends 3 miles to the southward from the west point of Quemoy island, and several patches, on which the sea breaks heavily, dry on it at low water. It is steep-to, and the lead will give no warning. Its western edge bears S.S.W. ½ W. from the west point of Quemoy; its southern end, named Quemoy spit, bears S. by W. ½ W. from the pagoda, and W. by S. ½ S. from Leeo-lu head.

A good mark to lead southward of Quemoy spit, is to keep Ki-sue pagoda open southward of Tae-tan island bearing W. by N. ⅜ N.; care must however be taken not to mistake the north division of Tae-tan for the island itself, for the pagoda seen over the low sandy isthmus between the north and south division of the island will lead across the bank; a mistake easily made.

LEEO-LU BAY.—The south-western face of Quemoy is composed of low sand hills. From the west point of the island the coast trends S.E. ¼ E., 3 miles to the south-west point, and from thence N.E. by E. 5 miles to Leeo-lu bay. Detached rocks extend 8 cables eastward of the south-west point, and great care must be taken to avoid them in foggy weather, as the tides here are so uncertain in their direction.

This bay is said to afford good shelter from N.W., round northerly, to East. Leeo-lu head (a low peninsula) will be known by a high peak rising 856 feet above the level of the sea immediately northward of it. The head may be rounded within a cable's length, and a berth picked up according to the vessel's draught, taking care to avoid a coral shoal on which the Dutch bark *Justina* grounded when running into this bay in 1856. It is said to be about a cable in extent, with only 9 feet on it at low water, and lies with Leeo-lu head bearing E. by S. ½ S., and Leeo-lu hill N. by E. ⅛ E.

HU-I-TAU BAY, formed between the eastern side of Quemoy and the mainland, will afford good shelter in the North-east monsoon. Hu-i-tau point, the eastern point of entrance, is 80 feet high, and on the hills north of it is a small fort, to the northward of which is a remarkable knob, 215 feet high. On the north-west side of the bay are two remarkable hills, which will serve to establish a vessel's position when in this vicinity. West peak, the highest of the two, is 1,714, and East peak, 1,390 feet above the sea. There is no doubt in the North-east monsoon that, besides the tide, vessels must calculate on a southerly set, and the same, but in a contrary direction, will most likely take place in the other season of the year; this set probably accounts for several vessels having mistaken this bay for the harbour of Amoy; the following remarks, however, will point out the difference in the approach :*—

Dodd island, in lat. 24° 26′ 16″ N., long. 118° 29′ 4″ E., may be known from Chapel island by a reef extending 3 cables to the N.N.E. of it, and on which the sea always breaks; the former also is uneven, gradually sloping to the eastward. Chapel island rises suddenly, and there is a difficulty in saying which is the highest part; it is 8 miles from the nearest land, whereas Dodd island is but 3 miles. The distance from Chapel island to the south point of Quemoy is 11 miles, but from Dodd island to Hu-i-tau point is only 5 miles. The rocks off the south end of Quemoy are peaked; the reef off Hu-i-tau point is flat. There are two pagodas on Quemoy point, which bear N.W. by W. and S.E. by E. of each other; on Hu-i-tau point is a small obelisk; and the land turns suddenly to the northward.

At 1¼ miles eastward of Hu-i-tau point is a sunken rock, on which 2¾ fathoms were found, but as in all probability there is less water its locality should be avoided. From the shoalest water obtained, the obelisk on Hu-i-tau point bore N.W. b. W., Dodd island S.W. ¼ W., and Scrag point Reefs N.E., which break heavily in bad weather, project 3 cables in a southerly direction from the point, and westerly a quarter of a mile from the first point inside the bay.

---

* *See* Plan of Hu-i-tau and Chimmo Bays, No. 1,959; scale, *m* = 1 inch.

HUI-TAU BAY.

Water may be obtained under the fort northward of Hu-i-tau point.

OYSTER ISLET and ROCK.—Oyster islet is a low flat rock bearing N.N.W. ¾ W. 2 miles from Hu-i-tau point. Oyster rock, which is awash at low water, lies S. ¼ W. 9 cables from the islet, with the obelisk on Hu-i-tau point bearing S.E. by E. ⅔ E.; the fort E.N.E., and the summit of Flak island in line with the left slope of a conical hill at the head of the bay, W. by N. ¾ N.

THALIA BANK occupies a central position in Hu-i-tau bay, and its east end bears W. ½ S. about 2 miles from Hu-i-tau point, and N. b. E. ½ E. from Dodd island; from thence it extends to the north-westward beyond the White rocks in the centre of the bay. The eastern end has 1¾ fathoms on it; the western end dries. The north-east part of the bank is steep-to, the lead giving no warning.

DODD LEDGE, bearing from E. b. N. to E.N.E. nearly 1¼ miles from Dodd island, has on it two patches of rock, one of which breaks and the other has 6 feet over it at low water; from the eastern edge of the ledge Scrag point, the east extreme of the land to the northward, bears N.E. ¼ N. There are two rocks, one with 3 feet, and the other with 6 feet on it, lying North three quarters and a mile respectively from Dodd island, and at half a mile N.W. by W. ¼ W. from the island, is a reef showing at half tide.

TIDES.—In Hu-i-tau bay it is high water, full and change, at 12h. 15m.; springs rise 16 feet.

DIRECTIONS.—Vessels requiring shelter during the North-east monsoon in this bay, will find good anchorage on its eastern side between Oyster islet and Oyster rock, taking care to avoid the latter which only shows at low water springs. There is also anchorage westward of Oyster islet in 5 fathoms, but the islet should not be brought to the southward of East, as a rocky ledge with only 6 feet on it lies 7 cables to the north-west of the islet.

Vessels requiring shelter in a southerly wind can run up the bay to the northward of the white rock and Thalia bank, and find anchorage in 5½ fathoms at half a mile to the north-east of Flak island. To avoid the northern edge of Thalia bank, do not bring Flak island to the northward of W. by N. ¾ N.; and by keeping Oyster islet open northward of the fort the bank will be avoided which extends from the north side of the bay.

There is a channel between the Thalia bank and Quemoy, but the ground is foul with several reefs; and it should not be attempted without some previous knowledge. To clear the south end of the Thalia bank, keep the chimney on the north end of Quemoy on a W. by N. ¾ N. bearing until the White rock bears North, then steer N.W. until the rock bears N.E. b. E., when shape a course to pass half a mile from the points of the bays on the Quemoy shore.

The channel north of Quemoy has a depth of 8 feet in it, and therefore might be used at high tide; but no vessel should attempt it without a pilot.

The COAST from Hu-i-tau point trends 9 miles to the north-east to Chimmo point, and is low, the sand hills rising about 300 feet. Near the coast are two walled towns, the southernmost of which has a small pagoda near it. None of the bays afford shelter. H.M.S. *Reynard* tried that under Scrag point, but was compelled to use her screw to get out of it.

CHIMMO BAY is between Chimmo and Yungning points, and its locality will be easily recognized when approaching from the eastward, by Mount Kusau and its pagoda, which is 760 feet above the sea, and 1¾ miles from the beach on the north side of the bay. The mount is the most conspicuous land on this part of the coast, and a fine land-mark in hazy weather. The shores of the bay, although barren, are very populous; the inhabitants bear a bad character, and it was here that the crews of the opium vessels were attacked in 1847. The walled town of Yungning stands on the northern shore.

On the south side of the bay, off Chimmo point, are two islets, named Sour and Pagoda, the channels between which and between Pagoda and Chimmo point are full of rocks. N. ½ W. 6 and 7 cables' lengths respectively from Sour islet, are the two Chimmo rocks, which show at low water. When on them the east end of Pagoda islet is in one with Flat reef, bearing S. ½ W. To pass northward of these rocks bring a large tree,* which stands half a mile from the beach in the northwest side of the bay, open westward of a remarkable shoulder peak 3½ miles at the back of it, bearing N.W., and when Yunguing islet (off Yungning point) is in one with Junk head (the first point to the north-east of it), the vessel will be to the westward of them; from these rocks to Yungning islet the distance is 1¼ miles.

This bay can only be termed a roadstead, and a dangerous one in the southerly monsoon. Yungning islet is steep-to, but a reef lies W. ¾ S. 3 cables from it, and covers at high water. Within the bay the depths shoal gradually, but vessels drawing 15 feet and upwards must not bring Yungning islet to the southward of E. ¾ S.

---

* This tree could not be made out.—*Henry McAusland, Master, H.M.S. Reynard,* 1850.

CHIMMO BAY.—CHIN-CHU HARBOUR.

TIDES.—It is high water, full and change, in Chimmo bay at 10h. 20m. and springs rise 16 feet. The tide sets with considerable velocity along the coast, between Hu-i-tau and Chimmo bays; but both the period and the rate vary considerably with the monsoon: the state of the tide will be known by the numerous fishing nets moored off the coast.

The COAST from Chimmo bay trends N.E. by N. 8 miles to Chungchi point, the southern point of entrance to Chin-chu harbour. Several sandy bays occur between these points, and afford shelter to junks, but from the number of rocks in and about them they cannot be recommended for square-rigged vessels. At $1\frac{1}{2}$ miles southward of Chungchi point is an islet with a building on it something like a bell.

CHIN-CHU HARBOUR.—Chungchi point is about 400 feet above the sea; sunken rocks extend 2 cables from it to the south-eastward. The entrance to the harbour is about 10 miles wide between this point and the town of Tongbu to the north-east, but its shores rapidly approach each other, so that its proper entrance may be considered to be not more than $4\frac{1}{2}$ miles wide between Chungchi point and the point North of it, and between which are the islands Tatoi and Seatoi, with the Hewen rocks, above water, lying half a mile S.W. of the latter; these all lie in a N.N.E. and S.S.W. direction, and between them are the navigable channels to the Lockyung river entrance. Seatoi is a low barren islet; Tatoi, 358 feet high, is the highest land in this neighbourhood. The Seatoi bank, with $2\frac{1}{4}$ fathoms on it, extends about 2 miles to the eastward of Seatoi island; and an extensive sand, named the Boot, runs westward from Tatoi island to the entrance of the river.*

PILOTS.—Chin-chu harbour is the only place where pilots can be obtained for Hai-tan strait or Hungwha sound, and it is advisable that all vessels bound there should take one, as the navigation is very intricate.

PASSAGE ISLAND lies N.E. $\frac{3}{4}$ N. $4\frac{1}{4}$ miles from Chungchi point, and to the eastward of it are three rocks, which cover at high water: the outermost rock bears E. $\frac{3}{4}$ S., half a mile from the island. A ledge also extends from the south-west point of the island, the outer rock of which is $1\frac{1}{2}$ cables from high water mark.

WHITE ROCKS.—N.E. $\frac{1}{2}$ N. about half a mile from Passage island are two White rocks, which are always partly uncovered; the channel between them is not safe. At three quarters of a mile northward of the White Rocks is Tahkut, an island at high water with a large town on

---

* See Plan of Chin-chu Harbour, No. 1,769; scale, $m = 1\frac{1}{2}$ inches.

it, and between them is a sunken rock, from which the highest part of the northern White rock bears S. by W. ½ W. half a mile, and the summit of Tatoi W. by S. ⅜ S.

LYNX ROCK, with only 6 feet water over it, lies E. by S. southerly not quite half a mile from the highest part of Seatoi, with Tatoi summit bearing N. b. W. ¼ W.; and Passage island N.E. by E. ½ E.

TAHEEN ROCK is 2 cables to the S. b. E. of the Lynx, and shows at low water; when upon it, Choho pagoda bears W. ¼ N., and Tatoi summit N. b. W. ¼ W. The bottom between the Taheen and Hewen rocks is rocky and uneven, and in several places there are only 6 feet at low water; a channel through, however, is sometimes used by the opium vessels when the wind is too far to the eastward to permit them to fetch between the Lynx and Seatoi; their leading mark is the highest part of the Hewen in line with Choho pagoda bearing W. ⅓ N.

MID-CHANNEL REEF.—Between Seatoi island and the Hewen rocks, rather more than a cable's length from the south-west point of the former, and a good half cable from the latter, is Mid-channel reef, three heads of which show at low water springs; it is about 2 cables in circumference, and from its centre the summit of Tatoi is in line with the west summit of Seatoi. Reefs also extend half a cable's length from Seatoi on its South, S.W., and eastern sides; thus rendering the channel between this island and the reef exceedingly awkward to a stranger.

CHOHO REEF.—A sandspit extends nearly 1¼ miles in an easterly direction from Choho pagoda, and from a reef lying on its northern edge, the pagoda bears S.W. ⅔ W., and is distant 6 cables, and the summit of Pisai island W. by N. ½ N.

OTA ROCK, which covers at high water, lies East half a mile from Pisai, and N.W. ½ N. from Choho pagoda.

TIDES.—It is high water, full and change, at Pisai island in Chin-chu harbour at 12h. 25m.; springs rise 17 feet.

DIRECTIONS.—Kusau pagoda, 760 feet above the sea (page 180), is an excellent mark for recognizing the locality of Chin-chu harbour when approaching it from the southward. From a position about 1¾ miles to the eastward of Chungchi point, steer North until Choho pagoda opens northward of Seatoi island bearing W. ¾ S., when it should be steered for on that bearing, and it will lead along the northern edge of Seatoi bank. The ship *Omega*, drawing 11 feet, struck upon a bank 1½ miles to the eastward of Seatoi, but not less than 2¼ fathoms were found upon the Seatoi bank in March 1844 ; the southerly monsoon may, however, cause the sand to accumulate.

If running for the harbour from the northward, and intending to anchor to the southward of the Boot sand, after passing about three quarters of a mile south of Passage island steer in with Choho pagoda W. ¾ S., until the peak on Tatoi island bears N. b. W. ½ W., and the eastern end of Seatoi island S.S.W. ½ W., then haul to the southward, and pass a cable's length to the eastward of Seatoi. Round the south side of Seatoi at half a cable's length, and when its western summit is in one with the highest part of Tatoi the vessel will be in the narrowest part of the channel, which is here barely a cable across.

Having passed Seatoi a W.N.W. course will lead to the anchorage above Pisai islet in mid-channel. By keeping this islet to the westward of W. by N. ½ N. the reef off Choho pagoda will be avoided; and the southern edge of the Boot will be cleared by not bringing Seatoi to the southward of E. by S. ¾ S. ; the outline of this bank, however, is generally visible. The opium vessels run in between the Lynx and Tahoen rocks with the south extremes of Seatoi island and Ota rock in line with north extreme of Pisai. The anchorage is North about 1½ or 2 miles from Pisai, where the channel is 3 cables wide.

If intending to anchor northward of the Boot sand steer to pass northward of Tatoi island, and if drawing less than 15 feet a vessel may run up until Choho pagoda bears S. b. W. ½ W., where she will have smooth water in any weather, as the Boot forms an excellent breakwater. The north edge of the Boot will be avoided by keeping the White rocks to the southward of East. There is a sunken rock lying 1½ cables from the northern shore, and N. b. W. ½ W. from the summit of Tatoi. There is good anchorage in north-east or northerly gales in 3½ and 4 fathoms, with the summit of Tatoi bearing S. E. b. S. ; but in a south-west gale the former anchorage is to be preferred. The Boot may be crossed by a vessel of light draught at high tide, but it should be sounded first, as the sands shift.

The entrance of the small river, leading to the town of Chin-chu, bears W. by N. ¾ N. 3 miles from Pisai islet. On the left bank near the entrance is a circular fort, 4 or 5 miles above which is the town standing on the north bank of the river. The channels to it are shoal and intricate, and the large junks have to wait in the neighbourhood of Pisai for tide before they can cross the flats, which are covered with artificial oyster beds.

PYRAMID POINT, at 3 miles eastward of Tongbu, is the southern point of entrance to Port Matheson, and when approaching it on a westerly bearing, it appears a bold black face of land, not in any way representing its name; but on a northerly bearing, or inside the point, it cannot be mistaken.

The Pyramid rock is connected with the point at low water, and to the S.E. of it is a rock which never covers. To the eastward of Pyramid rock are several reefs, from the outermost of which the Pyramid bears S.W. by W. ¾ W. 6 cables, the highest part of the land forming the north side of Port Matheson N. by E., and a cliff head at the head of the promontory (extending south-westerly from the above hills) is in line with a remarkable cone in the bay N. by W. ½ W.

ANCHORAGE.—Vessels requiring anchorage in the North-east monsoon will find it in the first bay westward of Pyramid point, where they will be sheltered to the eastward by the reef of rocks, mostly above water, extending south-east from the point, and forming a good breakwater: care must, however, be taken to avoid a sunken rock lying South a cable's length from the first point eastward of the walled city of Tongbu.

PORT MATHESON, called by the Chinese Gulai, is the next inlet to the north-east of Chin-chu, the isthmus near the city of Tongbu being only a mile across. The port is 4 miles wide at entrance, and will afford tolerable shelter to vessels of about 12 feet draught if the wind be to the northward of East; but it is only a roadstead, and that a bad one in the southerly monsoon. There are no dangers in it except a rock lying North 4 cables from the largest islet on the southern shore.

MEICHEN SOUND, the next inlet north of Port Matheson, is 6 miles across at the entrance, which may be recognized by the Ninepin rock, which lies nearly in mid-channel. A reef extends South from the Ninepin, and at the distance of a mile is a cluster of rocks, one of which, Square rock, does not cover at high tide; from thence the reef extends south-westerly 1½ cables, and its outer part dries at low water. A large spar* is moored about 1⅓ miles to the south-west of Square rock.

East, 6 cables from the Ninepin, is a flat patch of rocks awash at high water, and between this patch and Rogues point is good anchorage in the North-east monsoon. H.M.S. *Plover* rode out a gale to the westward of the Ninepin, without much strain upon the cable; but with an uneasy sea. Anchorage was therefore preferred under Rogues point; but since that period H.M.S. *Scout* found a rock here which renders this anchorage more difficult of approach. It lies midway between the Ninepin and the extreme of Rogues point, bearing from the former

---

* H.M.S. *Pique* stood into this sound in August 1858, but the spar could not be seen.

E. by S. ¾ S., and from the mound at the end of the sandy isthmus connected with Rogues point, South. H.M.S. *Comus*, in August 1856, anchored in 8½ fathoms with the Ninepin bearing South, Rogues point S.E. ¾ E., and a small white rock off Meichen village E. ¼ N.; but it was considered an unsafe anchorage during the southerly monsoon, and many rocks were seen in the sound, which are not noticed in the chart.

N. by E. ¾ E., one mile from the Ninepin, is a rock which shows at low water, and from it the highest part of Rogues point bears S.E. by E. ⅓ E. There is a passage between this rock and the Ninepin, but rocks extend a cable's length from the latter. Rogues point may be approached without fear except on its east side, where there is a reef lying rather less than a cable's length from the shore; a depth of 3½ and 4 fathoms will be found at the distance of 3 cables from the sandy beach. One and a half miles South of Rogues point is a patch with 4¾ fathoms on it.

Inner Harbour.—In the southerly monsoon vessels will find a good harbour to the north-west of Saddle island, which bears N.W. by N. 3¼ miles from the Ninepin. In approaching it, pass southward of the south islet off Saddle island, and haul to the northward round the western islet, giving it a berth of a cable's length at high water to avoid a ledge. The ground is very uneven hereabouts, and there are only 2½ fathoms water at a mile W.N.W. of the western islet.*

N. b. E., one mile from Saddle island, is a low Cliff islet, from the west point of which a sand bank extends 1¾ miles to the north-west; the south peak of Saddle island kept eastward of S.E. b. S. will lead westward of it. When Mound peak (which is on the mainland, and 3 miles northward of Saddle island, with a walled town and pagoda near it) bears East a vessel will be to the northward of this bank, and can haul in towards the town. W. by N. ½ N., 2½ miles from Mound peak, is a knoll with only 6 feet over it.

The junks use the channel between Mound peak and Cliff islet, and also pass between Mound peak and Meichen island. The former channel is deep, but requires personal knowledge; the latter is strewn with rocks, and in some places has not a greater depth than 9 feet. The sound runs back 10 miles to the northward of Mound peak, forming narrow isthmuses across to Pinghai bay and Hungwha sound.

Tides.—In Meichen sound it is high water, full and change, at 12h. 30m.; springs rise 17 feet.

Sorrel Rock, 60 feet high, bears E. by N. 3¾ miles from Rogues point. A rock lies three-quarters of a cable's length southward of it.

---

* *See* Chart, East Coast of China, Sheet 5, No. 1,761; scale, $d = 14½$ inches.

PINGHAI BAY, the next inlet north-east of Meichen sound, is 6½ miles wide at entrance, between the Rowan islands and Ping point, and carries a depth of 5 to 3 fathoms. Ping rock, 90 feet high and conical shaped, lies 4 cables to the southward of the latter point, and bears N.E. b. N. 9 miles from the Sorrel rock ; a sunken rock lies S.W. by W. a quarter of a mile from it.

The anchorage in this bay is in 3 fathoms, off the town of Pinghai, with the Ping rock bearing S.E. b. E. At 5 miles westward of the anchorage is a high range of hills, one of the peaks of which, named Marlin Spike, will form a good guide for this part of the coast. The bay runs back past the foot of the Marlin Spike range, but is shoal, there being seldom more than 2 fathoms to the westward of the range.

Reefs extend nearly a mile from the coast to the northward of the Ping rock.

LOUTZ ROCK lies E.S.E. about 5¼ miles from the Ping rock, and between them, at 1¾ miles from the Loutz, are two sunken rocks, named Loutz shoal, from which the Ping is in line with Marlin Spike peak bearing N.W. by W. ¼ W., and the islet lying north-east of the Loutz is in one with the South Yit E. ¼ N. There is a rock which shows at half tide, lying N.N.W. 2 cables from the above islet, and another S. ¾ W. 8 cables from the islet, and East from the highest part of the Loutz.

OCKSEU ISLANDS.—The Ockseu or Wokeu group consists of three islets, the centre one being a barren rock joining the eastern island. The western island, the largest, is in lat. 24° 59′ N., long. 119° 27½′ E., and elevated between 200 and 300 feet above the sea. It is round-topped with smooth sides, and bears from the Sorrel rock E. by S. ¼ S. 15½ miles, and from the South Yit S. b. W. ¾ W. 10½ miles.* The steam-vessel *Nemesis*, drawing 5 feet, anchored under the eastern island, which is low, rugged, and sandy, with a large fishing village on it, and detached rocks off its east and west points. It is doubted, however, if there is shelter sufficient in a breeze for a vessel of greater draught.

LAM-YIT ISLAND, the southern and largest of the archipelago called the Eighteen Yits, is 7 miles long in an E.S.E. and W.N.W. direction, and fronts the deep and extensive gulf named Hungwha sound. The eastern peak, High Cone, 565 feet above the sea, and the highest point of the island, is in lat. 25° 12′ N., long. 119° 35′ E.

---

* A strong tide ripple, or reef, appeared to break about 1½ miles W.N.W. of the western Ockseu island.—*Commander J. C. D. Hay, H.M.S. Columbine*, 1848.

CHAP. IV.] PINGHAI BAY. — LAM-YIT ISLAND.—HUNGWHA SOUND. 115

The south point of the island is a bold table land, off which and connected at low water is South Yit islet, to the north-west of which will be found a snug and excellent anchorage in the North-east monsoon. On rounding, give the South Yit a berth of a quarter of a mile and then haul up into the bay, being prepared to anchor directly the water shoals. N.W., 2 miles from the South Yit, is a flat rock which is always above water; and S. b. E. 4 cables from this rock, is a reef awash at low tide. This is the only danger in the bay, and it will be avoided by keeping within $1\frac{1}{2}$ miles of the South Yit, should the vessel not fetch up into smooth water after rounding it.

LAM-YIT CHANNEL is to the westward of Lam-yit island, and a vessel proceeding through it towards Hungwha sound from the anchorage on the south side of Lam-yit, must be careful on the flood to steer well to the south-west to avoid a sand-bank extending $2\frac{1}{4}$ miles in a southerly direction from the west point of Lam-yit. From its southernmost edge, in $2\frac{3}{4}$ fathoms, the South Yit bore E. $\frac{3}{4}$ S.; its western edge will be avoided by keeping Lam point (the west point of the island, which will be known by its three chimneys) to the eastward of North.

H.M.S. *Plover* examined this bank three different times, and on each occasion found a change. On one occasion a passage was discovered between it and the point; the outline of the bank, however, may be detected by discoloured water. On the western side of the channel there is also a rocky patch of $1\frac{3}{4}$ fathoms, the eastern edge of which bears S. b. W. 2 miles from Clam islet (the largest islet between Lam-yit and the main); from its southern edge, Lam point bore E. b. N.

ANCHORAGE.—The *Plover* rode out a strong N.E. gale between Lam point and Clam islet; but better shelter will be found to the southward of Lam point, where the junks anchor. The outer rock off the point always shows, and may be rounded close to; but it must not be brought to the westward of N.N.W., as the water shoals suddenly, and there is a sunken rock in the bay at 6 cables to the southward of it. The best position is as close up under the point as the vessel's draught will permit. There is anchorage for vessels of large draught in 4 or 5 fathoms, at $1\frac{1}{2}$ miles to the northward of the point.

HUNGWHA SOUND.—Besides Lam-yit island, which, as already stated, fronts Hungwha sound, there are many islands and rocks within the sound bordering its shores, the principal ones being near its entrance points. The only passages that must be used to enter it are, the Lam-yit and Hungwha channels, and Hai-tan strait.

DIRECTIONS.—If bound through the Lam-yit channel for the entrance of the Hungwha river, which flows into the western part of Hungwha

H 2

sound, steer northerly 7 miles from Lam point, when the vessel will be one mile to the northward of Knob island, and should then steer for Pitew point, which bears N.W., 7 miles from the Knob. A batch of rocks lies to the north-west of Knob island, the eastern one of which bear N. b. W. 8 cables from the island, and the north-westernmost N.W. ½ W. 1¾ miles; a part of them always show. There is another patch off Pitew point, the south-east end of which bears E.S.E. 2 miles from the south-east corner of the fort. There is good anchorage in 6 fathoms with the corner of the fort on Pitew point bearing E.N.E. The entrance to Hungwha river, leading to the town, bears W. b. S. from Pitew point; the depth shoals to 6 feet at low water at 5 miles from the fort. In 1844 there was a piratical establishment on the main, S.W. from Pitew point.

A vessel leaving Hungwha sound and intending to pass northward of Lam-yit island should use the channel north of the Passage islands, which are three in number, and bear N.N.E. 5 miles from Lam point. Between Lam point and these islands is Cliff island, in the vicinity of which are several reefs, rendering the channel between it and Lam-yit, and between it and the Passage islands, precarious. A ledge extends 2 cables in a westerly direction from the south-west point of the west Passage island.

The channel to the northward of the Passage islands* is 4 cables wide, and on its northern side is a rock, with a reef, which shows at low water, lying 1½ cables to the westward of it. North of the rock 1½ cables is a small islet; and 4 cables north of the islet is Rugged island.

The north-east Passage island is a bold bluff, steep-to on its north side; from thence a vessel may steer to pass either north or south of White island, which bears East 4½ miles from the Passage islands. If passing south of this island take care to avoid three rocks, named the Hung, which cover at first quarter flood and bear S. b. W. 1 1/10 miles from it. E. b. N. 2¼ miles from White island is the southern edge of a reef extending three-quarters of a mile from Kerr island;† having passed which, haul to the northward, and work up inside Chim island, to the westward of which there are no dangers, except a rock at the entrance of Vangan inlet, which may be avoided by keeping 1½ cables from the shore. Here vessels will have smooth water

---

* H.M.S. *Salamander* encountered a frightful race and chow-chow water in this channel, November 1851.

† The reef on the main, bearing E. b. N. from White island, is very dangerous, and extends nearly 2 miles off shore; it is quite covered an hour after low water. There is good anchorage in the bay to the N.E. of the reef.—*Commander J. D. Hay, H.M.S. Columbine*, 1848.

protected from the easterly swell by Chim island. On the south point of Vangan inlet there is a walled town and pagoda.

HUNGWHA CHANNEL is to the north-east of Lam-yit island, and its southern side is bounded by the Eighteen Yits, and its northen side by Sentry, Reef, Sand, and Chim islands.* On no account ought vessels to stand in among the Yits, as the ground is very uneven. Triangle Yit, with a reef to the eastward of it, lies $1\frac{1}{4}$ miles to the S.E. of the High Cone peak on Lam-yit island. Cap Yit, the south-easternmost of the group, lies E.N.E. 4 miles from High Cone Peak; and 2 miles S.E. from Cap Yit is a group of low rocks, named Scattered Yits, some of which are always above water. Double Yit, lies N.E. $1\frac{1}{4}$ miles from Cap Yit, and the channel between it and Sentry island is 3 miles wide. N.N.E. 4 miles from Double Yit there is a remarkable white island, named Sand island, with sandy beaches and detached hills.

Chim island, the highest island in this locality, rises with sloping sides into two peaks, one of which, 640 feet above the sea, has on it three chimneys, the usual pirate signal along the coast of the Fu-kyen province. At 3 miles to the S.E. of Chim island and $1\frac{1}{2}$ miles northward of Reef island are four rocks,† with reefs interspersed, called Chim bank.

DIRECTIONS.—Entering the Hungwha channel from the eastward, pass between Double Yit and Sentry island, and to the westward of Sand island and the rocky islets on its north-west face, off which there is anchorage, should daylight or the tide fail; but the best shelter is off Station island, to the northward of Chim island. On no account whatever pass between Sand, Sentry, Reef, and Chim islands, as this locality has not been sufficiently examined, and beware of the reefs to the eastward of Reef island.

HAI-TAN ISLAND.—This large and irregular shaped island lies near the mainland between the parallels of 25° 24′ and 25° 40′ N. Its

---

* A rock, small and steep-to, with only a few feet water over it, is said to lie in Hungwha channel, nearly midway between Vangan point and the N.E. Yit, with Vangan pagoda bearing N. b. E., and White island W. $\frac{3}{4}$ N. The Master of the opium vessel who discovered this danger, sounded on it with a boathook. There are also many dangers between Hungwha sound and Hungwha channel, and the chart of this part is not strictly to be relied on; for instance, the Cliffy islands and North Yit do not exist; there is but one Cliffy island with rocks detached off its south-east part, which may be the south-west Cliffy island marked on the chart. There are two dangerous rocks, awash at half tide, between Cliffy island and Red Yit, in line with the former and a little to the northward of the latter, and directly in the way of navigation. The Hung rocks are somewhat out of position, being more to the eastward of Red Yit.—*Charles G. Johnston, Master of H.M.S. Bittern*, 1856.

† The chart only shows two.

northern part is high, the peak of the Kiangshan hills rising 1,420 feet above the level of the sea, while the eastern and western shores are low, and indented by deep sandy bays. Numerous small islands and rocks occupy Hai-tan strait, the channel between the island and the coast, and although it is not to be recommended, being very intricate, yet the junks invariably use it; one was found lying there, having been detained 27 days waiting for an opportunity to get out at the northern end.*

Hai-tan point, the south extreme of Hai-tan island, is a rugged, sandy headland, with large boulders sticking up here and there. Off the point are several rocks a little above high water, and a sunken rock lies 7 cables eastward of them, and nearly 6 cables from the shore. The best mark to avoid this rock is not to haul into the South-east entrance of Hai-tan strait until the rocks off Hai-tan point bear E.N.E. Station island is $3\frac{1}{2}$ miles to the north-west of this point, and the south coast of Hai-tan, between, is shoal with detached reefs, and should not be approached within a long mile. The reef lying to the westward of Station island is covered at high tides.

From Hai-tan point the south-east coast of the island trends N.E. by E. $6\frac{1}{2}$ miles to Hae head, and between these points is a deep sandy bay, with several detached rocks, the most remarkable of which, Trite Island, forms in three peaks. S. $\frac{2}{3}$ E. 2 miles from this island is South reef, portions of which are visible unless the tides are very high, and the water smooth; from it Chim island bears W. by S. and Turnabout island N.E. by E. $\frac{3}{4}$ E.

Between Hae head and Tan point, 7 miles to the N. b. E. is Hai-tan bay, a deep sandy bight, with numerous rocks both above and below water. Tan point, which is a low cliff with a mound at the back of it, forms the south extreme of Kwing bay, and at $1\frac{1}{4}$ miles to the eastward of it are the Tan rocks, some of which are always visible. Kwing island lies a mile to the northward of Tan point, and reefs extend in a south-easterly direction a mile from its eastern side. The channel between the island and the point is much obstructed by reefs at its western end, and the swell rolls home to the Hai-tan shore. Between Kwing and Hai-tan is another islet; but the tide rushes through these channels with such velocity that they ought never to be taken.

DIRECTIONS.—A vessel approaching the south-east entrance to Hai-tan strait from the northward, after rounding Hae head, will avoid the South reef by passing about half a mile to the southward of Trite island. Junks occasionally take shelter under Hae head, and it is said that some

---

* See Plan of Hai-tan Strait, No. 1,985; scale, $m = 1$ inch

vessels have done so in the North-east monsoon; it will, however, be found much exposed should the wind haul to the southward of East.

If entering from the southward pass about a mile to the eastward of Chim bank, and when the northernmost of the rocks bears W.S.W. one mile, steer about N.W. b. W. until Junk Sail rock bears North, to avoid a sand spit which extends from the point north-west of Station island, and then haul up for Junk Sail, from which a reef extends half a cable's length both to the southward and to the westward. N.W. b. W. a mile from Junk Sail is Pass island, from which a sand-bank extends in a southerly direction, and its extreme end bears from the west point of Junk Sail S.W. by W. $\frac{1}{2}$ W., and the channel between is rather less than half a mile wide. A reef of rocks, showing at half tide, lies N.E. 3 cables from the summit of Pass island. Keep to the eastward of this reef, and between it and a small islet lying 4 cables to the N.N.E., having a mud spit with rocks extending S.S.E. 3 cables from it; nor can the islet be approached within a cable's length of high-water mark on its western side.

Having cleared this part of the channel, steer N. b. W. $\frac{1}{2}$ W., to pass eastward of Flag island, which has a spit extending South 3 cables from it, and a ledge of rocks off its north-east point, on which H.M.S. *Plover* lost her false keel. From thence bring the east end of Flag island in line with the west end of Pass island bearing S. $\frac{1}{2}$ E., and it will lead in mid-channel 5 miles above Flag island. Care, however, must be taken not to open them, as there is a reef, which shows at low water, lying 1$\frac{1}{4}$ miles northward of Flag island, and from it a chimney hill on Hai-tan bears E. b. N.; by keeping the chimneys on Chim island just open westward of the west point of the islet lying N.E. of Pass island, it will be avoided.

When Pillar rock (on the Hai-tan shore, N. b. E. 6$\frac{1}{2}$ miles from Flag island,) bears N.E. b. E., steer N.W. b. W. until Slut island bears N. b. W., when it may be steered for, passing westward of Tower rock, which lies N. $\frac{3}{4}$ W. 8 miles from Flag island, and has a reef 1$\frac{1}{2}$ cables westward of it. The summit of Slut bears N. b. W. $\frac{1}{3}$ W. 4 miles from Tower rock, and between them are several reefs; the west end of the reef (part of which always shows) nearest the rock, bears N. $\frac{3}{4}$ W. 8 cables from it.

N. b. W. 2$\frac{3}{4}$ miles from the Tower rock is a reef which only shows at low water, and when on it the Cows Horn (a remarkable peak on the main outside the strait) bears N. b. W. and is in line with the east end of Slut island, Pillar rock bears S.E. b. S., and Tower rock is in line with the south-west point of Hai-tan. The channel out of the strait is between this latter reef and a Black-peaked rock, bearing from the reef W. by N. $\frac{1}{4}$ N. three-quarters of a mile. Rocks visible at low tide extend from the Black-peaked rock south-easterly a quarter of a mile, and there is also a reef lying half a mile to the southward of it. Both these reefs on the western side of the channel will be avoided by keeping the summit of Slut to the northward of N. ½ E.; there are several reefs between Black-peaked rock and Chung island.

The best channel out of Hai-tan strait is to the eastward of Slut island, between Slut and Shingan islands. Reefs extend from both shores, narrowing the channel to 4 cables' lengths. When working through the narrows, the summit of Slut must not be brought to the southward of S. W. ⅓ S. as a rocky patch with only 9 feet on it lies 7 cables to the north-east of the island. N.N.E. ¼ E. 2¾ miles from the summit of Slut island is a sunken rock on which the sea breaks at low water; when upon it the Cows Horn bears N.W. ⅔ N. Shingan island, on the eastern side of the narrows, trends away to the N.E., breaking into detached fragments and giving a little more room for a board, but the main difficulty is the tide, which, after a vessel is through the channel, affords little or no help, so that unless there is a slant of wind she is liable to be driven among the small islets north of Hai-tan, and if a dull sailer, and unable to clear the dangers in one tide, she will be compelled to bear up before dark.

There are three other channels between Slut island and Hai-tan island, none of which, owing to the height of the islands and consequent liability to be becalmed, are so good as the one described. The flood tide enters through all these, but with great irregularity; it should, however, be observed, that while the *Plover* was employed on this portion of the survey, a very severe Typhoon occurred to the northward, which may in some measure have caused the difficulty experienced by her getting out at this end.

TESSARA ISLANDS are a group of four islets lying N.N.E. 6 miles from Slut island, and between them and the Cows Horn the depth is 6 fathoms. The long swell which set into the bay prevented the *Plover* anchoring, and giving these islands as well as the islets to the eastward that close investigation that could have been wished. The only conclusion arrived at was that there was nothing here sufficiently extensive to shelter a vessel in the North-east monsoon. A reef extends S.S.E. 3 cables from the easternmost islet.

RED ROCK is a small islet with reefs about it, lying S.E. by S. 3 miles from the Tessara islands. Vessels should not close the Hai-tan shore to the eastward of this rock, as the intervening space between it and the Warning rocks (which are about 80 feet high and lie 7 miles to the eastward) is strewn with reefs.

NORTON ROCK, about 50 feet high with a rock awash half a mile to the westward of it, lies East 6½ miles from the Tessara islands.

WHITE ISLAND.—At 7½ miles to the northward of the Tessara is the southernmost of a group of rocks and shoals, which extend hence all the way to Sand peak. Junks anchor under the largest, named White island, but there is almost always a heavy ground swell setting into this bay. A sandy beach extends from the Cows Horn to Sand peak, a distance of 16 miles, and a vessel may stand towards it until the group just described is reached, which it will be advisable to keep outside of, taking care to avoid a rock lying 9 cables eastward of White island.

SAND PEAK.—Under Sand peak the banks at the entrance of the river Min commence; 3 fathoms will be found at 2 miles from the shore, and boats may find their way into the Min by the channel between Sand peak and Woufou island, but the navigation even for them is difficult, and entirely impracticable to any but of such light draught as can go over sands that dry at low water. This, however, when the tide will admit, will be found the best channel for a vessel lying at the White Dogs to communicate with Fu-chau fu. There is a large fishing establishment under Sand peak.

TURNABOUT ISLAND, lying E.S.E. about 4 miles from Hae head, is in lat. 25° 26′ N., long. 119° 58′ 42″ E.; there are two small islets off it. A sunken rock, on which the sea breaks occasionally, lies 2 cables to the northward of the island.

WHITE DOG ISLANDS, called by the Chinese Pih-kuen, bear N.N.E. 23 miles from the peak of the Kiangshan hills on Hai-tan island, and N.E. ½ N. 15 miles from Norton rock. They consist of two large and one smaller islet, named Middle Dog, South Dog, and Tong-sha island.*

Tong-sha, the western island, and the largest of the group, has a reef of rocks running off its western extreme, terminated by a square islet called the Breakwater; and a half tide rock lies a cable's length from the western point of Village bay, on the south side of the island. The highest part of the island is flat topped, and 590 feet above the sea. Fresh water may be obtained in small quantities.

Rocks and reefs extend both northerly and westerly from the Middle Dog, but the outer ones always show; a rock on which the sea generally breaks lies N.E. by E. ½ E. 1/10 miles from its north-east point.

---

* *See* Chart of River Min, with views, No. 2,400; scale, $m = 1\cdot2$ inches.

The channel between the Middle Dog and Tong-sha is safe. The islands are inhabited by a few fishermen, and are occasionally visited by pirates; as on the point between the Breakwater and the village the officers of the *Plover* found a framework with six buckets suspended, each of which contained a human head.

ANCHORAGE in the North-east monsoon, for vessels of any draught, will be found under Tong-sha island. Small vessels will find good shelter in 18 feet, close under the breakwater, and here whole fleets of Chinese junks remain during foul weather. As the water decreases gradually towards Tong-sha, vessels of greater draught may approach as convenient, bearing in mind that the rise and fall is 18 feet. H.M.S. *Cornwallis* anchored here for five days, with strong north-easterly winds, and rode easy, with the Breakwater bearing N. ¼ W., the village N.N.E., and the Middle Dog E. ½ S.

DIRECTIONS.—The passage from Lam-yit to the White Dog islands may be considered as the most difficult portion of the coast that a vessel has to contend with in the North-east monsoon, and it is believed there are few men who know the coast of China but will allow that Turnabout island is well named. The attempt of the flood to force its way through Hai-tan strait forces the water back, and occasions a strong current off Kwing bay, at the north-east end of Hai-tan. It is a great misfortune that this bay does not afford shelter, as it would prove an uncommonly good half-way house; it is, however, one of the worst places on the coast of China the *Plover* dropt anchor in, being full of rocks, with a heavy swell. Sailing vessels have, therefore, no alternative but to stand boldly off and trust to a slant on the Formosa side, or to take the Hai-tan strait. The open sea is, however, preferable, notwithstanding that some vessels have got successfully through the strait; yet it requires local knowledge and a handy vessel to prevent great detention.

PESCADORES ISLANDS.

The Pescadores or Ponghou archipelago consist of twenty-one inhabited islands, besides several rocks, and extend from lat. 23° 11¼′ to 23° 47′ N., and from long. 119° 16′ to 119° 40′ E. From their basaltic formation the land is generally flat, and no part of the group is 300 feet above the level of the sea. The two largest islands, named Ponghou and Fisher, lie near the centre of the archipelago, and between them is an extensive and excellent harbour. The general depth of water on the western side of the archipelago is 30 and 35 fathoms; there are, however, some places where there are 60 fathoms. To the eastward of the group the depth is 40 fathoms, and the current strong.*

These islands contain (in 1843) a population of about 8,000, and are extensively cultivated, potatoes, maize, millet, and ground nuts being produced in considerable quantities, as well as a few other vegetables, but the soil is not good; owing to the violence of the wind there are no trees, but the islands are well supplied with fruits and vegetables from Formosa. Bullocks are numerous, being used to till the ground. Fresh water was abundant in the months of June and July, but it was stated that at some seasons it was scarce. H.M.S. *Plover* watered from a well at Ponghou island that yielded 3 tons daily. Dried fish forms the only article of export, and the imports are rice, sugar, fruits and vegetables from Formosa, tea, &c., from Amoy.

JUNK ISLAND, the southernmost island of the Pescadores, is 2 miles long, east and west, and 1½ miles wide; the depths of water in its vicinity being 15 and 16 fathoms. The highest part of the island is 260 feet above the sea, and from it, High island bears N.W. ¼ N. 8¾ miles, Reef island N.E. b. E. 5¼ miles, and East island E. b. N. 13 miles. A reef of rocks extends 6 cables from its south-west side, and within them is a small artificial harbour for junks. Its eastern face is fronted by bold cliffs; and its western extreme is a long shelving point.

REEF ISLANDS are three in number, one of which, Steeple island, is a remarkable pyramid. The other two are rather more than a mile each in circumference, and are connected at low water by a stony ledge; reefs extend half a mile to the southward of them, and South from the west end of the eastern island is a pyramidal rock rising 80 feet above the sea. There is also a low flat rock, nearly level with the water's edge, lying S.W. b. S. 1¾ miles, and a small peaked rock with a reef to the northward of it lying S.E. 2 miles from the east end of this island.

EAST ISLAND is 8¼ miles to the eastward of the Reef island, and between them and distant 5¼ miles from the latter is a smaller island, named Pe-ting, 1½ miles in circumference, with a reef extending in an easterly direction, not quite a mile from its north point. East island is 2½ miles in circumference, and has a small islet lying half a mile from its north-western shore.

NINE-FEET REEF lies N. b. E. ¾ E. 12¾ miles from the north end of East island, and from it Dome hill on Ponghou island bears W. by N. ½ N.

---

* *See* Chart of Pescadores Islands, No. 1,961; scale, $m = 0.8$ of an inch.

10¾ miles, and Three island N.N.W. ½ W. 4 miles. The lead gives no warning, but if there is any tide the ripple will be sufficient to point out its position.

ROVER GROUP consists of two large islands, Pa-chau and Tsiang, and several rocks, and are sufficiently extensive to afford shelter under their lee in either monsoon. The general depth is 7 and 8 fathoms on the southern, and 13 and 14 fathoms on the northern shore. From the highest part of the group, the lighthouse on the south-west point of Fisher island bears N. b. W. 10½ miles.

Pa-chau, the western island, is 2½ miles long, north and south, and a mile broad, and its summit, which is near the eastern shore, rises like a dome with a large pile upon it. S.W. ¼ W. 2$\frac{6}{10}$ miles from the summit is the end of a reef extending in a westerly direction from the south point of the island, and part of it shows at all times of tide. There is also a reef which covers at high water, bearing W? by S. ¾ S. from the summit, and lying 2 cables from the shore. The north-west point of the island is not steep-to; and a rock, which always shows, lies off the north-east point, having a channel 4 cables wide between it and the point.

Tsiang, the eastern island, is only 1¼ miles long N.E. and S.W., and about 1½ miles broad, and the channel between it and Pa-chau is barely a cable wide. The east point of this island is remarkable from an isolated cliff, called Rover Knob, 100 feet high, which forms the most striking feature in the group; and 7 cables eastward of the cliff is a ledge of rocks, part of which are always above water.

DIRECTIONS.—The channel between the Rover group being so narrow and intricate, the only excuse for a stranger using it would be his vessel being caught at anchor to the northward of the group in a breeze from the northward, and unable to fetch clear either eastward or westward. On the north-west face of Tsiang are two islets, under the southern of which a small vessel might find shelter in a northerly wind, taking the precaution not to stand too far into the bay, as there are only 6 feet water at 2 cables from the shore. On the west end of the island, which is a cliff, are three embrasures.

In the centre of the southern part of the channel is a small rock with a reef extending southerly half a mile from it. The passage out is to the eastward of this rock, and the channel is a quarter of a mile wide. E. by S. 4½ cables from the small rock is a reef which may always be detected from the mast head, as well as two other patches lying respectively 4 and 7 cables to the eastward of it.

HIGH ISLAND, bearing W. by S. ⅔ S. 9¾ miles from the highest part of Pa-chau, is dome-shaped, 247 feet high, and three-quarters of a mile in circumference. At one mile to the eastward of it is a low flat island, and between the two are several rocks, one of which has a remarkable gap in it, and rises 60 feet above the sea. A rock nearly level with the water's edge lies S.E. ½ E. 1¾ miles from the summit of High island.

YIH-PAN ISLAND, 158 feet high, 2 miles in circumference and uneven in appearance, is 4 miles to the northward of High island, and S.W. ½ S., 12 miles from the lighthouse on the south-west end of Fisher island.

TABLE ISLAND, bearing S.S.E. ¾ E. nearly five miles from the lighthouse on Fisher island, is aptly named, the summit being a dead flat 200 feet above the sea ; near its south-west end is a sudden fall nearly to the sea level, giving it at a short distance the appearance of two islands. The island is not quite 2 miles long, in an E. b. N. and W. b. S. direction, and is seldom 3 cables wide. The 2 fathoms line extends 2 cables from its eastern extreme.

Water.—There was a good run of water in the month of June towards the north-east end of Table island.

TABLET ISLAND lies about a mile to the northward of Table island, and between them the depth is 12 to 19 fathoms. A shoal, with only 9 feet least water upon it, extends N.W. ½ W. 1¼ miles from the north-west side of the island, and from its south-west edge, in 4 fathoms, the south end of the island bears S.E. ½ E. ; from its north-east limit the north point of the island bears S.E. by E. ; and from the north-western limit Dome island bears N.E. by E. ¾ E.

PONGHOU ISLAND, the largest of the Pescadores, is 9½ miles in extent, in a north and south direction ; it is, however, separated into three portions, by narrow channels, which have only 2 feet in them at low water, and are further blocked by stone weirs. The whole of the western face of the island is fronted by coral reefs. On its south-eastern side, between Hou and Leechin points, are two bays with fishing villages, either of which will afford anchorage in the North-east monsoon. The best shelter will be obtained in the northern bay of the two, as it is protected by some rocks, the reefs lying off which may be seen from the mast head, as the water is very clear. Dome bay, on the south-west side of the island, will also afford good anchorage in 6 fathoms.

MAKUNG HARBOUR is formed at the south-west part of Ponghou island, and although much confined by coral reefs it has sufficient depth for vessels of large draught. The town of Makung stands on the north side of an inlet, close to the north-east of the entrance, and will be easily recognized by a citadel and a line of embrasures. The large junks waiting for a favourable wind to take them to Formosa, anchor to the south-west of the town in 7 and 8 fathoms water, with Black rock lying midway between Fisher island and Makung, bearing N.E. by N. The junks belonging to the place lie close to the town, in a creek to the north-eastward of the citadel.

The harbour runs back 3 miles to the eastward from Chimney point, the south point of entrance, on which is an old Dutch fort. The southern shore is low, and on Dome hill, which is 154 feet above the sea, and the highest part of the land hereabouts, is a large pile of stones; the land between the hill and Chimney point is low and in two places less than a cable across. Dome hill overlooks Dome bay on the south-west face of the island in which there is a village and a fort. The isthmus immediately eastward of the village is low enough for the sea to break over it at high water during a south-east gale. The *Plover* anchored with Chimney point bearing N.W. ¾ W. distant 6 cables, which is also the width of the harbour here.

Within the harbour there are four coral patches, awash at low water springs, but they may always be detected from the mast head in time to avoid them. From the westernmost patch Chimney point bears N.W. by W. ¼ W. and Dome hill S. by E. ¼ E.; the next patch lies a quarter of a mile farther eastward, with the fort on Chimney point N.W. by W. ¼ W. and Dome hill South; from the next patch the fort bears N.W. ½ W., and the hill S. ½ W.; and from the fourth patch the fort bears N.W. ½ W., and the hill S.W. by S. They are all small in extent, and steep-to.

DIRECTIONS.—In running for Makung harbour from the westward pass about half a mile to the southward of Litsitah point the south extreme of Fisher island, and then steer E. ½ N. for the town of Makung, which, as before observed, may be recognized by a citadel and a line of embrasures. The only dangers to be avoided in entering this passage are, the shoal with 9 feet on it, extending N.W. ½ W. 1¼ miles from Tablet island; and a reef, just awash at high water, at half a mile westward of Dome island. Flat island, which lies 2 cables westward of Chimney point, is also surrounded by reefs to the distance of a cable's length from high water mark; and shoal water extends three-quarters of a cable in a northerly direction from Chimney point, on which is the old Dutch fort.

FISHER ISLAND, which in a collection of voyages in Dutch, published in 1726, is called D'Vissers island, lies to the westward of Ponghou, and between them is the excellent and extensive harbour of

Ponghou. The island is 5 miles long, north and south, and 3¼ miles broad. The south-east point, Siau head, is a bold cliff rising 170 feet above the sea. A reef, which breaks at low water, extends 7 cables from the western shore of the island, and its outer extreme bears N. b. E. ¼ E. from the lighthouse.

LIGHT.—A *fixed white* light is exhibited at 225 feet above high water, from a lighthouse standing on the south-west extreme of Fisher island; but as part of the windows are glazed with oyster shells, and the apparatus very rude, it will not be seen much farther off than a mile.

The lighthouse, 30 feet high, was built 90 years ago by subscription, and the expense of lighting is defrayed by a port charge of a dollar upon each junk entering Makung harbour.

ANCHORAGE.—Vessels seeking shelter in a north-east gale will find smooth water off the southern shore of Fisher island between the lighthouse and Siau head, where there are two sandy bays; in the eastern bay is a fort or line of embrasures, and in the western a run of fresh water, except during the dry season.

Niu-kung bay, between the north end of Fisher island and Pehoe island, will afford shelter in the South-west monsoon. The north-east point of the former island is a table bluff with reefs, which cover at high water, extending 2 cables in a north-easterly direction from it.

PONGHOU HARBOUR.—The eastern coast of Fisher island trends to the northward from Siau head, and forms several small bays which are steep-to to a cable's length of the beach until 2¼ miles north of the head when reefs extend nearly 3 cables from the shore. To avoid these reefs the fall of Siau head must not be brought southward of S. by W. ¼ W. after Makung citadel opens northward of the Black rock, which lies N.E. ¾ E. 1½ miles from Siau head, and part of it is always uncovered. When passing eastward of this rock, keep within 4 cables' lengths of it, as coral patches extend some distance from Ponghou island.

The *Plover* anchored at about 3 miles northward of Siau head, with Black rock bearing S. by E. ¾ E., and the highest part of Tatsang island E. ½ N.; in the bay abreast of her were two runs of good fresh water. In working up for this anchorage, to avoid the coral reefs which extend from the Ponghou shore, do not stand farther eastward than to bring the Black rock S.S.W. The harbour to the northward of this anchorage is much choked with coral patches. There is a passage out to the northward between Fisher island and Pehoe island, and it may be used on an emergency by vessels of 15 feet draught, but a local knowledge is necessary to render it available.

The archipelago, to the northward of Fisher and Pehoe islands, does not afford any inducement for a vessel to enter it. The external dangers therefore will only be noticed.

TORTOISE ROCK, 9 feet above high water and steep-to, lies about $2\frac{1}{4}$ miles from the north-west point of Fisher island, and N. by E. $\frac{1}{8}$ E. $7\frac{1}{2}$ miles from the lighthouse. There is a shoal patch of $1\frac{3}{4}$ fathoms at 6 cables S. $\frac{3}{4}$ E. from the rock, and N.W. $\frac{3}{4}$ N. from the north-east point of Fisher island.

SAND ISLAND, three quarters of a mile long, north and south, and a quarter of a mile broad, bears N.E. by E. $\frac{1}{4}$ E. $2\frac{3}{4}$ miles from the Tortoise rock, and will be known by a hummock which rises on the low land in the centre of the island and also by its yellow appearance; a rock lies off its south-west end and reefs extend north-westerly 3 cables from its north-west point. At half a mile eastward of this island is a flat black islet, and to the northward of it a cluster of stones, some of which are always above water.

BIRD ISLAND bears E.N.E. from Sand island, and a long sandy point, off which is a small sand island with a house upon it, forms its southern extreme. On the west point is a low hill connected with the rest of the island by a sandy isthmus.

Shoal water extends 3 miles to the northward from the north point of Bird island, and near its centre is North island, which has a house on it to shelter the fishermen, and upon a reef half way between them is another house. The northern edge of the shoal water uncovers at low tide, bearing from N.N.W. $\frac{1}{2}$ W. to N. $\frac{3}{4}$ W. from North island distant $1\frac{4}{10}$ miles; and from the reef at its west extreme, which is steep-to (for the lead gives no warning), Sand island bears S. by W. From the west point of Bird island to this reef are many reefs which will be avoided by not bringing Sand island to the westward of S. by W. until the west point of Bird island bears eastward of E. by S.

ANCHORAGE.—Shelter during a north-easterly wind might be found on the west side of Bird island; and from southerly winds, to the northward of the reefs extending from the north point of the island.

N.W. OUTLIER is a shoal patch of 5 fathoms, lying N. b. W. $\frac{3}{4}$ W. from Sand island and West from North island.

SABLE ISLAND, bearing S.E. by S. 5 miles from the north-east end of Bird island, is a small islet with a sand patch on its south cliff, and surrounded with rocks. It is nearly connected with the two islands to the southward of it, by reefs at low water; the southern island of the two has a large village on it.

ORGAN and RAGGED ISLANDS.—Organ island, bearing S. by E. ½ E. 3 miles from Sable island, has a reef lying N.E. ¾ N. one mile from it, and from which Sable island bears N.W. b. N. Ragged island is nearly a mile S.E. by E. from Organ island.

The whole of the east coast of Pehoe and the north coast of Ponghou abreast the above five islands is shoal.

ROUND and THREE ISLANDS.—Leechin point, the eastern extreme of Ponghou, is low and shelving, and at 1½ miles eastward of it is Round island, bearing S. by E. ⅔ E. 3$\frac{6}{10}$ miles from Ragged island; and S. ½ E. 1$\frac{3}{10}$ miles from Round is Three island. N.W. by W. ¼ W. from Three, and S.W. from Round island, is a reef which covers at half tide. Between Round and Organ islands are several overfalls.

TIDES.—It is high water, full and change, in Makung harbour at 10h. 30m.; springs rise 9½ feet, and neaps 7 feet. The tidal streams among the Pescadores run with great strength, but they are much affected by the prevailing winds. H.M.S. *Plover*, during the southerly monsoon in August, sometimes experienced a stream of 4 knots per hour on the flood running to the northward, whilst with the ebb, the current slackened for two and three hours, but seldom ran with any velocity from the northward. Vessels therefore navigating in this neighbourhood may safely allow that the effect of the current and tidal stream together will set them, according to the prevailing monsoon, 17 miles in one tide. Tide races are common, and overtop with great violence.

FORMOSA BANKS occupy a large space on the charts to the south-west of the Pescadores, and as they have not been surveyed and there is at present no account of them they should be approached with great caution. They appear to trend in the direction of the Pescadores channel and to carry 5 to 10 fathoms water. There is, however, probably less water over them, for Captain Livingstone, of the ship *Sea Star*, of Glasgow, reports that his vessel struck the ground in lat. 23° 19′ N., long. 118° 53′ E., and carried away part of her keel. The depth he considered to be about 15 feet, and High island bore E. ½ S. about 20 miles.*

---

* Nautical Magazine, p. 54, January 1858.

CHAPTER V.

EAST COAST OF CHINA—WHITE DOG ISLANDS TO NIMROD SOUND.

VARIATION 1° 00′ to 1° 40′ West, in 1861.

RIVER MIN.*—The entrance of this river is 8½ miles N.W. ½ W. from the anchorage at the White Dogs (page 121), and is formed between sandbanks which extend 7 miles from the land, and partly dry at low water. The northern range of banks terminates to the eastward in a detached rocky patch, named Outer Min reef, two peaked heads of which show at the last quarter ebb. The large island of Woufou, 6 miles long east and west, and 4 miles broad, is situated within the entrance, and near its north-east point is the little island of Hokeang, with its two contiguous islets called the Brothers.

The city of Fu-chau stands on the left bank of the river 34 miles within the entrance, and during the survey of 1841 the navigation of the river, 4 miles below the city, was obstructed by piles of stones and stakes, which had occasioned great detriment by preventing the flow of the tide, and causing the sandbanks to accumulate and shift; and, as it is one of those rivers where changes may be looked for each season, a stranger had better obtain a pilot. The usual anchorage is off the south point of Losing island at 9 miles below the city.

TIDES.—It is high water, full and change, at the White Dog islands at 9h. 0m., and springs rise 18 feet; at Temple point, river Min, at 10h. 45m., and springs rise 19 feet, neaps 14¼ feet; and at Losing island, river Min, it is high water at noon.

DIRECTIONS.—With a 16 feet rise of tide, the best time for entering the Min is from half-flood to half-ebb. The depth is 15 feet on the Outer bar, and 13 feet on the Inner bar, at low water springs. At low water neaps there are 19 feet and 17 feet respectively, and 27 and 25 feet at high water. At half-tide, both at springs and neaps, the depth is 21 feet over the Inner bar.

When the north sands of the entrance begin to dry, there are scarcely 16 feet on the bar. At low water springs they dry about 3 feet; at neaps they do not show. In fine weather, the North and South breakers appear

---

* The description of this river is by John Richards, Master, R.N., Commanding H.M. Surveying vessel *Saracen*, who re-surveyed it in June 1854. *See* Chart of River Min, with Views, No. 2,400; scale, $m = 1\cdot2$ inches.

from half-ebb to half-flood, and the Outer knoll, which has only 7 feet on it, seldom until after the last quarter; but in bad weather a line of breakers extends from the Outer knoll across to the north bank, and a continuous line from the South breakers to Black head.

The first of the flood-tide sets in from the N.E., and, running with great velocity through numerous small channels, and over the north banks inside of Rees rock, sets across the entrance of the river, passing Sharp peak direct for Round island, gradually changing its direction for Hokeang island, as the tide rises. The first of the ebb comes from the direction of Round island, and sets across the Sharp peak entrance over the north banks; as the tide falls, the stream takes the regular channel.

Outside of Rees rock the ebb runs strong to the eastward until nearly low water, when it changes its direction to S.E. The flood, now coming from the N.E. turns the stream off to the southward; and near the Outer knoll it runs strong to the S.S.W. for 3 hours, changing its direction to the westward as the tide rises. After half-flood, the stream sets towards Round island, and abates considerably in strength.

At Temple point, on the south side of Woga island, the ebb runs down for nearly 2 hours after it is low water by the shore, and the flood-stream runs for about $1\frac{1}{2}$ hours after high water.

OUTER BAR.—A vessel bound for the river Min, from the anchorage under the White Dogs, should steer about N.W. $\frac{3}{4}$ W. $8\frac{1}{2}$ miles for the entrance of the channel, south of the Outer knoll. This is the track in, for the channel north of the Outer knoll is not safe, and should not be attempted by vessels of large draught. If the weather be cloudy keep the Breakwater rock, off the west end of Tong-sha island, nearly in line with the south point of the Middle dog, about S.E. b. E. High Sharp peak, 1,232 feet high, open southward of Sharp Island peak, 616 feet high, N.W. $\frac{3}{4}$ W., is a good mark to lead in between the Outer knoll and the South bank,* till Triangle head comes open of the small black rocks off Sand Peak point, W. by S. $\frac{1}{4}$ S., or when the North breakers bear North, then haul up N.W. or N.N.W. (according as ebb or flood is running), and crossing the outer bar, gain the deep channel to the northward.

NINE FEET PATCH.—If intending to pass northward of the Nine Feet patch, Sharp Shoulder should be well open to the northward of Sharp Island peak, before Sand peak, 742 feet high, comes in line with the middle of the black rocks off Sand Peak point, S.W. $\frac{3}{4}$ S. If passing southward, Sharp Shoulder should be kept a little open to the southward before crossing that line of bearing.

---

* *See* views A. and B. on chart.

INNER BAR.—When Sand peak appears well open westward of the black rocks off Sand Peak point, Sharp Shoulder may be brought in line with Sharp Island peak, gradually opening the Shoulder to the southward as Serrated peak, 2,028 feet high, comes in one with the south-east extreme of Woufou, S.W. by W. ⅔ W., which now becomes the leading mark, until the middle of Brother A. islet comes on with the north high extreme of Brother B. bearing N.W. by W. ¾ W. ;* with which mark on, cross the Inner bar, steering a mid-channel course for the river when Round island comes on with the south-east extreme of Woufou, bearing S.W. b. S., and taking care to avoid a sunken rock with only 5 feet on it lying three quarters of a cable's length off Woga point. There is good anchorage in 5½ fathoms, stiff mud, outside the Inner bar, with Brother B. in line with, or a little open of, Sharp Peak point, and Rees rock in line with Black Head.

Vessels of small draught turning in over the Inner bar, will find the following marks useful: Stand no nearer the north bank than with Temple point in line with Sharp Peak point, nor nearer the south-east side of Hokeang bank than with Sharp Island peak on with the middle of Sharp Point bluff; nor to the north-east side of Hokeang bank than to bring the right high extreme of Brother A. in line with the left high extreme of Brother B.

SIX FEET ROCK.—To pass to the southward of this rock, which lies in mid-channel off Temple point, keep Sharp Island peak open of Woga point. The mud extends westerly a mile from Brother A., and on its northern edge is a patch of rocks which covers at quarter flood, and from them Brother A. bears E. by S. ½ S., and Temple point N. b. E. Sharp Island peak shut in behind the high land of Woga, will lead inside, or northward of the Temple Point rock. In the North-east monsoon, the high land of Woga in line with or a little open of Temple point is a good line to anchor on; in the South-west monsoon Woga creek is the best anchorage.

KINPAI PASS is dangerous to strangers, particularly at or near spring tides, for then the violence of the current produces eddies among the rocks, that occasionally cross the channel, and render the vessel totally unmanageable, even in a fresh breeze; it therefore should never be taken without a pilot or personal knowledge, and then at slack tide. On the flood a dangerous eddy extends from Kinpai point above it, in the direction of the Ferry; and for this reason, the passage north of the Middle Ground is considered the best. The Wolverine rock, with 13 feet over it, lies

---

* Beacons were being prepared to mark these spots, in 1854.

S.W. by W. ½ W. from the north extreme of Kinpai point, and 1½ cables from the shore. The Vixen spit, at the eastern end of the Middle Ground, lies S.W. 3 cables from the point, and the distance is about a cable from 1½ fathoms on its south edge to the southern shore.

After passing White fort, close with the northern shore, for it is steep-to, and may be approached with safety. The highest part of Pass island in line with White Fort bluff outer extreme is a near clearing mark for the northern shoulder of the Middle Ground. It is recommended to shut Pass island in altogether until past that point, opening it again immediately afterwards.

The danger of this passage is in passing the northern shoulder, which forms a sharp angle of the bank, with only one foot on it at low water springs, and 4 fathoms close-to; from this point to the opposite shore the distance is only 1¼ cables. After clearing this point, in passing either up or down, the tide will tend rather to set the vessel from the bank into the stream.

The high Serrated peak in line with the Ferry-house, S. ¾ W., leads through between the Middle Ground and the Quantao shoal, and is a good line for vessels to anchor on when coming down the river, and waiting for an opportunity of dropping through the Pass.

TONGUE SHOAL.—Passing the Ferry-house on the port hand the Tongue shoal is reached, steep-to, having 7 feet water near its northern extreme. This part is cleared by keeping the Ferry-house midway between Kinpai bluff and the tower, until the highest point of Kowlui head comes in line with Half-tide rock, seen ahead. Between Half-tide rock and Tintao, also ahead, the bottom is very irregular.

MINGAN PASS.—Proceeding upwards, the river narrows at the Mingan Pass. About three-quarters of a mile above Mingan, and on the same side of the river, is Couding island, off the east point of which H.M.S. *Scout* grounded on a rock at the end of a ledge projecting 25 yards from the islet, with 7 feet near its extreme.

At the upper or south end of the gorge are two islets, Spiteful and Flat islands, on the east bank of the river, and which must be left on the port hand. The Spiteful rock shows at low water: it is part of a rocky ledge projecting about 30 yards from the island.

To pass between the Spiteful rock and Losing spit, and avoid the latter, do not shut in Younoi head with Flat island until Black Cliff head, just passed (marked with a white spot), comes in line with the northern edge of Spiteful island.

The Pagoda rock, off the south point of Losing island, dries at low water springs. The best anchorage is between this rock and about half a mile above it; should this anchorage be full, a vessel should anchor near the south shoulder of Losing island, where she will be out of the strength of the tide. The river is navigable for vessels three quarters of a mile above the pagoda on Losing island; but the channel is narrow, the tides strong, and the latter anchorage is generally preferred.

LEAVING the RIVER MIN.—In dropping through the Mingan Pass with the ebb tide, it will be necessary to guard against a dangerous eddy setting from the point above Couding island on to the Scout rock.

On leaving the river, take care that the set of the tide across the channel between Sharp peak point and Rees rock does not force the vessel on the shoals on the north side of the channel. Fair anchorage in 6 fathoms, to stop a tide, will be found with Rees rock bearing S.S.E.

The junks generally use the Woga channel between Woga and Sharp Peak islands, but to the northward of the latter island there are several sandbanks which show at low tide, and there are not more than 6 feet water between the banks.

MATSOU ISLAND lies to the north-east of the entrance of the river Min, and North 10 miles from the western White Dog: and between the two and N. b. E. $\frac{1}{4}$ E. $6\frac{1}{2}$ miles from the latter is a precipitous black rock, 60 feet high, surrounded by reefs, named the Sea Dog.

S.W. b. S. one mile from the Sea Dog is a rock called Hobe reef which shows when there is a heavy swell and at low water springs; from it the west end of Matsou bears N.N.W. $\frac{1}{4}$ W., and the Breakwater rock at Tonk-sha island S. b. W. $\frac{2}{3}$ W.: the east end of Reef island (off the east point of Matsou) in line with Changchi peak N. b. E. $\frac{1}{2}$ E. will lead to the westward.

Between the Sea Dog and Matsou are two other rocks above water, named the Sea Cat and the Flat rock, but they should not be approached within the distance of 2 cables.*

CAUTION.—A dangerous rock, on which the sea breaks at low water, has lately† been discovered by the river Min pilots, lying East 3 miles from the Sea Cat, and N.N.E. from the highest part of the Middle Dog. Until this danger has been farther examined, the mariner should use great caution in approaching its locality, for its position is given by compass bearings, and therefore must be considered doubtful.

ANCHORAGE.—A good roadstead will be found on the western side of Matsou island during the North-east monsoon, and good shelter in the

---

* See Chart: East Coast of China, Sheet 6, No. 1,754; scale, $d = 14\frac{1}{2}$ inches.
† Commander T. Colville, H.M.S. *Camilla*, Dec. 1859.

deep bay on its northern face in the South-west monsoon. H.M.S. *Hornet* anchored in the latter bay in July 1857, and was well sheltered in 5 fathoms, muddy bottom, at a third of a mile from the shore, with the west extreme of bay bearing N.W. ½ W., east extreme E.N.E., centre peak of bay S.W., and Pastel rock N. b. E. ½ E. There are several villages around the bay, and fish, goats, and a small quantity of poultry may be procured; fresh water can be obtained in both bays.

CHANGCHI ISLAND.—At 1¾ miles to the north-east of Matsou is Changchi island, having two remarkable sharp peaks on it, the highest of which is 1,030 feet above the sea. On the northern face of the island are several islets, the largest of which, Gordon islet, bears North 2½ miles, but there is no safe passage between them. N.E. 1½ miles from Gordon is a small black rock with a reef lying westward of it.

At half a mile S.S.E. ¾ E. from the islet off the south point of Changchi are two rocks always above water; and West 1¼ miles from the south point is the Pastel rock.

N.E. b. E. ½ E. 2 miles from the north-east point of Changchi, and with a channel between them, are three peaked rocks named the Trio, 50 feet above the sea.

ANCHORAGE.—The bay on the south side of Changchi affords good shelter in the North-east monsoon. Vessels entering from the northward can round its eastern point close-to, and anchor within the point, in 6 fathoms. Either this or the anchorage on the western side of Matsou should be used by sailing vessels bound to the River Min during the North-east monsoon, as they may always get to the bar from hence to the precise moment they require it, but from the White Dogs a vessel will barely fetch.

ALLIGATOR ISLAND or Tungsha is a barren rock, about 40 feet above the sea, in lat. 26° 9′ N., long. 120° 26′ E. It lies East 22½ miles from Matsou island, and N.E. by E. ⅓ E. 26 miles from the south end of the White Dogs.

LARNE ROCK and ISLET.—N.W. b. W. 12½ miles from Alligator island is Larne rock, which is low and flat, with a reef lying 2 cables north of it. Larne islet, bearing N. b. E. 5½ miles from Larne rock, has ledges extending from its north and south ends. It is about 200 feet above the sea, with large boulders sticking up here and there; near its summit are three houses.

BLACK ROCK, 40 feet high, is 7½ miles to the W.N.W. of Larne islet, and the channel between it and Ragged point is 6 miles wide.

A reef which shows at low water lies E.N.E. 5½ miles from the Black rock, and midway between Larne and Cony islets, with Larne bearing S. b. E. ½ E. 5 miles, the north end of Tung-ying island E. ¾ S., and Cony islet N.W. ⅜ N.

TUNG-YING, the easternmost island on this part of the coast, bears E. ½ N. 13 miles from Larne islet, and its peak rises 855 feet above the sea. The appearance of this island is level and flat, with steep cliff shores, and a large village stands on the western side; off its south extreme is a ledge of rocks. There is another island half a mile to the north-westward of Tung-ying, appearing as part of it, except on a N.E. b. N. or S.W. b. S. bearing.

ANCHORAGE.—There is a good anchorage in the North-east monsoon, in 10 fathoms, at half a mile to the southward of the small island lying off the north-west point of Tung-ying.

CONY ISLAND is a remarkable conical island, lying W.N.W., 19 miles from Tung-ying; a reef extends 3 cables off its north-east shore, otherwise the channel, which is nearly 1½ miles wide, between it and the two islands north of it, is safe. There is a rock, awash at low water, lying East $1\frac{1}{10}$ miles from the cone, and another S.E. ⅓ E. $1\frac{4}{10}$ miles; from the latter the south end of Spider island bears W. ½ N.*

SPIDER ISLAND lies 3 miles to the westward of Cony island, and its highest part is 620 feet above the sea. There is a large village in a bay on its south side, a reef off its south-west point, and four islets off its north-east face. Between Spider island and the main, which is 5½ miles distant, there are three other islets; between the first and Spider island is a half-tide rock; the centre one, named Isthmus, has a sandy isthmus and a mud bank extending westerly from it, but the channel between it and the first islet is clear. The passage between Isthmus and Inside islet to the westward of it, is obstructed by half tide rocks. The channel between the latter islet and Cox point has 6 to 4 fathoms water, and is a mile wide. To the southward of Isthmus islet are the Larva rocks, four of which are above water; reefs however extend northerly from them, rendering the passage between them and Isthmus islet barely a mile wide.

ANCHORAGE.—There is good shelter from N.E. winds on the west side of Spider Island.

TING-HAE BAY, formed on the west side of a peninsula on the mainland 13 miles westward of Changchi, affords safe anchorage in 2½ to 3

---

* See Plan of Sam-sah Bay.

fathoms in the N.E. monsoon; there are the remains of a walled town here, but the place is nearly deserted.

Fronting this bay to the southward and south-east are many islets and rocks. The outermost (four islets above water, named Square rocks) lie 3 miles to the southward, with reefs extending northerly from them. To the north-east of the Square rocks is Crab islet, surrounded by reefs, which extend off its north-west part at least half a mile. In the channel between Crab islet and Ting-hae point are two islets.

WANKI BAY, 6 miles to the E.N.E. of Ting-hae, is frequented by junks, but although it affords them good shelter it cannot be recommended for larger vessels. There is a rock, which shows at low water, lying near the centre of the bay at 7 cables from the shore, with Pe-kyau point bearing E. ½ N., and the nearest Claret rock S.E. b. S.

CLARET ROCKS lie 1½ miles to the southward of the east point of Wanki bay. Three of them are from 20 to 30 feet above the sea, but they are all surrounded by sunken rocks, the southernmost of which lies S.W. ½ S., half a mile from the south Claret, with the hill over Ting-hae bay bearing W. ⅓ N., and the summit of Matsou S. b. E. The northernmost rock lies N.E. ½ E. a mile from the north Claret, with the north end of Gordon islet in one with a small islet beyond it bearing E. by S. ½ S. Pe-kyau point is half a mile to the northward of this rock; there is a channel between them, but the sunken rocks lying off the point narrow it to two cables; a stranger therefore should pass south of the Claret rocks, and haul up when the village in Wanki bay bears North.

RAGGED POINT is the extreme of a narrow peninsula, in some places only half a mile across, which runs 5½ miles to the E.N.E. of Wanki bay. Off the east end of the point, distant a quarter of a mile, is Diplo islet, with a reef three-quarters of a cable's length to the eastward of it. The junks use the passage between Diplo and the main, but vessels have no business in it, as the tides are strong.

SAM-SAH INLET.*—The entrance to this inlet, at 10 miles to the westward of Spider island, is 1¾ miles wide, with deep water and strong tides. On the eastern side, close to the entrance, is a small bay with a fort in it, and here the junks remain for a tide, but the water shoals too suddenly for vessels that cannot take the ground. A rock lies in mid-channel, with Castle point bearing E. ¼ N., centre peak of Cone island

---

* Sam-sah inlet is not known by that name to the natives or European coasters. The true Sam-sah lies farther north, between Fuh-ning and Nam-quan. Commander G. T. Colvile, H.M.S. *Camilla*, December 1859.

N. ¼ W. and Steep rock N. b. E. ¾ E. ; the west end of Cone island in line with the highest peak of Crag island will lead eastward of it.*

The *Plover* made a running survey of the interior of this inlet. In proceeding to the westward, she left a large island on the port hand, then hauled to the northward, and found anchorage on a middle ground, three-quarters of a mile from the shore, and 5½ miles above the island. The bay extended to the northward 13 miles beyond this anchorage, terminating in a sandy isthmus, over which Fuh-ning bay was seen. The bay also runs back to the west and south-west ; in the latter arm is the town of Nin-le-heen.

At 4 miles to the southward of the entrance of Sam-sah inlet is the opening into another inlet, which is ten miles deep ; there are 30 fathoms water at the entrance, but circumstances did not admit of its being examined.

RAG ISLANDS.—Off the entrance to Sam-sah inlet and 7 miles to the south-westward of Spider island there are three islets, named Rag islands, having the Bittern rock, which covers at high water, lying a mile to the northward of them. The *Plover* anchored to the westward of the western-most islet and found tolerable shelter. The tides here run with great strength, and a long swell rolls home into the bay with north-east winds.

TIDES.—It is high water, full and change, at Changchi island at 9h. 30m., and at Spider island at 10h. 0m. ; springs rise 17 feet. Inside Matsou and Changchi islands the tidal streams are very perceptible, there being a great indraught into Ting-hae bay and the northern entrances to the river Min with the flood, and the velocity off Ragged point sometimes amounts to 3 knots. There is also a great indraught into Sam-sah inlet.

To the northward of Changchi the flood came from the E.N.E. at the rate of 1¼ knots per hour, and the ebb from W. b. S. 1½ knots; also off Cony island the ebb averaged 1¼ knots from W. b. S. at neaps. At the anchorage inside Sam-sah inlet the ebb came from the N.W., and it ran 11½ miles in a tide ; the flood set E.N.E. for the first 3 hours, then S.E.

DOUBLE PEAK ISLAND is 3¼ miles long, N.N.E. and S.S.W., and near its northern end are two remarkable peaks, the highest of which rises to the height of 1,190 feet above the sea. It lies 3 miles to the north-east of Spider island, the only danger in the channel between being the rocks lying off the north end of the latter island.

There are two cone-shaped islets between Double Peak and Cony island, with channels between too narrow for sailing vessels, but there is

---

* *See* Plan of Sam-sah bay, No. 1,988 ; scale, *m* = 0·7 of an inch.

a good passage between the southernmost of these islets and Cony island; reefs extend 3 cables in a north-easterly direction from the latter, and the west point of the former is not steep-to.*

ANCHORAGE.—Good anchorage in the North-east monsoon will be found to the south-east of a small islet, with a rock above water on each side of it, lying three quarters of a mile to the westward of the west point of Double Peak island; the two cone-shaped islets to the northward of Cone island sheltering from the eastern swell.

FLAP ISLAND, at $1\frac{1}{2}$ miles westward of the north end of Double Peak, is a low flat islet, with a sunken rock off its southern point. There is no passage fit for vessels between this islet and the main land, but there is good shelter abreast the first sandy bay within the point westward of it. Here were found six piratical junks plundering part of a convoy they had captured.

BITTERN ISLAND.—To the northward of Flap and Double Peak islands the coast trends to the northward for $9\frac{1}{2}$ miles to Fielon island, and off it is Bittern island and several rugged rocks which it will be advisable for vessels of large draught to give a berth to, and not to close the shore under the depth of 6 fathoms. Bittern island is from 3 to 4 miles in circumference, and between it and the main there is a passage three quarters of a mile wide and a mile in length, affording good anchorage in $3\frac{1}{2}$ fathoms for small vessels in either monsoon. On the north-west side of the island is a sandy cove where fresh water will be found. H.M.S. *Bittern,* when in search of piratical junks, anchored in $4\frac{1}{2}$ fathoms with Goodridge point E. $\frac{1}{4}$ N., and the extremes of the island from S. by E. $\frac{3}{4}$ E. to S.W. by S.

FUH-NING BAY.—From Fielon island the coast falls back to the westward, forming a deep but shallow bay, in which is the city of Fuh-ning. In the northern part of the entrance is a group of islets extending 2 miles from the coast. The *Plover* anchored under the south-western, named Fong-ho, which is the largest, but the shelter was not good.

PIH-SEANG ISLANDS.—N.E. b. E. 10 miles from Double Peak island is the Pih-seang or Tsih-sing group. The northern islet, named Town island, is the largest, and at its south-west angle there is a little cove, which will afford shelter to one or two small vessels. Between the northern and southern islets of the group there is a channel free from rocks, but the intervening space is thickly studded with fishing stakes.

---

* *See* Chart : China, East Coast, Sheet 6, No. 1,754 ; scale, $d = 14\frac{1}{4}$ inches.

FUH-YAN ISLAND, 1,700 feet above the sea, lies North 12 miles from the Pih-scang group, and between it and the coast is a good roadstead, named Lishan bay. The anchorage in the bay is on the Fuh-yan side, abreast an islet and a joss house. The northern entrance to the bay is broad and open. To the southward are three entrances: the first, Fuh-yan pass, between Fuh-yan and Chuhpi island, is only a cable wide, and vessels using it are apt to get becalmed under Fuh-yan. The Chuhpi pass between Chuhpi and Angle island is 8 cables across, but there is a patch of low rocks (which must be left to the westward) to the S.W. of Chuhpi that narrows the channel to half a mile; but there is a sunken rock off the north-east point of Angle island. The third entrance, between Angle island and the main, called Little Pass, is only fit for small junks or boats.

Water.—Good water is plentiful and easily obtained at the anchorage in Lishan bay.

TIDES.—In Lishan bay it is high water, full and change, at 10h. 15m., and the rise at springs is 16 feet. The first of the flood comes from the E.S.E. at the rate of three quarters of a knot per hour, then from E.N.E. at half a knot; the ebb runs to the N.E. at three quarters of a knot.

DANGEROUS ROCK is in lat. 26° 53′ N., long. 120° 34′ 18″ E., and its summit is 8 feet above high water, or 24 feet above low water springs.

TAE ISLANDS.—E. b. N. 16 miles from the eastern point of Fuh-yan are the Tae islands, the easternmost of which, rising to the height of 618 feet above the sea, is the largest, and remarkable for its table top. Shelter can be had under this island as close as a vessel can safely go (say half a cable's length), but it is bad.

S.S.W. $\frac{1}{4}$ W. 3 miles from the easternmost Tae island are two rocky islets named Strawstack, about 100 feet high; they almost join. Close to the north-east point of the northern Tae island is a remarkable Mushroom rock 260 feet high.

Between the Tae group and Fuh-yan are the Incog islands, too small to afford shelter; they are low and flat, with steep cliffs. At 3 miles to the N.W. of these islands is Solitary rock, with a reef extending 2 cables in an easterly direction from it; the soundings between this rock and the main, from which it is distant $3\frac{1}{2}$ miles, vary from 7 to $5\frac{1}{2}$ fathoms.

CAUTION.—Vessels passing inside the Tae group should keep well to the westward, as the ground in their vicinity has not been well explored. Two reefs, which show at low water, have been found; one, with the rocks on it 8 feet above high water, lies with the Mushroom rock bearing

E.S.E., and the west end of the eastern Incog island S.W. b. W. ¾ W., on which bearing it is in line with the east end of Fuh-yan. The table top island of the Tae group bears from the other E. b. S. ¼ S. and the west rock of the group N.E. b. E. 1 1/10 miles.

SEVEN STARS are three small rocky islets with several rocks awash near them, lying N.E. by E. ⅓ E. 6½ miles from the eastern Tae island. At 3 miles to the N.N.W. of these is Cleft rock, 50 feet above water.

PIH-QUAN HARBOUR.—N.W. 14 miles from the Tae group is the entrance to Pih-quan harbour, to the northward of which is a remarkable high peak, Pih-quan peak, in lat. 27° 18′ 48″ N., long. 120° 28′ 45″ E. The harbour is formed between Ping-fong and Chin-quan islands, is 1½ miles wide, carries a depth of 3 fathoms, and affords good shelter in the North-east monsoon to vessels under 15 feet draught.

Ping-fong has three chimneys on its summit; off its south-east point is a low rock which is never covered, and between this rock and Ping-fong is a sunken rock. Vessels bound to this harbour from the northward may round this low rock within a cable's length, and then haul up for the south point of Ping-fong, giving it and also the south-west point a berth of 2 cables. The Pih pass, between the north end of Ping-fong and the main, is fit only for such junks as use sculls.

Water.—Fresh water may be obtained in the sandy bay at the foot of the three chimneys on Pih-quan.

NAM-QUAN HARBOUR.—The south point of Chin-quan island is a bold steep bluff, having under it a rock which may be passed close-to. Anchorage in 9 and 7 fathoms will be found on the west side of Chin-quan after a second rock has been passed. The soundings shoal suddenly to the northward in the north part of Nam-quan bay, where stands the walled town of Nam-quan.

Immediately to the westward of Nam-quan bay is the entrance to an inlet called Nam-quan harbour, which runs about 15 miles in a general N.W. direction, when it appears to expand into a wide basin called Gordon bay.*

On the point at the north side and a little within the entrance is a town. South of the town point is a small rock which never covers, having rounded which haul up to the northward, giving the western end of the town point a berth of 1½ cables, to avoid a sunken rock off it which shows till quarter flood. When within the point anchor in 14 fathoms, as the mud banks rise almost vertically. On the south side of the entrance is a

---

* *See* Plan of Nam-quan Harbour, No. 1,980 ; scale, *m* = 1·7 inches.

small fort with a few houses. The narrowest part of this channel is 6 cables wide, and the strong tides and baffling winds make it necessary to have a boat ready to tow the vessel's head round. The *Plover* traced the inlet for 15 miles to the N.W. from the town point, and had then a depth of 8 fathoms; the channel, which is, however, narrow and tortuous, is surrounded by high hills, and there was apparently little or no traffic.

NIMROD ROCK.—H.M.S. *Nimrod* when proceeding up Nam-quan harbour, January 1857, struck on a rock with only 9 feet of water on it, lying about 11 miles from the entrance, and $1\frac{3}{4}$ cables eastward of a small islet on the western shore.*

TIDES.—In Nam-quan harbour it is high water, full and change, at 10h. 0m.; springs rise 17 feet.

BOUNDARY.—The boundary line of the provinces Chi-kyang and Fu-kyen passes through Pih-quan harbour.

The COAST from Pih-quan harbour trends N.E. b. N. 19 miles to Ping-yang point; at 12 miles from the harbour is Tanue bay, which is too shallow to afford shelter to any vessel drawing over 10 feet water. A low rock, named Gap islet, lies $1\frac{1}{2}$ miles to the southward of Tanue point; and N.E. $\frac{3}{4}$ E. $4\frac{3}{4}$ miles from it is Farmer rock, which shows at low water, and lies $3\frac{1}{4}$ miles off shore, with Ping-yang point bearing N.N.W., and Nam-ki peak E. b. N.

From Ping-yang point the coast takes a north-westerly direction and is fronted by mud banks, which dry 3 miles from the land at low water, and on which are several small islets and rocks. At the distance of 11 miles from the point is the embouchure of the Shwin-gan river, by which the commerce of Wan-chu fu is maintained; there are only 9 feet on the bar at low water.

Off the entrance of the Shwin-gan are the Tsang islets, four in number, the southern of which is the largest. In the channel between this latter islet and the mud bank at the entrance of the river the depth is only 9 feet. Between the south islet and the one next it to the northward, there is a channel close to the latter with 4 fathoms in it; and inside the two central islands the depth is 3 fathoms, but the space is confined.

NAMKI ISLANDS lie N.E. b. N. 29 miles from the Tae group, and the largest, 740 feet above the sea, has a good harbour during the North-east monsoon on its south-eastern side, called Port Namki. Vessels should not pass among the islets forming the south-west part of this group, as

---

* Capt. C. C. Forsyth, R.N., H.M.S. *Hornet*, 1857.

there are many reefs which cover at high water. The westernmost islet, Turret island, makes like a cone, and has reefs to the northward of it. The southern islet is a castellated rock, and lies S.S.W. 5 miles from the rest of the group.

Water.—Good water can be obtained in Port Namki.

PIH-KI-SHAN ISLANDS.—N.N.E. 9 miles from Namki is another group, the largest of which is called Pih-ki-shan. There are four small islets lying close to its south-east side, which protect the anchorage on the south side of the island from the easterly swell. Vessels should not, however, choose this anchorage, unless from necessity. Fresh water may be obtained.

TUNG-PWAN and TAE-PIH ISLANDS.—West 11 miles from Pih-ki-shan, with five small islets intervening, is another group of one large and four smaller islets. The large islet, called Tung-pwan or Brass basin, has anchorage off its south-west face in 8 fathoms in the North-east monsoon, but the shelter is not so good as that on the south side of the Tae-pih islands, lying 3 miles to the N.W., under which the water will be smooth in 4 fathoms.*

In working up to the northward of the Tae-pih and Tung-pwan groups, shoal water will be found to extend 8 miles from the foot of the hills on the main; at which distance is the 2 fathoms' line of soundings. On the eastern edge of this line, at 6½ miles northward of Tae-pih, is the Pang-peto reef, which is visible at low water; from it the western of the Tae-pih islands bears S.S.W. ½ W., and the southern of the Tseigh islands E. by S. ½ S.

TIDES.—At the Namki and the Pih-ki-shan islands it is high water, full and change, at 8h. 30m., and the rise at springs is 17 feet. At the anchorage under the southern side of the latter group, the ebb runs to the N.N.W., and the flood to the S.E. by E.

FONG-WHANG GROUP.—The Tseigh islands, three in number and named North Tseigh, South Tseigh, and East Tseigh, lie N.N.W. 8 miles from Pih-ki-shan, and form the south extreme of a large and numerous group. Between the Tseigh and Pwan-peen island, the next island to the northward, is a navigable channel for vessels 3 cables wide. Fong-whang, the largest island of the group, is 6 miles long N.E. and S.W., 2½ miles at its extreme breadth, and its eastern face is high and precipitous; there is a channel for junks between it and Pwan-peen.

---

* *See* Chart: East Coast of China, Sheet 7, No. 1,759; scale, $d = 14\frac{1}{2}$ inches.

Coin island, the north-eastern of the Fong-whang group, has three rocks lying to the N.W. of it, and to the W.S.W. is a low flat islet, Flask island, with rocks off its southern end, and two rocky islets to the westward, between which there is a safe channel carrying a depth of 8 fathoms.

BULLOCK HARBOUR, the entrance to which is between the Tseigh group and a high island with bold cliffs, named Fakew, has excellent anchorage in 4 to 10 fathoms, sheltered from all winds. The distance is 2 miles between the Tseigh and Fakew, and on entering a vessel will have to pass over a bar with 4 fathoms on it, deepening to 6 and 8 fathoms, and then shallowing to 4 and 3 fathoms at the head of the harbour. The anchorage is in $5\frac{1}{2}$ fathoms off the west end of Pwan-peen island.

Supplies.—Water can be procured in this harbour, and bullocks of the best description.

TIDES.—It is high water, full and change, in Bullock harbour at 8h. 30m., and the rise at springs is 17 feet.

DIRECTIONS.—Vessels approaching Bullock harbour from the southward cannot pass between the Pih-ki-shan and the Tseigh islands, as there are clusters of rocks interspersed with reefs between them, but they should pass between Tung-pwan and Shroud islet, which may be recognized by its bluff; the islands near it are low. Care must be taken to avoid a sunken rock lying North of the rocks immediately westward of Shroud; and also the reef North of the islet lying N.N.W. $2\frac{1}{2}$ miles from Shroud.

In approaching the harbour from the northward, through the San-pwan pass, which may be taken by a vessel of 12 feet draught, pass to the westward of Fakew, bearing in mind that a rock with only a foot over it at low water lies N.N.W. $\frac{1}{2}$ W., rather more than a mile from its south-west point, with the west point of Fong-whang in line with the east extreme of Great San-pwan bearing N.E. $\frac{3}{4}$ N.

To the N.W., 4 miles from Fakew, is the island of Niaow; the channel lies between these two, and between Niaow and Fong-whang, where, from both shores being shoal, it is only 6 cables across. Great San-pwan is almost connected with Niaow, there being but a very narrow channel between them. Close to the south-east point of Great San-pwan is a bold perpendicular islet, and the channel is between this islet and Little San-pwan. The winds being variable and the tides uncertain, unhandy vessels will have difficulty in clearing this pass, especially if a strong northerly wind has been blowing, as there is usually a heavy swell at such times setting into it.

WAN-CHU RIVER.—N.W. by W. 8 miles from Niaow island is Wan-chu island, fronting the mouth of the Wan-chu river. A mud spit extends 6 miles to the south-eastward of this island, leaving only a shallow channel of 7 feet water between it and Niaow.

TIDES.—At the entrance of Wan-chu river it is high water, full and change, at 9h., and at 9h. 30m. at Wan-chu fu; and the rise at each place is from 15 to 16 feet.

DIRECTIONS.—When bound to Wan-chu river from the southward, after passing Coin island, steer N.W. ½ N., leaving the Cliff rocks to the north-east and the north rock of Great San-pwan island to the southward. Having passed the latter, edge away W. by N. for the south point of Hutau island, leaving a remarkably steep bluff island, called Hokeen, to the southward. Off the south point of Hutau, and abreast of Hokeen, is a sunken rock lying 1½ cables * from the shore, but it will be avoided by opening the south-west point of Hutau to the southward of a white rock in Hutau bay. South of the white rock there is a middle ground confining the channel to a width of 7 cables. There is good anchorage in 4 and 5 fathoms to the S.W. of the white rock, but the bay within it is shoal.

From the south-west point of Hutau the entrance of the river bears W.N.W. 5 miles, and it will be known by an isolated range of hills, with a square fort at the east, and a small walled town at the west end. The depth varies from 3 to 4 fathoms in the channel, which is more than a mile wide, but the mud dries upon either side, and it shoals suddenly. Having passed the range of hills keep the left bank or north shore of the river aboard, until the first hill on the flat island (Wan-chu island) on the south side of the river bears S.W. b. S., when the vessel will have cleared a middle ground at half a mile from the south shore, and 1½ miles to the E.N.E. of this hill; the highest part of Hutau in line with the south foot of the hills at the entrance bearing E. ¼ S. is the mark for its northern edge.

From abreast this middle ground edge over to mid-channel, passing a large walled town on the north side of the river, then gradually haul over to the first point on the south side, where the hills come down to the water's edge, passing a point with a circular fort, and a building like a large jar upon it close-to. Vessels ought not to go above 2½ miles beyond Jar point; they will then be in from 3½ to 7 fathoms water. From this anchorage the distance to Wan-chu fu is 5½ miles, but the channel is too intricate for a stranger. The water of the river contains a great

---

* Commander Vansittart, H.M.S. *Bittern*, states that this rock is within half a cable's length of the point, and may be passed close-to; and that a sailing vessel must be careful of the ebb tide, which sets with great strength to the E.N.E., across the flat between Hutau and Wan-chu point, and between Hutau and Junk island, especially as the flat seems to have grown to the southward.

[c.]

deal of sediment, and is not used by the inhabitants for culinary purposes. From the summit of Fort hill the canals with junks in them were traced to the westward, where they probably communicate with the Shwin-gan river, which appears to monopolise the commerce of the district, as but few junks were seen on the Ngau river, notwithstanding its capabilities for navigation.

LOT-SIN BAY.—Junk island is low and rocky, and lies on the north side of Hutau island. The channel between them, and between Junk island and the main, can only be used by small junks.

To the northward of Junk island is Lot-sin bay, which runs back to the northward 20 miles. There is good anchorage in its southern part, but its head is shoal except a narrow channel, named Hebe Lock, which makes Ta-ou-an island.

QUANG-TA ISLAND.—At 2 miles eastward of Hutau is Quang-ta island, under the west side of which H.M.S. *Plover* anchored, but the water was found to shoal very suddenly. There is a channel between Quang-ta and the Cliff rocks to the south-east, and also between Quang-ta and Ta-ou to the northward; take care, however, to avoid the islets and rocks off the north-east part of Quang-ta.

KEMONG HARBOUR.—Near the east point of Ta-ou island is a bight named Kemong harbour, with an islet off each point, in which the junks are fond of taking shelter. It is, however, confined, and vessels will find better anchorage to the eastward under either Taluk or Seoluk islands.

Captain Meier, of the Hamburgh barque *Kingman*, reports * the existence of a rock, awash, lying in the middle of the entrance to Kemong harbour. From the rock,—which was only seen twice, one or two feet above water, during the three weeks the vessel remained in the harbour,—the east extreme of the rocks extending from the north-eastern point of Quang-ta bore South, and the east point of Nam-pan S.W. Before the arrival of the *Kingman* no European vessel had brought a cargo to this port. The Chinese knew of the existence of the rock.

SEOLUK, TALUK, CHIN-KI, TOWAN, and PE-SHAN ISLANDS lie from 3 to 14 miles to the eastward of Ta-ou. The Seoluk consist of three islets lying north and south of each other. Taluk is a higher island, 770 feet above the sea, lying 1½ miles to the northward of the Seoluk, and in the channel between them the depth is 7 and 8 fathoms. West of Taluk is Chin-ki, a low flat island with a large village

---

* Nautical Magazine, page 277, **May 1860**.

on it; there is anchorage between these in 3 to 4 fathoms. The bay to the north-west of Chin-ki is shoal; at its head is the entrance to Hebe Lock communicating with Lot-sin Bay. At 8 cables to the north-east of Chin-ki is Towan island, with a channel of 4 fathoms' water between them; but as a sunken rock lies in the middle of this channel, and a reef runs out from the north point of Chin-ki, vessels have no business here. Between Towan island and the rocks off the north end of Taluk island, the passage is a mile wide.

Pe-shan, the easternmost islet of this group, is $1\frac{1}{2}$ miles long, east and west, and off its northern face are three rocks, and off its southern two islets. W. b. N. $1\frac{1}{2}$ miles from Pe-shan is a low level islet, named Flare island, and to the N.W. is Sugar Loaf island, with a small islet lying close to its north side. Between Sugar Loaf and Flare islands the depth is 5 fathoms.

TIDES.—At the anchorage between Chin-ki and Taluk islands it is high water, full and change, at 9h. 20m.; and springs rise 13 feet.

TAOW-PUNG ISLAND, bearing N.N.E. 9 miles from Pe-shan, is 7 miles long N.N.E. and S.S.W., and $1\frac{1}{2}$ miles broad, and to the westward of it is Yey-van bay, which is shoal and affords no shelter. The island is separated from the main by a narrow channel called Penetration pass, through which all the country trade passes. Near the north end of the pass, on the main, is the walled town of Song-men.

Song-men point forms the south end of Taow-pung, and to the south-west of it, at 2 and 3 miles respectively, are two flat rocks above water. To the south-eastward of the point are several islets; the nearest, named San-shi, has a reef to the westward; the outer islet of the three has a shoal off its north end. There is a navigable channel, a mile broad, between San-shi and the rocks off the point. At 3 miles to the N.E. of Sanshi, are the Stragglers and Shetung islets; the latter, the northern and highest islet of the group, has a reef lying 3 cables from its south-west point, and many rocky islets off its south end, between which and the Stragglers there is a channel carrying a depth of 6 fathoms. Indifferent shelter in the North-east monsoon may be found under Shetung island.

Between Shetung and Taow-pung island are two islets forming three channels, the eastern of which, between Shetung and the next islet westward, has $3\frac{1}{4}$ fathoms in it, but the other two are too narrow for vessels. Junks lie inside the inner islet, where there is a small village. To the north-east of these two islets are three rocks above water, the northern of which has a reef off its east end. Soudan, the eastern islet of this group, bears N.E. 15 miles from Pe-shan: it is flat-topped, and has a reef on its south side.

CHIKHOK ISLANDS.—Chikhok island lies North 6 miles from Soudan island, and as it rises abruptly to the height of 760 feet above the sea, and has a broad yellow stripe on its south-eastern side, it forms altogether one of the best leading marks on the coast. N.N.W. 1¼ miles from Chikhok is an islet named Low Chikhok with a half tide rock lying N.W. 3 cables from it. West 2 miles from Chikhok is Crookback island, with many rocks about it. H.M.S. *Plover* anchored to the south-west of Crookback in 2¾ fathoms, but a long swell sets in here, and the channel to the northward of it is too shallow to get through on that side. The same may be said of all the channels amongst the islands to the north-west of Chikhok.

TAI-CHAU ISLANDS.—East, distant 9½ miles from Chikhok island, is Hea-chu islet, the southernmost of the Tai-chau group; off its south side is a remarkable finger rock. The group extends 9 miles to the northward of Hea-chu, and consists of two large and ten smaller islands. Between the two large islands is an excellent harbour, the approaches to which, both from the eastward and westward, are free from danger. The southern large island, 750 feet high, is called Hea-ta, and the northern Shang-ta, which is well inhabited. Between Shang-ta and the Shang rock, to the N.N.E., there is a safe passage.

At about 2 miles to the south of the west point of Hea-ta are two rocks, the western of which shows at all times of tides, and lies S.S.W. 3¼ miles from the highest part of Hea-ta; the other, which bears N.E. ½ N. 4½ cables from the western rock, and S. by W. ¾ W. from the highest part of Hea-ta, covers at high water.

ANCHORAGE.—The best anchorage in the harbour formed between the two larger islands of the Tai-chau group, during the North-east monsoon, is to the south-east of the islet lying off the south-west extreme of Shang-ta.

Water.—Several watering places will be found on Shang-ta island, but the supply from any one of them is not abundant.

TIDES.—It is high water, full and change, at the anchorage at the Tai-chau islands at 9h. 0m., and springs rise 14 feet.

SQUALL ISLANDS.—At 6 miles N.W. b. W. from Shang rock, the northern islet of the Tai-chu group, are the two Squall islands, but so close together as to appear as one, except on an E.N.E. and W.S.W. bearing. Rocks lie off the north-east and north-west points of the northern island, and a reef extends from the south-east end of the southern island. Junks take shelter under the western point during strong north-east winds.

Crate island, a small cliff islet, lies 2½ miles to the eastward of the Squall islands, and the channel between them has 8 fathoms in it; but the western end of Crate is not steep-to.

TAI-CHAU BAY and RIVER. — Tai-chau bay, to the N.W. b. W. of the Tai-chau islands, is wide and shallow, and at its head is the entrance to Tai-chau river. On the right bank of the river is the walled town of Haimun, 4 miles above which the river separates into two branches, one taking a north-west, the other a south-west direction. The city of Tai-chau is on the north branch of the river, about 24 miles in a direct line from Haimun. There are only 8 feet at low water across the bay to the entrance of the river, but inside the entrance points the depths are 4½ and 5 fathoms. The inhabitants reported that vessels of 12 feet draught could not cross the bar, except at high water, and that the tide, which rises from 18 to 20 feet in this locality at springs, would carry them up to the city.

At 9 miles to the south-west of the Squall islands is the North Foreland, an islet lying off the southern side of Tai-chau bay, 1¾ miles from the coast, with a depth of 10 feet inside it. South of it are two other islets; and there is a half-tide rock which bears West southerly 12 miles from the north point of Shang-ta, S.S.E. 2½ miles from the North Foreland, and N.N.W. from Chikhok island, on which bearing Low Chikhok island is in one with it. In the channel between the Squall islands and Tai-chau bay, the water shoals gradually towards the main; but by not bringing the North Foreland to the eastward of South, a vessel will be in 2½ fathoms at low water.

CHUH-SEU ISLAND, lying N.N.W. 4½ miles from the Squall islands, is remarkable, having a sharp cone, 670 feet above the sea, over its southern point, and a beacon on its western summit. Between Chuh-seu and the Squall islands are four rocks; and S.E. b. E. ⅓ E. 2¾ miles from the former is a solitary rock named Fir Cone.

ANCHORAGE and WATER. — Good anchorage in 6 fathoms and a convenient watering place, with abundance of water, will be found under and to the south-west of the cone of Chuh-seu, between the south-west shore of Chuh-seu and an islet with a reef off its north-east point.

The channel between Chuh-seu island and Mud islet (a hill on the mud on the north side of Tai-chau bay) is shallow, with several rocks in it covered at high water. North 1⅓ miles from the western islet off the Chuh-seu group is a rock showing at low water.

TUNGCHUH ISLAND. — East, a little northerly, 5 miles from Chuh-seu is Tungchuh or Bella Vista island, 700 feet high, the easternmost of this group. The two Reef islands lie S.S.W. 2¼ miles from the south point of Tungchuh; a reef extends north-easterly from the southernmost of the two. Midway between Reef and Chuh-seu are a cluster of rocks.

The island of Gau-tau, remarkable for four barren peaks, lies 3 miles to the north-west of Tungchuh. The channel between them has not been examined; there is generally a heavy swell in it. The low north-eastern promontory of Gau-tau is an island at high water; a half-tide rock lies North 3 cables from its eastern end.

ANCHORAGE.—Shelter may be had in the North-east monsoon under the south side of Tungchuh, but there is generally a heavy swell, which renders riding there unpleasant, and vessels had better gain the anchorage under Chuh-seu island, or endeavour to reach Barren bay.

BARREN BAY, formed between Gau-tau and Kin-men islands, is 2¼ miles wide at its north-eastern entrance, and besides the half-tide rock just mentioned off the eastern promontory of Gau-tau, there are rocks off the eastern point of Kin-men, and a mud spit off the north-west point of Gau-tau. Immediately to the south-west of Kin-men, and separated by a deep-water channel rather more than a cable across, is Nine Pin island, divided near the centre by a sandy isthmus, on which is the rock from whence the island is named. Very poor shelter in 6 to 3 fathoms will be found between Gau-tau and this island, the deeper water being towards the latter.

There is a channel to the westward of Nine Pin, but it cannot be recommended, as there are depths 1¾ and 2 fathoms to the northward of Nine Pin, and between it and Pine Cone, an islet lying N.W. 2½ miles from it. South 2 cables from the west end of Nine Pin, is a rock which will be seen at half-tide.

FALL and CHAIN ISLANDS.—Fall island lies nearly 2 miles to the northward of Kin-men island, with two rocks above and one below water off its west end. The channel is safe between these islands, and also between Fall and Chain islands, but the latter are not steep-to.[*]

Chain islands, three in number, bear N.W. b. W. 4¼ miles from Fall island. South 2 cables from the centre island is a half-tide rock, and there is a rock awash and two small islets lying off the west end of the southernmost island. Between the Chain islands and Pine Cone island, to the southward, are four detached rocks.

CAUTION.—Vessels should keep to the eastward of the whole group just described, for the channel inside Chuh-seu, Kin-men, Chain, and San-mun islands is shallow, and has several rocks in it covered at high water.

---

[*] See Chart:—East Coast of China. Sheet 8, No. 1,199; scale, $d = 15$ inches.

CHAP. V.] BARREN BAY.—MONTAGU ISLAND.—SAN-MUN BAY.   151

HIESHAN GROUP, consisting of three inhabited islands and eight rocks, lie N.E. b. E. ½ E. 17 miles from Tuugchuh island, and occupy a space 5 miles in a north and south, and 2 miles in an east and west direction, but they are too small and too detached to afford shelter. The southernmost island, 320 feet above the sea, is the largest, and makes like a saddle. The inhabitants, who are Fu-kyen men (and most likely pirates), call the islands Ung-shan; they are all fishermen, and excellent fish may be obtained.

The rocks are steep, with remarkable cliffs. The sea has so much undermined the northernmost, named Mushroom, as to cause it to bear some resemblance to a large mushroom. N.E. ⅜ E. 1¾ miles from Mushroom is a sunken rock, with 8 feet water on it, from which the Cheng rock appears in one with the south-east end of Cliff or Sha-ho island, bearing S.S.W. ⅓ W. N.N.W. a quarter of a mile from the Mushroom, is a rock awash at low water.

MONTAGU ISLAND, or Tauto-Shan, 20 miles to the N.N.W. of the Hieshan group, is separated from the main islands by channels varying from 1 to 1¾ miles wide, the navigation of which is much obstructed by sunken rocks; shelter however in the N.E. monsoon will be found under its south and south-west extremes. The island is 740 feet high, and nearly divided into two parts, the connection being a low shingly isthmus; the northern portion is called Gore island.

To the southward of Montagu, and at the distance of 2 to 5 miles from the eastern coast of Nyew-tew island, are six islets; the southernmost, called the Twins, is 8 miles from Montagu, and the others are 1½ to 6 miles from it, with clear channels between them. A rock awash at low water was reported in 1851 to lie S.W. 3 miles from the eastern or larger Twin. Abreast the middle islet of the five (Dike islet), and which is the nearest to the main, is Nose islet, and vessels passing between them must bear in mind that neither are steep-to; Nose islet is nearly connected with Nyew-tew island at low water.

SAN-MUN BAY.—The entrance to this bay is 20 miles to the W.N.W. of the Hieshan islands, and it will be readily recognized by a remarkable thumb peak, 800 feet above the sea, called by the Chinese Tafou, and by the opium vessels Albert peak; it rises from the northern end of Tafou island, on the northern side of the bay.

Vessels wishing to stop a tide or driven in by bad weather, will find good shelter in the North-east monsoon in the bay immediately westward of Lea-ming island, which forms the north point of entrance of San-mun bay. In running for this anchorage, give a berth of 2 cables to the south-west point of the island, to avoid a reef lying off it. The soundings will shoal suddenly after the north peak of the island is brought to the southward of East; the bottom is soft mud.*

S.W. ⅜ S. 2½ miles from Lea-ming, is Sanchesan or Triple island, and the depth between them is 10 and 11 fathoms. West, distant 6 miles from Lea-ming, is a conical islet, named Cone island, with a reef off its south end; and N.W. b. N. 6 cables from Cone is a small islet with a rock off its south-east face. At 4 miles to the westward of Cone is a small islet.

Having passed to the southward of Cone island, St. George island will be seen bearing N.W. 4 miles; the bay shoals gradually as this island is approached, and the anchorage in 3 fathoms at half a mile South of it is secure in N.E. winds. There is a well of good water on this island, but it is not easily got at nor plentiful.

The bay northward of St. George island is shoal and full of rocks; it extends a considerable distance, leaving an isthmus 7 miles wide between it and Nimrod sound. There is an entrance into Sheipu harbour at 4 miles north of St. George island, and it is frequently used by the junks.

Westward of St. George island is a group of high islands, the largest of which is called Tinwan. There are several islets and rocks on the eastern face of this group, and between their western face and the main is a deep water channel a mile wide. S.W. from Tinwan island is the embouchure of a river, on the bar of which there are only 4 feet, but deep water inside. On the left bank of the river, 5 miles from Tinwan, is the walled town of Kien-tyau. W. b. N. from Tinwan is the mouth of the Ning-hau river, on the north side of which, at 6 miles from Tinwan, is Quarry island, and to the southward of this latter island there is good anchorage in 6 and 4 fathoms; a mud spit extends 2 miles eastward from Quarry island. Between Kien-tyau and Tau-tew point, abreast of Tinwan, the hills rise abruptly from the coast-line to the height of 1,000 feet; but the water shoals to 2 fathoms in some places, at the distance of 2 miles from the shore.

TIDES.—At the anchorage under St. George island, San-mun bay, it is high water, full and change, at 10h. 20m.; and the springs rise about 15 feet.

SHEIPU ROAD.—Vessels bound to the roadstead off the town of Sheipu may pass close to the northward of the islets off Gore island, the northern portion of Montagu island, and steer in West for the two forts

---

* See Plan of San-mun Bay and Sheipu Harbour, No. 1,994; scale, $m = 0.7$ of an inch.

standing on the summit of Tungmun island, which forms the southern side of the entrance to Sheipu harbour. North of the roadstead are three islands named Bangao, and South 3 cables from the eastern point of the centre island, Wangchi, are the Bangao rocks, which always show. There is deep water close to these rocks, except to the westward, where it shoals to 2¼ fathoms; to avoid which do not bring the higher fort to the southward of West.

Cliff island, or Seao-Seao, lying nearly in the centre of the roadstead, has anchorage off its north-west end in 4 fathoms, but with a strong wind a considerable swell rolls in. A reef of rocks extends westerly from Cliff, and the channel between it and the islands off the main carries 3 fathoms water. South of Cliff is an islet with foul ground between; and E. by S. 7 cables from Cliff is a flat rock, and between them a sunken rock. The channel eastward of Cliff island will be found very narrow; and in using it care must be taken to avoid another sunken rock lying S.S.E. ¾ E. 4 cables from Cliff island.

SHEIPU HARBOUR is between the main land and Nyew-tew island, and at high water it has the appearance of a splendid basin, but at low tide the mud dries off shore a long distance, giving it the appearance of a river. At the western end of the harbour is an entrance into San-mun bay, and another to the southward leading into the bay west of Lea-ming island. The town of Sheipu stands on the main, forming the northern boundary of the harbour, and derives its importance principally as a convenient stopping place for the coasting trade; the walls are in a dilapidated state, and the houses and shops are not good.

There are three very narrow entrances, with rapid tides and chow-chow water in them, leading from Sheipu head into Sheipu harbour. Two of these entrances are formed by Tungmun island. In the centre of the middle entrance, between Tungmun and Sin island, is a rock on which H.M.S. *Sphinx* struck in 1853. It lies in the narrowest part of the channel, and the least water on it was 10 feet, with irregular soundings around it, the deepest water being towards Sin island; it appeared very small, and is probably quite smooth. This passage is not recommended for large vessels, and if used they should keep well over on the southern shore.

The northern entrance between Tungmun and the main, although tortuous and narrow, is safe; there is also less chow-chow water than in the middle entrance. The south entrance, between Sin and Nyew-tew, is long and narrow, and near its mouth is a small flat islet with a reef extending eastward from it. Vessels pass to the north-eastward of this islet; but it is said the Chinese junks never use it, and they report rocks in mid-channel.

The **COAST** from Sheipu trends in a northerly direction about 25 miles to the entrance of Nimrod sound, and is fronted by several islets none of which are large enough to afford shelter, and the depth generally is under 3 fathoms.

HALF-TIDE ROCK lies N.E. b. N. 6 miles from the east point of Montagu island, with the Bear (an islet near the main with a sharp peak at its western end) bearing N.W. ½ N., and distant 11 miles. Should high tides and smooth water prevent this rock being seen, the east point of Montagu kept westward of S.W., will lead to the eastward.

KWESHAN ISLANDS are eleven in number, besides several rocks. The largest island is 3 miles long, and deeply indented, and its greatest breadth is 1¼ miles; in two places, however, it is not more than a cable or 1¼ cables across. It rises near its western end into a sharp peak 490 feet high; its coast line is steep, high cliffs, and, with the exception of six small sandy bays, the island is steep-to on all but its western side. The other islands are much smaller. The whole group is thickly populated, the inhabitants subsisting principally on fish; they have pigs, goats, a few fowls, and sweet potatoes.

Patahecock, the south-easternmost of the group, is remarkable from its flat and table-like appearance. It lies North 31 miles from Saddle island, the south-western island of the Hieshan group, and its summit is 450 feet above the sea.

The north-eastern island of the group is a narrow cliff islet uninhabited; to the westward are four small islets inhabited and cultivated; and North of them, at the distance of 3 cables, is a flat precipitous rock, the coloured appearance of which (it being composed of red porphyry) renders it remarkable. This face of the islands is free from danger, the depth being 7 or 8 fathoms near the shore.

The north-western island of the group is the second in size and attains an elevation of 400 feet; its northern extreme is remarkable, in consequence of several isolated masses of rock. The body of the largest island bears South of the north-west island, and between the two is a mud bank gradually shoaling towards the larger island. By keeping the west extreme of the north-west island to the eastward of N.N.E., not less than 3 fathoms will be found with good holding ground, and not much swell.

South of the large Kweshan island, and separated by a channel 1½ cables wide, is another island, which is also high, with steep cliffs; off its western point is a half-tide rock, and a reef runs off from its south end. The Holderness rock lies W. ¼ N. one mile from the highest part of this latter island, and having only 6 feet water over it, occasionally breaks: from it the highest part of north-west island bears N.N.E. ¼ E.:

a small peaked islet to the south-east S.E. ⅝ E.; and Patahecock table E.S.E., the reef of rocks lying off the south end of the nearest island being in line with it. Another sunken rock with only 5 feet on it lies S. b. W. ¾ W., three quarters of a mile from the summit of the same island; when upon it the east end of the large Kweshan is in one with the east end of the nearest island, bearing N.E. ½ E., and Patahecock table E. by S. ¾ S.

Between the Kweshan group and Bear islet to the westward, the depths vary from 6 to 3½ fathoms, gradually shoaling towards the latter.

TIDES.—It is high water, full and change, in the neighbourhood of the Kweshan islands at 9h. 30m., and springs rise about 14 feet. The ebb stream out of San-mun bay will be useful in working to windward, provided the vessel heads up to the northward of N.N.W.

Between the Hieshan and the Kweshan islands the flood against a strong northerly wind causes an angry sea. At the Kweshan the change in the direction of the stream does not take place until two hours subsequent to the change of depth. From hence the flood stream comes from the southward, and its rate seldom exceeds 2 knots per hour; it will, however, sensibly assist a vessel in getting into the Chusan archipelago.

MOUSE, WHELPS, and STARBOARD JACK ROCKS.—From the north extreme of the Kweshan group, the Mouse, a small rock, nearly level with the water's edge at high water, bears N. ¼ W. 4¾ miles; the Whelps, a cluster of four small islets, W. by N. ⅔ N. 8 miles; and a low flat reef, with two rocks off its eastern end, named Starboard Jack, bears N.W. 7¼ miles.

The CORKERS are several isolated patches of rock lying between the Whelps and Buffaloes Nose, an island lying 6 miles to the N.N.W. From the outer or eastern rock, which is occasionally covered, Buffaloes Nose bears N.N.W. ¾ W.; there are two islets lying a cable's length to the westward of it, which, should the rock be covered, will point out its position. The distance between the Corkers and Starboard Jack is about 3 miles, and the channel between has a depth of 6 to 5 fathoms.

The TINKER is a steep cliff rock, 80 feet high, lying N. b. E. ¾ E. 2¾ miles from Starboard Jack. The Buffaloes channel between them has 6 and 7 fathoms in it, and will be found the most eligible to take in entering the archipelago during the N.E. monsoon, as the vessel will be well to windward; in using it, however, recollect that a sunken rock lies S.E. b. E. 2 cables from the Tinker.

MESAN and LANJETT ISLANDS.—Four large and several smaller islets or rocks lie three quarters of a mile to the northward of the Tinker. The largest islet, named Mesan, is not quite a mile in circumference, and about 400 feet high; its barren summit forming one of the most remarkable features in the Buffaloes Nose channel. There are 7 and 8 fathoms water in the channel between it and the Tinker, but sunken rocks extend a short distance from both shores.

HARBOUR ROUSE.—Between the Mesan group and Front island, (which lies 3 miles to the E.N.E. and is the southernmost of the islets extending from the southern part of Beak island), is the entrance to Harbour Rouse, which will be found a convenient stopping place in the northern monsoon, for a vessel that has missed her tide through the Beak Head channel. The entrance lies between Front island and a castellated rock 2 miles to the westward, and the depth inside varies from $5\frac{1}{2}$ to $2\frac{1}{2}$ fathoms.

BUFFALOES NOSE ISLAND, lying N.W. $\frac{3}{4}$ W. 16 miles from the north-east extreme of the Kweshan islands, is $1\frac{1}{4}$ miles long north and south, and three-quarters of a mile broad. Its eastern shore is rocky, and an islet lies off its north-west end; its western side has several deep indentations, one of which nearly separates the island into two parts. There are three peaks on the island, the central one of which, 500 feet above the sea, is the highest. Near its northern end the island is perforated, from whence its native name (Niupi-shan) is supposed to be derived.

ANCHORAGE.—The anchorage between Buffaloes Nose and the Ploughman group is secure; during the North-east monsoon, however, the wind blows directly through, and occasional violent squalls are experienced.

Supplies.—Fresh provisions and water may be obtained at the above anchorage, but the supply of the latter cannot be depended upon.

PLOUGHMAN GROUP is composed of three islets, and the largest lies W.N.W. nearly a mile from Buffaloes Nose, the depths between varying from 5 to 18 fathoms. The largest is an even flat-topped islet with a reef extending from its north-east point; there is also a detached reef at 6 cables N.W. b. N. from the same point. The other two islets are narrow and small, and lie to the north-west of the larger one.

Junks usually pass inside the Ploughman and Buffaloes Nose, and to the westward of the Corkers; there are, however, many reefs, and the tides are strong, and vessels will do better to keep to the eastward of Buffaloes Nose. As before noticed, page 155, the channel between the Tinker and Starboard Jack is the best to take during the North-east monsoon, and a vessel will have better anchorage under Luhwang than under Buffaloes Nose.

BUFFALOES NOSE ISLAND.—NIMROD SOUND.

NIMROD SOUND.—The entrance to this sound lies 5 miles to the W.N.W. of Buffaloes Nose, and is fronted by the south-west islands of the Chusan archipelago. The sound is a deep inlet running 27 miles in a W.S.W. direction from the entrance, which is between the Hunter islands, (six in number), lying near the south point, and a small island named Bateman lying 4¾ miles to the northward.*

From about 2 miles to the southward of the latter island the course up the sound is W.S.W. southerly to abreast Castle rock, which is on the edge of the mud on the northern shore, N.W. ½ N. 3 miles from the Hunter islands, and should be given a berth of about half a mile. From Castle rock the edge of the mud bank, which is dry in most parts at low water and extends 2¼ miles from the north shore, trends 5 miles towards a small low islet named Barren island lying close to the shore.

Between Barren island and Nimrod point, on the south shore, the sound is 2½ miles wide. Between Nimrod point and the Hunter islands is First Cone point, with an islet off it ; and to the westward of this latter point is Cone rock and David island, with a half-tide rock lying a cable's length to the north-west of the latter. Nimrod point is high, and has several sunken rocks lying 3 cables off it. Four miles within Nimrod point is an islet, which, from its central position, is called Middle island ; and to the southward of this islet is the entrance to Medusa creek, which carries a depth of 4 to 6 fathoms.

Above Medusa creek the sound, between the southern shore and Parker island, contracts to three-quarters of a mile, and the water is deep, and the tides strong ; off the east end of this island are some dangerous rocks which are steep-to and only show at half-tide. S.W., 1½ miles from Parker Island, is the entrance of a small river for boats, leading up to a village 3 or 4 miles inland, having about 6 feet in it at low tide.

At 7 miles above Medusa creek the sound is separated into two branches by the Treble islands. Pass to the northward of these islands, keeping in mid-channel to avoid a half-tide rock on the northern shore.

To the north-west of the Treble islands, on the northern shore, is the village of Tung-ju, from whence there is a paved footpath communicating with the Fungwha branch of the Ningpo river, the distance from hence to Ningpo being 20 miles in a direct line. On the south side of the sound, at 3 miles to the south-west of the Treble islands, is also a paved footpath leading to San-mun bay. Having passed the Treble islands good anchorage will be found in 6 or 7 fathoms, mud, off the village of Tung-ju.

TIDES.—It is high water, full and change, in Nimrod sound at 10h. 30m. ; springs rise about 20 feet.

---

* See Plan of Nimrod Sound, No. 1,583 ; scale, $m = 0{\cdot}7$ of an inch.

CHAPTER VI.

EAST COAST OF CHINA.—NIMROD SOUND TO THE YANG-TSE KIANG, INCLUDING THE CHUSAN ARCHIPELAGO.

VARIATION 1° 40′ West, in 1861.

CHUSAN ARCHIPELAGO.—This large assemblage of islands, of which Chusan is the principal, lies near the mainland between the parallels of 29° 39′ and 30° 50′ N. The archipelago may be entered from the southward by the Buffaloes Nose, the Beak Head, the Vernon, and the Sarah Galley channels, among which the two former channels may be considered the best to enter by, and the Vernon to go to sea. The channel to the northward of Chusan between the chain of islands extending W.N.W. from Fishermans group and Chin-san island, is generally taken during the North-east monsoon by vessels bound to Ning-po fu and Chusan, and it appears clear of danger with the exception of the Mariner reef at its western entrance.*

LUHWANG, the largest of the islands in the south-west part of the archipelago, is 9½ miles long N.W. and S.E., and 6 miles wide at its broadest part, which is the western end ; near the centre it is not more than 2 miles across, and not much elevated above the sea. The south-eastern body of the island rises to the height of 865 feet, being a conical bare hill ; on the isthmus is an isolated peak 718 feet high, and on the north-west side of the island are five high peaks, one of which is 910 feet above the mean level of the sea. The western part of the island, forming the eastern side of Duffield pass, has several small bays with stone embankments extending from point to point. Cape Luhwang, the north extreme of the island, is high and bold. The island is well cultivated and maintains a large population.

The southern face of Luhwang has two deep indentations with sandy bays, and a reef extends 3 cables from the point abreast the Mesan and Lanjett group, described in page 155. Reefs also extend half a mile from the northern extreme of the latter group, narrowing the channel between them and Luhwang to less than a mile. The coast line of

---

* *See* Charts :—East Coast of China, Sheet 8. No. 1,199 ; scale, $d = 15$ inches : and Chusan Archipelago. South Sheet, No. 1,429 ; scale, $m = 0.8$ of an inch.

Luhwang immediately westward of the reef point trends to the northward, forming a deep bay with three islets in it, extending to Duffield pass. South one mile from the easternmost islet there is a mud bank, having $3\frac{1}{4}$ fathoms on it, to avoid which a vessel may keep the islet aboard, giving a berth to a rock lying half a cable from its south extreme. Between this island and Duffield reef, which lies off the eastern side of the entrance to Duffield pass and consists of three rocks above water with a sunken rock between them and Luhwang, there are from 9 to 5 fathoms, good holding ground.

FU-TO ISLAND, to the westward of Luhwang, is about $2\frac{1}{2}$ miles long, north and south, and a mile broad, and its southern extreme, forming a narrow point, is connected at low water to St. Andrew island. A spit runs off the north extreme of Fu-to, to the north-east of which are three islets, with a rock lying a cable's length to the north-west of the northernmost, named Chloe island.

Tree-a-top island lies $3\frac{1}{2}$ cables to the southward of the south extreme of Fu-to, with a deep water channel between. This island, 180 feet high and about 4 cables in circumference, has a pile of stones on its summit, but no tree; the old name, however, given it in the chart by Thornton in 1703 is still adhered to.

DUFFIELD PASS, between Luhwang and Fu-to, is $1\frac{1}{4}$ miles wide at the southern entrance (where the water suddenly deepens from $5\frac{1}{2}$ to 40 fathoms), and half a mile in the narrowest part, which is near the centre. On the Fu-to island shore are several islets; among them the water shoals to $4\frac{1}{2}$ and 5 fathoms, and a vessel may anchor and stop a tide if necessary. Off the fourth point on the Luhwang side is a reef extending a cable from the shore; otherwise this side of Luhwang is very steep-to, the depth being 35 fathoms within a cable of the mud. Two small islets, named the Notches, lie in the centre of the pass, abreast this reef, and between them and Fu-to is a half-tide rock; unless this rock shows, vessels should not tack inside the Notches so as to pass westward of them.

At the north end of the pass there is a rock with only 16 feet over it at low water, lying 2 cables eastward of Hebe island; when on it, the north extremes of Hebe and Chloe islets are in one bearing N.W. b. W. and the east extreme of Fu-to is in one with the west extreme of Tree-a-top island, seen over the mud connecting St. Andrew with Fu-to. On the Luhwang side, to the north-east of Hebe island, and a cable from the shore, is the Bird rock, which formerly had a stone pillar on it, but it was either thrown down or removed in 1846. Two islets lie 2 cables to the southward of Bird rock. Beyond the rock the coast-line of Luhwang turns suddenly to the north-east to Cape Luhwang.

GOUGH PASS, formed between Fu-to island and the central isles, is 1½ miles long and half a mile wide, and is far preferable either to Duffield or Roberts pass, for both shores are steep-to, and the lead, if hove quickly, will give warning of approach to the shoal which extends half a mile to to the S.S.W. from the southern islet of the Central islands.

The south-western of the Central islands is a small islet connected at low water with the largest of the group by a reef and spit. At half a cable's length to the northward of the northern island is a reef.

ROBERTS PASS is to the westward of the Central islands, between them and the mud which dries one mile from the embankment on Mei-shan island. This channel is 2 miles long, N.E. and S.W., and 4 cables wide, but as the lead will give no warning, its boundary on the Mei-shan side will not be known except at low water; the depths in it vary from 6 to 40 fathoms. Mei-shan island appears formerly to have been eight islands, now, however, united by substantial stone walls, one of which, on its northern face, is 1½ miles in extent. The mud dries 1½ miles from its south, and a quarter of a mile from its north end; on its east side the bank is steep-to.

On the north-eastern side of Mei-shan are the two Damson islets, from the northernmost of which, named Cliff islet, the 3 fathoms line extends nearly a mile to the northward. By keeping the Central islands open of the Damson islets until the vessel is three-quarters of a mile past the Cliff islet, this shoal will be avoided, and the Ketau shore can be approached. The course for Ketau point, after clearing this pass and Gough pass, will be N.E. 9½ miles.

JUNK CHANNEL, between Mei-shan island and the Ketau shore, is 2¼ cables wide, and carries a depth of 5 and 6 fathoms except at the southern entrance, where it shoals considerably, and not more than 10 feet water was obtained; some parts, however, may be deeper, as only one line of soundings was taken across the bar. On the mainland, near the centre of this channel, is a custom-house, and the entrance to a canal which communicates with the two populous villages. Two miles to the northward of Mei-shan is the walled town of Kwokeu, where the mate of the *Lyra*, merchant ship, was kidnapped, and attempts made to interrupt the surveying operations in 1840.

ANCHORAGE will be found anywhere along the Ketau shore, between Mei-shan and Ketau point, until abreast of Sing-lo-san island, where the water deepens.

CAUTION.—As there is no anchorage besides the above, but in very deep water, until that (page 170) under Elephant island is reached, it

CHAP. VI.] CHUSAN ARCHIPELAGO; BEAK HEAD CHANNEL.   161

would not be prudent for sailing vessels to proceed farther unless the wind and tide will ensure their gaining that position.

TIDES.—In the above Passes, at full and change, the first of the flood often comes from the northward, and runs sometimes for 3 hours before it takes the direction of the ocean tide.

BEAK HEAD CHANNEL (Taou-sau-mun of the Chinese) is the next passage north-east of Buffaloes Nose channel, and considered one of the best to enter the archipelago by from the southward. The entrance is between Beak head, the east extreme of Beak island, and Vernon point, the east end of Vernon island, which bear N.N.E. ¼ E. and S.S.W. ¼ W. from each other, distant 2¾ miles. Beak Head island is nearly 5 miles long, in some parts very narrow, and remarkable for two hummocks near its west end. Off Beak head are three islets; and to the south-westward of the head are several islets and a rock, which together with Luhwang island form Harbour Rouse (page 156), which will be found a convenient stopping place for a vessel that has missed her tide through the Beak head channel. The channel between Luhwang and Beak island has 3½ fathoms water; but there would be no object in using it while there are passages so superior.

Off the north-east face of Beak island are two reefs, lying 3 cables' and half a cable's length respectively from the shore. Off the north end of the island are Gull, Shag, and Puffin islands, with a reef of rocks above water between the two former; a reef also extends 3 cables from the north-west end of Puffin island. Near the west end of Beak island the channel narrows to half a mile between the reef of rocks, the northernmost of which is always above water, and two small islets lying off the south side of Conical Hill island. This island is midway between Beak and Vernon islands, and between it and the latter are two islets, the reefs off which render the channel between Conical Hill and Vernon islands more intricate.

DIRECTIONS.—A N.W. by W. ¼ W. course for 8½ miles from the eastern entrance of Beak Head channel will lead to the southward of Conical Hill and Conway islands, and from thence a N.W. course will clear the channel; care must, however, be taken in light winds to give the Pai rock, the last islet on the north side of the channel, a wide berth, as the flood sets directly towards it. Good anchorage in 9 and 10 fathoms will be found on the north-west side of Conway.

To the northward of Conway island is a group of islets and rocks, through which there is a passage into the Vernon channel; but owing to the rapidity of the tides, it should not be attempted without local experience. On the Luhwang side of Beak Head channel is a reef, and an

[C.]   L islet with a small pinnacle on it; the reef, which is generally uncovered, bears S.E. ¾ S. 2 miles from Cape Luhwang, and by keeping the cape to the westward of N.W. ½ N. it will be avoided. The mud dries 7 cables from the Luhwang shore, in the bight to the southward of this reef. Landing is difficult on this side of Luhwang, except at high water.

VERNON CHANNEL or South-east passage (Hea-che-mun of the Chinese), to the northward of Boak Head channel, is formed by Vernon island on the south and Taou-hwa island on the north. This will be found a convenient passage from Chusan during the northern monsoon, the distance from Elephant island to the open sea being only 17 miles; it should not, however, be attempted with light winds, as vessels are liable to be becalmed and experience flaws under the high lands of Taou-hwa, and in some parts the depths are 60 fathoms, and the tides strong.

Vernon island is 5¾ miles long in a W.N.W. and E.N.E. direction, and on its north-east side is a wide bay, with two islets and a reef in it, where vessels may anchor in 4 and 5 fathoms, and procure water from Taou-hwa island; there are several cascades, and the water may be obtained without removing the casks from the boats. The east end of Vernon island is rugged, with large boulders of granite; at this end there is a cove, which runs back three-quarters of a mile to the westward and affords shelter for boats.

The eastern entrance to the Vernon channel is 1½ miles wide, but 5 miles within it is divided into two passages by John Peak island, which has a rock lying half a cable's length from its north-east extreme and uncovers at the last quarter ebb. The passage on the north-east side of John Peak is only 3½ cables' wide between this rock and two small islets and some rocks which bound its north side. The passage between John Peak and Vernon islands is half a mile wide, and good anchorage will be found on the south side of the former. The Taou-hwa shore is bold and precipitous, and the peak of the island rises to the height of 1,680 feet above the sea. Near its western end the land becomes low, rising, however, again, and surmounted by a peculiar perpendicular crag, called Millers Thumb, 606 feet high, which will be recognized nearly throughout this part of the archipelago.

SARAH GALLEY CHANNEL, the next passage to the northward of the Vernon, is by no means so eligible as those just described. Near the entrance, at 4 miles N.E. b. E. ¾ E. from the south point of Taou-hwa island, will be seen the Jansen or Laoush rock, a steep cliff islet, with rocks extending 1½ cables from its south end; there is also a half-tide rock lying W. by N. ¾ N., 1¼ miles from the north extreme of the Laoush, with the highest part of Ousha island bearing N.N.E. ¼ E. 1¾ miles.

CHAP. VI.] CHUSAN ARCHIPELAGO; SARAH GALLEY CHANNEL. 163

The coast line of Ousha island is steep cliffs, and off its north-west end is a ledge of rocks; the southern end of the island is the highest, and rises in a round peak. The channel between the north-east point of Taou-hwa island and Peak island is not navigable, owing to reefs and strong tides; neither is there a fit passage between Peak island and Tang-fau. Vessels may pass between Peak island and the two patches of rock lying westward of Ousha; but there are some rocks off the north end of Peak which must be avoided.

The channel, named Cambrian pass, between Ousha and the large island of Chukea, or Chus Peak, is 2 cables wide, but, from the violence of the tides, it should not be used without a commanding breeze.

DIRECTIONS.—Vessels entering the Sarah Galley channel from the southward generally pass westward of Laoush rock and Ousha island, and from thence the channel is between the latter island and the two patches of rock to the westward, which are almost covered at high water; they lie N.N.E. and S.S.W. of one another, 2 cables apart, and the distance between them and Ousha is half a mile. After passing these rocks the course is North $2\frac{1}{4}$ miles, leaving two small islets, named Teen and Yung and a reef between them, to the westward; and Hut island (so called from a house on its summit) with a reef of rocks off its south extreme, to the eastward. The channel here is three quarters of a mile wide.

From thence steer N.W. b. N. for $1\frac{3}{4}$ miles, leaving an island with two hummocks to the southward, and Druid island to the northward; but be careful after passing Hut island, that Flat or Liwan island (at the west entrance to the channel), is not brought westward of W. by N. $\frac{1}{4}$ N., as the water shoals suddenly on the north side of the channel, and the mud dries nearly all the way from Druid island to South Chukea island, leaving a small boat channel.

When in the vicinity of Liwan island the east end of Chusan will be seen, having on it a small temple composed of large stone slabs. Between Liwan and Chusan is Lokea island, the southern shore of which is not steep-to; and this is the case with the whole of the islets on the south side of Chusan, between this and Pih-lou, after which they become steep-to. After passing the smaller islets south of Ta-kan the shoal water will be avoided, when standing northward, by not bringing the rocks off the southern part of Pih-lou on with Trunk point on Elephant island. Liwan has two rocks off its south end; the anchorage in its vicinity is noticed in page 177.

CHUKEA ISLAND is about 7 miles long, north and south, and on its western side are many deep indentations, some of which are enclosed

L 2

from the sea by stone walls. Near its south extreme are four remarkable peaks, and near its centre is a smooth-topped cone, 1,164 feet high, named Chukea peak, which is one of the most prominent objects in making this part of the archipelago.

There are several indentations on the eastern side of this island, and the southern one, Wolf bay, affords anchorage in the North-east monsoon, and was resorted to in 1842 by H.M. ships from Chusan, for water. On the north side of the bay is a black islet, with rocks extending southerly and easterly from it. Fronting the bay, and $1\frac{1}{2}$ miles from the shore, is a peaked rock, named Pillar, off which, at 2 cables to the N.E., are two reefs, showing at half tide. In the small inlet north of Wolf bay is a reef visible at low water, and it will be avoided by tacking outside the headlands; Nob rock lies 3 cables from the north point of the bay, and is always above water.

TONGTING and PIHTING ISLANDS, and PELICAN ROCK.—To the eastward of Chukea, at the distance of 5 and 8 miles, are two islets named Pihting and Tongting. Tongting, the outer one, about 40 feet high, has detached reefs to the south-west of it. Pihting is a similar islet.

The Pelican rock lies $2\frac{1}{4}$ miles from the Chukea shore, and only shows at low water springs; but the disturbed water over it, when covered, will generally indicate its position. From the rock Yangsi islet, off the north-east end of Chukea, is in line with the summit of Putu island, N.N.W. $\frac{1}{2}$ W.; Pihting islet bears E. $\frac{1}{4}$ N.; and Chukea peak N.W. $\frac{2}{3}$ W.

PUTU ISLAND lies $1\frac{1}{2}$ miles from Whang head, the east extreme of Chusan, and the channel between is called by the Chinese Leenhwa-yang, or the sea of water lilies. The island is $3\frac{1}{2}$ miles long, north and south, and in one place only half a mile across. The temples on it are numerous, but the two largest, which are on its eastern side, are falling into decay. A narrow projecting point extends from the eastern side of the island, forming to the southward a deep sandy bay, in which there are 3 fathoms water; the islet off the point has a sunken rock lying on its eastern side. The western face of Putu is shoal, the $2\frac{1}{2}$ fathoms line of soundings being 3 cables from the shore. A smaller islet with rocks to the northward of it lies off the north end of Putu; vessels may pass between the rocks and the islet.

Water.—A stream runs into the above bay, on the eastern side of Putu, and it might be used should the well at the south side of the island prove dry. This stream runs in a small sandy bay to the westward of a hill with three chimneys on it, and may be known by a small joss house. The landing place of the pilgrims is at a causeway east of the well bay.

ANCHORAGE.—There is anchorage off the eastern side of Putu in 12 and 14 fathoms water, but several vessels have had a difficulty in purchasing their anchors; it is also much exposed, and by no means desirable in bad weather.

ISTHMUS ISLAND is three-quarters of a mile from the north-east point of Putu, and the channel between has deep water. A half tide rock lies 4 cables from the south-east point of the Isthmus, with the east and south-east extremes of Putu in one bearing S.W. ⅝ S., and the south summit of Isthmus W. ½ N.

To the eastward of the south point of Putu, and off the north-east end of Chukea, are four islands named Loka, Pih-sha, Lakeah, and Lakeati. There is a passage between them and Chukea, and a good channel between them and Putu.

N.E. ISLAND and NINEPIN ROCKS.—N.E. island is a conical rock, in form something like a haycock, lying N.E. ¼ E. 2 miles from the north-east end of Isthmus island. The Ninepin are four pinnacle rocks with reefs around them, lying a mile to the south-eastward of N.E. island, and N.E. by E. ½ E. 5 miles from the summit of Putu island.

EAST ISLET and EAST ROCK lie respectively 6 and 7½ miles to the eastward of Putu. The islet is 30 feet above the sea, and from it, Loka, the northernmost of the islands on the north-east face of Chukea, bears W. by S. East rock, which is nearly awash at low water, lies E. by S. 2 miles from East islet, with Tongting islet bearing S. by E. ¼ E. 7 miles, and the summit of Putu (which will be known by a look-out house on its summit, and the high land of Chusan forming a table top at the back of it) W. by N. ¼ N. This rock forms the southern horn of Lansew bay; Video island, Fishermans group, and the chain of islands to the westward, bounding the bay to the northward.

A description will now be given of Chusan island, with directions for approaching Ting-hai; after which the north part of the Chusan archipelago.

CHUSAN ISLAND, so called from its supposed resemblance to a boat, is 51¼ miles in circumference; its extreme length in a N.W. and S.E. direction being 20¾ miles, and its greatest breadth 10½ miles. From the beach at Ting-hai on the south side of the island to the northern shore, the distance across is 7 miles; towards the east end it becomes narrower. The island is beautifully diversified with hill and dale and well cultivated. Of the numerous small streams which run from the mountains, the most considerable is the Tung kiang which falls into Ting-hai harbour. The products are rice, millet, wheat, sweet potatoes, and yams; the tea plant is found everywhere, but is treated with little or no care. The cotton plant is largely cultivated near the sea. Besides the harbour of Ting-hai there are three other commercial ports, viz., Chinkeamun at the south-east end of the island, Ching Keang on the north-west side, and Shaaon at the north end.

The town of Ting-hai is 1¾ miles in circumference and is surrounded by a wall 14¾ feet high and 13 feet wide, surmounted by a parapet 14½ feet high and 2 feet wide. The southern face runs east and west; the west face north and south; and the eastern face north 350 yards, and then north-west. A canal, 33 feet wide and 3 feet deep, nearly encircles the city, and enters it near the south gate, which is about half a mile from the shore of the harbour. Canals form the principal means of transportation, the roads being merely footpaths on the stone embankments which prevent the encroachment of the sea on the rice fields. Every large field has its canal for the purpose of carrying away the produce.

The population of the town and suburbs at the commencement of 1843 was about 27,500, but in 1846 it had increased to 35,000; the population of the entire island was estimated at 200,000. The principal exports are, fish, coarse black tea, cotton, vegetable tallow, sweet potatoes, and some wheat.

Water.—The water is not good at Ting-hai, and is sometimes scarce, the tanks in the rice fields near the sea being the only supply, excepting wells, which afford but a limited quantity; no running streams were found. The place latterly adopted for watering by the squadron during the China expedition in 1840–1843, was in the bay westward of Chuh or Guardhouse isle.

WINDS and WEATHER.—The following is a meteorological abstract deduced from monthly registers, kept at Chusan during the period the island was occupied by the English troops in 1840:—

The climate of Chusan is subject to a range of temperature similar to that in the same latitude upon the coast of North America; the thermometer in the shade standing at 103° in September, and at 25° in February.

September was generally fine, only four rainy days for short periods; 1·8 in. of rain fell. The barometer generally standing below 30 inches; falling in strong south-easterly, and rising with northerly winds; height of the cistern above the sea 72 ft. 7 in. Very strong breezes were not experienced during this month. Winds easterly 10 days, south-easterly 6 days, north-easterly 8 days, and north-westerly 6 days. Range of thermometer 103° to 65°.

The first 10 days of October were fine, the remainder of the month overcast; weather squally, much rain during the last week. Except the four first days of this month, the barometer was never below 30 inches, and rose as high as 30·335 in., rising with fresh winds from the north-west. The winds variable, changing frequently during the 24 hours; they were from the North 6 days, N.E. 12 days, N.W. 9 days, and 4 days from S.E. to S.W. On the 29th the meteorological instruments were removed to the suburbs, where the height of the cistern of the barometer above mean tide level was 24 feet. Range of thermometer 92° to 51°.

November was generally overcast with rain, the barometer in easterly winds fell below 30 inches. Winds were N.E. 2 days, N.N.W. 8 days, N.W. 4 days, northerly 4 days, westerly 4 days, S.S.W. 2 days, and calm 4 days. Range of thermometer 74° to 40°.

In December the weather was finer than last month; the barometer kept very high, being 30·588 inches on the 10th; winds light from the N.W.; the mercury generally rose as the wind freshened from that quarter, and during calms fell to 30·02 inches. Winds south-westerly half a day, westerly 2½ days, north-westerly 15 days, north-easterly half a day, northerly 5½ days, easterly 1 day, and calm 6 days; much rain during the last week. Range of thermometer 77° to 27°.

During January the weather was misty with much rain; barometer ranging from 30·606 to 30·084 inches, falling previously to south-easterly winds. Snow the last two days. Winds fresh with squalls; from the N.W. 20¼ days, West 2 days, S.W. 1 day, S.E. 1 day, North 2¾ days, and calm 2¾ days. Range of thermometer 60° to 28°.

February was generally fine; winds N.W. 5½ days, North 2½ days, S.W. 1 day, S.E. 2¼ days, calm 5 days. Range of thermometer 60° to 25°.

The greatest range of temperature during 24 hours was 28°. During January, the barometer was at the height of 30·606 inches, and generally fell in light or easterly winds. A few days south-easterly winds occurred in September, but the northerly monsoon could not be said to have commenced until the beginning of October. The following are the number of rainy days in each month: September 4 days, October 3 days, November 12 days, December 7 days, January 11 days, February 3 days.

TING-HAI HARBOUR, formed on the south side of Chusan, is fronted by many small islands, between which are the several channels leading to it. The outermost and westernmost island is Ta-maou or Tower-hill; east of which and distant 1 and 4¼ miles respectively are Teijo or Elephant island, and Pih-lou. Within, or to the northward of these, reckoning from the westward, are the islands called Ha-tse or Bell, Pwanche or Tea, Seaou-keu or Deer, and Ao-shan. The two small islands Tawu or Trumball, and Wae-wu or Macclesfield, lie inshore or to the north-east of Tea island, and there are many small islands and rocks among those larger ones just named.

The harbour is difficult of access in all its approaches, owing to the strong tides and sunken rocks. The best approach is through Tower-hill and Bell channels, between Tower-hill and Bell islands, and between the latter island and Tea island, in which no hidden danger has been found; the tides, however, are strong, and sailing vessels in light winds must be careful that they are not set by their influence between Tea and Elephant islands, where the ground is foul and the narrow channels deep.*

DIRECTIONS through TOWER HILL CHANNEL.—The best approach to Ting-hai harbour for large or unhandy vessels is through Tower Hill channel. Unless favoured by a commanding breeze and neap tides, they ought not to take the channel between Roundabout island and Ketau point, as the tides run there with great strength. After passing eastward of Roundabout steer to pass a convenient distance from the south extreme of Tower Hill island. Should the tide fail, anchorage will be found under the islands to the eastward of Tygosan island; for which purpose pass 3 cables to the southward of Square Stone islet, to avoid the reef lying $1\frac{1}{2}$ cables to the S.W. of it, and anchor before the channel between Little Tygosan and Chuen-pi islands opens, as the water shoals suddenly off the east end of Entrance island, the islet to the south-westward.

Having rounded Tower Hill island, haul up, steering first for Bell island then for Tea island. The soundings in Bell channel, between Bell and Tower Hill islands, vary from 30 to 40 fathoms, except off the north-west end of the latter, where there is a mud bank with 3 fathoms over it, extending $1\frac{1}{2}$ cables from the shore.

Good anchorage will be found in 10 and 12 fathoms water between Bell and Tea islands, but vessels intending to remain here should not open the channel between Bell island and Chusan, as the tides are stronger and the ground loose. On proceeding from hence to Ting-hai harbour, take care to avoid the Nab, a sunken rock with 14 feet over it at low water, lying $2\frac{1}{4}$ cables from the Chusan shore, and South of a small hillock in the valley near the shore; the marks for it are Taching point, the west extreme of Tea island, in one with the east side of Taewang or Bell rock, S. $\frac{1}{4}$ W., and the south point of Guardhouse isle nearly in line with the summit of Trumball island. A 3 fathoms patch also lies about $3\frac{1}{2}$ cables W.S.W. of the Nab, and E. by N. $\frac{1}{4}$ N. nearly 4 cables from Ap-tan-shan island.

The anchorage, named Spithead, on the Chusan shore, between the Nab rock and Guardhouse isle, will be found a convenient place for

---

* See Chart:—Chusan Archipelago, North Sheet, No. 1,969; scale, $m = 0·8$ of an inch; and Ting-hai Harbour. No. 1,395; scale, $m = 4$ inches.

watering; the anchoring ground is steep-to, and the tides are irregular, and off the entrance to the watering creek is a mud flat, having 3 fathoms on it at low water. With light winds, vessels should avoid the strength of the ebb when passing through the channel between Tea and Guard-house islands, which otherwise is liable to set them through the southern or Melville channel. A ledge of rocks covered at high water, extends a cable's length from the high water mark at Kouching point, the north extreme of Tea island.

In proceeding towards Ting-hai harbour, and being abreast of Guard-house isle, steer towards Macclesfield island, taking care to avoid the Middle Ground, which has only 2 feet on its shoalest part. Tower Hill in line with the slope on the southern rise of Tea island will lead along the southern edge of this shoal, in 4 fathoms. The Wae-wu channel is only 2¾ cables wide between the 3 fathoms line on the edge of the Middle Ground and Wae-wu island. The usual anchorage is abreast of Taotau, the suburb of Ting-hai, but vessels must moor as the eddies are strong. The channel between Chusan and Guardhouse isle is only fit for boats.

CAUTION.—Spring tides set at the rate of 3 and 3½ knots per hour in the Tower Hill channel, and with light winds and a strong flood vessels have been swept away to the westward, and carried by the tide beyond Just-in-the-Way, and even through the Blackwall channel; and after rounding Tower Hill and entering the Bell channel many have been borne by the ebb amongst the islands between Tower Hill and Elephant island, or between the latter and Tea Island, where the channels are narrow, the water deep, and the ground foul. In these cases the bower anchors and chains should not be used, but a good kedge and stout hawser, which (as the holding ground is good and if care be taken to con the vessel and not break her sheer) will bring a vessel up and prevent her being driven into these narrow passages, where some have been brought up in from 30 to 40 fathoms water, with two anchors down and three or four round turns in their hawse.

Having rounded the north end of Tea island with a strong ebb, it is necessary to guard against its taking the vessel through the Melville channel, and if not able to pass to the northward of Macclesfield island, send the boats a-head and endeavour to keep the vessel to the northward of Takeu and Sarah islands, where the water is not so deep.

Through MELVILLE CHANNEL.—The Melville or southern passage to Ting-hai harbour is between Elephant and Deer islands, but as two sunken rocks lie in the centre of the channel and narrow it to 1¾ cables, it should not be attempted unless there is a commanding breze, and the mariner has a thorough knowledge of their position. Its navigation is rendered more difficult in the neighbourhood of these dangers by the tidal streams, which, rushing through four different channels into this, form eddies which render a vessel unmanageable even with a good breeze at the springs. In taking it at the neaps, a boat will be found useful a-head.

The entrance to this channel will be easily recognized by Elephant island, which is remarkable from a curious crag near the summit; and by a cone-topped island, named Pating, to the N.N.E. of it. There is anchorage in the southern part of the channel between Elephant and Tung islet, in 16 and 18 fathoms; but the holding ground is not good. Beyond Round island, which lies 4 cables from the north-east point of Elephant, the water deepens to 28 and 40 fathoms to the southern rock, on which H.M.S. *Melville* struck in 1840.

The Melville rock, which has only 10 feet water over it, lies S.E. b. E. ¾ E. 2 cables from the Black rock, and E. by N. ⅓ N. 1¾ cables from the rocky ledge extending towards it from Ledge island, and which covers at half tide; the marks for it are the Cap rock in line with the saddle of Kintang island, bearing W. by N., and the Joss house on the hill near the suburbs of Ting-hai showing between Trumball and Sarah islands, N. ½ E. The northern or Dundas rock, is a small patch, about 30 feet by 20 feet in extent, lying N. ¼ W., 1¾ cables from Melville rock, and N.E. by E. ⅔ E. 2 cables from the Black rock, and the least water on it is 9 feet; the marks for it are, a bushy tree on the eastern slope of Ta-keu island, in line with the middle beacon on Tsingluy Tau or Beacon hill, N.E. ¼ E., and the north end of the Black rock on with the south side of Cap rock W.S.W.

From Roundabout island a N.W. ½ N. course for 4½ miles will lead to the entrance of the Melville channel. Pass on either side of Round island, and when to the northward of it keep its east extreme touching Trunk point, bearing S. ½ W., and this mark will lead between the Melville rock and Ledge island, and between the Dundas rock and Black rock, rather westward of mid-channel. When clear of the Dundas keep in mid-channel, and when abreast the south end of Sarah island steer for the west end of Macclesfield island, which should be rounded rather close—to avoid the Middle Ground, the southern edge of which, in 3 fathoms, is only 2¾ cables distant. A rock, covered at high water, lies barely a cable's length from the northern face of Macclesfield.

Through DEER ISLAND CHANNEL.—Ting-hai harbour may also be entered from the eastward by passing between Deer and Takeu islands, which are 1½ cables apart. The Melville and Dundas rocks will be avoided by keeping Deer island aboard, but it must be borne in mind that neither shore of the channel is steep-to. The Beacon rock, awash at high water, to the north-east of Takeu, may be passed on either side; and from thence steer for the Chusan shore, keeping a **cable's** length to the eastward of Grave island until the harbour beacon opens north of it, when it can be steered for, passing between it and the Chusan shore, keeping the latter aboard, until Takeu island is shut in by Trumball island.

This passage, although narrower is superior to the Melville channel, as vessels have the tide in their favour all the way. The principal objection to its use is the liability to flaws of wind under Deer island; but the main point to be guarded against is the flood from the eastern channels carrying them so far westward as not to fetch far enough to the eastward of Grave island. There is convenient anchorage between Trumball and Takeu islands, in 8 and 10 fathoms. A spit extends from the south-east end of Trumball, the 3 fathoms line being 3 cables distant from the shore; the south end of Macclesfield island, open of the summit of Tea island, will lead south of it.

The **CHANNEL between BELL ISLAND and CHUSAN** is not recommended, owing to the tides, which attain the rate of 5 knots at springs. Nearly mid-channel is a half tide rock, named Kwa-fau, with a stone beacon on it; and to the S.W. of the beacon is a 9 feet patch lying with the south end of Kwo-kan, the westernmost of the two islets on the Chusan shore, in line with the south end of Kiddisol island. Neither is the north end of Bell island steep-to, consequently, should a vessel use this passage, the channel between the beacon and the Chusan shore should be preferred to that between the beacon and Bell island.

KIDDISOL ISLAND, which has a patch of $2\frac{3}{4}$ fathoms off its south-west end, lies 2 cables to the southward of Yanglo point, the south-west extreme of Chusan, with a deep water channel between, but the eddies are violent at the springs. From hence to Sinkong point, 4 miles to the N.W. by N., the coast line of Chusan is mud, with the exception of a small hillock at the edge of low water.*

ANCHORAGE in 10 and 12 fathoms will be found all along the Chusan shore between Yanglo and Sinkong points, but in standing towards the shore bear in mind that the water shoals suddenly after 10 fathoms.

CHING KEANG HARBOUR, on the western side of Chusan and distant 7 miles in a direct line from Ting-hai, is formed by the islands Wa-teo, Lin, and Latea, (that is to say Outer, Middle, and Inner Hook) and Chusan. Upon the islands, and on the point near the entrance, are extensive stone quarries. A white rock lies off the south-west point of Wa-teo, and a mud spit extends from the island nearly to the rock.

---

* *See* Chart of Kintang Channel, No. 1,770; scale, $m = 1\cdot2$ inches.

Between Wa-teo and Chusan the entrance to the channel is 6 cables wide with 7 and 8 fathoms water in it, forming a snug anchorage much frequented by the junks as a stopping place, and defended from pirates by a fort. Abreast of Lin the channel is less than a cable wide, with 7 fathoms water. The town stands on the banks of a stream on the Chusan shore, which at high tide is navigable for boats. Here the channel is also less than a cable wide, and the depth from 5 to 4 fathoms.

STEWARD ROCK, 50 feet above the sea, lies in the middle of Blackwall channel, between Chusan and Kintang island. The depths are 45 fathoms in its vicinity, except at 2 cables to the eastward, where there is a rocky patch on which the least water that has been found is 9 fathoms.

KINTANG or SILVER ISLAND lies between the west end of Chusan and the entrance of the Yung or Ning-po river. Near its south-east extreme is a remarkable saddle hill 1,432 feet high, which with the Cap rock forms one of the marks for the Melville rock (p. 170); there is also another remarkable peak, 1,520 feet high, rising at 1¼ miles to the northward of the saddle hill.

Alligator point, the south end of Kintang, has a reef, which covers at half tide, extending 2 cables to the southward; and off Algerine point, the south-east extreme of the island, is an islet connected at low tide by a mud flat, from which a ledge of rocks extends S.S.E. 2 cables, the south end covering at high water. The eastern face of the island is bold-to, without any anchorage along it.

TA-OUTSE HARBOUR.—Off the north end of Kintang there is a group of seven islets, amongst which there is anchorage; and off the northwest end is Taping island, to the southward of which is the small harbour of Ta-outse, which affords good anchorage in 7 to 10 fathoms. The entrance is between Kintang and Ta-outse island, and the channel is barely 2 cables wide. Between Ta-outse and Taping there are not more than 8 feet at low water.

* "Ta-outse harbour is small, but affords good anchorage, and may be recommended as a sanitary station for vessels obliged to make a lengthened stay in the Yung river. Supplies of all kinds can be readily obtained by native boats from Ning-po fu. Kintang is well cultivated and produces abundant supplies, but they all appear to be sent to Ning-po."

BLACKWALL CHANNEL.—Tsih-tze or Blackwall island, which gives its name to this channel, is about six miles in circumference, and divides

---

* Captain C. F. A. Shadwell, H.M.S. *Highflyer*, 1858.

the northern entrance into two passages, one between Blackwall island and the north-east end of Kintang, called Blackwall pass, and the other between the four islets on the Chusan shore and Blackwall, named Ketsu pass. The southern entrance, between Kiddisol and the south-east end of Kintang, is nearly 5 miles wide, and just within it is the Steward rock described above.

From the anchorage off Sinkong point the distance through the Blackwall pass is 6 miles, and no anchorage will be found until to the northward of Blackwall island near Cliff islet, but this is exposed to northerly winds. This pass is three-quarters of a mile wide, and in it the eddies are so strong that vessels have been turned round in a double-reefed topsail breeze. Rondo, a small islet, lies off the south-west end of Blackwall island, and there is deep water between them, but the Kintang side will be found the best to border on. There is a long bay on the Blackwall side, from the north end of which, Blackwall point, a reef extends westerly 1½ cables; to avoid it do not open the Steward rock to the eastward of Rondo islet.

Between Blackwall and Chusan is a flat island named Ketsu, and the channel between it and Blackwall is 3 cables wide, but not recommended as the tides are strong, and a sunken rock lies 1¼ cables from the north-east point of Blackwall. Between Ketsu and Chusan the channel is only a cable wide, and neither shore is steep-to.

BROKEN ISLAND, which is connected at low water to the north-west extreme of Chusan by a mud bank, is steep-to on its north-eastern side. Crack islet lies about half a mile from its north point, and between them is a narrow channel carrying a depth of 5 to 8 fathoms, but it is not calculated for vessels of large draught, as a bank with 6 to 18 feet on it extends a mile from the north-west point of Broken island. A mud spit runs off north-westerly 4 cables from Crack islet.

N.W. 3½ miles from Broken island is a group of low islets called Dunsterville, which may be approached as convenient, the depth of water between them and Crack islet varying from 5 to 4 fathoms. The tides are strong in this neighbourhood, the flood running to the west, and the ebb to the east.

SHAAON HARBOUR, or North bay, formed between Chang-pih or Fisher island, and the north end of Chusan, is 2 miles long, 1¾ miles wide, and has a varying depth from 5 to 9 fathoms. Broken island, as before stated, is steep-to on its north-east side, but shoal water extends half a mile from the west end of Chang-pih. The southern shore of Chang-pih is an extensive mud bank, a considerable portion of which has been enclosed from the sea by embankment; the water is shoal off its south-east end, the 3 fathoms line being half a mile from the shore.*

The shore of Chusan is bordered by a mud bank, which renders landing, unless at high water, difficult, except in one place near the eastern end of the harbour, where there is a causeway. Near the causeway are some houses, but the principal village is situated some distance up the valley. A small islet lies off the north end of Chang-pih, and a group of islets, named Cluster or Midway islands, off the north-east end.

DIRECTIONS from SHAAON HARBOUR through KWEI CHANNEL.—Vessels bound to the eastward from Shaaon harbour may pass either through the Kwei channel, between Lan-sew or Sheppey island and Chusan, or to the northward of Lan-sew, which is the better channel of the two, but both are difficult for a stranger. A sunken rock lies on the Chusan shore S.E. 2 miles from the south-east point of Chang-pih, and from it the south extreme of Chang-pih bears W. $\frac{1}{2}$ N., the largest of the Cluster islets N. $\frac{1}{2}$ W., and the summit of Lan-sew E. by N. $\frac{1}{4}$ N.

The Houbland islets are between Chang-pih and Lan-sew, but nearer to the latter; the Kwei channel is between them and two rocks lying on the Chusan shore off Ma-aou point, and then south of Grain or Sewshan islet, which lies 2 cables from the south point of Lan-sew. The channel here is 2 cables wide, being formed between the small islet with a reef off its south-east end, lying south of Grain islet and Kanlan point on the Chusan shore, and it should not be attempted during the strength of the tide. There is another channel, named the Kwimun, closer to the Chusan shore; it is, however, crooked and there is a sunken rock near the centre.

The island of Lan-sew appears formerly to have been two; the intervening space having been gained from the sea by embanking; it is now called by the Chinese Lan-shan and Saw-shan, and is $3\frac{1}{2}$ miles long, and $2\frac{1}{2}$ miles broad.

Through CHANNEL NORTH of LAN-SEW.—To pass north of Lan-sew when leaving the anchorage in Shaaon harbour by the Chang-pih channel, steer about N.E. b. E. for Kwi-si, a barren island with a round peak upon it. The south-east side of this island is steep-to, and the distance between it and the north-west point of Lan-sew is $1\frac{1}{2}$ miles; a mud bank dries $1\frac{1}{2}$ miles from the western side of the latter, and is steep-to, the lead giving no warning, but its northern edge will be avoided by keeping the north end of Mo-uu islet (the largest islet off the north end of Lan-sew) open northward of the north extreme of Lan-sew.

Having passed Kwi-si steer for Kwan island, which must be kept close aboard, to avoid a reef which lies half a mile to the southward and covers

---

* *See* Plan of North Bay, No. 1,744: scale, $m = 1\cdot 2$ inches.

at high water; from the reef Kwi-si hill bears W. b. N., and the highest part of Lan-sew S.S.W. ½ W.; the ground between this reef and Lan-sew is foul. Although the channel is half a mile wide it is difficult to shoot owing to the eddy tides and flaws off Kwan. When the reef is passed take care to avoid a ledge of rocks extending a short distance from the north-west point of Mo-un, which bounds the channel to the southward.

To the eastward of Kwan are nine islands lying off the south-east end of Tae-shan; a reef lies off the southern end of the first. From thence an East course may be steered along the southern coast of Keu-shan island and the Fishermans group.

ANCHORAGE.—Vessels wishing to anchor on the east side of Lan-sew island may haul to the southward after passing the first islet east of Mo-un, running between it and Gan-ching, a cluster of rocks to the eastward. At the east end of Lan-sew is a low cliff, named Harty island, which may be passed at a cable's length, and anchorage will then be found in 5 fathoms; the water shoaling gradually towards the shore. H. M. S. *Pylades* anchored here in 5½ fathoms, with the east end of Harty island bearing N. ¾ W. and distant 6 cables, and Grain islet S.W. b. W. In the northerly monsoon there is a better anchorage at 7 miles to the north-eastward in Peaked Rock bay, on the southern shore of Keu-shan.

CLIFFS and DOUB ROCKS.—To the eastward of Lan-sew, at the distance of 2 miles and 5 miles respectively, are two cliff islets, called Cliffs and Doub rocks. South 2 cables from Cliffs, the western islet, is a ledge of rocks nearly awash at high water, and in its neighbourhood the ground is foul; there are rocks, also, which show at low water, lying 1½ cables from the north-east point of the same islet.

N.E. and EAST COASTS of CHUSAN.—The north-east coast of Chusan, eastward of Lau-sew, trends to the S.E. for 11 miles to Whang head, a low peninsula forming the east end of Chusan. At the distance of 3 miles is Thornton island, with a narrow passage between it and Chusan, and a deep bay westward of it, in which the mud dries out a long way, rendering it difficult to land except at the extreme points; an islet and rocks lie off the north-east face of Thornton. At 2¾ miles farther to the S.E. is a larger island, named Tsae, with a remarkable fall in the hills near its centre; a small islet lies a mile to the westward of the islets off its north end. The Chusan shore in this locality is shoal-to, there being only 1¼ fathoms between this islet and the coast.

To the eastward of Tsae island are three islands at the distance of half, 1½, and 3¼ miles. The nearest, named Meih-yun, the largest of the three, has a patch of rocks lying N.N.W. 4 cables from its north point. Meih-ting, the central islet, has a pinnacle rock lying E. b. N. half a mile from it, and a rocky patch at 2 cables to the westward of its north extreme. The outer islet, Jow rock, is a narrow cliff with a rock lying a cable's length from its north side.

From Tsae island to Whang head the distance is 4½ miles; half way between the two is a low island, named Ta-chen, and the depth of water in its vicinity is 3 fathoms. A reef lies three quarters of a mile to the south-east of Ta-chen, and a quarter of a mile from the Chusan shore, with the north-east point of Ta-chen in one with north-east point of Tsae island N.N.W., and the north end of the Putu group E. b. N.

The north-west and west face of Putu island (described in page 164) is shoal-to, leaving, however, a channel between it and Whang head nearly a mile wide. The northern part of this channel has only 4 fathoms in it, and in working through, when southward of Whang head, do not bring the head to the eastward of North, as the Chusan shore is shoal.

The channel off the south-east end of Chusan is 2 cables wide, and in the centre of it is a reef with a stone pillar on it. The flat extending towards Putu island has only 1½ fathoms on it at low water, and some hard casts; therefore vessels drawing over 12 feet should not attempt this passage, but use the Sarah Galley channel, page 162. In working up from the southward between Lokea and Kin-ho island, bear in mind that the shoal water extends 3½ cables from the former, and 6 cables from the latter; the above pillar or beacon in one with a cliff islet to the northward of it, is a good mid-channel mark. After passing westward of the beacon bring the cliff islet in line with a building on Whang head; this will lead over the flat in the deepest water, and when the south end of Putu bears East it may be steered for.

CHINKEAMUN HARBOUR is at the south-east end of Chusan, and carries on a considerable fishery to the eastward of Putu island; about 35 junks, each having from 30 to 35 men, and 250 smaller boats, averaging 5 men each, are employed for this purpose, and the proceeds are carried principally to Ning-po, the fish being preserved in ice during the summer. The harbour, formed between the island of Lokea and the Chusan shore, is 1½ cables wide with 4 and 5 fathoms water in it abreast the town. The south-west entrance to the harbour, between Lokea and Maoutze island, has not more than 2¼ fathoms in it at low water; the mud extends westerly 4½ cables from the former island, and a rock lies S.S.E. a cable's length from the east end of Maoutze.

H.M.S. *Pylades* anchored between Maoutze and Chusan, in 5 fathoms, the width of the channel being 2¼ cables; the high land, 600 feet high, on the Chusan side occasioned the squalls at times to be very violent. H.MS. *Conway* anchored eastward of Lokea, with Liwan island, which has two rocks off its south end, (page 163,) bearing West three-quarters of a mile in 5 fathoms at low water.

The distance from Chinkeamun to Ting-hai harbour is 11½ miles. The Shei-luh channel along the southern shore of Chusan has deep water in it, but in some places it is so narrow as to be practicable only for small steam vessels or boats.

The principal islands bounding the south side of this channel are, (reckoning from the eastward,) Maoutze, Ta-kan, Yin-gar, and Ao-shan. Between Ta-kan and Maoutze there are not more than 6 feet at low water, and the same depth between the two latter; between Ao-shan and Deer island there is a deep water channel, but it is confined by mud banks and obstructed by reefs.

LAN-SEW BAY is formed between the islets and rocks off the north-east face of Chusan and the extensive chain of islands running in a W.N.W. direction from Video island. The navigation of the southern part of this bay, from the north-west point of Chusan to Putu island, has been noticed in page 174; the northern part, beginning at Video, will now be described, and also the anchorages which may be useful to a vessel proceeding to Ning-po or Chusan in the northerly monsoon.

The **NORTHERN PART** of the **CHUSAN ARCHIPELAGO** consists of numerous islands and rocks, extending a considerable distance to the northward of Chusan, and fronting the northern part of Hang-chu bay. All of them are inhabited, with the exception of the Barren isles and Leuconna, and small supplies may be obtained, but the natives, except at Tae-shan island, are in a very miserable condition, owing to the constant visitation of pirates. Many good anchorages will be found among them; the depths on the surrounding banks varying from 5 to 17 fathoms, deepening to 25 and 30 fathoms on the outer part of the bank.

As vessels bound to the Yang-tse kiang pass to the eastward of this archipelago, and as, in the northern monsoon, they endeavour to make the island of Video if they cannot weather the Barren isles, we shall commence with these latter islands, and then continue the description to the westward.

VIDEO ISLAND, in lat. 30° 8' N., long. 122° 46' E., bears E.N.E. 21 miles from the summit of Putu. It is about 500 feet high, nearly square, and has a bold precipitous appearance and a remarkable white cliff, which shows when the island bears N.W. by N.; when first seen from the south-west it appears flat and shelving.

E. by N. ½ N. 5 miles from Video are four rocks called the Four Sisters; and E. by N. 9 miles from Video are two rocks named the Brothers. The depth of water in this vicinity is above 30 fathoms; any cast, therefore, below that depth will, in thick weather, point out that a vessel is among the chain of islands.

LEUCONNA ISLAND bears N.N.E. ¼ E. 18 miles from Video, and when seen from the southward it makes like three abrupt round topped hummocks.

BEEHIVE ROCK, 35 feet high, has a rock awash lying 3 cables to the eastward, and a depth of 14 and 16 fathoms around it. From it Leuconna bears E. by N. ¾ N. 12¼ miles, and Video S. by E. ⅓ E. 12¾ miles.

BARREN ISLES, three in number, are three-quarters of a mile in extent, east and west, and about 50 feet high, and at 2 cables to the south-eastward of the eastern isle is a reef* awash at high water. They lie E. ¼ N. 16 miles from East Saddle island, and N.N.E. ¾ E. 20 miles from Leuconna, and their position is lat. 30° 43′ N., long. 123° 7′ 14″ E.

FISHERMANS GROUP.—From Video island a chain of islands extends W. by N. ½ N. 45 miles, terminating in the Volcano islands facing Hang-chu bay. Between Video and the Fishermans group, the first islands westward, there is a channel 2 miles wide; but among the Fishermans group (consisting of four islets and several rocks) vessels ought not to go. Perhaps the best channel through this chain is close to the westward of this group, S.S.W. ½ W. 9½ miles from the Beehive. The channels between any of the intervening islands ought not to be attempted, as, from the character of the land, there are, no doubt, many sunken rocks.

ANCHORAGE.—Shelter will be found under Hall island, at 7 miles westward of Fishermans group; but a vessel had better go on to Keu-shan island, and anchor in Peaked Rock bay to the westward of Eden point (page 175), bearing in mind that the head of the bay is shoal. Along the southern side of Keu-shan are several islets and rocks, to which a berth of 2 cables' lengths should be given.

TAE-SHAN CHANNEL.—The channel between the Doub and Cliffs rocks (page 175) and the west end of Keu-shan island is 1½ miles

---

* This reef appeared to extend three-quarters of a mile to the south-east of the isle.— *Edward H. Hills, Master of H.M.S. Highflyer, June* 1859.

wide. From the Cliffs the southern entrance to the Tae-shan channel bears North, and is formed by the islets of Pou-no and Pou-ti to the west, and Funing island, with the Cliff islet south of it, to the east; off the west end of the latter is a reef, covered at high water. N.W. b. W. 6 cables from Funing are two low rocks, and the channel between them and Funing is shallow. North, 3 cables from these rocks, is the south point of Chang-tau island, which is not steep-to; but north of the rocks there is a narrow channel, named Chang-tau strait, carrying 5 fathoms.

The Tae-shan channel, formed between Chang-tau and Tae-shan islands, is a mile wide. Both shores are shoal-to, and a sunken rock lies S.S.E. 2 cables from the projecting point on the Tae-shan shore. A mile to the N.E. of this point is Gan-su island, which has a double peak on it, and there are two islets on each side; the channel lies between it and Chang-tau, under the north head of which is a low rock. Chang-tau peak rises to the height of 920 feet above the sea over the west side of the island, rendering it one of the most conspicuous objects of the chain.

The directions for passing south of Tae-shan island, between Kwan and Lan-sew islands, have already been given in page 174; but it remains to describe Tae-shan and the channels between it and the Volcano islands.

TAE-SHAN ISLAND, 8 miles long and 5 miles broad, and the third in point of size in the archipelago, those of Chusan and Luhwang being larger, is very populous, and carries on an extensive salt manufactory from sea water. The centre of the island is an extensive plain, with many villages; the hills also separate near the eastern extreme, leaving a level plain across the island. Off the south-east end of the island are nine islets, among which vessels have no business to go. There is a passage close to the eastward of Kwan island; but owing to strong tides and the flaws under the bluff land of this island, vessels had better pass south of it and between Kwan and Kwi-si islands, where there is a channel a mile wide; the mud dries 3 cables from the west end of Kwan.

To the northward of Kwan and Kwi-si islands are three islets; the best channel is between Ning and Kwi-si, after which a vessel can haul up for the Tae-shan shore and anchor in 4 or 5 fathoms off Wou-hou creek, observing that there is a reef which covers at first-quarter flood, lying with the summit of Kwi-si bearing S. by E. $2\frac{1}{2}$ miles, and Ellicott isle W. by S. $\frac{1}{2}$ S. $2\frac{1}{4}$ miles; the north end of Peshan islet in line with the north point of Kwan island bearing E. by S. $\frac{1}{2}$ S., will lead south of it.

The mouth of Wou-hou creek bears N.E. 6 miles from the summit of Chang-pih island; it was here that the Chinese forces assembled in 1841 for the retaking of Chusan. The creek runs through the centre of the island, but is not accessible to large boats at low water. There is another creek near a village farther westward, but with these exceptions the whole face of this side of Tae-shan is difficult of access in consequence of the mud drying a long way from the shore.

At Tautau point, the west end of Tae-shan, the hills come down to the water's edge, and midway between it and Chang-pih are Miles and Ellicott isles, with 5 and 7 fathoms in their vicinity. The Show islands, one of which is high, lie 6 cables westward of Tautau point; the channel between having 4 fathoms at low water. On the north side of Tae-shan are four islets, which are too small to afford much protection in the North-east monsoon, but during the summer good anchorage will be found off the town near the centre of the island.

VOLCANO ISLANDS.—The East Volcano, which has four peaks on it and is the largest of the group, lies 6 miles westward of Tautau point, and is 4 miles long north and south. East of its south point is an islet; and between it and the Show islands are two islets lying close together, with steep cliffs, named Becher islets. North 1½ miles from the latter islets are two low rocks.

ANCHORAGE.—Vessels passing between the Show islands and the East Volcano should be careful not to stand too close to the latter, as the water shoals to 2 fathoms at 1¼ miles from the shore. East 3 cables from its north point is a half-tide rock.

There are many sunken rocks among the islets off the west face of this group, among which vessels ought not to go, but they will find shelter from northerly winds on the south side of the group, to the northward of a flat rock, lying westward of the south point of East Volcano. The northernmost islet of the group has a reef lying 1½ cables' lengths to the northward of it.

TIDES.—The tides in the vicinity of the Volcano islands will be found to have increased their velocity, the flood setting W.N.W. and the ebb E.S.E. It is high water, full and change, at 11h. 30m. and the rise at springs is 15 feet. In light winds a wide berth should be given to all the islets hereabouts.

SKEAD ISLET lies 4½ miles to the northward of the Show islands; on its north-west and south-east sides are smaller islets. The depth of water between it and the Rugged islands to the northward varies from 5 to 7 fathoms.

MARINER REEF.—A notice was published in the "North China Herald," in February 1857, of a rock or reef lying directly in the route of vessels running between Tae-shan and Chin-san islands towards Ning-po, and on which the merchant brig *Mariners Hope* struck. The vessel was 12 hours on the reef, which was stated to be about a third of a mile long in an east and west direction and 2 cables broad, and had 7 fathoms at her bows, with only 5 feet under her stern at low water. Skead islet bore S. ¾ E., distant 3 miles; south extreme of Chin-san E. ¾ S.; large Volcano S.W. westerly; and extremes of Rugged islands from N.N.W. ¼ W. to N.N.E. ½ E.

CHIN-SAN ISLAND, 8 miles long in an east and west direction, lies W. by N. 13 miles from the Beehive rock (page 178), and 5½ miles to the north-east of Tae-shan. The channel between the chain of islands extending W.N.W. from the Fishermans group and this island is sometimes taken during the northerly monsoon by vessels bound to Ning-po or Chusan, and it appears preferable to that through Lan-sew bay, and clear of danger, with the exception of the Mariner reef just described. There are several small islets lying off the eastern and northern face of Chin-san; the best anchorage in the northerly monsoon is to the westward of the south-eastern islet, between it and Chin-san; and there is also tolerable shelter on the western side of Chin-san off Pennell point.

SADDLE GROUP.—The southern islands of this group, South Saddle and East Saddle, bear W. ⅓ S. 16 miles from the Barren isles, and N.N.W. 17 miles from Leuconna island (page 178). South Saddle island is rugged, the highest part, at the north-east end, rising 680 feet above the sea. A rock, which shows at low water, lies in the bay on the east side of the island, with the highest part of the rocky islet close to the eastern point of the bay in line with a conical hill over the west point of East Saddle island.*

Eight miles to the north-west of South and East Saddle islands is North Saddle island 780 feet high. Between these is False Saddle island; and S.W. of North Saddle are the Side Saddles, two narrow islets which will afford shelter, but not as good as that under South and East Saddle islands. North Saddle forms the north end of the Chusan archipelago, and from it the Amherst rocks at the mouth of the Yang-tse kiang bear N.W. ¾ N. 26 miles, the soundings gradually shoaling from 12 to 6 fathoms. The tides are regular; the flood setting to the N.W. and the ebb to the S.E., it being high water an hour before noon on full and change days, and the rise 14 feet.

Water can be obtained at the east end of East Saddle island.

ANCHORAGE.—The most convenient anchorage in the northern mon-

---

* *See* Plan of South and East Islands of Saddle Group, No. 1,418; scale, m = 2·2 inches.

soon amongst the Saddle group is under the East Saddle, and in the event of being caught in a southerly wind vessels might run through between them, taking care to keep South Saddle close aboard, as there is a patch of 3 fathoms lying in the centre of the channel, and three rocks wash north of it.

CHILDERS ROCK, which uncovers at low tides, lies 4¾ miles to the southward of East Saddle island, with the Barren islands bearing E.N.E., Leuconna island S.S.E. ¼ E., and the summit of Senhouse island W. b. N. The lead will give no warning of approach to this danger, the depth being 24 fathoms close to.

PARKER ISLANDS.—West 11 miles from South Saddle island are the Parker islands, of which Raffles is the largest. About 4 miles westward of South Saddle is the Bit rock, not much elevated above high water. At half a mile from the north-east point of Raffles island is a sunken rock. An island with steep cliffs, named Senhouse, lies 1¾ miles to the southeast of Raffles; there is a good channel between them, and anchorage will be found on the south side of Raffles in the northerly monsoon.

Brooke island lies a mile to the south-west of Senhouse island; the channel between them should not be used, as the wind is liable to fail under the latter; there is, however, a good passage 2 miles wide west of Brooke, between it and the Bonham isles. Off the north-west end of Raffles island, and distant from it 1¾ miles, are the Elliot islets, on the south-west side of which H.M.S. *Plover* anchored, and found fair shelter, with the wind blowing hard from the northward. From these islets Gutzlaff island bears W. by N. ½ N., 10½ miles.

CAIRNSMORE ROCK.—This dangerous pinnacle rock, not more than thirty or forty feet in diameter, and on which the ship *Cairnsmore* was wrecked in 1858, rises almost perpendicularly from soundings of 12 fathoms at about 2⅓ miles eastward of the east end of Raffles island. When examined * the precise depth on the pinnacle could not be ascertained, as the wreck, with her foremast standing, quite covered it, but there cannot be more than 11 feet over it at low water springs.

The position of the rock is lat. 30° 42' 10" N., long. 122° 34' 40" E., and from it the south-east point of Senhouse island bears South; a small rugged rock lying close to the south-east point of Raffles island, and in line with the point, bears S.W. by W. ¼ W.; and the northern rock of the group lying off the north part of Chesney island, N.W. by W. ¼ W.

CAUTION.—Vessels navigating the channel between the Saddle group and Raffles island are cautioned for the future to keep well over towards

---

* By Lieutenant J. Ward, R.N., H.M. Yacht *Emperor*, 1858.

the Saddle islands to avoid the above danger, as the lead will give no warning when approaching it. In sailing north, when the Bit rock opens south of the South Saddle they will be to the northward of the Cairnsmore; and in sailing south, when the same rock opens north of the South Saddle they will be to the southward.

RUGGED ISLANDS lie W.S.W. 15 miles from Raffles, and between them are the Morrison islands, the largest of which is very precipitous. The Rugged group affords shelter in both monsoons, but the tides set through them with considerable velocity. Tayung, the largest and highest of the group, is 660 feet above the sea, and differs from the rest by being round topped, whereas the others are, as their name denotes, rugged Under the south side of an islet west of Tayung is Pirate bay, which will afford snug anchorage during the northerly monsoon, and a better shelter than that within the S.W. and N.W. Horns of the group. A reef, which generally breaks, lies off the east side of Pirate bay.

The largest island on the north side of this group is Tripoint, remarkable for its triple peak; and east of it is Spire islet, on which is a curious pinnacle.

HEN and CHICKS.—N.E. by N. $3\frac{3}{4}$ miles from the N.W. Horn of the Rugged islands is an islet with several rocks north-west of it, called the Hen and Chicks.

A shoal with only 10 feet water over it has been reported to lie S.W. 7 miles from Gutzlaff island, which would place it E. by N. $\frac{1}{2}$ N. not quite 2 miles from the Hen and Chicks.

GUTZLAFF ISLAND, 210 feet above the sea, bears N.E. 12 miles from the N.W. Horn of the Rugged islands; a small rock lies off its north side. * "It has been reported from many sources, that a bank with only $2\frac{1}{2}$ fathoms on it, extends a mile from the western side of Gutzlaff."

TIDES.—In the South-east or Vernon channel, at the south end of the Chusan archipelago, it is high water, full and change, at 9h. 40m., and springs rise 14 feet; in Ting-hai harbour at 11h. 0m., springs rise 12 feet, neaps 9 feet; at Putu island at 8h. 15m., springs rise 12 feet; in Lan-sew bay at 10h. 0m., springs rise 13 feet; at the Volcano islands at 11h. 30m., springs rise 15 feet; and at East Saddle island at 11h. 0m., and springs rise 14 feet.

Under Luhwang island the flood sets to the N.W. at the rate of 2 knots per hour, and the ebb to the S.E. at $1\frac{1}{2}$ knots. In Duffield, Gough, and Roberts passes, the first of the flood, at full and change, often comes from

---

* Commander C. M. Mathison, H.M.S. *Mariner*, 1850.

the northward, and sometimes runs in that direction 3 hours before the tide through the Buffaloes Nose channel overcomes that through the south-east channels. In Duffield pass the tide sometimes runs 5 knots per hour; in Gough and Roberts passes it is not so strong; in Beak Head channel 4 knots is about the maximum; and in Vernon channel it has been known to run 6 knots. Off Roundabout island the tidal streams are not so violent, but the eddies take command of a sailing ship at springs.

In the southern entrance to Sarah Galley channel, between Laoush and Ousha islands, the flood ran W. b. S. 2 knots per hour, the ebb E.S.E., 1½ knots; the moon was then 18 days old. In the Cambrian pass between Ousha and Chuken islands, H.M. steamer *Vixen*, with the *Cambrian* in tow, could not stem the ebb.

In the Tower Hill channel, as before stated (page 169), with a strong flood, vessels have been swept away to the westward, and carried by the tide beyond Just-in-the-Way, and even through the Blackwall channel; and after rounding Tower Hill and entering the Bell channel, many have been borne by the ebb, between Tower Hill and Tea islands. Having rounded the north end of Tea island with a strong ebb, it is necessary to guard against its taking the vessel through the Melville channel, and if not able to pass northward of Macclesfield island, send the boats a-head and endeavour to keep the vessel to the northward of Sarah island, where there is shoal water to anchor. In the channel between Bell island and Chusan, the tide at times runs with great strength, so much so that on one occasion the *Madagascar* steamer had great difficulty in stemming it.

In the Blackwall channel, the eddies are as strong as they are off Roundabout island, taking a sailing ship round against both helm and sails. In the Kintang channel, between Kintang island and Deadman island, the rate of the tides is sometimes 4 knots.

In the northern part of the Chusan archipelago in Lansew bay, with Lansew island bearing West 5 miles, the flood ran to the W.N.W. the first hour, then N.W.; total amount of tide 11 knots. The ebb, S.E. b. S. the whole tide; total amount 5¾ knots.

KINTANG CHANNEL, between the south point of Kintang or Silver island and the main land, is about 2½ miles wide, but it is narrowed to 1½ miles by an extensive mud bank which borders its southern shore, and by the reef which extends 2 cables from Alligator point (page 172), and covers at half tide. This mud bank dries upwards of three-quarters of a mile from the shore, is steep-to, and the lead gives no warning; there are some small islets lying on its outer edge, near the easternmost of which is a boat creek, from whence there is a paved footpath leading to Tein-tung and so on to Ning-po; the whole distance being about 18 miles; the last 6 miles may be performed by canal.*

JUST-IN-THE-WAY.—This small rock, 20 feet high, with rocks extending $1\frac{1}{2}$ cables from its S.S.E. side, lies in the eastern entrance of the Kintang channel, and to the south-east, between it and Tygosan island, there is fair anchorage in 12 to 16 fathoms, which will be found a convenient stopping place should there not be sufficient tide to take a vessel to Chin-hai, the anchorage outside of which is much exposed.

The DEADMAN is a square island lying W. $\frac{1}{2}$ N. $2\frac{1}{2}$ miles from the south extreme of Kintang and $4\frac{1}{2}$ miles westward of Just-in-the-Way. The channel between it and Kintang is rather less than 2 miles wide, with deep water and rapid tides. The Ko channel, between the Deadman and the main, is half a mile wide, but it is not recommended, as the tides are violent, and the limit of shoal water on the south side is not well marked.

BLONDE ROCK lies a short half mile to the northward of the Deadman, at the western entrance to the Kintang channel, and shows at low water springs. The marks for it are, the easternmost islet off the north-east point of the Deadman in one with San-shan islet S.E. $\frac{1}{4}$ S.; Taping point, the north-west point of Taping island, N. $\frac{1}{2}$ E.; and the west extreme of Dumb islet S.W. $\frac{3}{8}$ W. Beacon hill at the east side of the entrance to the river Yung, in line with the citadel bearing W.S.W., will lead to the northward.

TSE-LE or SQUARE ISLAND bears N.W. b. W. $2\frac{3}{4}$ miles from the Deadman; there is a patch with $2\frac{3}{4}$ fathoms water on it lying S.E. by S. 6 cables from its north end. H.M.S. *Conway* anchored to the westward of this island with it bearing E.N.E., and Pus-yew, the western of the Yew islands, South. This anchorage in the summer season is safe, but during the autumn and winter violent gales with thick weather rise rapidly, causing an uneasy sea, in which a vessel will have difficulty in weighing her anchor; consequently, the anchorage at Just-in-the Way or that in Ta-outse harbour, at the north-west end of Kintang, should be resorted to at this season.

YUNG RIVER.—The entrance to this river is 15 miles to the westward of the Bell channel leading to Ting-hai harbour, and is fronted by three islets, called the Yew islands or Triangles, which form three en-

---

* See Chart of Kintang Channel, No. 1,770; scale, $m = 1\cdot 2$ inches.

trances into the river. The town of Chin-hai is built immediately to the south-westward of the Citadel hill, on the western side of the entrance to the river, of which it is the maritime town. From Chin-hai the river trends in a S.W. and West direction for 11 miles to Ning-po fu, and is about 2 cables wide, with depths varying in mid-channel from 5 to 2 fathoms. Vessels of 17 feet draught can proceed up to the city from Chin-hai at half tide.

The Yung separates into two branches at Ning-po fu ; one, the Yuao and Tsie-kie branch, taking a north-west, the other, the Funghwa branch, a S. by W. direction ; the latter is barely a cable wide, and is crossed by a bridge of boats at a quarter of a mile beyond the junction. As the turning at this junction, from the river Yung to the Tsie-kie branch, is very sharp and difficult to take, owing to the crowded state of the river and the flood tide setting towards the Funghwa branch, large vessels should anchor below it and wait until it is cleared. The British Consulate is on the left bank of the Tsi-kie branch opposite the city.*

† " Ning-po fu was occupied by a detachment of marines and British troops during the winter of 1841, and the vessels of the squadron anchored off the north-east and eastern parts of the city. To prevent an attack from fire junks an expedition of troops and blue jackets, under the command of Vice-Admiral Sir W. Parker and General Lord Gough, embarked in December 1841, in H.E.I.C. steamers, *Sesostris*, *Phlegethon*, and *Nemesis*, and proceeded with the boats of the squadron up the Yuyao branch, clearing the river of suspicious junks, and making the Chinese troops evacuate the city of Yuyao. The *Sesostris* anchored about 3½ miles below the city, her draught being 17 feet."

TIDES.—At Chin-hai it is high water, full and change, at 11h. 20m., and springs rise 12½ feet. At Ning-po fu it is high water at 1h. 0m., and springs rise 9 feet.

DIRECTIONS.—A vessel bound from Ting-hai harbour, Chusan island, to the river Yung, should, after clearing the Bell channel, steer W. b. S. for Just-in-the-Way, bearing in mind that the south-east face of that islet is foul, and that a reef extends a cable's length from Insular point, the north extreme of Tygosan island. As before stated (page 185), if the tide should fail, there is fair anchorage to the south-east of Just-in-the-Way. From hence the peak of Tower-hill island in line with Insular point will lead southward of the rocks off Alligator point, after which keep over towards the Kintang shore, until well past the Deadman, or until Beacon hill at the east side of the entrance to the river Yung is in line with the

---

* See Chart of Yung River, No. 1,592 ; scale, m = 3 inches.
† John W. King, Master of H.M.S. *Modeste*, 1841.

citadel W.S.W., which will lead northward of the Blonde rock, and to the southward of the 2¾ fathoms patch lying S.E. b. S. 6 cables from the north end of Tse-le island.

It will be prudent for a stranger, before entering the river Yung, if unable to obtain a pilot, to examine the entrance in his boat, for since the survey of this river the stakes and sunken junks which blocked the channel between the citadel and Peak islet have been removed, and this may have caused some change in the mud banks and soundings outside.

The Yew islands, as before stated, form three entrances into the Yung river, the easternmost of which is between the islands and Look-out hill on the eastern side of the entrance to the river. The first danger in this channel is the Nemesis rock, which is covered at half flood and lies E. by N. ¾ N. a quarter of a mile from the summit of Ta-yew, the eastern Yew. By keeping Pas-yew, the western Yew, open of the south point of Ta-yew, this danger will be avoided.

Having passed the east point of Ta-yew, keep it and Seaou-yew, the middle Yew, aboard, to avoid the Sesostris rock, with only 8 feet on it, which lies in mid-channel with Friendly islands (lying 7 miles north-west of Chin-hai) in one with Talung island (a high bluff island beyond it) bearing N.W. ½ W. ; Peak islet (a remarkable rock on the east side of the river abreast the citadel) in line with Cone Hill bearing S.W. ½ S., will lead westward of it.*

Having cleared the Sesostris, steer so as to pass between half and 1½ cables to the southward of Pas-yew, and then for the point under the citadel, taking care that the tide does not set the vessel over to the eastern bank of the river, where the water shoals to 2 fathoms at half a mile from the shore.

The middle entrance, or that between Seaou-yew and Pas-yew, is probably the best of the three. A mud spit extends north-westerly 1½ cables from the west end of Seaou-yew, but it will be avoided by keeping the citadel open westward of the west end of Pas-yew ; then steer as before so as to pass to the southward of Pas-yew.

The channel between Pas-yew and Chung or citadel point carries a depth of 2 fathoms at low water, and is the broadest and best for small vessels when the tide has risen sufficiently high for them to enter it ; the only danger being the Tigers Tail rock which covers at high water, and lies rather more than a cable's length N.W. ½ N. from the summit of

---

* The merchant barque *Moltan* is said to have struck on a rock having 9 feet on it and 18 feet close to ; when on it Friendly island was just showing northward of Pas-yew, and the northern extreme of Look-out hill bore East.—*Nautical Magazine*, 1852, *page* 395.

Pas-yew, with the south-east foot of the citadel hill in line with Cone peak bearing S.S.W. ¾ W. Chung point is steep-to on its east side, and vessels will find good shelter under the fort.

The COAST from Chin-hai trends in a north-west direction, and is fronted by a mud bank which dries at low water for nearly three-quarters of a mile from the embankment, and is steep-to. At the distance of 7 miles from Chin-hai, and three-quarters of a mile from the shore, are a group of five islets, named the Friendly islands, inside of which there was shelter in a depth of 3 fathoms at the time of this survey, but the water is said* to be shoaling fast. Care must be taken in rounding the west end of the largest islet to avoid a spit which extends 3 cables to the S.E. from it.

At 4 miles farther to the N.W. is a high bluff, named Talung island, rising to the height of 920 feet, and forming the southern horn of the Tsien-tang estuary, or Hang-chu bay.

CAUTION.—From Talung the coast trends to the westward, and for upwards of 30 miles is fronted by a dangerous mud bank which, at the distance of 8 miles from Talung, dries 7 miles from the shore, and on its edge are some small hillocks. The *Kite* transport was lost upon this bank in 1840, the tide, which here begins to increase its velocity to 6 knots at the springs, turning her over the moment she tailed on it.

MIDDLE GROUND.—N. b. W. 3½ miles from Tse-le island is a Middle ground with less than 2 fathoms on it, to avoid which vessels in proceeding to the northward from the river Yung, must keep over towards the Kintang shore, and if drawing 18 feet water, should not bring Tse-le island to the eastward of South. There is a passage to the southward of this Middle ground for vessels of 15 feet draught, but there are two patches, on which H.M.S. *Contest* grounded, lying in a N.W. direction from Tse-le island, one with 12 feet on it at 9 cables, and the other with only 5 feet at 2¼ miles from the island.

NANHO or North island, bearing N. ¾ W. 14¼ miles from Tse-le island, is the largest and easternmost of the first group of islands met with when steering to the northward from Chin-hai; it is flat-topped, 216 feet above the sea, three-quarters of a mile in extent east and west, and cultivated. As the water deepens close around this island to 26 and 32 fathoms, vessels cannot anchor near enough to get shelter, but the holding ground is good. North about half a mile from it is a small rock which always shows.

---

* Commander T. H. Mason, R.N., H.M.S. *Medea*, 1849.

CHAP. VI.] HANG-CHU BAY; SESHAN ISLANDS.—CHAPU BAY. 189

WEST STORK is a small islet lying W. ¼ N. 3¾ miles from Nanho island, and there are 8 and 9 fathoms water between them.

SEVEN SISTERS lie North 9 miles from Talung island, and although small, will afford shelter from northerly winds. The channel between this group and the dangerous mud bank just described, is 4 miles wide, and the depth in it varies from 6 to 2 fathoms, shoaling towards the bank. A reef, which shows at low water, lies N.N.W. half a mile from the western islet of the group.

SESHAN ISLANDS form three distinct groups. East Seshan, the easternmost group, lies North 18 miles from Nanho island; the largest islet is about 400 feet high, and has six small islets around it. The Middle Seshan group lies 6 miles to the W.N.W. of East Seshan, and consists of one large and eight smaller islets, the southernmost of which is a small rock nearly level with the water's edge, lying nearly 4 miles to the southward of the highest; the western islet, House islet, is an abrupt cliff with a house on its summit. Neither of these two groups are sufficiently large to afford shelter; but fair anchorage will be found in the neighbourhood of the western group, named West Seshan, which consists of three islets, and lies W.N.W. 10 miles from the Middle Seshan.

CHAPU BAY, formed on the northern shore of Hang-chu bay, is 9 miles wide, in a N.E. by E. and S.W. by W. direction, and 3 miles deep. The position of the roadstead off the city of Chapu will be readily known by the hills in its vicinity, as well as by the islets which protect the road from the eastward; on the eastern of these islets is a remarkable white house.

Vessels steering for this anchorage should round the southern islet at about a quarter of a mile and haul up for the houses which will be seen to the westward of the hills. The anchorage is sheltered from E.N.E. to S.S.W.; but the velocity of the tide at springs is 5 knots, and the rise and fall 25 feet. The mud dries half a mile from high-water mark, is steep-to, and the lead gives no warning. At 4 miles to the southward of the southern islet is a shoal on which the ship *Bentinck* tacked in 3 fathoms, and where there is probably less water; should the tide therefore set vessels in this vicinity, it will be prudent to anchor.*

Off the southern horn of Chapu bay the tide runs 7½ knots, and the rise and fall is 30 feet. In this bight of the bay are some islets and a pagoda; the latter most likely is in the neighbourhood of Hai-ning. At 13 miles from Chapu there is a bay, protected in some measure by a small islet, in which several boats were lying aground. On the hill over

---

* *See* Plan of Chapu Road, No. 1,453; scale, $m = 2$ inches.

it was a four-gun battery and a numerous garrison; this place, answering to the name, is supposed to be the Canpu of Marco Polo. Tseenshan, which is 24 miles westward of Chapu, appeared to be an islet connected with the main by a causeway under which boats were lying; on it was a four-gun battery and a small pagoda, and, assuming the geographical position which the Jesuits assign to Hang-chu fu to be correct, this place is 60 miles from the city. It was in this neighbourhood that the H.E.I.C. Steamer *Phlegethon* experienced a tide of 11¼ knots at springs and 8 at neaps; the depth of water across the estuary at low tide was found to be less than 1½ fathoms.

DIRECTIONS.—Vessels bound to the northward from the river Yung should endeavour to leave with the first of the flood, and when northward of Tse-le island, if drawing more than 18 feet, they should not bring that island to the eastward of South to avoid the Middle ground. In working up for the East Seshan group some casts of 3½ and 4 fathoms were obtained with the eastern islet bearing N. by E.; it will, therefore, be advisable that a vessel of large draught should not stand into Hang-chu bay unless bound for Chapu road, in which case pass about three miles to the southward of East Seshan, and steer for the southernmost islet of the Middle Seshan group. After passing the West Seshan the low land on the north side of Hang-chu bay will be seen, and to the southward the Fog islets, a group of five low rocky islets bearing W. by S. ⅓ S. 14 miles from the Middle Seshan, the depth of water about them being 5 and 6 fathoms. The position of Chapu road, as stated above, will be readily known by the hills in its vicinity.

If bound for the Yang-tse kiang keep to the eastward of the Seshan islands, steering between the East Seshan and Rugged islands. The tides in the vicinity of the Volcano island will be found to have increased their velocity, the flood setting W.N.W. and the ebb E.S.E. The Rugged islands affords shelter in both monsoons, but the tides set through them with considerable strength. From the Rugged islands, steer to pass on either side of the Hen and Chicks in 6 and 7 fathoms water, recollecting the 10 feet shoal (page 183), and also on either side of Gutzlaff island; it will be as well, however, if the vessel is of large draught, to pass eastward of the latter island, as a bank with only 2½ fathoms on it is said to extend a mile from its western side.

TIDES.—It is high water, full and change, at the Seshan islands at 11h. 45m., and springs rise 14 feet; at the Fog islands in Hang-chu bay at the same time, and springs rise 17 feet; and in Chapu road at noon, and springs rise 25 feet.

The velocity of the tidal streams increases as Hang-chu bay is approached; in the neighbourhood of Nanho island and the Volcano group, the flood runs W. b. N., and the ebb E. b. S. sometimes at the rate of 3 knots, and in light winds, unless great care is taken, vessels are liable to get entangled among the Dunsterville or Volcano groups. At the Fog islands in Hang-chu bay, the velocity increases to $4\frac{1}{2}$ knots per hour; at Chapu to 5 knots; and in the south-west part of Chapu bay to 7 knots, with a rise and fall of 35 feet. Twenty-five miles above Chapu, the tide at springs was found to run $11\frac{1}{4}$ knots per hour, and at neaps 8 knots, with a rise and fall of 40 feet. In the vicinity of the outer Seshan islands, and the Rugged group, the flood runs $2\frac{1}{2}$ and 3 knots; to the southward of Gutzlaff island the first of the flood makes to the southward of West.

YANG-TSE KIANG.

The Yang-tse kiang, the largest of the Chinese rivers, leads to Nanking, which is about 220 miles from its entrance, and then on to Hankau, which is 384 miles above Nanking. The entrance is wide, and divided into two channels by Tsung-ming island, which is 32 miles long in a W.N.W. and E.S.E. direction, and 5 to 10 miles broad, and is said to be the largest alluvial island in the world, containing a population of about half a million, although in the fourteenth century it did not exist above water.*

On the southern shore of the Yang-tse, nearly abreast of Bush island, and about 40 miles within the entrance, is the mouth of the Wusung river, with the village of Wusung on its left bank. The great commercial city of Shanghai is 12 miles higher up on the same bank of the river, and is one of the most active trading cities in the empire. On each side of the mouth of the Wusung there is a fort about $1\frac{1}{2}$ miles below the village.

YANG-TSE CAPE, forming the south point of entrance to the Yang-tse kiang, is 17 miles W. by N. from Gutzlaff island, and on it is a beacon 25 feet high and 35 feet above the sea. The land hereabouts is very low and quite level, having been entirely gained from the sea, and the mud dries out half a mile at low water from the embankment. There is anchorage in $4\frac{1}{2}$ fathoms under the beacon, and fair shelter from northerly winds, unless the wind draws to the eastward. The coast for 20 miles to the northward of the cape is fronted by a mud bank, which commences at the cape, and its eastern edge, in $2\frac{1}{2}$ fathoms, is 14 miles from the shore, and 12 miles N. $\frac{1}{2}$ W. from Gutzlaff.

---

* *See* Chart of the Yang-tse kiang, No. 1,480; scale, $m = 0.2$ of an inch, corrected to 1859.

SHAWEISHAN ISLAND, bearing N. ½ E. 37½ miles from Gutzlaff, is a small peaked islet 196 feet high, fronting the entrance to the Yang-tse kiang.*

AMHERST and ARIADNE ROCKS.—The Amherst rocks, 10 feet above high water, lie N.N.E. 25 miles from Gutzlaff island, and S.S.E. ⅜ E. 16 miles from Shaweishan. The Ariadne rock has only 5 feet on it at low water, and from it the Amherst bear E. by N. 7¼ miles; Shaweishan, North a little easterly 15½ miles; and Gutzlaff S. ¾ W. 21½ miles.

The SOUTH ENTRANCE to the YANG-TSE KIANG is bounded to the northward by the rocks just described and the southern edge of the Tungsha banks, and to the southward by the extensive mud bank which fronts the shore to the northward of Cape Yang-tse. The following are the positions of the light-vessel, buoys, and beacons, placed in the river by the Chinese authorities at Shanghai, and by George L. Carr, Master of H.M.S. *Pique*, March 1857:—

The LIGHT-VESSEL, painted *red* with two masts, each surmounted with a ball, at the entrance of the Yang-tse, is moored in 4½ fathoms at low water, one mile from the southern edge of the Tungsha bank's, and N. by W. ¾ W. 23 miles from Gutzlaff island. A gun is fired from her to attract attention, when a ship is observed running into danger, and the signal, by Marryat's code, of the course that should be steered, is then exhibited. A light from a common lantern is shown on board from sunset to sunrise.†

PILOTS.—Competent pilots, English and American, will be found cruising in the neighbourhood of the Saddle islands during the summer months, and just outside Gutzlaff island in the winter. No sailing directions can do away with their usefulness to the stranger, where the safety of the vessel depends so much upon a correct knowledge of the tides. The signal of the authorized pilots is a flag, half red and half white horizontal, with the number of the boat in black.

KIU T'OAN BEACON TOWER (a plain structure of brick, painted *red* and *white* and 70 feet high) is erected on the southern shore of the,

---

* Where the Admiralty chart gives a depth of 3½, 3¼, and 7 fathoms on each side of Shaweishan island, there is now only 2¼ fathoms.—*John Thomas, Master of H.M.S. Contest*, 1852.

† This light is so faint, that it cannot be seen at a greater distance than 3 miles, and may be easily mistaken for a light on board a fishing boat or junk. The light-vessel and the Kiu T'oan beacon tower are exceedingly useful, and so are the buoys, but unfortunately the latter are not large enough for the distances they are required to be seen.—*Captain Sir F. Nicolson, H.M.S. Pique*, April 1858.

Yang-tse kiang, at Kiu T'oan, near a spot known as the Three trees. It bears from the light-vessel N.W. by W. ⅜ W., distant about 16 miles, and from the trees upon Blockhouse island S. by E. ⅓ E. 8 miles.

Buoys.—The following eight iron Nun buoys are moored between Gutzlaff island and Wusung, upon the most projecting points of the southern edge of the Tungsha banks, and upon the northern projecting points of the south shore bank:—

Fairway Buoy (*black and white horizontal stripes* and surmounted with *staff* and *vane*) lies in 4¾ fathoms at the entrance of the river, with Gutzlaff island bearing S. by E. 18 miles, and the light-vessel N. W. ½ W. 7½ miles.*

Buoys on South Shore Bank.—The edge of this sand is marked by *three* buoys coloured *black*, each surmounted with *staff* and *vane*, and numbered respectively 1, 3, and 5.

No. 1 is in 18 feet at low-water springs, with Kiu T'oan beacon tower bearing N. W. by W. ; light-vessel East 3⅓ miles ; buoy No. 3 N.W. ½ W. 8⅓ miles ; and *red* buoy No. 2 on southern edge of the Tungsha banks, N. by W. 3¾ miles.

No. 3 is in 18 feet on the edge of the bank extending south-east of the beacon tower, with the tower bearing W. by N. ⅜ N. westerly ; light-vessel S.E. by E. ¼ E. ; and *red* buoy No. 4 on southern edge of Tungsha banks N. by W. ¼ W. 2¼ miles.

No. 5 is in 20 feet, with Blockhouse island bearing N. by E. ⅝ E. ; the beacon tower S.E. ½ S. ; and *red* buoy No. 6 on southern edge of the Tungsha banks N.E. 1¾ miles. At about a ship's length to the S.S.W. of this buoy, the depth is only 2 fathoms.

Buoys on Tungsha Banks.—The edge of this sand is marked by *three* buoys coloured *red*, each surmounted with *staff* and *vane*, and numbered respectively 2, 4, and 6.

No. 2 is in 19 feet at low-water springs, with the light-vessel bearing S.E. ⅓ E. 5½ miles ; Kiu T'oan beacon tower W. by N. ½ N. ; and *black* buoy No. 1 S. by E. 3¾ miles. It shoals suddenly inside the buoy, and the nearest point of the sand, which is dry at low water, bears N. ½ W., distant about one cable's length.

No. 4 is in 20 feet, with the beacon tower bearing S.W. by W. ⅜ W. 3½ miles ; Blockhouse island N.W. ½ N. ; and *black* buoy No. 3 S. by E. ¼ E. 2¾ miles.

No. 6 is in 19 feet, with Blockhouse bearing North 2 miles; the beacon tower S. by E. ½ E. ; and *black* buoy No. 5 S.W. 1¾ miles.

---

* The Fairway buoy disappeared in the gale of September 1857.—*Captain Sir F. Nicholson, H.M.S. Pique, April* 1858.

The numbers of the buoys begin from seaward, and in entering the river, when within the Fairway buoy, leave the *red* buoys on the starboard hand, and the *black* buoys on the port hand. When standing towards *black* buoy No. 5, tack the first shoal east, for the bank is steep-to : close outside the buoy the depth is 5 fathoms.

ANCHORAGE.—It will be generally safe to anchor in 4 to 6 fathoms water off the entrance of the Yang-tse kiang, outside Gutzlaff island. It is recommended that a vessel should not anchor at night under the Saddle islands, during the North-east monsoon, unless there are appearances of bad weather, as it will frequently take all the daylight of the next day to work up to the entrance. In the summer season, if bad weather is approaching, which the barometer usually foretells, a stranger should not attempt to run in unless certain of getting within the bar ; but an anchorage should be sought either under these islands, or the vessel kept at sea, as it is dangerous to enter the river when a gale is coming on. It will be preferable to anchor than to stand out to sea, as the weather is sometimes thick and foggy, the tides strong and uncertain, and the vessel's position may not be quickly ascertained.

TIDES.—It is high water, full and change, in the vicinity, and eastward of Gutzlaff island, between 11 and 12 o'clock; and the rise at springs is about 15 feet.

At the entrance of the Yang-tse kiang, it is high water at about noon ; springs rise 15 feet, neaps 10 feet, and neaps range 5 feet ; its rate, which is from $1\frac{1}{4}$ to $4\frac{1}{2}$ knots, and its direction, is affected by the prevailing wind. At the entrance of Wusung river, it is high water at 1h. 30m., and the rise is about the same. The depth on the outer bar of Wusung at the lowest springs is 16 feet, and on the bar above Wusung about 12 feet. At Shanghai it is high water at 1h. 40m. ; springs rise 10 feet, neaps 7 feet, and neaps range 4 feet. Vessels drawing 16 and 17 and even 18 feet can cross the Inner bar at any high water ; if of larger draught they must wait for spring tides. The greatest draught ever brought up to Shanghai was between 21 and 22 feet, but a vessel of that draught will have to wait for the springs to pass either up or down the river.

From the Saddle islands to Wusung the streams generally set N.W. by W. and S.E. by E. when fully made, if north-east gales or heavy rains do not interfere. The flood makes first to the southward, then S.W., and gradually round to N.W. at half-flood, which is its direction at the strength of the tide.

The first of the ebb sets to the northward, over the Tungsha banks, and in like manner, changes round to the eastward, gradually running the strongest when S.E.

During the survey of the Yang-tse in October 1842, off the mouth of the river with Shaweishan island bearing N.E. by E., on the first day of the moon the ebb ran to the S.E. 20 miles on the whole tide; the flood, commencing at N.N.W., then W.N.W., afterwards N.W., ran only 10 miles. On the following tide the amount of ebb amounted to 21 miles, and the flood to 16 miles.

It is at the turn of both streams that most caution is necessary to avoid being set out of the channel. Round the south-east edges of the south bank the flood sets W.S.W., and the ebb the contrary way. Leaving the position off Gutzlaff, at a quarter ebb, a sailing vessel will carry the flood to Wusung, if there is any wind.

DIRECTIONS.—Vessels bound to the Yang-tse kiang from the southward, and not intending to call at Chusan or the river Yung, should pass eastward of Chusan, and enter the archipelago to the northward of that island. In the northerly monsoon they should endeavour to make the Saddle islands, as being the most weatherly land-fall, but if they cannot fetch so far to the northward, and they have reached the parallel of 30° N., the high-domed shaped island of Video, 500 feet high, will then be a conspicuous object, for it may be seen in clear weather 50 or 60 miles; it has a remarkable white cliff, which shows when the island bears N.W. by N., and in thick weather any cast below 30 fathoms will point out that the vessel is in its vicinity. The most remarkable land to the southward of Video is the island of Chukea, on which there is a round-topped peak 1,164 feet high. There are several islets (page 164) to the eastward of Chukea; Tongting, the outer one, is about 40 feet high, with detached reefs to the south-west of it.

If unable to turn to windward, anchorage will be found on the southern side of Ousha island, in the entrance to the Sarah Galley channel. If able to weather the north end of Chukea, the south side of Putu island will be found the best stopping place; the anchorage in 12 fathoms is under the hill, with three chimneys on it; the mud bank from the shore is very steep, shoaling quickly from 12 to 2 fathoms. From this position, in a handy vessel the best route will be through Lansew bay, and through the channel between Lansew and Tae-shan islands, described in page 174; but large vessels had better pass eastward of Video, and enter the archipelago farther to the northward. If unable to fetch to windward of the Barren islands, a convenient anchorage can be found among the Saddle group, should the tide or weather be unfavourable for entering the river.

During the South-west monsoon, endeavour to make the island of Video, as it is the highest islet to the southward. If late in the day an anchorage may be sought under the Saddle islands, which afford shelter in both monsoons, and it may here be noticed, that as the entrance of the Yang-tse kiang is somewhat difficult for a stranger to make in fine weather, no vessel should attempt it in bad, without a good departure either from Gutzlaff island or the Amherst rocks, and strict attention to the course and distance made good.

Leaving the Saddle island keep North Saddle bearing about S.E. by E. until Gutzlaff island bears South, distant 15 to 16 miles, recollecting that if Shaweishan island shows plainer than Gutzlaff, the vessel is too far to the northward, and in danger of entering the *false* channel to the northward of the Tungsha banks. Gutzlaff island, 210 feet high, when first seen, will appear like a small round lump. Shaweishan island, which is a little larger than Gutzlaff, and 196 feet high, is not often seen when a vessel is in the right position for approaching the Tungsha banks. Thus far the tide sets N.W. by W. and S.E. by E. from $1\frac{1}{2}$ to $3\frac{1}{2}$ knots; but it is affected greatly, both in direction and velocity, by the prevailing wind.

With Gutzlaff island on the above bearing and distance, if a clear day, the light-vessel will be seen, when steer for her to cross the bar, passing her as most convenient, taking care, however, not to bring her westward of W.N.W.* when eastward of her, nor southward of S.E. by E. $\frac{1}{4}$ E. when westward of her, and to make due allowance for the set of the tides over the Tungsha banks.

From the light-vessel steer about N.W. by W.; when about dipping her hull, the Beacon tower at Kiu T'oan will be seen, and when it bears about W.S.W. the vessel will be in 6 fathoms at low water, with the south shore plain in sight. Continue the above course, passing the Beacon tower at about 2 miles, when the dry north bank will in all probability be visible on the starboard hand, as it is only covered at the highest springs. Blockhouse island† will soon rise, having at first the appearance of a cluster of fishing boats, and gradually showing itself a low island covered with bushy trees. When the large house on this island bears N.E. by E., the vessel will be in the narrowest part of the channel, which here is only $1\frac{1}{4}$ miles wide.

After passing Blockhouse, the south shore should be gradually closed to about a mile, and kept at that distance until the marks and buoy for

---

* Duncan J. Louttid, Master of H.M.S. *Assistance*, 1859.

† Blockhouse and Bush islands have much increased in size since 1842. There are many places where the water has shoaled since the last survey: there is a patch with $2\frac{1}{4}$ fathoms on it at half flood with Gutzlaff bearing S. by E. $13\frac{1}{2}$ miles.—*George B. F. Swain, Acting Master, H.M.S. Pilot*, 1850.

Wusung spit are seen. As the south shore bank is steep-to, that shore should not be approached nearer than three-quarters of a mile. The second clump of large bushy trees on the low point, open half a point of the square and well-defined outer point of Paushan, will lead clear of the Wusung spit, if the buoy should at any time be removed.

Great attention must be paid to the set of the tidal streams at the entrance of the Yang-tse, and also to the lead. So long as the weather is clear Gutzlaff island forms an admirable mark, and it has only to be kept to the westward of South until it is distant 16 miles, when a vessel may steer N.W. by W. for the Fairway buoy or the light-vessel; but in thick weather and a working breeze with a variable tide under her lee, it is difficult to ascertain when 16 miles have been made, and she will be liable to be horsed over to the Tungsha banks, where several vessels have been wrecked. These banks should always be approached with caution, as their southern edge gives no warning, unless it is by the lead indicating hard bottom; and, as the tide sets across and not into the river, it will be as well to ascertain the vessel's true rate over the ground by using the deep sea lead instead of the log-ship, and taking the opposite to the bearing of the line as the course. The break on the head of the Tungsha banks will sometimes be seen after passing the Ariadne rock, but in thick weather the southern side of the channel is no doubt the one to border on. Should the lead indicate hard bottom, and there be much sand amongst the mud on the arming of the lead, the probability is that the vessel is on the north shore.

In working up from the Saddle islands, do not bring Gutzlaff to the eastward of South, until 15 or 16 miles to the northward of it, when it may be brought to bear S.S.E. The vessel will then be on the edge of the south bank. She may now stand to the westward, nearly into her own draught, bearing in mind that the flood sets W.S.W. round the south-east edge of it, and the ebb contrary. All vessels should keep as near as possible to this bank, and not wait for a shoal cast to tack, when standing to the north-eastward.

After passing the light-vessel, do not, when standing to the northward, bring her to the southward of S.E. by E. $\frac{1}{2}$ E., and tack in $3\frac{1}{2}$ fathoms when standing towards the south bank. The deepest water is near and along the southern edge of the north bank, but in standing towards it do not wait for the second shoal cast to go about. Generally the edge of the north bank is lined with heavy fishing stakes, planted in 4 and 5 fathoms, with only a few feet water a ship's length inside them. As before stated, when standing towards *black* buoy No. 5, tack the first shoal cast, for the bank is steep-to.

The foregoing directions are for vessels of about 18 feet draught; small craft may close with the south bank when Gutzlaff island bears South, distant between 12 and 15 miles, and steer up with the lead for their guide. The south shore is not to be depended upon all the way; after passing the Beacon tower the bank is steep-to, and should not be approached within three-quarters of a mile.

DIRECTIONS for WUSUNG RIVER.—The north spit at entrance to this river is marked by a *red* buoy, surmounted with *staff* and *cage*, and numbered 8. It lies in 17 feet at low-water springs, with Paushan point bearing N.W. by W. ¾ W.; leading marks into Wusung river S.W. ¼ W., and mud fort S. by W. westerly.

Three poles, each 60 feet high, stand on the inner angle of the stone fortification (Fort A. on the chart) on the left bank of the Wusung river, and are used as leading marks for entering. The two rear poles have crows' nests built around them, and are painted *red*. The pole in front has on its summit a bull's eye or target, and is painted *white*.

It would be imprudent for a stranger to enter the Wusung without a pilot, who is always in attendance at the entrance, for the banks within are constantly undergoing changes from the alluvial deposits. In approaching the entrance, a peaked tower near the town of Paushan will be seen to the westward; and on the embankment, in front of it, a beacon, which must be kept a little open southward of the tower until the leading mark for entering the river is on, viz.: the *white* pole kept in one, or a little open southward of the southernmost or inner *red* pole. In entering pass close to the eastward of the *red* buoy, which marks the north spit; the deep-water channel here is narrow, and the spit is composed of hard substances. After passing the poles keep the western shore aboard. On the opposite shore, a little higher up the river than the town of Wusung, a sharp point—Pheasant point—juts out, off which there is shoal water, while on the Wusung side the water is deep, especially at the elbow of the river.*

Above this elbow, the opium vessels are moored along the western shore. Abreast the northernmost of these vessels, but near the eastern shore is the Inner bar, from which a shoal extends about 3 miles up the river. This shoal, known as the Middle ground, is rapidly increasing in height, and will soon form an island; a great portion of it is visible at

---

* See Chart of Wusung river, surveyed by Commander J. Ward in 1858, No. 1,601; scale, n = 3 inches. The directions from the outer bar to Shanghai are by Captain Sir F. Nicolson, H.M.S. *Pique*, April 1858.

half tide. Between it and the eastern bank of the river is the narrow ship-channel leading to Shanghai.

The Inner bar is marked by two boats painted *white* and *red*, which carry a flag and ball of corresponding colour by day, and a light of the same colour (*white* and *red*) at night; the red light is very faint. The white boat lies in 14 feet at low-water springs, on the western extremity of a shoal spit extending from the eastern shore. The red boat is in 12 feet on the northern extremity of the Middle ground.

To cross the Inner bar, pass close to the southward of the white boat; then after keeping well to the northward of the red boat, steer boldly towards the eastern shore until within two or three ships' lengths of it. After the Inner bar is crossed, steer close along this shore to avoid the Middle ground, until the junk anchorage is reached, about 4 miles above the opium vessels. The vessel will then be above the Middle ground and about a mile below Black point, which juts out from the eastern shore, and has trees and houses on it. There is deep water near this point and generally along the eastern shore, which should be kept aboard until the houses of the foreign settlement at Shanghai are seen. Then edge over to the opposite shore, steering for the British consulate, easily distinguished by its jack and flag-staff, and as being the most northerly large building.

As vessels of large draught are obliged to cross the Inner bar while the flood tide is running strongly, care must be taken to sheer over towards the white boat in good time. This caution applies with peculiar force to sailing vessels, for the flood sweeps up the river and towards the Middle ground with great strength. If intending to wait for high water, be careful not to anchor too near the bar. If the vessel is of large draught, it will be better to anchor below Wusung so as to give plenty of time and room to swing the ship; for with a strong flood a vessel may be abreast the bar before her head is the right way. No vessel of any size should attempt to pass through the junks or across the bar in light winds if the tide is running strongly; and it should be borne in mind that both flood and ebb streams continue to run for some time after the time of high and low water on the shore.

As the bar boats are not unlikely to be drifted away, and may perhaps be entirely removed, owing to the negligence of the Chinese authorities, the following directions may be useful to a stranger in crossing the Inner bar:—

After passing Pheasant point, to which a good berth should be given, steer across for the second creek on the eastern bank above this point. There is a wooden bridge across this creek, about 150 yards inshore from the bank of the river. By keeping this bridge in sight and the mouth of the creek open, the bar will be crossed in the deepes water ; then proceed as before.

When approaching Shanghai* with the flood, it will be advisable to anchor below all the shipping, unless a good berth has been previously selected. The space in front of the British consulate, at the entrance of the Su-chau creek, is generally clear of vessels and always looks inviting, but it should be avoided, as the chow-chow water renders this part of the river very insecure as an anchorage ; the holding ground also is indifferent, and a vessel is constantly swinging round and round. The best berths are abreast the Chinese custom house, along the west bank of the river ; the tides here run regularly and with less strength, and a fairway along the eastern bank is left clear. Vessels ought to be moored taut with at least 36 fathoms on each cable, and a mooring swivel should be invariably used. A heavy fine is imposed on vessels neglecting this precaution.

WINDS.—By a meteorological register kept at Shanghai, the prevailing winds from 1848 to 1854 appear to have been as follows :—

January	- N.E. to N.N.W. and generally N.N.W.	May	-	- E.S.E. to S.S.E.
February	- N.E. to N.W. and generally N.W.	June	-	- S.E. to S.S.E.
		July and August		S.S.E.
March	- N.E. to S.E. and variable.	September		- N.E. to E.
		October	-	- N.E. to N.W.
		November		- N.W. and variable.
April	- E.N.E. to S.E. chiefly S.S.E. and variable.	December	-	- N. to N.W.

TEMPERATURE.—The temperature by day and night taken by a self-registering Fahrenheit's thermometer in the open air in the shade at Shanghai, from 1848 to 1854, gives the following as the *extreme* ranges and the average *mean* temperature of each of the months for those years :—

	Maximum by day.	Minimum by night.	Average monthly mean.		Maximum by day.	Minimum by night.	Average monthly mean.
January	- 67	- 18	- 41	July -	- 100	- 64	- 85
February	- 65	- 19	42	August	- 100	- 63	- 84½
March	- 75	- 28	- 50	September	- 92	- 51	- 77¼
April	- 79	- 33	- 59	October	- 90	- 37	- 67
May -	- 87	- 37	- 69	November	- 80	- 25	- 56
June -	- 99	- 58	- 75	December	- 77	- 19	- 46

The mean average height of the barometer in the spring and winter months is above 30 inches, and in the summer months below it ; viz., from January to April 30·25 inches ; from October to December, 30·34 inches ; from May to September, 29·83 inches ; ranging lowest with southerly winds and during the season of the North-east monsoon.

---

* Shanghai signifies, upon the sea.

WEATHER.—January is in general fine at Shanghai. In February, thick fogs occur. March is damp and disagreeable. April has more rainy days than any other month, except June, which is the wettest month. In May there is but little rain, and that little occurs in heavy showers. July is hot, dry, scorching, with considerable rain in the form of evening thunder-showers. July and August are the hottest months. In September the South-west monsoon is wholly broken up, and the temperature is very changeable. In November the winter fairly sets in, the first frost appearing from the 12th to the 20th. December is the driest month of the year, and the weather clear and freezing, though fogs are of occasional occurrence. In May, June, and July fogs also occur.

The summer gales are strongest from the S.E., and generally give good notice, the barometer beginning to fall sometimes as much as 24 hours previous. The rules for judging the barometer on the Chinese coast generally hold good for the neighbourhood of Shanghai; a rapid fall betokens a gale, and a high range the continuance of northerly winds.

WUSUNG TO HANKAU.

Until re-surveyed, buoyed, and local pilots established, the Yang-tse kiang, above the red buoy at the entrance of Wusung river, must be navigated with extreme caution; the constant accumulation of a very fine kind of sand having created banks, where, at the period of the last survey in 1842, deep water existed, and shoals, which then were at all times covered, are now, from the increase of alluvial deposits, at low water in many places exposed to view.*

BLONDE SHOAL, which in 1842 had 2¼ fathoms on it, and a knoll marked to the S.S.W. of it, appear to have formed a junction, and the bank made by them now shows at low water in several spots.

In November 1858, the vessels forming the escort to the Embassy, *Retribution, Furious, Cruizer, Dove,* and *Lee,* passed to the westward of this bank, but at low water not more than 15 or 16 feet will be found in the channel. The *Retribution,* drawing nearly 20 feet, had to wait until half-flood, and then had but six inches to spare.

WEST BLONDE CHANNEL.—This channel was chosen in preference to the one on the eastern side of the above bank, merely from its having

---

* See Chart of the Yang-tse kiang, from the sea to Nanking, corrected to October 1859, No. 1,480; scale, $m = 0·2$ of an inch.

The description of the navigation of the Yang-tse kiang above Wusung is by Commander J. Ward, R.N., who accompanied his Excellency the Earl of Elgin in the expedition to Han-kau, in November 1858.

been examined and partially buoyed; want of time preventing the examination of the other, which will be found to be the deeper, straighter, and more desirable channel of the two. The course from the red buoy off Wusung to the entrance of West Blonde channel is N.W. ¼ W., about 7 miles; when that distance has been run, it will be prudent, if drawing not more than 15 feet, to approach the land (a large and rather conspicuous clump of trees bearing South), touching, with the lead, in from 3½ to 4 fathoms on the edge of the bank off the shore of the mainland, it being less abrupt, and affording more warning, than the edges of the Blonde shoal, or rather bank.

As the monotonous embankments of this part of the river afford no landmark that could be recognized from description by a stranger, their height hiding the houses, &c. in the rear, the lead must be mainly depended on. The course through the channel is about N.W.

A quick eye may possibly detect two *small* Joss poles which occasionally may be visible, through the tops of the trees, over the embankment. If seen, when brought to bear West, a push should be made to the northward, clearing the north end of the Blonde shoal; should the Joss poles not be visible, a boat should be used to sound, and to mark the N.N.W. end of the shoal, clear of which the channel should be crossed steering N.N.E., on which course from 8 to 10 fathoms will be obtained, until the bank extending from Tsung-ming is approached. When in from 5 to 6 fathoms, steer so as to keep on the edge of this bank in that depth, the course being about N.N.W. ½ W.; by this means the Doves Nest, a very dangerous collection of banks, will be avoided. The *Cruizer* and *Dove* both grounded on the 7th November, and the *Furious* on the 9th on these shoals, from which the following bearings were obtained :— village on Harvey point N. ¾ E.; single tree W. by N. ¾ N.; great bush W.N.W.; and left extreme of Mason island N. ¾ W.*

The Captain of the steamer *Confucius*, in the pay of the Chinese Government, reports that there is a good channel close along the main shore; but he acknowledged that his vessel, drawing about 8 feet water, often grounded in it; it was therefore not considered desirable to try that channel until time permitted of its examination.

HARVEY POINT may be passed at about three-quarters of a mile, and when a conspicuous clump on it bears S.E. by E. ½ E., steer

---

* The *Cruizer* kept along the south shore of the river, in 3½ and 4 fathoms, and when Harvey point bore North steered directly for it, by which means she grounded on the south edge of these shoals.— *William D. Strong, Master H.M.S. Cruizer*, 1858.

Mason island and Tsung-ming are now connected by a sand bank, dry at half tide. The channel off Harvey point has narrowed, and the point is no longer steep to the bank. —*Captain S. Osborn, H.M.S. Furious*, 1858.

N.W. by W. ¼ W. From 6 to 7½ fathoms may be expected on this course, until the conspicuous Single tree marked on the chart bears S.S.W. ¾ W.; then steer W. by N. ½ N., making due allowance for the tides, which here run N.W. and S.E.

PLOVER POINT may be known by the village on it, and a small fort or breastwork. A number of junks are generally at anchor in a river or creek opening at this point. When the fort bears S.W., the dangerous banks and shoals known as the Lang-shan crossing may be said to commence.

LANG-SHAN CROSSING.—The squadron navigating by the chart of the former survey in 1842 passed along shore, and was stopped abreast Fu-shan by a long sandy spit effectually barring the passage, and for three days the *Dove* and *Lee*, and the boats of the squadron, were engaged in searching for a passage through, until at length the tail of the spit was rounded by the *Dove*, but as low down as Plover point, and the squadron had to return 13 miles.*

The fort on Plover point, which is low and will require a good glass to make out, must not be brought to the southward of S.W., until Fu-shan hill, if seen, bears West; then steer for the hill. If the weather be clear, Fu-shan will be made in the shape of a hummock, crowned by trees, and a few white houses; a small fort, like a mortella tower, standing on the slope, may also possibly be seen. If unable to get a bearing of Fu-shan, when Lang-shan pagoda, a very conspicuous object situated on the summit of the highest of three hills, bears N.N.W. ¾ W., and a white house on left bank of river, if distinguishable, N. 27° E., steer West, until the pagoda bears North, the depth being from 7 to 9 fathoms; then haul more northward, W. by N. ½ N., until Lang-shan pagoda bears N. 13° E., when the course becomes N.W. ½ N. until Lang-shan pagoda bears E. by N.; the course may then be again altered so as to approach closer the left bank of the river, which may now be, for some distance, kept aboard.

KU-SHAN POINT, which when seen from the south-east is wedge-shaped, the thick end 90 feet high being outwards and very conspicuous, should not be approached in passing nearer than 2½ to 3 miles.

Great alterations have taken place in the vicinity of this point since

---

* Capt. S. Osborn also remarks, that "the tail of this spit alters its position almost weekly; for between November 14th, 1858, and January 10th, 1859, its bearings varied from the fort on Plover point, from N. 55° E. to N. 22° W. The old channel across the river, between Fu-shan and Lang-shan, is almost obliterated, and only 13 feet could be found in it at high water; the shoal water also between Fu-shan hill and Ku-shan point has several islands of recent formation rising from it."

the survey of 1842. An entirely new island, cultivated and inhabited, has risen up; the small islands, marked in the chart, appear to have formed a junction; and extensive banks, occasionally showing, have come into existence, and are constantly being added to, by the earthy matter brought down the river; doubtless, eventually, perhaps shortly, to become islands themselves, and add their quota of rice, &c. to the support of the teeming population of China, while at the same time the river appears to be scooping out the land on the opposite or left bank.

After passing Ku-shan point, the most anxious and dangerous part of the navigation of the Yang-tse may fairly be said to have been accomplished; in no other portion of the river do we find the same rapid alterations in the bed, especially in the vicinity of Fu-shan and Lang-shan, where the strong tides appear to be actively and constantly engaged in removing some banks while others are being formed. Until a good local pilotage has been established, vessels, especially sailing vessels, proceeding up, would act wisely by always having a boat ahead, showing the soundings. The time lost in this slow mode of progress is not to be compared with the loss of time consumed, exclusive of damage, in heaving a ship off a sand bank.

HWANG-SHAN.—After passing the islets formed and in course of formation abreast Ku-shan point, the river becomes pretty clear, and a mid-channel course may be safely pursued, steering for the high land about Hwang-shan, (which, from Ku-shan point, looks like an island in the centre of the river,) the lead giving no bottom at 8 fathoms. Anchorage, if required, can be had in Hwang-shan bay, but the water is deep. The squadron anchored there, in 12 fathoms close in shore. The hills in the vicinity range from 250 to 300 feet high. The rise and fall of tide in Hwang-shan bay was from 4 to 6 feet.

STARLING ISLAND.—The river narrows to a mile between Hwang-shan bay and Kiang-yin, but immediately afterwards becomes wider. A mid-channel course is still to be steered, following the trend of the river, until approaching Starling island, when the left bank must be neared, and may be kept pretty close (about a quarter of a mile), passing eastward of the island.

A dangerous shoal bearing from Keun-shan or Chu-shan pagoda S. 63° E. is forming in the river, and will doubtless before long become an island. It shows at low water, and lies close to the left or eastern bank of the river, with, at present (November 1858), a boat channel inside it.*

---

* Captain S. Osborn says that "this bank in January 1859 was dry to the eastern shore."

CHANG-SANG-CHAU.—The left bank of the river should still be kept aboard, passing eastward of a long low island, which appears to have grown considerably since 1842. No other banks appear to be forming, until abreast the islet named Chang-sang-chau, where from the right bank a dangerous shoal stretches nearly half way across the river, to avoid which the island of Seau-sha should be kept aboard.

The river banks are excessively monotonous between Kiang-yin and Keun-shan—the left bank quite flat—the only rising ground, through the whole extent, being Ku-shan hill, on which are some houses.

TIDES.—H.M.S. *Styx*, experienced a strong ebb all night when at anchor off Ku-shan point in June 1854, and it was supposed that the flood stream had no existence 10 or 15 miles below this point; this, however, must be considered to refer merely to that period of the year. From the fact of the river then having a large body of water in it, the downward current would naturally have greater weight, and check the flood stream; but in November 1858 the level of the river was lower, and the influence of the flood was felt much higher, and on 29th December, the level of the water being still lower, the flood stream was sensibly felt as high as Nanking. The *Retribution*, *Dove*, and *Lee* were occasionally compelled, on that day, to move their engines to enable them to preserve a position off the forts, during the time some negotiations were being proceeded with on shore.

SILVER ISLAND.—In passing Silver island, the southern channel should be taken, being careful to keep a *mid-channel* course to avoid the Furious rock, having 14 feet on it, lying about a third of the passage over from Silver island;* and a rock terminating some broken ground which extends from the right bank. A quick helm will be required in passing through, to avoid being at the mercy of the whirling eddies caused by the check these rocks oppose to the stream.

Silver island is evidently destined to form a junction with the small island called Ia-sha. A spit now runs out from the low flat tongue of ground which has accumulated to the north-eastward of the high ground of Silver island towards Ia-sha; while another appears to be working its way from Ia-sha towards Silver island, and the channel between the two is clearly filling up, as the chart of 1842 shows 12 fathoms; but no such depth is to be found now, and the passage has become so narrowed,

---

* The bearings from the *Furious* when aground on the reef extending to the southward from Silver island were:—Golden island pagoda W. by S. ½ S.; Keun-shan pagoda E. ¾ S.; and extremes of Silver island from N.E. ¼ E. to N.W. by N. At two ships' lengths to the south-east of the reef there is a rock with only 16 feet water on it.—*Captain S. Osborn, R.N., November* 1858.

that it was not deemed prudent to take the *Retribution* through in November 1858.

GOLDEN ISLAND.—After clearing Silver island, the left bank of the river should be gradually approached, and passing Chin-kiang fu, kept close aboard, to avoid a rock, said to have 10 feet on it, lying nearly in the centre of the river, to the north-west of Golden island; in the vicinity of which island several rocks appear to exist.

Golden island is now connected with the mainland by a low isthmus well covered with grass.

PIH-SIN-CHAU.—Both channels, north and south of Pih-sin-chau, appear to be safe and clear. The south channel was taken by the squadron; but as a Chinese Imperial steamer was observed passing through the north channel, it is very probable therefore that it is navigable: the Chinese steamer drew 9 or 10 feet.

In proceeding through the south channel, Pih-sin-chau should be kept aboard; and, after passing it, a mid-channel course steered, until abreast Yen-tse-ki, when the left bank must be closed in order to avoid an outlying rock said to exist near the right bank, after passing which a mid-channel course may be again steered, giving a mud flat, recently formed at the north-west side of Tsau-hia island, a berth. After passing this flat, the river appears to be quite clear.

NANKING.*—Theodolite point and the Nanking forts may be approached within pistol shot. The (two) forts on the right bank of the river are erected on a detached tongue of land in front of the city walls. Twelve 24-pounders, mounted on solid wooden carriages, were on the mud bank, in front of the wall of the lower fort, not on it, consequently no shelter was afforded the gunners. The upper fort on right bank mounted three 24-pounders, and six 6-pounders, in the same exposed position; some gingalls and light guns were on Theodolite point, where armed junks were also stationed. A fort on the left bank mounted a few light guns and gingalls.

Vessels forcing a passage should keep close in to the right bank of the river, and steady rifle practice will effectually keep down any fire, especially as the elevation of their guns is great, and they never alter it. After passing Nanking, a mid-channel course may be steered. At Sanshan the Admiralty survey of 1842 terminated.

---

* Nan signifies South, and Peh, North; thus Nanking, South capital; Pehking' North capital. Pak is thus pronounced in the Peking dialect, Pih in the Nanking dialect, and Pak in the Canton dialect.

ELGIN REACH appears to be clear. A mid-channel course may be taken, until the centre of some rising ground, about 150 feet high, and 3 miles S.W. of a remarkable rebel stronghold on left bank of river, bears West; when the right bank should be closed to avoid a spit, running to the north-east of what appeared to be a small island.*

Wade Island.—The squadron took the channel east of Wade island, and generally found no bottom at 8 fathoms. Some rebels in three small stone forts, at a village called Tsai-shih-ki, had the temerity to fire a few matchlocks and gingalls, but a couple of well-directed shots from the *Retribution* knocked the forts down, and they were still in ruins when the squadron returned at the end of December.

The channel west of Wade island is stated to have been used by the U.S.S. *Susquehanna*, and to have 8 fathoms in it; both channels may therefore be said to be navigable. The western one, if clear, is decidedly the preferable of the two, as by using it some shallow ground $3\frac{1}{2}$ miles W.S.W. of Tai-ping pagoda, and abreast the small village of Tang-tu, will be avoided. Should the eastern channel be used, do not approach the right bank of the river near Tang-tu, but pass close to the south-west end of the small island south of Wade island, as a dangerous shoal stretches out from the village before mentioned; by keeping Tai-ping pagoda nearly touching the south side of the small island, the depth was not less than 17 feet.

The Pillars.—When the river, as in January, may be considered to be at its lowest level, close the left bank to within half a mile, keeping Tai-ping pagoda as before, and steer a mid-channel course, passing between the East and West pillars, which are two rugged eminences strongly fortified; one, the eastern, is supposed to contain the treasures of the Tai-pings.

From the Pillars the course up the river is southerly. Morton point may be kept close aboard, passing eastward of a small flat island. A rock just showing in November, but dry 6 feet in December, lies on the right bank of the river, 3 miles south of Morton point and about a cable's length from the shore, and would be covered earlier in the autumn.

WUHU REACH.—Off the rebel city of Wuhu a shoal lies about a cable's length from the right bank, and dries in December; a mid-channel course clears it, and may be steered with safety through the Wuhu reach, gradually closing the right bank on approaching a range of hills

---

* *See* Charts :—The Yang-tse kiang, from Nanking to Tung-liu; and Tung-liu to Han-kau, Nos. 2,678, 2,695, scales, m = half an inch: surveyed by Commander J. Ward and Officers of H.M.S. *Actæon* and *Dove*, November 1858.

700 feet high, abreast which are a small islet and some mud patches at about 1½ cables from the shore of the left bank. These mud banks were covered in November, but dry for nearly 2 miles in December. After passing these banks a mid-channel course may be again followed, passing southward of Barker island, from the north-east point of which Kicu-hien pagoda will be seen.

Kicu-hien.—As the difficulties in the navigation appeared to be increasing, it was determined to leave the *Retribution* at Kicu-hien, where it was understood liberal supplies were to be obtained. Good anchorage in from 5 to 8 fathoms was found off the city, which stands on the right bank of the river, about 80 miles above Nanking; some conspicuous hills, ranging from 1,500 to 2,000 feet high, rise some 3 or 4 miles to the southward of it. The left bank of the river is a complete flat. Some mud banks are accumulating about the south-west end of Barker island. A mid-channel course should be steered, the least water found being 3¾ fathoms; this part of the river must be approached with caution. The channel north of Barker island is supposed to be clear, and, if so, would avoid the shoal water before mentioned.

TIDES.—From the 24th November 1858, the day the *Retribution* anchored off Kicu-hien, there was a daily rise and fall of 6 inches, but a steady decreasing of the level of the river until the 18th December, when the fall—since the 25th November—had amounted to 8½ feet. From the 18th December, when there commenced a week's constant fall of rain, with fresh N.E. and easterly winds, the river rose gradually 3 or 4 feet and the vessel swung occasionally to a flood stream. The influence of the flood was sensibly felt off Nanking on the 29th December, and very slightly off Tai-ping and Wuhu.*

OSBORN REACH is clear. After passing Tcih-kiang keep close to the right bank of the river until clear of Osborn reach, and approaching a large village built on the left bank, cross over, keeping that bank aboard until after passing the village.

WILD BOAR REACH.—A mid-channel course may now be steered through Wild Boar reach, which trends to the southward. Some high land will here be seen on the left bank of the river, which was the first rising ground met with on that bank after passing through the Pillar hills, a distance of more than 50 miles. Keep the left bank of the river in view, to prevent being enticed into a wide channel opening in the right bank, and which at first has the appearance of being the main

---

* William F. Hains, Master of H.M.S. *Retribution*, 1858.

stream; it is about 2 miles to the northward of a walled village which stands on the left bank. On nearing this walled village, close the left bank slightly to avoid some shallow ground lying abreast it off the right bank; after passing which, Wild Boar reach is quite clear, the course still southerly. After passing a ruined temple, which stands on a very conspicuous bluff about 100 feet high, edge over to the left bank, to avoid some shallow ground on the right bank, where the channel takes about a W.S.W. course for about 23 miles.

Fitz-Roy Island.—Some shallows lie off the left bank of the river east of Fitz-Roy island, to avoid which keep the right bank aboard until Chichau pagoda bears South; then edge over towards the east part of Fitz-Roy, passing through the channel north of it, keeping pretty close to the left bank, as some mud flats lie on the north side of the island.

After passing Fitz-Roy island the river is again clear, and a mid-channel course may be steered, about S.W. by W. The country about this part of the river is hilly.

Dangerous Shoal.—About 8 miles to the south-west of Fitz-Roy island, and about $1\frac{1}{2}$ cables from the left bank, is a very dangerous shoal, dry in December, but covered a month earlier. To avoid it, pass within a cable's length of a conspicuous rocky islet about 30 feet high, named Tai-tzu-chi, which lies nearly in the centre of the river, and cannot be mistaken,

Lang-kiang-ki.—From thence until past Lang-kiang-ki, or Hen point numerous rocks lie in the bed of the river. From Lang-kiang-ki a dangerous cluster extends for more than half way across the river. In November the outer rock was marked by a small bush sunk on it; it was, however, dry in December. To clear these dangers keep the left bank of the river aboard, the course becoming again southerly for 5 miles, when steer to the westward through Nganking reach.

NGANKING REACH.—Approaching Nganking, keep the left bank of the river aboard, and pass *close* under the walls of the city, to avoid extensive shoals and mud flats which exist on the other bank.

Nganking is an extensive walled city standing upon the left or north side of the river. The river face or south side of the walls is fully $1\frac{1}{2}$ miles long. Extensive suburbs, which once existed, were now in ruins. The city in November was in the hands of the rebels, but closely besieged by the Imperialists, who, however, appeared in December to have raised the siege.

After passing Nganking the river is again clear, steering south-westerly until passing a sandy point, when the course becomes West, keeping on the left bank.

CHRISTMAS ISLAND.—After rounding a small islet called Rover island the course is again to the southward, gradually approaching Christmas island, the southern point of which should be passed at about 1½ cables, to avoid some mud flats lying on the left bank, and only dry in December.

TUNGLIU REACH.—Having passed the above flats, a mid-channel course may again be steered, until approaching Tungliu, a third-class city, with rather formidable looking walls, built on the right bank of the river, and abreast which, on the left bank, an extensive flat is in course of formation. It will be advisable to give the point, on which stands Tungliu pagoda, a good berth, as, although shallow water was not obtained there, a great commotion was observed in the stream, apparently caused by some rock, or other check to its even course.

After passing this pagoda, keep on the right bank of the river, thereby avoiding some banks in the centre, dry in December. The squadron, ascending the river in November, crossed over to the left bank, and became entangled among these shoals, the *Furious* grounding on one; they lie abreast three brick-kilns, looking like mounds of earth and stones. In January the *Furious* and *Cruizer* cleared them by keeping close to the right bank, and had deep water. When abreast Hwayuen-chin, where there is a custom-house, having Joss poles, these dangerous banks will have been passed.

On the right bank of the river here are some high ranges, but the left is quite a flat, and although the river banks were in November and December from 25 to 30 feet high, the country showed evident signs of being frequently inundated; sampans were found at most of the farm-houses as far inland as 3 or 4 miles, affording a very significant hint as to the state of the country, when the river is at its high level. It must be borne in mind, that the left bank, probably both, would then be covered, the river becoming a large lake; under these circumstances, it would be advisable to keep in the most rapid part of the current, as it always runs strongest in the deep water.

BULLOCK REACH.—At Dove point the river takes a sudden bend at right angles to its former trend for a short distance, the course being about W.N.W.; keep the left bank aboard, until entering Bullock reach, when a S.S.W. course is gradually obtained.

Little Orphan.—Near the southern termination of Bullock reach is the Siau-ku shan or Little Orphan, a most remarkable small rocky islet, rising almost perpendicularly out of the river, and nearly 300 feet high. It has some Joss houses and temples on its summit; half way up its southern face some houses are perched, probably the residence of the officiating priests ; if it were not for these convincing proofs to the contrary, it might well have been deemed inaccessible. In November it was separated by a very narrow belt of water from the left bank, but in December its base was connected to it by mud.

Immediately abreast the Little Orphan, a bold rocky head crowned by forts and look-out houses, rises abruptly to a height of 400 feet ; at its southern base is situated a fortified town (name unknown). The right bank of the river is still rich in hills, which about here are very rocky and uneven. A mid-channel course may be steered in passing the Little Orphan ; no bottom at 9 fathoms was obtained, until nearly abreast Siah-kia-kau, when the left bank was approached, to avoid some shallow sand banks near the opposite shore, the course becoming about West for 5 miles, when, still keeping to the left bank, it takes a more southerly direction into Blackney reach.

BLACKNEY REACH.—About half way down this reach is a shallow, apparently extending right across the river, and over which, in December, the greatest depth appeared to be about 14 feet. After passing a village on the north edge of a small creek or stream about a mile, steer S. by W. for a low point, near which are some houses (Chang-kia-kau), until the water deepens to 5 fathoms, when the right bank may be followed, gradually deepening the water to 8 and 10 fathoms. Off Becher point a sharp helm will be required, the eddies here being very rapid.

Oliphant Island, lying westward of Becher point, is about $5\frac{1}{2}$ miles long, and divides the river into two branches, which are both shallow. The southern branch was used in November, when $3\frac{1}{2}$ fathoms was the least water obtained; the northern branch was examined, but not approved of, as although more water was found in it several dangerous banks were sounded on (page 215). On 22d December the water had fallen 7 feet since the examination of these channels in the preceding month, consequently it was found necessary for the *Furious* and *Cruizer*, when descending the river, to wait a rise in the river before attempting either channel. They were anchored off the imperial city Kiu-kang, and were detained there some days ; when the shoals having been carefully buoyed, and a rise of water having fortunately taken place, the north channel was passed. The *Furious* grounded.

Opposite Becher point are several sand hills on the right bank of the branch which conducts the tributary waters of the Poyang lake into the main stream. A fortified ? temple, built on a steep cliff, will also be seen on the same bank.

SEYMOUR REACH.—After passing the west end of Oliphant island the course is S.W. by W., past the imperial city of Kiu-kiang, which stands on the right bank of the river, and has most imposing looking walls, enclosing desolation and ruin. After entering Seymour reach, the trend of the river is more northerly, and a mid-channel course may be safely steered.

Hunter Island.—In November the channel south of Hunter Island was passed through, but much difficulty was met with in getting the *Furious* over a flat extending right across the river, and on which are several sand banks. This shallow ground lies abreast some very conspicuous red cliffs, from 40 to 60 feet high, situated on the right bank of the river. When opposite the west end of these cliffs, cross the river carefully, feeling the way by the lead.

' In December the channel north of Hunter island was taken. The *Furious* grounded, but after some hours worked a passage for herself through the mud, and got into a vein of deep water very close to the left bank.

Court Reach.—Steer in mid-channel through Court reach, about W. by S., passing the town of Wuhiutsun, which stands on the left bank. A lively trade in timber appears to be existing here. Three miles west of this town some hills, about 600 feet high, occur, the first break to the dull monotony of the left bank since leaving the vicinity of Nganking.

Abreast these hills the course becomes a little more northerly.

Futz-kau.—Opposite the town of Futz-kau some shallows are forming in the river. The right bank appears to have most water.

Ke-chau.—The course up the river is now about N.N.W. No shallows appear to exist after passing Futz-kau until approaching Ke-chau, when a remarkable ruined fort, standing on an isolated rock, must be closed, in order to avoid mud flats lying off the right bank. On passing two small hills about 2 miles below Ke-chau on the right bank of the river, steer for the ruined fort, passing it a hundred yards outside, and the shore at Ke-chau at the same distance; by this means the mud flats which extend some 4 miles parallel to the bank will be avoided.

Ward Reach, which trends about N.N.W. and S.S.E., is now entered, and appears to be quite safe and clear; the left bank of the river was kept aboard by the squadron, and no check whatever was experienced.

Ke-tau, or Cocks Head, may be passed close to. It is a remarkable bluff rising perpendicularly to a height of 300 feet on the right bank of the river. and cannot be mistaken.

Lee Rock.—A dangerous rock or rather a collection of rocks, on which the *Lee* struck when descending the river, lie abreast some limestone quarries, at a placed called Shih-wuy-aou, on the right bank; in December there were only 6 feet water on them. From Ke-tau steer West, being careful not to approach the right bank until Cocks Head is touching the low point of the opposite shore (left bank), when the Lee rock will have been passed to the northward. The right bank may now be kept aboard, passing close to the densely populated little town of Hwang-shih-kang, when the left bank should be gradually closed, taking the channel east of Collinson island. A small rocky hill, 70 feet high and about 2½ miles north of Hwang-shih-kang, lies on the left bank, and marks the commencement of this channel; in navigating which, the left bank is to be kept aboard.

Collinson Island.—Off the north end of this island is an extensive flat extending across the river; 4 fathoms was the deepest water found about mid-channel. A careful lead is the best guide here. The right bank may now be gradually closed, and, passing the small village of Yang-ki, kept close aboard, to avoid a bank on the opposite shore, and which was dry in December.

PAHO REACH.—After passing a small ridge of hills, on one of which is a remarkable and conspicuous boulder, cross over to the left bank to avoid some shallow ground lying off the small village of Tzko-kang; 3¼ fathoms was the most water found at this crossing. On obtaining 5 fathoms on left bank, steer boldly up the Paho reach, passing northward of two rocks, one 18 and the other 10 feet high; the latter lies north of the third-class city named Wu-chang-hien. In the summer these rocks would be covered; to avoid them, keep the left bank aboard.

Bythesea Channel.—After passing Wu-chang-hien, a mid-channel course may be steered, until abreast Hwang-chau pagoda, on the left bank of the river, when edge over towards the Bythesea channel. The squadron ascending and descending the river was compelled to use this channel, there not being sufficient water for the *Furious* in the eastern channel.

The Bythesea channel must be navigated with extreme *caution*, keeping the right bank aboard; it is so narrow that a vessel touching on either side, and swinging across, would ground on the opposite bank, and have the whole weight of the stream pressing her down. At an earlier period of the year the eastern channel would have plenty of water in it, and, if so, the Bythesea channel should be avoided.

The course now becomes North, and the river clear. Gravener island should be kept aboard, to avoid sand banks on the opposite shore.

WASHINGTON REACH.—After passing Gravener island, a sudden bend of the river leads into Washington reach, through which a mid-channel course (about W.S.W.) may be taken; 4½ fathoms was the least water found in this reach in December, and the river was then nearly at its lowest level. The right bank is to be approached on nearing some rising ground about 300 feet high, which lies on that side of the river; from thence the same bank is to be followed, as there appears to be shallow water off the left bank after passing the Pih-hu shan, or West Tiger hill, a prominent elevation about 400 feet high. The river again takes a northerly course (about N.N.W.), and appears to be clear, and free from any impediments.

Yang-lo, a small town on the left bank, may be approached close to; a ruined temple standing on the spur of a hill, one mile south of Yang-lo, is a conspicuous object.

PAKINGTON REACH.—A mile north of Yang-lo, Pakington reach is entered, the course taking rather a sharp turn to the westward, gradually turning to the southward into Hankau reach.

HANKAU REACH trends nearly S.W. and N.E., becoming still more southerly at Hankau. The two last-named reaches are, by keeping on the left bank, free of any impediments. North of a remarkable bluff (200 feet high), called Kin-shan, which is on the right bank of Pakington reach one mile inland, lies a sand bank which dries in December; it is easily avoided by nearing the left bank.

Opposite Hanyang, just above the entrance to the river Han, lies an extensive mud bank, dry 4 feet in December. A spit, gradually deepening, stretches to the northward from this, and affords good anchorage in from 3 to 7 fathoms.

HANKAU.—At Hankau, 384 nautical miles above Nanking, the river still maintains the same characteristics, showing no signs whatever of a decrease either in breadth or depth. The water under the walls of Wuchang fu is just as deep as at Nanking; no bottom at 9 fathoms was obtained.

The cities of Hankau and Hanyang stand on the left bank of the river and the city of Wuchang on the right bank. Hankau, signifying the mouth of the Han river, is flourishing, and fast recovering from being burnt down by the rebels. Hanyang and Wuchang are walled cities, the former is utterly and entirely ruined, and nothing remains of it but a heap of bricks. Wuchang is very extensive, and said to be much larger than Canton.

The season of the year at which the squadron ascended afforded good opportunities of observing banks and shoals which would be covered at an earlier period, but no just estimate could be formed of the force of the constant downward stream. A rate varying from $1\frac{1}{2}$ to 4 knots was observed; the latter only obtained in certain localities. In the summer the stream is said to obtain a constant rate varying from 5 to 7 knots, a circumstance which will effectually deter sailing vessels from attempting the voyage up.

An immense fleet of junks are always at anchor at the mouth of the Han, discharging and taking in cargoes; and large fleets appear to be constantly moving up and down the Yang-tse and the Han, telling tales of populous cities in regions still farther inland.

The position of Hankau is lat. 30° 32′ 51″ N., long. 114° 20′ E. (approximately); variation 0° 13′ E. No observations for dip were obtained.

Supplies.—Among the staples observed at Hankau were specimens of excellent iron, quite as good as any British or Swedish metal; the best quality was selling at about the rate of 14*l.* per ton.

Coal was to be had in any quantity from 2*l.* 5*s.* to 2*l.* 15*s.* per ton. The best quality appeared to burn very fairly when mixed with a little Welsh coal or assisted with wood. It left a white ash, but seemed to give quite as good results as Formosa or Labuan coal.

Flax, tea, insect wax, raw silk, copper, tin, and manufactured cottons were also to be obtained. Considerable quantities of British cottons were exposed for sale, as well as American drills; all the woollens were Russian.

HANKAU to WUSUNG.*—The squadron left Hankau on its return voyage on the 12th Dec. 1858, and on arriving at Hwang chau, where the deep water channel was about the width of the *Furious* and very tortuous, it was discovered that the river had fallen fully 7 feet since the 3rd of Dec. This difficult pass (page 213) was cleared without accident on the 13th, but the *Furious* was brought up by a bar at about noon; the water at this point had diminished to 10 feet in the channel used in the way up.

On the 14th an intricate channel with 17 feet water in it was discovered and buoyed, and the squadron passed through, the *Furious* grinding over a 14½ feet patch of sand, which the strong current prevented the boats from finding. The headway of the vessel carried her over the sand, when she was again anchored to buoy another channel with a sharp twist and a current through like a rapid. On the 15th, the vessel was taken through this passage, though the risk, had she touched, would have been great. In the afternoon the *Lee* struck heavily on a patch of rocks (page 212) with only 6 feet water over them, lying nearly in mid-channel to the westward of Ke-tau point, and over which the current was running 3 knots.

---

* The description of the descent of the Yang-tse is by Capt. S. Osborn, H.M.S. *Furious*, 1858.

The red cliffs (page 212) were reached on the 16th. The squadron remained here the 17th and 18th looking for a channel deep enough for the *Furious*; at last one was found with 15 feet in it, but turning sharp to the right and left. After marking it carefully with buoys it was entered ; the *Furious*, however, caught some knoll, fell athwart the current, and as her three anchors would not bring her up, she drifted through, the rolling sand of which the bed of the river is formed, yielding to her pressure.

On the 20th, the squadron anchored off Kiukiang (page 211), and boats were sent to examine the channels north and south of Oliphant island. Only 4 feet were found in the south channel ; 10 or 11 feet in the middle channel, and the bar wide and composed of hard sand ; and the same depth in the north channel, the current running like a mill-stream. As the *Cruizer* was drawing 14 and the *Furious* 15 feet, it was determined that they should remain at anchor off Kiukiang, and that the Ambassador and suite should proceed to Shanghai in the *Lee* gunboat accompanied by the *Dove*.

From the 24th of December it rained heavily with but slight intermission until the 28th, when a 15-feet channel was found north of Oliphant island, the water having risen about a foot. Throughout the 30th of December and 1st January the river was rising steadily. The weather improved, and the sun acting upon the snow-covered hills, added by a thaw to the quantity of water flowing on all sides into the river, so that on the 2nd it had risen 2½ feet.

On the 3rd of January there being 17 feet in the channel north of Oliphant island, the *Furious* and *Cruizer* proceeded by it, but when halfway through the *Furious* grounded on a 13-feet bank. The anchor coming home, the vessel fell athwart the current, and before another anchor could be laid out, the pressure of the vessel had formed a sand bank under her lee bilge with only 9 feet water on it ; fortunately, however, towards midnight the action of the current began to loosen the sand, and at 3 a.m. on the 4th she floated. By buoying and sounding, the bar at the eastern end of the channel was eventually crossed, carrying 16½ feet water ; the tide running like a sluice. On the 18th the vessels passed Nanking, and the next day arrived at Wusung.

CHAPTER VII.

EAST COAST OF CHINA.—WHANG HAÏ OR YELLOW SEA; GULFS OF PE-CHILI AND LIAU-TUNG; AND WEST AND SOUTH COASTS OF KOREA.

VARIATION 2° 0′ TO 3° 30′ WEST IN 1861.

The WHANG HAÏ, or Yellow Sea, is bounded on the west by the deep bight of the coast formed between the Yang-tse kiang and the Shan Tung promontory, and on the east by the coast of Korea. It is mostly muddy, and of a yellow colour near the land, and has been little frequented by European vessels, nor has any part of the coast been explored between the Yang-tse and the Shan Tung promontory, although it is known to contain several excellent harbours, and to possess an extensive coasting trade. The following are a few of the best known points of the coast.*

WHANG HO, or Yellow river, the entrance to which is said to be in lat. 34° 2′ N., long. 119° 51′ E., is almost unknown to Europeans, but it is stated to be little inferior to the Yang-tse kiang in magnitude. The whole of the low coast between these great rivers is fronted by extensive flats and shoal banks, projecting in some places above 60 miles from the land, and rendering the approach dangerous for vessels of large draught until better known, although there may probably be channels among these banks in the neighbourhood of the coast frequented by the native trading vessels. H.M.S. *Highflyer,* July 1859, had soundings of 12 fathoms, abreast of and 100 miles from the mouth of the Whang ho.

KYAU-CHU, or Glue city, said to be in latitude 36° 17′ N., long. 120° 12′ E., stands at the north-west part of the head of a deep bay, and is bounded by a peninsula on the eastern side. It has a spacious harbour, and is the principal emporium of the province of Shan Tung. Teih-mei-heen, or Black Ink city, about 24 miles to the eastward, on the bank of a river which runs into the north-east branch of the same bay, is said also to be a place of considerable trade.

---

* *See* Chart: China from Hong Kong to Liau-tung, No. 1,362; scale, $d = 2$ inches.

URH TAO, or Ear island, also called Staunton island, is in about lat. 36° 47′ N., long. 122° 16′ E. It is of middling height, and lies near the south point of the peninsula of Shan Tung.

ACTÆON SHOAL.—A dangerous shoal, lying to the southward of the Shan Tung promontory, was sounded on by H.M.S. *Actæon*, on the 19th February 1860. The least depth obtained on it was 22 feet, in lat. 36° 31½′ N., long. 122° 28′ E.; but less water probably exists.

Approaching the shoal from the southward, the depth gradually decreased from 12 fathoms at 8 miles south of the shoalest part, to 10, 8, 7, and 5 fathoms, and then rather suddenly to 22 feet; it then rapidly deepened to the northward. The land was in sight occasionally through the haze, but not sufficiently distinct to get bearings of its extremes.

CAUTION.—Until an opportunity offers of ascertaining the dimensions and features of this shoal, vessels approaching its vicinity should keep a careful lead going. It will be prudent not to make the land until nearly on the parallel of the promontory.

SHAN TUNG PROMONTORY, in lat. 37° 25′ N., long. 122° 45′ E. and the easternmost land in China, is the eastern extremity of the Shan Tung peninsula. The promontory is high and bold with a rugged termination near the sea, and has a small pagoda near its end. The soundings are 16 and 18 fathoms about 9 miles from the promontory, but increasing fast to 30 and 40 fathoms when it is approached within 3 miles.

About 2 or 3 miles to the north-west of the promontory there is a small but high island, named Alceste: it appeared to have a reef extending about half a mile around it, and there are some rocks above water on the reef. At about 7 miles westward of Alceste there is another round island at some distance from the main land, which here forms a deep curve or bay, and is mountainous.*

Close under the promontory, in about lat. 37° 23′ N., is Sang-kau bay, having in it an island called Le-tau, where the coasting junks anchor; and there is said to be a spacious and deep harbour, surrounded by rocks, with extensive shoals on the left side of the entrance. Another large harbour, called Toa-sik-tau, or Ta-shih-tau, frequented by the Chinese junks, is also said to be near the promontory.

WEI-HAI-WEI or Oïc-haï-oïc harbour, at about 25 miles westward of Alceste island, is formed between Leu-cung island, 517 feet high, and a deep bight of the coast, and is the most eastern anchorage on

---

* *See* Chart of Yellow Sea and Gulf of Pe-chili, No. 1,256; scale, $d = 3.9$ inches, corrected to April 1861.

the north shore of the Shan Tung province.* It is easy of access and capable of affording shelter to a considerable number of large vessels. Moreover, it has two entrances, one on the west, the other on the east side of Leu-cung island, thus affording a facility for access or departure with almost any wind.

The western entrance, although much narrower than the other, has the deepest water, and should be used by all vessels drawing above 18 feet. The soundings in it are 10 and 12 fathoms, but when abreast Observatory island (a rocky islet near the north-west side of Leu-cung), they increase suddenly to 17 fathoms, and decrease again rapidly to 5 fathoms; after which the depth gradually decreases to the southern shore, and into the bay to the westward, where the town is situated.

Round island and three or four adjoining rocks lie off the northern point of the western entrance; the outer rock, scarcely a mile E.N.E. from the point, is 10 or 12 feet high, and steep-to. A rocky patch, which covers at high water, lies between this outer rock and Round island: no other hidden dangers are known.

The best anchorage is close to the west point of Leu-cung island, in 5 to 7 fathoms on excellent holding ground of mud, the island protecting the anchorage from the north-east. At half a mile E.S.E., from the eastern end of the island, is a reef of rocks, steep-to, but as a portion of them always shows above water, they may be easily avoided. H.M.S. *Actæon*, April 1860, anchored in 5 fathoms, with the apex of Leu-cung bearing N.E. by N., small Gingall fort N.W. ¼ W., left extreme of Observatory island N.N.W., and Channel island, centre, S.E. ¾ E. This position is only open 2½ points to the sea, from S.E. by E. ¼ E. to E. ½ S., while to the westward the main land is well overlapped by Observatory and Leu-cung islands.

Watering the ship was a tedious operation, but it must be remembered the above period is the driest time of the year, and most probably a better supply would be forthcoming at a later season.

An extensive wall surrounds the small town of Weï-haï-weï, and continuing up the side of a hill, encloses, as well as the town, a considerable space allotted to gardening purposes. No guns were mounted on the wall. The population appear to be fishermen and agriculturists. Large fleets of junks come from Lai-chau and that locality at this time of the year to fish in the more favoured waters of this neighbourhood.

Agricultural operations were going on, all over the country; every inch of available land was being placed in a fit state to do its work towards

---

* *See* Plan of Oïe-haï-oïe harbour, by Lieut. J. Crawford, 1816.

supplying the wants of this thickly-populated province; the grain grown is buck-wheat, millet, kau-liang, &c.

The place of observation at Weï-haï-weï was at high water mark at the east end of Observatory island on the north-west side of Leu-cung, and is in lat. 37° 30′ 19″ N., long. 122° 07′ 00″ E.; var. 3° 41′ W. in 1860: high water, full and change, at about 9h. 30m.

Supplies.—No bullocks could be obtained at Weï-haï-weï; the usual answer, to the request for some, all along the coast invariably being, that they had none, except the miserable two or three occasionally seen at work in the fields. This answer, however, was intended to mislead, as at several parts of the coast, a little way inside the first range of hills, numerous droves were seen. Some 80 head were discovered concealed in a sequestered glen within three or four miles of Weï-haï-weï; in fact, the province of Shan Tung appeared rich in cattle. A few ill-conditioned sheep, and a small supply of poultry were obtained, and some pigs; fish (herring and cod) were in great abundance, and there was a fair supply of shell fish. Of vegetables only a few onions were obtained.

Water in small quantities was found at the well of a village on Leu-cung island. A small stream waters the eastern wall of the town of Weï-haï-weï in the rainy season, but in the months of March and April, with the exception of two or three little pools, into which trickled a very small stream, it was quite dry. Another stream in precisely a similar condition was found running through the first village south of the town.

Fuel is not to be obtained, the natives having barely enough wood and straw for their own immediate culinary necessities. A small quantity of charcoal is consumed by them. The wood they burn is an oak scrub, some few patches of which are occasionally to be found on the islands and adjoining main, but the larger portion is brought by sea from some better wooded district. No coal was seen, nor did the peasantry appear to know of it.

Directions.—When bound to Weï-haï-weï harbour from the east-ward, after rounding the Shan Tung promontory, and giving Alceste island a berth of 2 miles, the course for the Channel islet in the eastern entrance of the harbour is W. $\frac{1}{4}$ N., and the distance from Alceste 21 miles. This will lead about $1\frac{1}{2}$ or 2 miles to the northward of Coast island, and clear of all known danger, up to Channel islet, a small round rocky islet about 20 feet high, which may be safely approached and passed to a quarter of a mile. Vessels drawing 17 feet and less may pass on either side of this islet, but those of 18 feet should pass to the southward, rounding the islet close to, and steering for the west end of Leu-cung island, anchoring as above in from 5 to 7 fathoms. After passing Channel islet the soundings will decrease to 3½ and 3 fathoms over an extensive flat stretching across from Leu-cung to the main shore, but they will increase as the west end of the island is approached.

In working in through the eastern entrance the lead may be safely trusted, there being no hidden dangers as yet known. The shore of the mainland may be approached to a mile, and that of Leu-cung to 3 cables.

Vessels of large draught running for this harbour from the eastward should pass outside of Leu-cung island. An offing of a mile will clear all danger, and when a small gingall fort, on the mainland, on the west side of the bay, bears W. ½ S. the western entrance will be open. The vessel can then steer for the fort; just to the southward of it is a village, off which a fleet of junks are generally to be seen, and they will assist in showing the position of the fort. The course should be then gradually altered to the southward, and when Observatory island comes on with the left extreme of Leu-cung steer for the anchorage.

Approaching Weï-haï-weï from the westward, Round island and the adjoining rocks are conspicuous marks for the entrance. The outer rock is steep-to, and its north and east sides may be passed at a cable's length. The right extreme of Observatory island should then be steered for until the gingall fort bears W. ½ S., then keep a mid-channel course until Observatory island comes on, as before, with the left extreme of Leu-cung, then steer for the anchorage.

CHI-FAU HARBOUR.—Cape Chi-fau, about 60 miles westward of Alceste island, is high and bold, and when seen at a distance appears like an island. Chi-fau harbour is formed by a receding coast-line between White rock and Cape Chi-fau, and is sheltered on the north by the Kung-kung-shan islands. The best anchorage, spacious and sheltered from all winds, is under these islands in 4 to 5 fathoms.*

There is also anchorage in Village bay on the south side of the cape in 2 to 4½ fathoms, but a northerly wind sends in an unpleasant swell, and a sailing vessel, with a south-easterly wind, would find a difficulty in leaving it.

The approaches to this harbour according to our present information are clear of all danger.

---

* *See* Plan of Miau-tau strait and Chi-fau harbour, No. 1,260; scale, $m = 0·2$ of an inch; and Plan of Ki-san-seu harbour by Lieut. D. Ross, 1816. Chi-fau is the name of the Cape, and as applied to the harbour is evidently a misnomer, its proper name being Yen-tai, and it is so marked in all Chinese maps; it is known to the Chinese navigator only by the latter name, and he would probably not know what was meant if asked to be directed to Chi-fau. The mistake doubtless arose from the first European visitors landing at the Cape, who asking the name of the place, were told the name of the Cape. —*Commander J. Ward, H.M.S. Actæon,* 1860.

Supplies.—Only one or two horned cattle were seen at Chi-fau; the rest, if there were any, having probably, on the first appearance of the foreign ships, been driven out of sight, or possibly, at this season of the year (March and April), they may take them to better watered districts: but in general, as at Wei-hai-wei, the people ignored the existence of any except a small number for agricultural purposes. A few ponies were seen, and donkeys and mules in abundance. Two or three sheep in a miserable condition were brought off, also a small supply of poultry and eggs. Fish,—herring and cod,—became abundant about the end of March, affording occupation for the whole population, and were freely parted with for dollars.

Water is the great want along the whole of this part of the coast of Shan Tung, and appears to be more keenly felt at Chi-fau than anywhere else. In March the Natives were bringing water in casks, or rather deep tubs, slung between two mules, from some considerable distance in the interior, the village wells not affording sufficient water to supply even the very moderate demands of the Chinese population, whose only consumption of it is in the shape of tea.

Fuel of no kind can be obtained here. The natives burn a small quantity of a dwarf oak, mixed with a considerable portion of straw and reeds. They *had seen* coal, but it was brought from the north, probably from Fu-chau, in the province of Liau-tung.

KUNG-KUNG-SHAN ISLANDS.—This group, as before stated, shelters Chi-fau harbour from the northward. The North rock or island of the group bears E. $\frac{3}{4}$ N., distant $7\frac{1}{2}$ miles from Cape Chi-fau. When approaching it from the eastward it appears round, with a smooth top sloping southward, but when seen from the northward and westward it is wedge-shaped. A small rock just awash at high water, and therefore nearly always visible, lies N.E. by E. $\frac{1}{2}$ E. 3 cables from the North rock, and is steep-to, there being 9 fathoms close outside it.

Double rock bears from North rock S.W. by W. distant $2\frac{1}{4}$ miles, and seen from the eastward appears, as its name denotes, to be a double island, the northern part like a wedge; the southern part, which is much higher, being about 150 feet, is an irregular mound, rather elongated to the westward.

S.E. island is 60 feet high, and bears from North rock S.W. $\frac{1}{2}$ S. $4\frac{1}{4}$ miles. This and the two islets just described are safe of approach, and except the small rock lying off North rock, appear to have no detached dangers. Three high rocks lie between South-east island and Kung-kung, the largest island of the group, but no hidden dangers have as yet been discovered.

A spit, which shows at low water, extends southward and westward from Kung-kung island: its extreme end has 4 fathoms close to, and from it the highest summit of the island bears N.E. ¾ N. northerly, distant 1 4/10 miles, and the Mound N.N.W. ¾ W. This spit shelters the anchorage from easterly and south-easterly winds.

DIRECTIONS.—When bound to Chi-fau harbour from Wei-hai-wei, or from the eastward, after rounding Cape Cod and Eddy island, the course and distance to the Kung-kung-shan islands is West 25 miles. The high hill over Knob point, kept on a W. by S. ½ S. bearing, will lead eastward of these islands, giving S.E. island a berth of half a mile. If intending to anchor, this mark must be followed until Stick-up rock comes on with the eastern part of the Mound, bearing N.N.W. ⅛ W., when the end of the spit will have been passed, and the course may be altered for the Mound, until Finger rock, which is conspicuous, comes on with the west extreme of Kung-kung island, N.N.E. ¾ E., then haul up more to the eastward, anchoring with the centre of the island bearing about E.N.E. in 4 to 4½ fathoms. The bottom is mud, and good holding ground, and there is room for a large number of vessels.

If wishing to run farther on for the anchorage in Village bay on the south side of Cape Chi-fau, when the mark for clearing the spit has been reached, Chi-fau peak bearing N.W. will readily be distinguished. Steer N.W. ¾ W. for the head of the bay, and anchor in 4 to 5 fathoms, with the extreme of the cape bearing about N.N.E. or N.E. by N.: the bottom here is also mud. H.M.S. *Actæon* in 1860, anchored in 3¾ fathoms at low water, with Chi-fau peak bearing N.W. by N.; Sentry rock N.E. by E. ½ E.; the summit of Kung-kung E. by S. ½ S.; and Knob point S.S.E. On the south side of the peninsula, which is connected with the mainland by a low neck of sand, is a village and a small square gingall fort.

In working for this harbour to the eastward of the Kung-kung-shan islands, North rock, Double, and S.E. islands may be safely approached to half a mile, on the one side, and the mainland until the soundings decrease to 4½ fathoms on the other; the depth gradually lessens as the shore is approached. Between the islands and Knob point is a mud bank, from a mile to 1½ miles wide, east and west, having in one or two places 4 fathoms at low water springs, but the general depth is 4½ and 4¾ fathoms.

In working towards Village bay, as the spit extending from Kung-kung island is approached, remember the bearing of the hill over Knob point, W. by S. ½ S., and do not go to the northward of that bearing until the clearing mark, Stick-up rock and the Mound, comes on. A longer stretch may then be made on the port tack, having care, however, not to approach the Mound nearer than to bring S.E. island just in sight to the left of the western part of Kung-kung island, when it will be seen bearing E. ¾ S. over the sandy flat between the two portions of the island. This line will clear the west sand spit, the south extreme of which bears from centre of Mound S. by E. ¼ E. nearly, three-quarters of a mile, and W. ¼ N. from centre of Kung-kung.

Approaching the harbour from the westward, Chi-fau peak, which is 980 feet high, and the land in its immediate neighbourhood forming the cape or peninsula, shows out conspicuously, appearing from a distance like an island; the low sandy isthmus connecting it with the mainland not being visible. There are no hidden dangers known at present in the vicinity. Three or four detached rocks are dotted along the shore, but they are all well within half a mile of it, and above water; a course a mile off, and parallel to the shore, clears everything. Sentry rock lies S.S.E. of the cape, and may be rounded at 2 cables' distance in 7 fathoms, and the anchorage under the cape steered for.

If intending to anchor under the Kung-kung-shan islands, after rounding the Sentry rock, steer for Knob point until the clearing mark for the west sand spit (S.E. island touching the left side of the west part of Kung-kung island, bearing E. ¾ S.) comes on; then run in on that line and anchor as before directed.

Tides.—It is high water, full and change, in Chi-fau harbour at 10h., and the rise is about 8 feet.

The COAST westward of Cape Chi-fau falls back to the southward, forming a sandy bay, terminated by Sloping point, bearing N.W. by W. ¼ W. from the cape, distant 11 miles. Two small bays are also formed between Sloping and Low points, the latter of which is distinguished by a conspicuous nipple, or small mound upon it, 250 feet high.

TENG-CHAU is a city of the second class, governed by a Chifu. whose name is Li, and whose authority extends over the neighbouring villages. The inhabitants were civil, evincing no other feeling than that of curiosity. The city is commanded on three sides; the rising ground of Teng-chau head, which is 250 feet high, commands it on the west. A small detached fort also stands outside the walls to the west, but it appeared to be unarmed. The city is surrounded by rather a formidable-looking wall, but without guns; a break in its sea face forms the entrance to a small camber, in which a fleet of junks lie closely packed and sheltered from all winds. The entrance is so shallow that a very moderate sea breaks right across it.

The only trade going on appeared to be rather a brisk one in grain. The little camber was a scene of bustling activity, some junks taking in cargoes of buck-wheat, kan-liang, &c., and others discharging the same grain, rice, and fuel, consisting of wood and millet straw. Coal is occasionally imported from Fu-chau. The shops in the city appeared to have little else for sale but the usual description of grain and dried peas.

Water.—A small stream of water empties itself into the camber at Teng-chau, but its purity may be doubted, as it seems to run through a large and populous part of the city.

DIRECTIONS.—The course and distance from Cape Chi-fau to the anchorage off Teng-chau fu is first N.W. by W. 23 miles to abreast of Low point, and then W. ½ N. 10 miles to the anchorage. If intending to anchor off Teng-chau after rounding Low point, steer W. ½ N. until Spit point, the south extreme of Chang-shan island, comes on with Island head, immediately to the northward of it (on the eastern side of the island), bearing N. ¼ E.; then steer for the town, taking up the anchorage on the same bearing, in 3 to 6 fathoms.

But in running westward be careful not to bring the nipple on Low point to the eastward of E. by S. ¾ S., to avoid a dangerous rocky ledge extending 2¼ miles east of Teng-chau head and nearly a mile off shore. This reef partially protects the anchorage from the eastward, as Teng-chau bank does from the westward, but it is entirely exposed to the northward, and these winds send in a heavy breaking sea, which renders the anchorage unsafe, and communication with the shore impossible, the Miau-tau group being too distant to afford any shelter.

The Teng-chau bank projects in a W.N.W. direction 6½ miles from Teng-chau head. A depth of 2¾ fathoms was obtained on it at high water, with the head bearing E.S.E., and the west point of Ta-hi-shan island N. ¼ E. The bank has a general depth of 3 to 4 fathoms on it, but there are some patches of only 3 to 6 feet.

The **MIAU-TAU** or Meih-shan group, consisting of fifteen islands, exclusive of two or three small rocks, extend in a northerly direction from Teng-chau fu to within 15 miles of Liau-tie-shan head, (named the Regents Sword by Sir Murray Maxwell in 1816,) and separate the Yellow Sea from the Gulf of Pe-chili. The peak of the northernmost island is in lat. 38° 23′ 37″ N., long. 120° 52′ E. There are several passages through these islands. Miau-tau strait, between the south part of the group and the mainland, has generally been used by vessels bound into the Gulf of Pe-chili; but if not intending to anchor off Teng-chau, or among the southern islands of the group, there are much better channels north of Chang-shan island.*

The Chang-shan channel, between the north side of Chang-shan and Hou-ki, is decidedly the best, and may be taken at night if the islands be seen. In fact, with the exception of the Hesper and Fisherman rocks, and a reef extending a mile to the southward from Sha-mo island, the whole of the entrances northward of Chang-shan appear to be remarkably clear of danger. A small rock, which dries 6 feet at low water, lies in mid-channel in the eastern part of the deep narrow passage between North and South Hwang-ching islands, and there is another of the same height lying three-quarters of a mile from the north-east shore of North Hwang-ching.

Besides the rock in mid-channel between North and South Hwang-ching, there is also a reef with a flat rock on it extending a quarter of a mile from the north-west point of South Hwang-ching.

The Liau-tie-shan channel, north of the Hwang-ching islands, is the most northern entrance into the gulf. It is supposed to be clear of all hidden dangers, with the exception of the small rock (before mentioned) which dries 6 feet at low water, lying three-quarters of a mile from the north-east side of North Hwang-ching.

Supplies.—On the first appearance of the surveying squadron in the southern part of the Miau-tau group in June 1860, all the cattle in Chang-shan and the islands in the immediate neighbourhood were concealed, and eventually, during the stay of the vessels, removed either to the mainland or the northern islands. Three days after the arrival of the ships nothing in the shape of bullocks was to be seen, and on being questioned the natives denied having any except what were in the fields at work. A few pigs were procured; also a small quantity of poultry. H.M.S. *Cruizer,* during her stay at Miau-tau island, obtained a few sheep.

To-ki, which lies nearly in the middle of the chain, although not the largest island, appears to be the most productive. At the anchorage on the south side of this island, H.M.S. *Wellesley* in 1840 obtained about 50 bullocks, and a supply of eggs, poultry, and vegetables. The *Cruizer,* in 1859, was supplied with 11 bullocks, some vegetables, principally cucumbers; but in 1860 no cattle of any description nor vege-

---

* The description of the Miau-tau group, and the Sha-lui-tien banks at the entrance of the Pei-ho, is by Commander J. Ward, H.M.S. *Actæon,* in 1860, and from the remark books of H.M.S. *Squadron,* 1840-1860. See Chart of Miau-tau Strait and Islands, No. 1,392, scale, $m = 0\cdot4$ of an inch; surveyed by Commander J. Ward, Lieut. C. Bullock, R.N., and Assistants, in 1860.

tables were to be had there, the inhabitants declaring that the last was taken from them by the foreign ships last summer.

At South Hwang-ching, although a small island, some 70 or 80 head of cattle were seen; they were small but in very fair condition. So many on such a remote island gave rise to the idea that they had been transported there from the more southern islands, to avoid the observation of the foreigners.

The villages on the south part of the Mia-tau group appear to have a better supply of water than is usually found along the coast of the mainland; the village on the north part of Temple island has four wells. At To-ki the *Wellesley*, in 1840, procured 30 tons of water in one day from the wells of the village on the south side of the island. In 1860 it was with great difficulty that the *Actæon* obtained 5 tons with two pinnaces in one day. A better supply may probably be found at a later season.

ANCHORAGE.—There are two or three good anchorages among the islands forming the southern extreme of the Miau-tau group, but Hope sound is the best, where ships of any draught of water, and in almost any number, may lie quite sheltered from all winds, so that even boat work would be seldom interrupted. The sound is on the west and northern side of Miau-tau or Temple island, and is sheltered on the east by that island and Chang-shan, on the north by Chang-shan and Siau-hi-shan and some rocks between them, and on the west and south-west by Ta-hi-shan and a reef extending to the south-east of that island. Having several entrances even sailing vessels under all circumstances of wind and tide may freely run in and out of it. If drawing under 14 feet they may anchor between Miau-tau and Chang-shan, and if this be too open to the southward, they can anchor south of Chang-shan in 4 to 6 fathoms, sheltered from all but westerly winds. The *Actæon* anchored in $4\frac{1}{4}$ fathoms, with Chang-shan peak bearing N.E. by E. $\frac{1}{4}$ E., the west extreme of Chang-shan N.W. by N., and the temple in the rear of Teng-chau-fu South a little easterly.

There is anchorage in 6 to 9 fathoms in Chief bay on the south side of To-ki island; it is well protected from the northward and westward, but quite open to southerly winds.

H.M.S. *Wellesley* in 1840 anchored in 12 fathoms under Kao-shan or Quoin island during a strong northerly wind, with the island bearing from North to N.N.E. $\frac{1}{2}$ E. about a mile distant.

TA-CHU-SHAN, or Great Bamboo island, the easternmost of the Miautau group, is 480 feet high, and can be seen at a distance of 30 miles. Although of a barren appearance it has a village on its south-eastern side, and cattle was observed on the sides of the hills. The island has a white shingly beach around it, and appears bold-to.

p 2

CHANG-SHAN, or Long island, the largest of the Miau-tau group, has a sandy spit named Chang-shan Tail, extending South a long half mile from Spit point, its south extreme, with irregular soundings of $4\frac{1}{2}$ and 2 fathoms to the southward, the latter depth being nearly $1\frac{1}{4}$ miles from the point. The Tail shows at low water; a tidal overfall is very perceptible on it, and continues so for a considerable distance across the straits like breakers far to the southward of real danger. H.M.S. *Furious,* April 1858, grounded at $1\frac{1}{4}$ miles from Spit point, with the east extreme of Chang-shan just shutting in with the south extreme, bearing N. $\frac{1}{4}$ E., and the western end of Ta-hi-shan island N.W. by W. As night was approaching there was no time for examining the shoal, but the vessel appeared to have grounded on its southern limit, having $2\frac{1}{2}$ fathoms at her bows and amidships, and 5 fathoms under her stern.*

A small round hill, with a heap of stones on it, forming the extreme of the land to the north-eastward of the village on Miau-tau island, kept open of Ship point (a low bluff of a reddish colour forming the western extreme of the southern part of Chang-shan), N.N.W. $\frac{3}{4}$ W., will lead in 5 fathoms water to the south-west of the spit. The above hill is low, and to the north-east of the village is a higher hill, having also a heap of stones on its summit.

TA HI-SHAN and SIAU HI-SHAN, or Great and Little Black islands, lie to the westward of Chang-shan, and between them is a small island named Miau-tau or Temple island, 310 feet high. Hope sound, on the north-west side of Temple island, as before stated, is the best and most sheltered anchorage among the Miau-tau group.

TO-KI ISLAND, about 10 miles to the northward of Chang-shan, may be readily distinguished by its peak, 613 feet high, and is in the form of a right-angle triangle, the shortest side facing the south and west. There are four villages upon the southern side of the island, and one or two on the north-east side. It is well cultivated, and fresh provisions and water may be procured.

The whole of the southern part of To-ki appears clear of danger. The small rock off its south-eastern point, and Mochang-shi islet off its southwest end may be passed at a cable's length.

KAO-SHAN is a remarkable little island, lying nearly 5 miles W.S.W. of To-ki. Its form is like a gunner's quoin, with the highest part (650 feet high) to the southward. The island to the southward, named Hou-ki, 310 feet high, has a reef extending some little distance from its northern side, and another off its eastern end.

---

* Captain S. Osborn, C.B., H.M.S. *Furious*, 1858.

NIMROD ROCK.—H.M.S. *Nimrod*, June 1859, whilst steering for the passage between To-ki and Kao-shan, passed a small rock just above water, and distant about a quarter of a mile. The following bearings were taken when abreast the rock:—North extreme of To-ki, N.W. ¼ W.; the rock in line with the eastern extreme of Hwang-chin island, N. by E. ¼ E.; and the rock in line with the eastern extreme of Ta-kin island, N. ½ W. This rock, we have every reason to believe, is identical with the Hesper; for in the survey of these islands by Commander Ward, in 1860, there is nothing less than 9 fathoms in the position assigned to it.

HESPER ROCK.—This danger was discovered, July 1859, by J. Loane, Master, R.N., Commanding H.M.S. *Hesper*, when endeavouring to find the Nimrod rock. It dries from 4 to 6 feet at low water springs, and is scarcely covered at neaps; in fact at the highest tides a break or mostly a ripple, visible in daylight and clear weather, shows its position. From the rock the west extreme of Ta-kin island, which is 590 feet high, bears N.N.W. ¼ W.; the summit of Kao-shan (which is conspicuous and quoin-shaped), W. ¼ N.; and the highest part of Ta-chu-san, 480 feet high, S. by E. The rock is only about 30 yards in extent, east and west, and 8 or 10 yards wide, and when first seen, bearing E. ¾ N., it had the appearance of a wreck or abandoned vessel with her timbers showing above water. Great caution should be used in approaching this locality at high water.

The *Hesper* passed the south and south-east sides of the rock at the distance of 3 cables, and carried 12 fathoms water. When it bore N. ½ E. it was in line with the east end of Hwang-ching, and when W.N.W. it was in line with the north side of To-ki; attention to these two bearings will lead either eastward or southward of it.

FISHERMAN ROCK is nearly in the middle of the channel between To-ki and Ta-kin islands, and is seldom visible, being only just awash at low water spring tides. A ripple generally shows its position during both flood and ebb streams when the sea is smooth, but when either stream has ceased, no signs of it appear. From the rock, the east extreme of Ta-kin island appears just touching the west extreme of North Hwang-ching island, N.N.E. ¾ E.; Quoin island is just seen over the north extreme of To-ki, S.W. by W.; and the western side of Sha-mo island is in line with the centre of Siau Chu-san, S. by E., easterly.

DIRECTIONS.—Vessels bound through Miau-tau strait from the eastward, should not bring the south point of Chang-shan in line with Island head, bearing N. ¼ E., until the north point of Miau-tau island is seen clear of Ship point (the western point of the southern part of Chang-shan) bearing N.N.W. ¾ W. This latter line of bearing clears Chang-shan Tail, when the course may be altered to the northward for the anchorage on the south side of Chang-shan. Or should the anchorage in Hope sound, on the north side of Miau-tau island, be preferred, after rounding Chang-shan Tail, steer N.W. by W. ¾ W., until Ellis island is just seen clear of Club point, bearing about N. by E., then run in on that line and anchor, with Cairn hill, the northern summit of Chang-shan, bearing N.E. by E., the temple on Miau-tau, E. ¾ S., and the summit of Siau-hi-shan, N.W. ½ W., or as near to this position as circumstances will admit. The bottom, as is generally the case on this coast, is stiff mud, and therefore holds well.

If intending to pass through the strait without anchoring, after clearing Chang-shan Tail, keep on the north side of the strait in 6 or 7 fathoms, and be careful of getting into 10 and 12 fathoms, as the deepest water borders the Teng-chau bank projecting from Temple point, on the southern shore of the strait, to avoid which, Teng-chau point should not be brought eastward of S.E. by E. until Ta-hi-shan island bears N. by E. ¼ E., when edge to the southward, or if necessary haul round into Temple bay, between the shoal and the rocks which extend nearly 1½ miles off between Temple and Hwang bays.

On leaving the anchorage in Temple bay keep to the westward, to avoid the rocks just noticed; and if bound into Miau-tau strait, in proceeding to the northward, the point off which they lie should not be brought to the westward of S. b. E. until Teng-chau point bears S.E. b. E.

Vessels bound to the Pei-ho or other ports in the Gulfs of Pe-chili and Liau-tung, are recommended to use the channel on the north side of Chang-shan island, the course and distance from 2 miles outside of Alceste island to the middle of which, is W.N.W. 99 miles. As before stated, with the exception of the Hesper and Fisherman rocks, and the reef extending a mile to the southward of Sha-mo island, the whole of the entrances to the northward of Chang-shan appear to be clear of danger.

The channel between To-ki and Ta-kin islands cannot be recommended to a stranger on account of the Fisherman rock; but if compelled to take it and intending to pass northward of the rock, do not bring the south end of Ta-kin to the northward of N.W. by W. ½ W., until Kao-shan island opens West of To-ki. In passing to the southward of the rock do not bring the northern point of To-ki to the southward of West until its eastern point bears South.

There is a narrow deep channel between the North and South Hwang-ching islands, but at its east entrance, nearly in the centre, there is the rock which dries 6 feet at low water, and, therefore, nearly always visible. There is also the reef with a flat rock on it, extending a quarter of a mile from the north-west point of South Hwang-ching.

TIDES.—It is high water, full and change, at the anchorage off Teng-chau fu at 8h. 0m., and the springs rise is about 7 feet. At Miau-tau island, it is high water at 10h. 35m., and the rise is about 6 feet. Between the Shan Tung promontory and the neighbourhood of Miau-tau strait the flood tide sets to the westward, and the ebb to the eastward; but within the strait, a few miles westward of Teng-chau fu, the flood will be found setting to the eastward, and the ebb to the westward. This is probably the effect of the water from the Yellow Sea flowing between Shan Tung promontory and the Korea into the Gulf of Pe-chili, and being repelled from the Liau-tung coast westward, around the circular shores of the Gulf of Pe-chili, has, when it reaches Teng-chau fu, sufficient strength to resist and overcome the feeble efforts of the eddy tide setting round Shan Tung promontory to the westward.

From Teng-chau fu the coast takes a W.S.W. direction for 25 miles, to a projecting point, on which stands a village; it then trends south, curving gradually round to the westward, and forming the southernmost shore of the Gulf of Pe-chili.

GULF OF PE-CHILI.

ASPECT of COAST.—From Miau-tau strait the southern coast of this gulf trends first in a south-westerly direction for 50 miles; it then bends round to the west, north-west, and north to the mouth of the Pei ho. The shore is low and flat, and shoal water extends some distance from the land.*

Between Miau-tau strait and Lai-chau the shore is exceedingly dangerous, and should be approached with caution. Chi-ma-tau promontory is a hill, about 250 feet high, joined to the mainland by an isthmus of sand; the sea face is abrupt, but reefs extend from it nearly 1½ miles, with 10 and 11 fathoms close to. Sang-tau island is low and flat, with a large village on it; the island is surrounded by extensive reefs, and should not be approached within 2 miles; the outermost reef has a sand island on it. Lutai bay is full of shoals. Sanson or Saddle hill, 300 feet high, forms a point in a sandy plain. Fuyung Quoin is an island resembling a quoin; a rock lies one mile outside it.

Lai-chau fu, or Edible plant city, said to be in lat. 37° 13′ N., long. 119° 50′ E., stands near the eastern point of the mouth of its con-

---

* The south coast of this gulf was surveyed by Lieut. C. Bullock, R.N., in December 1860.

tiguous river. There is a fort and high craggy cliffs a little to the eastward.*

The LAI-CHAU BANK, of hard sand, and exceedingly dangerous, extends 11 miles in a N.W. by N. direction from a low point between Fuyung Quoin island and Saddle hill. The Saddle bearing S.E. by E. leads in 7 to 8 fathoms close to the eastward of its north extreme; and Fuyung Quoin in line with the high sharp peak of Mount Elias, S.S.E. leads to the westward.

LI-TSIN HO.—The vicinity of this river may be known by the singular nature of the bottom—a yellow clay, into which the lead sinks 4 to 6 feet. Its bar is well marked by the Chinese, the estuary taking an easterly direction through the banks.

From Lai-chau to this river the coast is very low, and skirted by sand banks. From the Li-tsin ho to the Ta-san ho the shore is irregular and broken by large openings; the sand banks extend out in some places 3 or 4 miles. The Ta-san ho is smaller than either the Li-tsin ho or the Pei ho; the bar takes a northerly direction.

Between the Ta-san ho and the Chi-kau ho the sand plain is somewhat higher, and the beach steep at high water; at low tide it would dry out a mile.

The CHI-KAU HO is a salt water creek, which enters the sea through the banks by a narrow tortuous channel, having a bar nearly dry at low water. It runs up about 3 miles to some villages, is 60 to 70 yards wide, carries 15 to 16 feet water, and boats could lie in it in 2 fathoms at low tide, and land troops or stores on a hard mud, not more than 300 yards from *terra firma*. The springs rise about 9 feet and neaps 7 feet.

The anchorage off this river is open from North to South. The water is very shoal, there being only 4 fathoms at 8 miles, and 2 fathoms at about 2 miles from the entrance. There are shoals of 7 feet at about 4 miles from the mouth of the river; a clump of trees bearing S.W. $\frac{1}{2}$ W. clears the north shoal. Small vessels can close the shore at half-tide on that bearing to about $1\frac{1}{2}$ miles, in 12 feet water. The passage over the bar should not be attempted without buoying.

COAST between the **CHI-KAU HO** and the **PEH-TANG HO.**— Between the Chi-kau ho and the Pei ho the soundings are still shoal, the depths being only 4 fathoms at 7 or 8 miles from the coast. The sands, which dry out at low water to a distance of $1\frac{1}{2}$ miles, are hard,

---

* Horsburg, vol. ii., seventh edition, page 497.

and men can walk on them without inconvenience. At about 8 miles south of the river there is an inlet which may be mistaken for a river, and into which the water flows at half flood. At two places between the Chi-kau ho and Pei ho the sea overflows at very high tides, but only to a depth of a few inches; the country inside is a plain of sand, apparently dry, except at places at the top of the tide, and is almost entirely uncultivated. There appears to be almost an unbroken line of sandy beach at the high water level, raised sufficiently to be above the influence of ordinary tides.

An extensive flat runs out between the Pei ho and Peh-tang ho, dry land, appearing to run in about a north and south direction. The mud at the mouth of the Pei ho appears to be soft only where it is thrown up on the banks from the force of the stream.

SHA-LUI-TIEN ISLAND and BANKS.—Sha-lui-tien island, distant 120 miles to the N.W. by W. of Teng-chau fu (page 224), lies at the south-east extreme of an extensive range of sand banks, which should be approached with caution, particularly in thick or foggy weather. The island is low, but it has a joss house on it, which, standing alone and upon an elevated spot, is conspicuous. Some of the banks dry at low water.

There are passages between these banks, through which small junks go, and shoals innumerable, over which nets are spread, but there appears to be no open channel between the banks and the mainland; there is a junk passage in some part, available only at high water.

TIEN-TSIN HO or PEI HO.—* The *Pique* anchored off the entrance of the Pei ho, in 5 fathoms, with the entrance bearing W.N.W. distant about 7 miles, and the beacon on the bar W. ½ N. 4 miles. As it was then nearly high water and spring tides, the vessel was expected to touch the ground at low water; not less than 23 feet, however, was obtained alongside during her stay, which is the least depth a vessel drawing above 20 feet should attempt to anchor in. The forts were occasionally seen in clear weather from the ship; at times the flags could be made out, but these were rare occasions.

The holding ground at this anchorage is excellent. A heavy gale would bring in an unpleasant sea, yet with good ground tackling and plenty of cable out it was considered that a sailing vessel ought to ride out a summer gale. The anchorage seems to be a wild one in winter, but if the gales are off shore, the sea would not be heavy.

---

* Capt. Sir F. Nicholson, H.M.S. *Pique*, April 1858.

Bar.—The bar at the entrance of the Pei ho is about 2 miles in length, in a N.W. by W. and S.E. by E. direction, and consists of hard mud. It presents less difficulty than the mud banks on either side of the river entrance, for the passage across the bar is wide, while between the banks the deep water channel is much contracted. Neither are the banks easily distinguished; at high springs the ripples over them are not visible.

The river is very tortuous, as might be expected from its running through a flat country, but no peculiar difficulties were met by the gun-boats in its navigation in 1858 from its entrance to Tien-tsin. A vessel of sufficiently shallow draught to cross the bar would reach Tien-tsin without much trouble. Some of the straight reaches are shallow and must be passed at high water in a vessel drawing more than 8 feet. At the bends of the river the water is always deep—as much as 6 fathoms;—it need scarcely be remarked that all points must be avoided, and the vessel steered round the elbows of the river.

The river is 200 feet wide at Tien-tsin, above this it soon contracts and becomes too shallow even for gun-boats. The *Kestrel*, of $6\frac{1}{2}$ feet draught, ascended it about 6 miles, and found a reach with only 4 and 5 feet in it at high water; the rise and fall there being 3 to 4 feet. From Tien-tsin there is a water communication to Tung-chu by means of large boats and rafts to within 10 miles of Pekin.

Supplies.—At Tien-tsin and along the river ample supplies of bullocks, sheep, and poultry were obtained. Sheep are very plentiful, and fatten to a great size on oil cake. Vegetables rather scarce. At Tung-ku, as the village about $1\frac{1}{2}$ miles above the forts at the entrance is named, a junk for watering the ship was filled, and the water after being allowed to settle, proved to be very good. One junk load was about 70 or 80 tons, and the Chinese were glad to load her and bring her off to the *Pique* for a trifling sum.

Directions.—* "Having entered the Gulf of Pe-chili by the channel between Chang-shan and To-ki islands, Miau-tau group, the course and distance from Kao-shan island to the anchorage off the Pei ho is W.N.W. 138 miles, with regular soundings of 12 and 14 fathoms. With a strong S.E. wind caution is necessary, lest the vessel be driven too near the Sha-lui-tien banks.

"The south-western part of these banks is very steep-to, the *Pylades* having shoaled from 10 to 8, 6, and 3 fathoms, rocky and shingly bottom.

---

* George Norsworthy, Master of H.M.S. *Pylades*, 1840.

Good anchorage was found with smooth water in lat. 39° 2' N. off the western end of the banks, particularly in N.E. gales when the anchorage off the river is much exposed.

In running for the anchorage off the Pei ho, having sighted Sha-lui-tien island, do not come to the northward of lat. 38° 54' N., on which parallel the vessel will, when past the island, soon shoal to 12 fathoms, and will carry that depth until the west end of the banks bears North, when the soundings will decrease gradually towards the river to 8 and 7 fathoms, when she may either haul up for the anchorage off that place, or proceed farther north to the anchorage before mentioned, under the west end of the shoals."

* "In running for the entrance of the Pei ho should the joss house on Sha-lui-tien island be sighted, when it bears North and just visible from the deck, steer W. b. N., and a run of 30 miles will reach the outer anchorage off the river. In approaching the bar, bring the joss house at Tung-ku on a N.W. b. W. $\frac{3}{4}$ W. bearing, keeping it well open to the left of the southern fort.† A vessel should anchor and ascertain the height of the tide on the bar before attempting to enter.

To cross the bar, weigh at three-quarters flood, which sets strong to the northward across the flats; stand in, keeping the joss house on the above bearing until the mouth of the river begins to open, then haul up N.W. $\frac{3}{4}$ N. for a long house, still keeping the joss house to the left of the southern fort, behind which it must not be shut in till well within the river. The least water will soon be crossed, and when it deepens 2 feet and the joss house bears W. $\frac{1}{2}$ N., haul short in for it. The position of the banks will then be seen by the ripple on them, and the course is in mid-channel."

TIDES.—During the period the *Pique* remained at the anchorage off the mouth of the Pei ho, from 14th April to 10th July, it was high water, full and change, at the bar at 4h. p.m., and springs rose 9 to 10 feet, and neaps about 6 or 7 feet. The usual depths on the bar at high water springs were 10½ and 11 feet; occasionally there were 12 feet, but it was very rare. At low water the lead in several places was barely covered.

In October 1854, the time the U.S. surveying squadron remained off the river, it was high water, full and change, at 2h. 39m., and springs rose 8 feet, and neaps 6¼ feet.

---

* Capt. Sir F. Nicholson, H.M.S. *Pique*, April 1859.
† The forts at the mouth of the Pei ho have been blown up since the above directions were written.

The tides are irregular. North and N.W. winds retard the flood and diminish its rise; East and S.E. winds increase the rise and retard the ebb. Slack water sometimes lasts 3 to 4 hours at the neaps. The flood sets North; the ebb S.S.E. The tide in the river runs 2 to $3\frac{1}{2}$ knots per hour.

Near the Sha-lui-tien banks the flood takes a W.N.W. direction along their edge at the rate of $4\frac{1}{4}$ knots at the springs, and the ebb to the S.E. at the rate of 3 knots; on their west side it sets to the northward, but its velocity is not so great. The tide at the entrance of the river is subject to great irregularities, the stream in the river having a motion more or less towards the sea, except when the prevalence of strong southerly winds swells the gulf, and thereby augments the depth of water in all the adjacent rivers equally with the Pei ho.

A strong north-west wind drives the water out of the gulf of Pe-chili, reducing the depth several feet along the coasts; but a southerly wind forces the water into it, between the Korea and the Shan Tung peninsula, thereby augmenting the depth considerably all over this shoal gulf, which is gradually subject to a decrease in depth occasioned by the accumulation of soil, deposited by the Pei ho and other rivers.

WINDS and WEATHER.—During the period the *Pique* remained off the Pei ho the weather was fine, but sudden changes of wind were frequent, and as a breeze from seaward brings in a heavy sea, much caution is necessary to avoid accidents to loaded boats.

From 14th April to 7th May the changes of wind were constant, and rarely was it smooth enough for boatwork throughout the whole day. Latterly the sea was much smoother, and boat operations were not often interrupted. On the 7th June a very heavy squall came on from the northward, and it blew hard from that quarter until next day.

CLIMATE.—* The climate in the Gulf of Pe-chili appears generally very good. The weather, from the 11th of July to the 8th of September, was exceedingly fine, and the wind moderate, the thermometer ranging from 72° to 80°, and the barometer steady at about 29·50 inches.

Although the rainy season is said to be during the months of July and August the rain was distributed over the earlier summer months, and very little fell in August and September.

The winter begins at the commencement of November, and ends early in April, during which period the rivers are frozen, and the sea to a

---

* Commander J. Bythesea, R.N., 1858.

distance of 3 or 4 miles from the shore. Snow falls from 2 inches to 2 feet; the latter is considered severe.*

GULF OF LIAU-TUNG.

The shores of this extensive gulf were almost a *terra incognita* to Europeans until the year 1793, when H.M. Ships *Discovery* and *Alceste* navigated its southern portion and anchored in Hulu Shan bay. In August 1855, H.M.S. *Bittern* sailed along the eastern coast and anchored in Fu-chu bay and off the port of Niu-chwang. Subsequently, in July 1859, a survey was made by Commander J. Bythesea, H.M.S. *Cruizer*, and Major A. Fisher, Royal Engineers, of part of the western coast of the gulf, from the Great Wall of China to the Chi ho, at 25 miles south of the Pei ho.

The remaining shores of the gulf were surveyed in the fall of the year 1860 by Commander J. Ward, Lieut. C. Bullock, R.N., and Assistants. They have discovered and surveyed a fine harbour, named Port Adams, on the east coast of the gulf. It is formed at the head of a deep indentation (Society bay) of the coast line, and its entrance is in lat. 39° 16′ N., long. 121° 32′ E. It affords secure shelter for a large number of vessels, and at high tide offers a passage of 23 feet water. No large towns or symptoms of trade were observed in its vicinity.

HULU SHAN BAY, on the eastern coast of the Gulf of Liau-tung, affords excellent shelter from north or north-easterly gales. It is about 7 or 8 miles wide, and its north point, when bearing N.N.E. ½ E., has an abrupt aspect, sloping to the northward and vertical towards the sea; and has a reddish appearance. The land here is moderately high, and may be seen at the distance of from 24 to 27 miles. Between 2 and 3 miles within the point is the watering place, which it is not prudent to approach nearer than 3½ fathoms, at low water. H.M.S. *Blonde* in August 1840 anchored in 8½ fathoms, with the north point bearing N.N.W. ¼ W.; village E. ¼ N.; a remarkable red hill E. ½ S.; watering place E. by N. ½ N.; and south point of bay S. ¼ W.†

The *Discovery* and *Alceste* anchored in the northern part of this bay, August 16th, 1793, and by observations taken at a mile to the eastward of the north point, the lat. was 39° 31′ 35″ N., long. 121° 19½′ E. The former vessel obtained water easily from the second stony beach to

---

* For description of coast northward of the Pei ho, see pages 241–246.
† *See* Plan of Hulu Shan Bay, No. 1,393; scale, *m* = half an inch.

the eastward of the point; but the *Alceste* filled water farther to the eastward, where there was a better stream, although, on account of a flat, not so easily obtained.

When approaching this anchorage, the soundings continued regular until passing the first point about a mile, when they began to decrease fast, so that 2 miles within it there were but 3 fathoms water. When at anchor in 5 fathoms, the north point of the bay bore N.W. $\frac{1}{2}$ N., the southern point S. $\frac{1}{3}$ W., a remarkable red hummock East a little northerly, a village E.N.E., distant about one mile off the nearest shore to the northward. Whilst at anchor here, numerous junks were seen passing to and from the northward, many of which appeared deeply laden. The inhabitants were civil, but from their being totally ignorant of the value of dollars, the ships were unable to procure any refreshments. From the summit of a hill extensive lakes were seen to the eastward.

TIDES.—It is high water, full and change, in Hulu Shan bay at 2h. 30m., and the rise is about 9 feet.

FU-CHU BAY.—The land on the north side of Fu-chu bay, the next inlet to the northward of Hulu Shan, is of singular formation, and bears such a resemblance to extensive fortifications that at first sight there is a difficulty in believing they are not forts. The westernmost hill having this appearance is the largest, and has a small conical projection above its regular surface; two others with flat summits are near this to the eastward, and they all have a remarkable appearance from north or south.*

When standing into the bay the soundings will decrease gradually from 13 and 14 to 8, 9, 6, and 5 fathoms; the latter depths being carried some distance before they decrease. The *Bittern* carried 3 fathoms for a considerable distance, when endeavouring to close a fleet of piratical vessels, and at length hauled out in a few inches more than her draught.

The city of Fu-chu, said to command no trade, is some little distance up a river which runs into this bay. It produces coal, a sample of which, as also one of a manufactured article of combustion, were procured, and both found to be of little worth.

DIRECTIONS.—A vessel, after avoiding a reef, which will be seen off the south-west point of Fu-chu bay, may stand boldly in to the north-east and find capital anchorage in 5 fathoms water, with an island in the bay (which, from its appearance, was called Flat Top isle), bearing from E. by N. to E.N.E.; north-east extremity of the bay about N. by E.; and a projecting point (Flat cape) S.E.

---

* Capt. E. W. Vansittart, H.M.S. *Bittern*, August 1855.

In leaving this anchorage for the northward or coming in from that direction, care must be taken to avoid a spit which extends about 2 miles to the S.S.W. from a point to the westward of Flat Top isle, and upon which the sea sometimes breaks.

From Fu-chu bay the *Bittern* steered to the north-east, keeping generally about 10 miles off shore, in regular soundings, which decreased, as she proceeded, from 12 to 9, 7, 6, and 5 fathoms to lat. 40° 12' N., when the depth decreased suddenly to 3½ fathoms on a bank, said by the Chinese to extend from the shore. An island, which was named Saddle,* then bore E. b. N. ¼ N., and hauling out to the north-west, when Saddle bore East the water deepened to 14 and 17 fathoms. The depth was soon found steady at 12 fathoms, and she again kept to the north-east, the water gradually shoaling until she anchored for the night in 5½ fathoms.

The land on this part of the coast has a barren and irregular appearance, but not very low.

In a bay 10 or 15 miles to the northward of Fu-chu, a remarkable rock was observed, resembling a fore-and-aft schooner with gaff topsails set.

KAI-CHU FU, in about lat. 40° 30' N., long. 122° 25' E., and 10 miles inland, is surrounded by a high wall; the houses are low and ill built, but thickly inhabited, and it has an extensive trade. The *Sylph*, opium trader, in November 1832, was obliged to anchor here at a great distance from the land, there being only 2¼ fathoms water about 6 miles off, so flat is this part of the gulf. Not being able to communicate with the shore, which was fronted with ice, and having no shelter from strong north winds, this vessel proceeded from hence towards King-chu fu, a place of considerable trade, about 20 miles inland, on the bank of a river that falls into the northern part of the gulf, where, it is said, vessels may anchor in lat. 40° 37' N., about 6 miles off shore.

There are several dangerous shoals in the upper part of the gulf; for the *Sylph*, after weighing from the coast at Kai-chu fu, deepened gradually to 4, 5, and 6 fathoms, then grounded on a shoal in lat. 40° 34' N., long. 121° 48' E., about 24 miles from the land, and narrowly escaped being wrecked, the vessel striking hard for a considerable time, until the wind changed from the north-eastward to the southward, which raised the water in the gulf, and floated her clear of the shoal.

---

* This Saddle is a peninsula and not an island.—*Commander J. Ward, September,* 1860.

NIU-CHWANG CITY and PORT.—H.M.S. *Bittern* anchored in 4¼ fathoms in the north-east part of the gulf in lat. 40° 38′ N., long. 122° 00′ E. The land, although only 7 or 8 miles distant, had become so low as not to be visible from the deck, but from the mast head its extremes bore S. ¼ E. and N.W. by W.; the latter being the bearing of a hummock detached a short distance from the land to the eastward of it. To the north-east a large town was observed and found to be 2 or 3 miles within the bar of a large river,* which leads about 20 miles up to the city of Niu-chwang, reported to be of considerable extent and commercial importance. In this, its seaport town, there were evident signs of the existence of an extensive and thriving commerce in quiet times. The houses are mostly of one story high, built of stone and unburnt brick, with substantial roofs to them; light is admitted through oiled paper in the windows. The streets and roads were masses of mud and filth. The inhabitants had not before seen Europeans; they bear a strong resemblance to the purer Chinese, but are perhaps of smaller stature.

The channel used by the large junks and the deepest into this river is on the southern side of the entrance, where 4 and 5 fathoms water were found over the bar, but the officers of the *Bittern* were unable to carry these soundings out to the ship, or determine if a channel of any such depths extended so far.

Supplies are plentiful at the port of Niu-chwang, and large stores of grain were shown in spacious enclosures. Fish were in abundance and cheap. Coal is procurable from Niu-chwang, but the 30 tons the *Bittern* obtained proved of very inferior quality. A large supply of hemp, it was said, could be always furnished; a sample of white hemp rope was procured, costing by retail 5 or 6 cents per pound, and which is in various sizes from 1 to 6 inch.

TIDES.—During the *Bittern's* stay off this port the depth of water varied from 5 to 3½ fathoms, which was about the period of neap tides, showing a greater rise and fall than is experienced on the coast of Shan Tung; but the tides in the Gulf of Liau-tung must be greatly influenced by seasons and local circumstances. Upon one occasion in Fu-chu bay the depth varied only 2 feet.

The flood tide sets to the eastward and the ebb out of the gulf.

DIRECTIONS.—The Chinese pilots state that islands and shoal water will be found in the upper and centre parts of the Gulf of Liau-tung, and that large trading junks bound from Niu-chwang and neighbouring ports

---

* The Liau ho.

to the Pei-ho keep on the eastern shore of the gulf to Hulu Shan bay, or even farther south, before they steer across to the westward.

The *Bittern*, when proceeding down the gulf, kept a greater offing and carried regular soundings increasing from 5 to 16 fathoms off Fu-chu bay. The pilots stated there were other shoals to the westward of the spit on which she nearly grounded on her way up.

If vessels should be obliged to visit these regions they ought to be supplied with Chinese pilots; these, however, can only be trusted to a certain extent. A stranger should start with an offing of about 10 miles from the western head of Hulu Shan bay, and steer N.E. by N., preserving a distance of 15 miles off shore to avoid the dangerous sand spit extending 10 or 12 miles from the shore in lat. 40° 12′ N., and when Saddle island bears E. by S. edge to the eastward; soundings gradually decreasing from 14 fathoms will then be carried to the anchorage off the Liau ho.

WINDS.—The winds and weather were variable in the month of August 1855, during the *Bittern's* visit to the Gulfs of Liau-tung and Pe-chili, and the wind seldom lasted long from one direction. The monsoons are said to be less felt here than farther south; trading junks make three or four voyages in the year between Ning-po and Niu-chwang, or other northern ports in the Gulf of Liau-tung.

The barometer ranged between 29·70 in. and 30·10 in.; the thermometer between 70° and 80° of Fahrenheit.

The GREAT WALL of CHINA abuts the sea on the western shore of the Gulf of Liau-tung, in lat. 39° 58′ N., long. 119° 51′ E., originating within 100 yards of the beach, and having a masonry pier jutting out into the sea. The Wall rises generally from 20 to 30 feet, in sections similar to the walls of Chinese cities, and with a thickness of 15 to 25 feet. Running round and inclosing a portion of ground close to the seaside, and thus converting it into a fort, it then runs obliquely inward to the west, and at a distance of about 1½ miles from the beach embraces the city of Ning-hai; then striking over the highly cultivated plains at the foot of the mountains, it runs up one of the ridges, and apparently to a great extent along the higher portion of the chain, the different towers marking at intervals its course, after it has itself ceased to be visible. These mountains, about 2,000 feet high, approach to within about 4½ miles of the beach, and though to a certain extent covered with vegetation they are devoid of all cultivation; not so, however, the plain at their foot, which rises gradually from the sea shore to a height of about 450 feet up the side of the hills.

This part of the country appears to enjoy considerable prosperity, and is in a high state of cultivation; wheat, millet, and maize are mainly grown, and it is dotted over with villages and trees.

[c.]

From Ning-hai runs (along the plain at the foot of the mountains) the great high road to Peking, along which a great traffic exists. From accounts gathered from the country people the distance is said to be 680 li* or 223 miles English, though, according to the longitude of Peking as generally received (116° 32′ E.), the distance would not appear to be more than 150 miles. The road is described as being good, and fit for the passage of the country carts. As a rule it passes along the foot of the mountain ranges, though at times it runs over some hills of no great elevation. It does not lead through any woods or forest, though groves of trees exist in its vicinity in certain localities. The road is crossed at intervals by rivers, but there are none of any size; they are not bridged over, but are forded, except after heavy rains, when there are ferry-boats for travellers.

NING-HAI.—The anchorage off Ning-hai is near the extremity of the Great Wall. It is open from N.E., round easterly to West. With the pagoda bearing N. by W. the depth is $4\frac{1}{2}$ fathoms at $1\frac{1}{2}$ miles, and 2 fathoms at a quarter of a mile, from the shingle beach; inside the latter depth the bottom is rocky and unsafe. A shoal with only 3 feet on it and steep-to, extends about a mile off Shoal point.

Supplies.—The land in the vicinity of Ning-hai is pasture and cultivated, and cattle and corn are abundant.

TIDES.—It is high water, full and change, at Ning-hai at 12h., and the rise is about 6 feet.

CREEK POINT.—The sandy bay to the westward of Ning-hai, between Shoal point and Creek point, a distance of 8 miles, appears clear of rocks; the beach is steep, and the 5-fathoms line of soundings about 2 miles from the shore. A cultivated plain extends from the mountains (4 to 5 miles distant) almost to the water's edge; horses and cattle abound, and a large portion of the country is good pasture land.

The passage into the creek to the westward of Creek point is shoal, serpentine, and nearly dry at low water; small junks go in at high water. The rise and fall is 6 feet.

SHALLOW BAY, between Creek point and Rocky point is about a mile deep, clear of rocks, and the shore is sufficiently steep to allow large boats to land easily. The depth is 2 fathoms at half a mile, and $4\frac{1}{2}$ fathoms at 2 miles from the shore. A reef of rocks, which generally breaks, encircles Rocky point at half a mile distant.

---

* A Chinese li is about one-third of a geographic mile.

LIU-SIA-KWANG.—The anchorage off Liu-sia-kwang is open from N.E. by E. to S.W. The depth is 4½ fathoms at 2½ miles, and 2 fathoms at a quarter of a mile from the beach, which is of sand.

The passage into the beach, near Liu-sia-kwang, is between two sand banks, the one running out from Rocky point, the other from the mouth of the river Tai-cho. The depth in the passage is 2 fathoms at three-quarters of a mile, and 4½ fathoms at 2¼ miles from the shore; the beach is steep, and the landing good. The rise and fall of tide is 6 feet.

The land about Liu-sia-kwang is cultivated from the water's edge to the foot of the mountains, which are 4 to 5 miles distant. Horses and bullocks are abundant.

Water.—There are two wells of good water at the village near the beach.

TAI-CHO HO and YANG HO.—The river Tai-cho enters the sea at about 1¾ miles to the south-west of Liu-sia-kwang. This river is described as short, and arising from the low hills at the back of Liu-sia-kwang, and running in a westerly direction, it does not cross the great road to Peking. The bar at the river entrance has only 1½ feet on it at low water. Between the bar and Liu-sia-kwang the soundings are shoal.

The river Yang, which enters the sea at 2½ miles to the south-west of the Tai-cho, is very shallow, and though a few junks pass a short distance up the river at high water, the greater number discharge their cargo just within its mouth, whence it is carried into the interior in carts. The depth is 1½ feet over the bar, and the rise and fall 6 feet. The beach is composed of sand and mud.

The anchorage off the Yang is open from N.E. by E. to S.W. The water is shoal, the depth being 4½ fathoms at about 4 miles and 2 fathoms at about 1½ miles from the river's mouth.

The land adjoining the sea to the southward of the river Yang becomes less fertile, and apparently at high tides partially covered. A line of sand-hills runs along the beach, at the distance of 300 yards from the water's edge, and extends inwards in places for about a mile. Near the Yang is a small earthen battery for four guns, with a musketry parapet, but without any huts or accommodation for troops.

TIDES.—It is high water, full and change, at the entrance of the rivers Tai-cho and Yang at 0h. 15m.; and the rise is about 6 feet.

The PU HO, which enters the sea at 7 miles south of the Yang, though shallow and of no great length, is made use of by junks at high water. They discharge their cargoes near a dilapidated fort on the north bank, about a mile from the entrance, mounting six or seven guns, besides having a parapet for gingalls. The bar is nearly dry at low water. The rise and fall is about 6 feet.

The anchorage off the river Pu is open from N.N.E. to S.W. The depth of 4½ fathoms cannot be carried nearer the river than 5 miles East of the entrance, and 2 fathoms at a mile to the S.E.

The Pu river, running from the south-west, drains a flat of rather a swampy nature, and, indeed, appears to originate in rather an extensive marsh. Sand hills, 30 to 40 feet high, extend for several miles along the beach to the southward of the river, and the ground is swampy behind them.

At the extremity of these sand hills the formation of the coast changes. A bar of sand lies about half a mile from the coast, and forms a protection for junks, which enter at high tide through one of the breaks in it, and unload at low water.

Where the breaks in the sand exist, a sort of river seems to form, in which there may be 2 feet at low water; this sometimes extends inland for several miles, and occasionally joins the sea by a circuitous route some miles distant; the intermediate space of soft mud, covered with a thin layer of sand, dries at half tides.

LAU-MU HO.—The anchorage off this river is open from N.N.E. to S.S.W. The depth is 4½ fathoms at 1¾ miles to S.S.E. of the entrance, and 2 fathoms at 1½ miles. At the entrance there is a narrow bar, with 3 feet over it at low water. Having passed the bar, 15 feet may be carried close to the west point, and 12 to 13 feet up to a village or rather a series of storehouses, about a mile up the river on the right bank, where many of the junks discharge; a breastwork affords protection to this spot. Some junks ascend higher, and there is said to be a fort to protect the upper anchorage.

Water.—A strong stream of fresh water runs from the Lau-mu into the sea, discolouring it to some distance. A vessel might anchor off the bar, and pump in fresh water during the ebb, for though of a muddy colour, it rapidly settles, and is wholesome for drinking. The water in the river is exceedingly good.

TIDES.—At the entrance of the river Lau-mu it is high water, full and change, at 1h. 30m.; ordinary springs rise 5 feet.

HSIN-SHAI-KAU, about 2 miles to the south-west of the Lau-mu, is a bar creek, into which junks sometimes run for protection in bad weather; it trends to the westward, and eventually returns to the sea.

The adjoining country on either side of this creek is an extensive flat swamp, more or less covered by the tide. To the south-west, large plains of sand and hard mud exist for 6 or 8 miles inland, and present a desolate appearance.

The mirage on this coast is very deceptive, giving an appearance of water to the dry sand, and distorting the objects on shore considerably, small huts sometimes appearing, when first seen, to be large forts.

CHING HO.—At 16 miles to the south-west of the Lau-mu is the entrance to the Ching, which is a considerable river, but apparently conveying a smaller flow of water from inland than the Lau-mu. The anchorage of the river is exposed from N.E. to S.W. The depth is $4\frac{1}{2}$ fathoms at $2\frac{1}{2}$ miles to the S.E., and 2 fathoms at one mile, to the east of the entrance.

The passage to the river is through a break in the shoal, and across a bar, on which there are only 2 feet at low water. Inside the bar there is good anchorage for a large number of small vessels. The river has two entrances: one from the eastward through a creek which dries at half-tide; the other from the westward, which is nearly as deep as the main entrance. Mud flats, covered at high water, extend for some miles in all directions. No good landing can be found before arriving near the village of Ta-ching-ho, where large quantities of grain are landed and stored, and which stands on the left bank at about 5 miles from the entrance. Any vessel that can cross the bar will find sufficient water to enable her to reach the village. Junks ascend the river in considerable numbers, but apparently not farther than the village.

To the westward of the Ching ho is a mud flat, formed into an island by the main and western branches of that river, a few miles west of the western entrance; a sand pit, which is covered at high water springs, joins the Sha-lui-tien banks.

TIDES.—It is high water, full and change, at the Ching ho entrance at 1h. 20m., and springs rise about $6\frac{1}{2}$ feet.

HAI-YE-TSE and CHIANG HO.—Coasting round the east, south, and west sides of the Sha-lui-tien banks the village of Hai-ye-tse is reached, but it cannot be approached, even by boats, except at high tide, the shore drying a mile out at low water.

To the northward of Hai-ye-tse is the village of Chiang-ho. There is a small creek here, in which junks unload, and from which the village derives its name.

Both these villages are very poor, the country desolate in the extreme, barren and uncultivated; a desert of dry mud, sand, and salt, with here and there a stunted shrub. The inhabitants have little or no subsistence but fish, and have to send 12 miles for drinking water. Small whirlwinds are frequent which raise the dust in clouds. Poor as both these villages are, they both pretend to batteries for their defence, which, however, are unarmed and insignificant. There are three other villages to the westward of Chiang-ho, and to the south-west of them is the entrance of the river Peh-tang.

The PEH-TANG, though a smaller river than the Pei ho, and apparently navigable only for a short distance above its mouth, where stands the town of Peh-tang, has a deeper, and perhaps more easy channel of approach. The passage up the river is defended by two forts,* one on either bank, and which resemble those at the mouth of the Pei ho, but on a smaller scale.

The bar of the river has 5 feet over it at low tide; with the north fort bearing N.W. $\frac{1}{4}$ W., 14 feet may be carried across it at high water; the passage up the river is then about N.W. Vessels are recommended to cross the bar at high water, and wait for the banks to show themselves before proceeding up; the channel is from 1 to $1\frac{1}{2}$ cables wide between the banks.

The anchorage off the Peh-tang is open only from S.E. to South. The depth is $4\frac{1}{2}$ fathoms at 8 miles from the shore.

TIDES.—It is high water, full and change, at the bar of the Peh-tang at 10h., and the rise and fall is about $9\frac{1}{2}$ feet.

NORTH COAST OF YELLOW SEA.

From Liau-tie-shan head, the south extreme of the province of Liau-tung, the northern coast of the Yellow Sea extends upwards of 180 miles in an easterly direction, having many islands fronting it in some parts, but it is as yet little known to Europeans. Off Liau-tie-shan head are violent tide ripples, of alarming appearance to a stranger. After trending first in a north-easterly, and then in an easterly direction, the coast line takes a south direction, near the meridian of 125° E., forming a great concavity between Liau-tung and the western coast of the Korea.

In September 1840, H.M. ships *Blonde* and *Pylades* visited this part of the coast,† and determined the position of several points on their route. The south head of Liau-tie-shan, is a high, bold promontory: with the head bearing E.N.E. 15 miles, the *Pylades* anchored

---

* These forts, and those at the entrance of the Pei ho, were demolished during the operations in 1860.

† The description of the northern coast of the Yellow Sea (with the exception of Ta-lien-hwan bay, which is by Commander J. Ward, H.M.S. *Actæon*, 1860), is by George Norsworthy, Master of H.M.S. *Pylades*, 1840.

in 15 fathoms, mud, the ebb tide setting strong to the S.E. From this to the head, the water deepened to 20, 25, and 30 fathoms, and when the head bore N.W. by W. 6 miles, discoloured water was seen bearing North, having the appearance of a long dangerous spit running out from the land to the southward ; three boats were sent to examine it, but after sounding every part, had nothing less than 30 fathoms off shore, from 3 to 5 miles—the change in the colour of the water being occasioned, it is supposed, by the muddy bottom or the meeting of the tides.

The coast from Liau-tie-shan head trends to the N.E. by E., and is high and bold, with deep sandy bays, affording shelter for junks with the prevailing northerly wind. The *Pylades* anchored in a small bight, called in the chart, Scou-ping-tao, about a mile off shore, with the head bearing W. by S. ¾ S. about 16 miles. The bight is well sheltered from north-westerly and easterly winds, but exposed to the southward and south-westward. From the anchorage in 16 fathoms, the west point of a rocky island, which forms the bay, bore E. by S. ⅝ S., centre of the town N.E. by E. : the bottom is irregular, but the holding ground good. Good water may be procured in small quantities, N.W. by N. from the anchorage. Wood appeared to be scarce ; cattle were seen in considerable numbers.

The CAP.—E. by S. from the above bay is a small island, which on this bearing appears round, and much like the Cap in Sunda strait, but in other directions it resembles a gunner's quoin ; it appeared steep-to, and has a rock off it to the southward. The *Pylades* passed between the Cap and the coast running along shore to the eastward, having no bottom with 20 fathoms. Running from the Cap to the E.N.E., at the distance of 6 miles, passed close to two other islands, one of which resembled a ship under sail. These islands appeared steep-to ; no bottom was obtained with 25 fathoms at half a mile from the shore.

ENCOUNTER ROCK, which was found by H.M.S. *Encounter* in May 1860, lies in lat. 38° 33′ 50″ N., long. 121° 37′ E., a little west of the direct course from Shan Tung promontory to Ta-lien-hwan bay.

It is about 70 yards in length east and west, has 24 and 26 fathoms close to, and seen from the north or south appears like a patch of small rocks, but is in reality only two. From the eastern or highest, which is 12 feet above high water, the Cap bears N. ½ W. ; the summit of San Shan tau N.N.E. ¾ E. ; a prominent peak N. by W. ¾ W. ; Sampson peak N. by E. ½ E. ; and Liau-tie-shan summit, N.W. by W. ½ W.

TA-LIEN-HWAN BAY is at the southern end of the peninsula which forms the south part of the province of Liau-tung. It is an extensive inlet, 10 miles deep north and south, and its principal entrance, formed between West Entry point and the San-shan islands, is 5 miles wide. Between the two San-shan islands there is also a passage 1 mile wide, and another 2 miles wide between the inner San-shan and the east point of the bay. Both of these channels, and the main entrance, appear to be quite clear of danger; but only one cast, of 7 fathoms mud, has been yet obtained in San-shan-tau channel.

To a vessel making the land in clear weather, Sampson peak, a noble mountain, 2,213 feet high, would be visible, and when it is brought to bear North, the main entrance to Ta-lien-hwan bay is open.

The flat country at the foot of the hills surrounding Ta-lien-hwan, appears to be good arable land well cultivated. Large quantities of a kind of dwarf Indian corn, millet, and kaou-liang are grown on it, and with sheep, exported to Shan Tung. Vegetables are scarce, and from the latter grain above mentioned a spirit is distilled.

The hills afford grazing for sheep and cattle, and hay is preserved for winter consumption. In exchange for their grain and sheep, the natives bring cloth, tea, sugar, &c., from Shan Tung. About the end of February, a fleet of junks cross over to the Korean coast to catch salmon. Fish, in general, appear to be scarce in the bay, but shell fish, oysters, and large mussels, are very plentiful, and from the quantity of shells observed around the dwellings of these people, appear to form a considerable portion of their animal food.

SAN-SHAN TAU.—These islands, which extend in a southerly direction from the eastern point of Ta-lien-hwan bay, are 500 feet high, and may be approached to within a mile. They lie nearly North and South of each other, and when seen east or west of the former bearing, appear to consist of three, in consequence of the two high portions of the outer island being connected by a low isthmus of small boulders and shingle, about 10 feet above high water. This isthmus was chosen as the place of the Observatory, and is in lat. 38° 52' 54" N., long. 121° 49' 14" E.

VICTORIA BAY (so named by Captain Bourchier, H.M.S. *Blonde*, in 1840) is in the north-west angle of Ta-lien-hwan bay, bearing from the outer San-shan island N.W. by W. It is formed by the coast line trending N.W. by W. ½ W. 4½ miles from West Entry point, then inclining to the southward, and gradually curving round until its eastern point bears from West Entry point N. by W. ¾ W., distant 6 miles.

This bay affords good anchorage in from 5½ to 3 fathoms, protected from all winds, excepting those from E.N.E. to E.S.E. An easterly gale would have a clear fetch of 10 miles, but it is not quite evident that one has ever occurred in the spring and summer months. The inhabitants on the north side of Victoria bay stated that these gales were prevalent, and sent in a heavy sea, while on the other shore of the same bay it was said that they were not known; and the shores of the bay certainly show no evidence of ever having been visited by one. The same people who stated that easterly gales were so prevalent, also added that Victoria bay was full of sunken rocks; but none as yet have been discovered that are not visible at low water.

The conclusion come to was, that in spring and summer the prevailing winds are south-westerly, southerly, and south-easterly, occasionally easterly, and if the wind from the latter quarter blows with any force a heavy swell must necessarily set in. The *Actæon* anchored with Sampson peak bearing N.E., Bay rock W. by S. ¾ S., and the small San-shan island E. by S. ¾ S.

HAND BAY, which is formed at the north extreme of Ta-lien-hwan bay by a peninsula jutting out from the mainland, affords excellent anchorage, quite land-locked for small vessels of 10 feet draught, and within signal distance of vessels lying in Victoria bay. The middle of its entrance bears N. by W. ¾ W., distant 10 miles from the outer San-shan island.

A reef, 9 cables long, and dry, or nearly so, at low water, extends to the S.W. from the north side of this bay. Its south-west end bears North, distant one mile from the east point of the above peninsula.

DANGEROUS REEF.—E.N.E. about 12 or 15 miles from the San-shan islands at the entrance of Ta-lien-hwan bay, is a dangerous reef of rocks lying a considerable distance from the coast, about a mile in extent north and south, and nearly level with the water's edge. The *Pylades* passed 2 miles to the southward of them, having soundings of 35 fathoms. The weather being squally and rainy, no observation could be obtained.

BLONDE ISLAND,—From the above reef the *Pylades* steered E. by S. with a strong S.W. wind, for the south extreme of a group of islands, and found good shelter, in 17 fathoms, mud, on the east side of Blonde island, in lat. 39° 2′ N., long. 122° 49′ E.; the anchorage is sheltered from all but northerly winds. Stock of every description and vegetables were abundant.

Four miles to the eastward of Blonde island are two islands, lying north and south of each other, having a deep water channel between them. There is a remarkable rock bearing S.S.W. from the south point of Blonde; it is high, appears like a junk under sail, and can be seen 12 or 15 miles off. The *Pylades* passed inside this rock, and had no bottom with 30 fathoms.

From the anchorage off Blonde island the *Pylades* steered N.N.E. and at noon in lat. 39° 16' N., long. 122° 54' E., and in 22 fathoms water, the eastern point of a group of islands bore N. $\frac{1}{2}$ W. 3 or 4 miles. Proceeded to the northward, and entered an inlet formed by the above group and other islands to the eastward, the high coast of the Korea distant 12 miles; shoaling the water from 15 to 9 fathoms, hauled to the eastward and anchored under the last-mentioned islands. The southernmost islands of the group are barren, with sharp pointed rocks like the Needles. Observed something like a fort or town on the main, at the distance of 15 miles, the line of coast trending to the N.E. The flood tide here set strong to the northward, $3\frac{1}{2}$ knots an hour, and the ebb faintly to the eastward. High water, at full and change, at about 8h. 30m.

In lat. 39° 12' N., long. 122° 56' E., some patches of sand were found, with depths of 15 and 17 fathoms on them, and on which the *Pylades* anchored for the night: the main land at this time 18 to 21 miles distant. At noon the next day, in lat. 39° 2' N., long. 124° 39' E., lost sight of the main land at 21 or 24 miles distance, the ship in 24 fathoms. There was at this time a patch of low islands in sight, bearing N. by W. 10 miles, and a number of high islands, the eastern extreme of which bore S.E. by E. 16 or 18 miles. After steering S.S.E. from noon, at the rate of 7 knots, at 12h. 50m. the water suddenly shoaled from 15 to 7 fathoms, rocky bottom; hauled off immediately W.S.W., and soon deepened again to 22 fathoms; altered course again as before, and in a short time again shoaled to 17, 10, 7, 6, and 4 fathoms, when the ship was hauled off. From the broken water and the number of birds, it was supposed that there must be much less than 4 fathoms on this shoal; it appeared to extend in a N.N.W. and S.S.E. direction, in lat. 38° 56' N., long. 124° 37' E.

WEST AND SOUTH COASTS OF KOREA.

CHODO ISLAND.—The *Virginie*,* which had several times been hindered, in lat. 36° 55', and 37° 30', and 38° 7' N., from nearing the shores of the Korea by banks which gave warning of a decreasing depth of from 44 to 14 fathoms, made a last attempt on 31st August 1856, in lat. 38° 27' N., long. 124° 27' E., a little distance westward of Chodo island.

This island, the south point of which is in lat. 38° 27' N., long. 124° 34$\frac{1}{2}$' E., lies in a bight of the coast, on the southern part of which are two

---

* The description of the west coast of Korea, north of the Conference group, and of the south coast between Mackau and Aberdeen islands, is partly from Admiral Guérin's report of the voyage of the French frigate *Virginie* in 1855-56.

villages and a harbour for junks. Its eastern end is about 7 miles from the coast, but it is connected to it by a bank which partly uncovers, and appears to extend to the north. About a mile to the southward of the west point of the island, near which is a little islet, is the beginning of an extensive flat, which projects 2 miles to the southward from a point where there is a village, with a depth of more than 8 fathoms in the channel which separates them.

In the channel formed by the north point of the bank, and the little islet, the soundings are 14 to 16 fathoms, 5 fathoms near the islet, and 8 and 9 fathoms at a short distance south of Chodo island. This channel leads to the eastward into a bay of the coast, which is encumbered by a flat carrying about 3 fathoms water; but anchorage may be found in the channel in 8 and 10 fathoms quite sheltered by Chodo to the north-east, and by the bank to the west. The shore around this bight is studded with villages and well cultivated.

TIDES.—On the 3rd and 4th September 1856 (three and four days after the change of moon), it was high water in the bay on the south side of Chodo at 8h. 40m., and the tide rose 10 feet; the approximate establishment may be assumed at 6h. 20m., and the rise 11 to 12 feet on springs.

JOACHIM BAY.—In steering for this bay, the *Virginie* passed 3 miles north of Tas-de-Foin or Haycock islet, off the entrance of Shoal gulf, and had 11 fathoms in the middle of the central islands of the archipelago. These are not, as those to the southward, isolated rocks or islets of no importance, but they have much high land susceptible of cultivation, and a large population of fishermen. Between the Clifford islands and the coast the depth was 33 fathoms. The frigate anchored in 9 fathoms off the entrance of Joachim bay, about a mile from the west point.

Joachim bay, in lat. 36° 53$\frac{1}{2}$' N., long. 126° 17$\frac{3}{4}$' E., is a large bight which extends to the south-east, and looks from a distance as if it afforded excellent anchorage; but part of its entrance is barred by two reefs, the summits of which are never covered. Having passed eastward of these reefs, through a channel carrying 16 feet at low water, a sheltered anchorage was found in 7 to 9 fathoms soft mud, abreast a village containing about 4,000 inhabitants.

From hence the remainder of the bight is nothing but a large lagoon, forming to the south-east and east two bays 4 or 5 miles deep, having not more than 3 feet in them at low water. Having found that no important river flowed out here, the frigate continued her voyage to the northward.

The COAST from Joachim harbour trends to the north-east, where it forms a large bight or gulf, in which are numerous islands. Caroline and Deception bays are on the south coast of this gulf. The head of the bight was named the gulf of Prince Jérôme.

CHASSÉRIAU BANK.—The south point of this bank was found by the *Virginie* to be in lat. 36° 59' 20" N., long. 126° 18' E.; from thence the bank followed the trend of the coast as far as lat. 37° 6½' N., long. 126° 31½' E. It is about three-quarters to a quarter of a mile wide, and the soundings over it are irregular on all parts. There are only 6 to 9 feet on its two extremes, but both its sides appeared steep-to, as the depth was 12 to 15 fathoms near its western, and 11 to 12 fathoms near its eastern edge. It shelters both Caroline bay and Deception bay, and forms with the coast a long channel about 2 miles wide, in which the depth is more than 12 fathoms, and the anchorage secure in every part.

CAROLINE BAY, the west point of entrance to which is in lat. 37° 1¼' N., long. 126° 25' E., has a safe outer roadstead, carrying about 9 fathoms water, between Alfred peninsula and the Cormorandière rock, lying off the west point of the bay. There are numerous villages on the coast. No river flows into the bay. The tide rises about 26 feet.

The entrance to this bay, little more than a mile wide, is much contracted by a tongue of sand which uncovers, and beginning at the eastern point of the bay extends three-quarters of a mile to the westward athwart the channel; but between the western extreme and the opposite shore there is a channel, carrying 11 fathoms, which leads to a large lagoon, resembling that in Shoal gulf and Joachim harbour.

DECEPTION BAY, where the *Virginie* anchored after some difficult navigation between the land and the Chassériau bank, is (the entrance) in lat. 37° 3' N., long. 126° 33' E. It owes its name to the deceitful indications of a river which seems to open into the deep and narrow channel which terminates it to the south, and in which the depth is about 5¼ fathoms.

There is excellent sheltered anchorage off the entrance of this channel, but the outer passages to it present great difficulty of approach from being in the midst of islands and surrounded by banks; the greatest obstacle is the strength of the current, which in the inner channel runs not less than 5 knots. The last of the summits of the Table chain overlook the shores of Deception bay and those of Prince Jérôme gulf, before which the frigate anchored in 15 fathoms, in lat. 37° 11' N., long. 126° 35' E., after passing between Fernande island to the south, and Adolphe island to the north, carrying 21 fathoms.

PRINCE IMPERIAL ARCHIPELAGO.—To the northward of the above anchorage are the numberless isles of the Prince Impérial archipelago, and to the south-east the shores of Prince Jérôme gulf. In this last direction, to within 1½ miles of the coast, is a channel of 11 fathoms, breaking off suddenly into irregular depths.

To the S.E. by S. was an indentation of the land about 10 miles deep, with an average width of one mile. At that distance the shores approach each other, and there were all the signs of a considerable river; many populous villages, a well cultivated soil, junks at anchor in all the creeks in which are the villages, and in the mouth of one four junks of 130 to 150 tons (the largest seen) were waiting for the turn of tide to go up into the interior; lastly, two chains of mountains, one to the north, the steep cliffs of which seem to form the right bank, the other the table mountain to the south, separated by a large plain, point out the direction and length of the basin of these watercourses. Unluckily time failed the officer of this exploration; he was 20 miles from the frigate, with an adverse tide, and he only arrived on board at 10 p.m. The false indications of the maps leading Admiral Guérin to believe in the existence of larger rivers, that he still had to explore, determined him to rest contented with the information collected during this day.

The direction of this river passing only a few miles from Great Mandarin point in Shoal gulf, seems to be (judging from the trend of the northern heights), first S.E., then East in going upwards. A Korean sailor, on being questioned, indicated the direction of Séoul as E.N.E. The presence of the mandarins who followed the *Virginie* from her anchorage to Caroline bay, their sudden arrival from the capital, whence they declared they came, and in the district of which they said the frigate was, the crowd gathered in a few hours at Great Mandarin point, all give weight to the opinion that this river reaches very near Séoul.

The current in the gulf has the same strength as in Caroline and Deception bays, or from 4 to 5 knots an hour. The tide rises from 22½ feet to 26 feet at springs.

On the 26th of August, in the morning, the *Virginie* made sail. The breeze being variable from S.S.W., forced her to tack in order to weather the easternmost islets of the archipelago. Soundings very regular. At noon, doubled Chapeau or Hat rock in about lat. 37° 15′ N., and sought a passage to the north-west between the Adolphe and Marolles islets, when about 3h. 30m. p.m. the lead showed successively in a few minutes 27, 16, and 8 fathoms. Notwithstanding every precaution being taken, the frigate touched a minute after the last sounding in 14 feet, sandy bottom, on a shoal of which there was no indication.

This bank lies N.N.W. ¼ W. from the summit of Fernande islet, and

N. ¼ W. from the summit of Alfred islet. The next day was passed in surveying the bank and in repairing damages. M. Montaru places Fernande islet in 37° 9′ N., and 126° 30′ E.; Adolphe islet, S.E. point, in 37° 12′ 10″ N., 126° 24′ E., and the shoal the vessel grounded on in 37° 13′ N., 126° 27′ E. of Greenwich.

CHWANG-SHAN or DANIELS ISLAND.—On the 17th July 1832, the ship *Amherst* * made the land a little to the northward of Sir James Hall group. She stood towards a high bluff point of a large island, distinguished by a large detached mass of rock close to the point; on rounding this an extensive bay opened out, exposed to the north. The summits of this, as well as the other islands to the southward, were clothed with luxuriant vegetation and high trees; the lower part near the sea was cleared and cultivated, and numerous cattle and several villages were seen.

This island was called Chwang-shan by the natives. In Horsburgh, Vol. 2, 7th Edition, page 505, it is named Daniels island, and it there states that the *Amherst* anchored on its west side in lat. 38° 17′ N., long. 124° 56′ E., and that the island is 12 miles north of the northernmost islands of Sir James Hall group.

SIR JAMES HALL GROUP, consisting of three islands, was discovered by H.M. ships *Alceste* and *Lyra*, 1st September 1816. The west end of the northern island was ascertained by Captain Basil Hall to be in long. 120° 44¼′ E., and the south end of the southern island to be in lat. 37° 44½′ N. There is a white rocky islet off the west end of the middle island. The vessels had 20 to 30 fathoms in rounding the south-west end of the southern island, and anchored in 7 fathoms, black sand, in the middle of a bay on its south side.

This bay affords good anchorage, sheltered from all winds, except between W.S.W., round south, and S.E. There are two villages here; the inhabitants were suspicious and unfriendly. From the top of the highest peak on this island, about 750 feet high, the mainland of the Korea could be seen, high and rugged, stretching N.N.W. and S.S.W. distant about 8 or 10 leagues. Along the coast abreast were seen many islands. The channel between the middle island of the group and the southern island appeared clear and broad; but the northern and middle islands seemed connected by a reef above water at several places.

MARJORIBANKS HARBOUR, and SHOAL GULF.—The *Amherst* left her anchorage under Chwang-shan island 20th July, and steered to the

---

* From the voyage of the ship *Amherst* to the northern ports of China, 2nd Edition, page 215; also Gutzlaff's third voyage in 1832.

southward. The next day, after passing numerous islands, she anchored under the Loktaou islands, about 2 miles from a large village built on the brow of a steep hill, westward of Table mountain. On the 25th, with the assistance of a native pilot, the ship was moved 10 miles farther eastward. From the Loktaou she steered N.E. 7 miles for another group, and then N.E. towards a deep bay (named Marjoribanks harbour*) or rather passage, among numerous islands, where she found sheltered and convenient anchorage near a large village.

On the 7th August a party proceeded to the northward in the *Amherst's* long-boat to explore a deep and wide bay or gulf, which was named Shoal Gulf†, the wind fresh from S.W. and a strong flood tide. They coasted along the western side of the gulf, which was studded with numerous beautiful verdant islets, mostly cultivated and inhabited, and thickly wooded with fine fir timber, many of the trees of which are fit for spars, and the wood of the best quality, close grained, and full of turpentine. In advancing up the gulf it became much wider, at the entrance 5 miles, and 10 miles up nearly double that breadth; the soundings were variable, but 8 to 12 fathoms were carried in the fair channel.

The boat still proceeded along the western shore of the gulf, the object being to ascertain whether the land forming it was an island or not. The island to the eastward was 2 miles in length, fertile, and with several large villages on it. Having proceeded about 6 miles farther, and the head of the gulf appearing at a great distance, they abandoned reaching its extremity, and landed on the promontory in the centre of what was found to be an island, and named Lindsay island.‡

On ascending a high hill, the open sea was seen to the westward, and Lindsay island was observed to be separated from the mainland by a narrow passage. The head of the gulf was divided into two parts, the western of which extended about 7 or 8 miles to the northward, but the termination of the eastern one, which trends to the N.N.E., could not be seen; all the boats with mandarins came from the latter direction. In returning to the ship, the boat anchored for the night abreast a cluster of islands in the centre of the gulf, about 10 miles from the entrance.

Marjoribanks harbour and Shoal gulf were also visited by the French

---

* After Mr. Marjoribanks, late President of the Select Committee of Supercargoes at Canton.

† There is a discrepancy between the chart of the *Amherst's* voyage by Capt. Rees her commander, and the account of her voyage given by Mr. Gutzlaff and Mr. Lindsay. Marjoribanks harbour of the chart is called Gankeang in Gutzlaff's Voyages, page 330; and Shoal Gulf of the chart is named Marjoribanks harbour in Lindsay's Report, 2nd Edition, p. 240.

‡ After Mr. Lindsay, the H. E. I. Company's supercargo in the *Amherst*.

frigate *Virginie* in 1856. The harbour is described as lying 1½ miles south of Lindsay island, fronting the mouth of the channel leading into Shoal gulf, and is formed by three islands of moderate size; one to the north-west, one to the east, and one to the south, and by a small islet surrounded by reefs to the west. The depth is 8 to 12 fathoms at low water, in the middle of the harbour, abreast the south point of Ko-tian, or the east island; but north of this, and in line with Mauzac islet, lying off the south point of Chang-kon-yam, or the north-west island, the water shoals rapidly to 20 and 12 feet, near some rocks awash, which seem to encumber the north side of the harbour. To the southward the depth decreases less rapidly; there are 12½, 6, and 5 fathoms as far as the line which joins the northern point of the south island to the rock above water, which lies one mile to the W.N.W. of this point. Mauzac islet is in lat. 36° 26¾′ N., long. 126° 28′ E.

There is a valley in a bay with a sand beach on the middle of the west coast of Chang-kon-yam, and another village on the north-west coast of Ko-tian. At a mile eastward of this latter island is a large island named Roussin, on which there are three villages; it occupies nearly all the entrance of Shoal gulf. The channel which it forms with the south end of Lindsay island is 2 miles long and 1½ miles wide, but it is difficult to navigate, being encumbered with islets and rocks.

The approaches to Marjoribanks harbour are well pointed out by the Table mountain, 3,280 feet high, which rises E. by N. ⅜ N. from the anchorage, but they are studded with islets and rocks, amongst which it is difficult to navigate.

In running for the harbour, the *Virginie* had 27 fathoms water at 2¼ miles south of Guérin island, 951 feet high, and in lat. 36° 7′ N., long. 126° 1′ E.; from thence she steered to pass 4 miles east of Wai-ian-do island, elevated 1,050 feet, the soundings being 14 and 16 fathoms. After sailing 7 miles eastward of this position she steered to the northward, passing some distance east of Camille, Pellion, and Charner islands, in soundings of 8 and 10 fathoms, until abreast of Tas-de-Foin islet, in lat. 36° 24½′ N., long. 126° 24′ E., and from thence to Marjoribanks harbour.

Shoal gulf is described as extending 30 miles in a northerly direction from its entrance. Although its name implies that it is shallow, yet there is an average depth of about 11 fathoms to 25 miles north of Roussin island, which forms the western part of its mouth. All the western part of the gulf, as far as the Mathilde group, in lat. 36° 44′ N., was sounded over by the boats of the *Virginie*, the depth varying from 6 to 10 fathoms. In the bay North of the group, the soundings were 4 and 5 fathoms, and in the bay E.N.E. of the group 22 to 26 feet near the shore.

The water in the gulf rises about 26 feet. The tides produce strong eddies, which, although they favour the junks, would no doubt be an obstacle in its navigation by a sailing vessel of large draught, particularly with the variable winds of these latitudes. The coasting trade is considerable, and its importance to the Koreans is pointed out by the beacons, which are kept up with great care on all the points, and on the shoals which cover at high water.

The general aspect of the shores of the gulf is uniform. Lindsay island is a long range of wooded summits united by chains of rocks, covered at high water. In all the creeks are hamlets and villages, the principal of which are on Roussin island and at Port Lourmel. The eastern coast of the gulf, on the contrary, is an uninterrupted series of bare hills, with abrupt declivities, throwing out into the sea their granite bases. Behind these hills are the high mountains of the Table chain, which run in a N.W. and S.E. direction. There are only two or three villages on this coast, but there appeared to be a large population in the interior.

Three channels lead into the gulf. The two to the westward of Roussin island are only practicable for junks. The third must be sought for to the southward of the Marjoribanks islands, and which, bounding them to the east, carries $6\frac{1}{2}$ to 22 fathoms water, as far as the south-east point of Roussin, where a circular bay, formed by this point and Bouët island, offers good anchorage. All the southern part of the gulf between the Parseval group, Roussin island, and the mainland, affords fair anchorage near the shore, especially to the south of Penaud island, where the currents run with less strength than in the gulf. The Parseval are 3 miles north, and Penaud $1\frac{1}{2}$ miles north-east of the east end of Roussin.

In proceeding up the gulf the soundings were 9 to 16 fathoms, sand and shells, as far as the Dénudées islands, which lie 10 miles within the entrance. At a mile to the southward of the centre island of this group, in 8 fathoms, sand, a convenient shelter may be found abreast a creek, where the junks await the change of tide. At $1\frac{1}{2}$ miles to the N.N.W. of the Dénudées there is a flat 4 cables in extent, and carrying 26 feet water, which must be left to the westward, and the channel taken between it and Lindsay island.

The Mathilde group will now be seen at 6 miles to the N. by E. They divide the head of the gulf into two deep bays, one trending in a north, the other in a north-east direction. At half a mile south of the most southerly island of this group there is a small islet distinguished by a large white pagoda covered with tiles standing in the midst of trees. At 3 miles S.W. by W. of this islet, and at the same distance from the nearest point of Lindsay island, there is another flat, of less extent than the

[c.] R

former, but carrying the same depth, 26 feet; in proceeding north this also must be left to the eastward, so as to keep in the channel.

From thence, steering towards the north part of Lindsay island, a bight, named Port Lourmel, about 1½ miles deep, will be seen to the W. by N., and a large village on the declivity of a steep hill. This is one of the most commercial points of the gulf. A channel, carrying 6½ to 9 fathoms, leads to it and forms a sheltered roadstead, the only inconvenience being the currents, which set strongly from a lagoon leading to the westward. This lagoon is only navigable at high tides to the junks, which penetrate some distance into the interior.

Port Lourmel is the limit of the deep soundings, for as soon as the small islet to the north is passed, the depths are irregular, the channel becomes contracted, and not more than 26 feet will be carried for 6 or 7 miles. Here one of the most remarkable mountains of the coast is to be seen; remarkable, inasmuch as it may be recognized in all directions, and that it rises from the centre of three large bays. Its peak is conical, and has three little paps on its summit.

To the eastward of Port Lourmel is the entrance of the north-east bay of the gulf, which is open to the south-west. Its high shores seemed to announce deep soundings; not more, however, than 32 feet were found, and at low water a bank, more than 1½ miles in extent, was left dry surrounding Great Mandarin point.

TIDES.—At Marjoribanks harbour, it is high water, full and change, at 3h. 30m., and springs rise 29 feet. In Shoal gulf the rise is 26 feet.

BASIL or CHU-YING BAY, discovered by the *Alceste* and *Lyra*, 4th September 1816, is formed on the east side of a curved tongue of land, on which is a peaked hill. The south end of this tongue, by Capt. Basil Hall, is in lat. 36° 7′ 38″ N., long. 126° 42½′ E. The bay is about 4 miles wide, but it is open to the south, and has not depth enough to allow vessels of moderate draught to enter, there being only 2½ to 3 fathoms inside the points. The *Alceste* and *Lyra* anchored outside in 5 fathoms.

To the south-west of the western point of entrance, and at the distance of 5 miles, is a peaked islet named Helens island, 930 feet above the sea, which forms a good mark to lead to the bay. The bay is skirted by large villages, ornamented with trees, and surrounded by cultivation, both on the main land and on many of the neighbouring islands, which form a part of the Korean archipelago.

TIDES.—In Basil bay, September 4th, 1816, the rise of tide, 2½ days before full moon, was 15½ feet; low water at 8h. p.m., and high water at

2h. 30m. a.m., showing an approximate establishment of 4h. 15m., and a rise of 17 to 18 feet on springs.

KOREAN ARCHIPELAGO.—From Basil bay the *Alceste* and *Lyra* proceeded to the southward among the Korean archipelago, carrying soundings of 7 to 17 fathoms, and usually anchored during the night. On 8th September they came to in 11 fathoms, soft bottom, in Murray sound, about 2 cables from the east side of Thistle island, in lat. 34° 22′ 39″ N., long. 126° 3′ E. The islands here are so situated as to form a capacious and secure anchorage, with passages among them in all directions. The tides run at the rate of $3\frac{1}{2}$ knots at the springs; the flood to the north-east; the rise and fall is 15 feet. From the summit of a high peak, elevated 800 feet, of an island to the south-east, upwards of two hundred islands were counted, many large and high, and almost all cultivated.

The KO-KUN-TO GROUP are the only islands tolerably known in the interior of this archipelago. They are high, mountainous, and 9 miles in extent, east and west, and about the same north and south, if Peak islet be included. They are separated by narrow channels, and form a mass of bare peaks and arid cliffs. The sea breaks with violence on the rocky points during S.W. gales, and the flood stream runs violently into the passage between the islets. In the interior the parts are less abrupt and covered with trees; there are some sandy beaches between the rocky ground, and at the foot of the hills many houses surrounded by cultivated fields.

Camp islet, the easternmost and largest of the group, is 2 miles long, east and west, and one mile wide. Its eastern point, in lat. 35° 48′ 8″ N., long. 126° 31′ E., is high, and covered with stunted trees; it falls towards the west, where it terminates in broken rocks, and is low; it is not inhabited. On a shingle beach, between two points of rock forming a small bay, the crews of the French ships *Gloire* and *Victorieuse* disembarked on the 11th of August 1847. The anchorage on its south side is safe during the North-east monsoon, sheltered as it is by the lofty mountains of the mainland; but in a S.W. gale, the swell must be heavy, and the anchors would perhaps drag on account of the slope of the bottom from south to north.

The resources of these islands are trifling. The inhabitants are poor; they live on the product of their fishing, and some little barley and rice that they gather in; and they are very mistrustful of strangers. Ships can water at about 2 miles from the anchorage. The water there is good and abundant.

TIDES.—It is high water, full and change, at Camp islet at 2h. 25m.; the tide rises 20 feet at equinoctial springs, and 10 feet at neaps. The tide streams follow the direction of the channel, and average perhaps 3 knots an hour.

DIRECTIONS.—When bound towards the Ko-kún-to group from the westward, it will be necessary first to sight Ile de la Selle or Saddle island. It has a peak at either end, which, seen from the offing, nearly join and give it the appearance of the hollow of a saddle. It may also be known by an islet lying 1½ miles from its north side. The west peak of Saddle island is in lat. 35° 36′ 50″ N., long. 126° 6′ E. In passing the north or south side of the island, the two peaks separate, leaving between them a hollow, which has the appearance of a Turkish saddle. About 10 miles farther east are two isles lying close to each other.

When 5 miles north of Saddle island and steering East, Peak islet will be seen ahead, and to the north the islands of the Ko-kún-to group will rise one after the other. The flood tide, which runs strongly to the northward, must be allowed for. The soundings will vary from 10 to 3¼ fathoms. At 4 miles north of Peak islet is a point remarkable for a hole pierced in the rock a few feet above water, and which, when approaching it from the westward, will be easily perceived. When between Roche Percée and Peak islet steer N.E. by E., or for the east point of Camp islet, and anchor in 9 fathoms, over a bottom of soft mud, at one mile off shore the middle of Peak islet bearing S. by W. ¼ W., and the eastern end of Camp islet N.E. If working in, which can only be done with the tide, care must be taken in approaching the east bank off the mainland, for the bottom varies suddenly from 6 to 3 fathoms.

WINDS.—The west coast of Korea is subject to periodical winds from the N.E. and S.W. From August to the end of September, the monsoon changes; this is the period of rains, variable winds, and sometimes storms.

MODESTE ISLAND seems to form the north-west extremity of the southern group of the archipelago; it is 3 miles long, N.N.E. and S.S.W., and consists of two peaks, of which the northern is the higher, reaching to 1,214 feet. Its summit is in lat. 34° 42½′ N., long. 125° 16′ E.

MACKAU ISLAND, 8 miles eastward of Modeste, is 6 miles long, north and south. Its height is 1,443 feet, and its centre is in lat. 34° 40′ 5″ N., long. 125° 28½′ E. Between Modeste and Mackau isles, the soundings were 52 fathoms.

ALCESTE ISLAND, in 34° 6′ N., 125° 11′ E., seems to form the south-west limit of the archipelago. It is 1,935 feet high, about 3 miles long, north and south, and a village stands on the slope of the high ground;

some rocky islets lie near its north-west end. The soundings were 48 fathoms, broken shells, at 5 miles to the north.

SOUTH COAST of KOREA.—This coast and the islands lying off it are but little known. Lyra island, the southernmost of the Korean archipelago and of the group forming Murray sound, is supposed to be in lat. 34° 8′ N., long. 126° 7′ E. The French frigate *Virginie* in 1856 vainly tried to penetrate into the eastern part of the archipelago. On the 2nd of August she sailed from the Port Hamilton group, and sought a passage to the N.W. towards the continent. Strong and opposite currents retarded rather than aided her progress, the winds were also variable and light. In 34° 11′ N., and 126° 27¼′ E., it was hoped that a passage was found across a gulf that seemed to belong to the mainland. The bottom was regular, the depth more than 16 fathoms, and the frigate had nearly reached the end when she was brought up in lat. 34° 32′ N., long. 126° E., by a bank stretching ahead from N.W. to the land. The boats could find no passage, and the vessel was obliged to return; on the 6th August her position was lat. 34° 10′ N., long. 126° E., and she sighted Mackau island and the Hydrographers group.

H.M.S. *Nimrod*, on the 28th April 1860, passed about 2½ miles to the eastward of Modeste island, and from thence steered S.E., passing various clusters of islands. Having run 44 miles on this course, a thick fog came on and she endeavoured to find anchorage under a cluster of islands (about 500 feet high), but was unable to obtain less than 19 fathoms water, coarse sand and shells, close to the shore and in an exposed position. She therefore steered for another group, distant about 8 miles to the N.E., and anchored in a bay formed by the junction of two islands (probably between Montreal island and the island westward of it), open to the S.W., with a channel apparently safe but narrow, to the N.E.

The *Nimrod* here rode out a gale from the E.S.E., which lasted nearly two days, in 14 fathoms, over a bottom of sand and shells, with rocky patches. There are several villages in the bay, and the inhabitants were civil and obliging. On the west side of the bay are two smaller bays, where tolerable shelter for small vessels may be found in 5 fathoms, muddy bottom. Good water was obtained, and fish by hauling the seine. At about 7 miles to the S. by W. of the anchorage there is a remarkable high and craggy island, about 500 feet high (most probably Lyra island), something like a cock's comb; no other island in this locality bears any resemblance to it.

This anchorage cannot be recommended as secure, but may be preferred to the risk of navigating among these islands at night or in thick weather. The flood tide sets to the N.E., the ebb to the S.W. High water, at full and change, at about 8h. 30m.

From the above anchorage the *Nimrod* steered E.S.E. 24 miles, and then E. ½ S. for Port Hamilton, passing about 4 miles north of a group of islands, on the east side of the largest island of which the ship *Remi* was lost during a fog on the 25th April 1860, in lat. 33° 55½' N., long. 126° 22' E., on her passage from Hakodadi to the Gulf of Pe-chili.

QUELPART ISLAND, extending about 40 miles in an E.N.E. and W.S.W. direction, and 17 miles wide, is of considerable height, and detached from the Korean archipelago and smaller islands which face the south-west coast of Korea.

The general outline of Quelpart* is that of an oval, with few deep indentations to affect its regularity. Its general appearance, as viewed from the sea, is inviting, there being a pleasing variety of hill and dale, and on the northern and eastern sides much cleared land, cultivation rising probably to the level of 2,000 feet. Above this all appears to be buried in thick forests of pines and other northern trees, even to the highest peak of the island, Mount Auckland, which is 6,544 feet above the level of the sea. Towards the northern and eastern parts some of the cones, 500 to 800 feet high, are so smooth and circular, that, with their little batteries or watch towers on the summit, appear almost to be the work of art. This probably results from their method of cultivating the sides, as all the furrows appear to be made horizontally, which in process of time, by the constant falling down of the ridges, would effect such a regular outline.

The space on which the city stands is a broad valley, situated about the centre of the northern coast of Quelpart, having a conspicuous flat eminence on its eastern side, and a small river or copious stream on the western; the country immediately surrounding it on all sides being barren. The city wall, on the sea face, occupies a line of about 500 yards, containing seven bastions, apparently with embrasures throughout, but no guns were noticed.

The eastern extreme of Quelpart is a peninsular promontory, named Cape Dundas, 2 miles north of which is Bullock island. Between this island and the city on the north coast, the depths in-shore vary from 6 to 27 fathoms. There appears a second city nearly opposite the former, on the south shore of the island, with a small bay on its west side, fronted by two small islands, named Hooper and Burnet, the latter being outside the former. West of Burnet is Richardson island, and east of it the islands of Mahon and Barrow, the latter surrounded by a reef. These four islands occupy a space of about 8 miles in an east and west direction

---

* The description of this island is by Captain Sir E. Belcher, H.M.S. *Samarang*, who surveyed it in 1845.

parallel with the coast, and not more than one or two miles distant from it. Towards the west end of Quelpart, on its south shore, is a projecting promontory named Loney bluff, on the east side of which is a deep bay, with an island in it, Marryat island, surrounded by a reef, lying about 2 miles N.E. of the bluff. South of the bluff are Barlow and Giffard islands, at the respective distances of 1 and 4 miles from it; they are both low and flat, and the former is surrounded by a reef. The Samarang rock lies W. b. N. about 2 miles from Barlow, and S.W. about the same distance from Loney bluff.

Supplies.—The productions of Quelpart do not appear to be at all equal to the wants of the population and are in very small variety. Rice, wheat, barley, sweet potato, large Russian radish, maize, and small garden produce, comprise all that was noticed, either in the grounds, under cultivation, or among the people. This does not appear the result of any deficiency in land fit for cultivation, but rather in the poor nature of the soil.

Water appears to abound on the southern side of Quelpart, but only on Hooper island could it be procured easily. It can be easily obtained on Barlow island, but there is not safe and convenient anchorage near it. Wood was purchased from the authorities, but in such small portions that it did not repay the trouble of sending for it; it is, however, abundant on the mountains, and on two of the off-lying islets it can be obtained by slight labour.

The manners of the population, excluding the superior class, are filthy in person and habit.. Their fishing vessels are few, and of the most miserable construction. It is highly probable that Quelpart occupies the position of one of the penal settlements of Korea, and viewing it in this light, accounts for the gross manners complained of, and it will readily account for the variety in the races of beings which were found assembled, and for the low state of cultivation.

Anchorage.—Quelpart, throughout its extent, has but one safe anchorage, and that is off the southern bay of Bullock island, which here forms a channel with Quelpart, about 2 miles wide, and through which the current sets strong to the southward. The second temporary roadstead is off the city on the northern shore of the island, but a vessel would be compelled to seek an offing at the first symptom of a N.W. breeze.

The third anchorage, at the western extreme of Quelpart, within Eden island, affords shelter from North, round east, to N.W., and offers an escape to leeward if requisite. A fourth temporary but dangerous anchorage is off Hooper island, near the city on the southern shore; but this is open from West to S.E., and is too confined to admit of beating out, should wind and sea come in suddenly.

PORT HAMILTON. — The Port Hamilton group, lying about N.N.E. ½ E., 38 miles from the north-east end of Quelpart, consist of three islands, two large and one small, deeply indented, which form a well sheltered harbour within, named Port Hamilton, the entrance to which is at the south-east part of the group. These islands may be readily distinguished from the numerous clumps of islets and rocks in the neighbourhood, by their greater size as well as their peculiar position.*

Supplies.—Within Observatory island a vessel may be safely hove down for repair. Wood is scarce—fresh water is plentiful and good, and easily embarked. Fish may be caught with the seine. Although the natives were friendly, H.M.S. *Saracen* in 1856 could not obtain fresh stock of any description; the crew roamed over the island as they pleased, but the inhabitants would not allow them to enter their houses.

DIRECTIONS.—The Port Hamilton group, so far as their examination went, is clear of danger on all sides, but is best approached from the south-east. On entering the port, the only danger that does not show is the Saracen rock, with 7 feet on it, lying at the entrance, 2 cables from the east end of Observatory island. It is a small pinnacle, steep-to on the outside, but may be easily avoided by opening the Gap tree, on Aberdeen island, a little northward of Observatory island, before bringing Triangle peak in line with Nose point, bearing S.S.W. Vessels may anchor anywhere within the port; the holding ground is so good as to render it difficult to trip the anchor after a few days.

When working in, it should not be forgotten that the north shore of the entrance nearly up to Shoal point is as steep as a wall, and that the Observatory island shore is quite safe, although rather shelving. The 3-fathoms edge of soundings from Shoal point extends more than half way across towards the west point of Observatory island, and there are only 2 fathoms water for nearly one-third of this distance.

TIDES.—It is high water, full and change, in Port Hamilton at 8h. 30m., and springs rise 11 feet; but the rise is very irregular, the sea standing at a higher level by 2 or 3 feet with southerly winds than with those from the opposite quarter.

CASTLES GROUP.—There are no islands near the Port Hamilton group to the westward, nor nearer than 9 or 10 miles to the northward, but to the eastward a chain of small islands and rocks form a continuous line for the distance of 15 miles. The easternmost is named the Castles, from whence the islets and rocks appear to take a north-easterly direction

---

* See Plan of Port Hamilton, No. 1,280; scale, $m = 3$ inches.

towards the coast of Korea, the high land of which is well in sight in clear weather. The Castles are a group of islets four or five in number, each about 2 cables in diameter, and about 200 or 300 feet high. The south-easternmost islet bears E. ¼ N. 14½ miles from the south extreme of Aberdeen island, the largest of the Port Hamilton group.

PINNACLES GROUP.—N.N.W. 2 miles from the Castles are the Pinnacles, a group of about the same size and height, but, as their name implies, of a different form; the largest of them has a conical peak.

CONE and MOSQUITO ISLANDS.—About 3½ miles to the north-east of the entrance to Port Hamilton is an island, about half a mile in diameter, having a remarkable cone in the centre of it about 350 feet in height; and between this island and the port is the small high group called the Mosquito islands, which are steep-to and safe of approach.

CHAPTER VIII.

PRATAS ISLAND AND REEF; NORTH COAST OF LUZON; AND BABUYAN, BASHÍ, FORMOSA, MEIACO-SIMA, AND LU-CHU ISLANDS.

VARIATION in 1861, Pratas shoal, 0° 30′ E.; Bashí islands, 0° 30′ W.; Meiaco-sima group, 1° 15′ W.; Lu-chu islands, 2° 0′ W.

HAVING in the former Chapters completed the description of the eastern coast of China, we will now return to the southward and describe the islands and dangers fronting that coast, then the Japan and Kuril islands, the Sea of Japan, the Gulf of Tartary, and the Sea of Okhotsk.

PRATAS ISLAND and REEF.*—Pratas island, the north-east end of which is in lat. 20° 42′ 3″ N., long. 116° 43′ 22″ E., rises from the west side, and near the middle of the sunken part of the Pratas reef. It is about 1½ miles long, E. by S. and W. by N., half a mile wide, and 40 feet high, of which elevation the scrubby bush, with which it is covered, forms about 10 feet. It is composed of sand, not a particle of mould or earthy matter could be found on it, and its shape is that of a horse-shoe, enclosing a shallow inlet or lagoon, which runs into its western side for about half a mile, and must afford shelter to the Chinese fishermen who come here to fish in the early part of the year. Brackish water can be obtained by digging a few feet into the sand. Gannets are numerous, and may be knocked down with sticks.

The island is visible at a distance of 9 or 10 miles, in clear weather, from the deck of a large vessel; from the westward it will make like two detached but contiguous islets, the centre being lower than the ends. It is visible when near the south extreme of the reef, but more conspicuous when approaching it from the westward or northward.

The Pratas reef, the north-east point of which is in about lat. 20° 47′ N., long. 116° 53′ E., is a coral barrier of nearly circular form, encircling a lagoon with 5 to 10 fathoms water in it, and thickly studded with coral knolls round its margin, but comparatively clear near the middle. The reef is about 40 miles in circumference, one to two miles broad, and slightly flattened on the northern side. Nearly two-thirds of it, or the north, east, and south sides, are just dry at low-water springs, the re-

---

* Surveyed by John Richards, Master Commanding H.M.S. *Saracen*, April 1858. See China Sea, Sheet 4, Mindoro Strait to Hong Kong, No. 2,661; scale, $m = 0.05$ of an inch.

mainder, or western side, forms a sunken barrier, across which there are two channels leading into the lagoon, one on each side of Pratas island. The north channel is about 3 miles wide, between the island and the edge of the breakers, and 3 fathoms may be carried near the middle at low-water springs. The south channel is by far the best of the two, from its being wider, a little deeper, as well as its comparative freedom from coral knolls.

TIDES.—During the survey of the Pratas reef by H.M.S. *Saracen*, April 1858, it was high water, full and change, at about 4h. a.m., and the rise was about 5 feet. There was only one perceptible ebb and one flow in the 24 hours at the springs. The highest tide occurred on the third day after the full moon, but the tides were very irregular.

ANCHORAGE.—Although the Pratas reef is steep-to in most parts, there are several spots, where, in case of necessity, a vessel might find anchorage outside the breakers, particularly on the west side, abreast the middle of the channels, through the sunken part of the reef, and at the distance of about $1\frac{1}{2}$ or 2 miles on either side of the island. At each of these spots there is good anchorage in the N.E. monsoon, in from 20 to 10 fathoms, but the position abreast the south channel is considered the best, the sunken reef at this part being deeper and the bottom more even than in the channel north of the island. A vessel of light draught might even anchor in safety *on* the reef, in the middle of the south channel in $3\frac{1}{2}$ fathoms at low water, or cross it and take up a berth inside the lagoon in 10 fathoms fine sand.

Captain Ross, of the Indian Navy, visited this reef in the *Discovery*, with the *Investigator* in company, in August 1813. The first soundings obtained were 74 fathoms, fine coral, about $1\frac{1}{2}$ or 2 miles from the north-east point; from thence the former vessel steered along the north side, about three-quarters of a mile from the breakers, in soundings of 31 to 38 fathoms; the *Investigator* keeping about a quarter of a mile off, had great overfalls of 10 to 24 fathoms. After rounding the north-west part of the reef about a mile off in 35 fathoms rocky bottom, they anchored in 24 fathoms, at about $1\frac{1}{2}$ miles from the west end of the island, with the island bearing from S.E. $\frac{1}{2}$ S. to E.S.E. About half-way between this position and the shore the depths were 4 and 5 fathoms, and then very shoal water.

H.M.S. *Highflyer*, in May 1857, anchored about 8 cables from the west end of the island, in 20 fathoms coral and clay, the extremes bearing S.E. $\frac{3}{4}$ E. and E. by S. She also anchored, with stream anchor, at half a mile from the south-east edge of the reef, in 32 fathoms, white mud, with the centre of the island bearing N.W. $\frac{1}{2}$ W. distant 10 miles; there were 13 fathoms water at 2 cables from the edge of the shoal, and 7 fathoms at a short distance from the edge. In April 1859 H.M. Steam Gunboat *Leven* anchored three-quarters of a mile off shore in 5 fathoms, with the centre of the island bearing E. by N.

CAUTION.—In beating against, or running with, the strength of the monsoon up or down the China Sea,* vessels should always endeavour to pass to leeward of the Pratas reef on account of the invariable set of the current to leeward; for there are no soundings to indicate a near approach, and the weather is frequently thick and hazy in this vicinity.

The safest quarter to make the reef is from the north-west, the island being on its western side, and the currents in the neighbourhood invariably running in a N.E. or S.W. direction according to the monsoon.

In approaching the reef a vessel should be conned from the fore-top. The sun should be well above the horizon, and if possible astern or on the beam, as the bottom can then be easily seen in 10 fathoms.

NORTH COAST OF LUZON.

CAPE BOJEADOR, which forms the north-west extreme of Luzon, is a low point with a reef of breakers projecting off it about 1½ miles. From hence the coast takes a north-east direction to Bangui point, distant about 15 miles.

Cavndian point, distant about 21 miles to the north-east of Cape Bojeador, has a reef projecting about a mile out. The intermediate coast, at one part, forms a bay, with some rocky islets near the shore. There is also anchorage in the small bay adjoining the small port of Bangui, the entrance to which port is between two points, with reefs extending from them; but it is said that this port has been long shut up by an earthquake.

Caravallo point is a bluff, steep point of white cliffs, bearing about E. b. S. 11 or 12 miles from Cavndian point, and having a mass of high mountains contiguous, which go by the same name. Close to the point there is an islet, and other islets lie near the shore, about 1½ or 2 miles to the eastward. To the eastward of Caravallo point there is a round hill of middling height, called Pata point. The whole of the coast from Cape Bojeador to this place is steep, without any soundings until near the

---

* The Macclesfield bank, lying in the fairway track from Singapore to Hong Kong, is said to be growing up. The Truro shoal, in lat. 16° 19′ N., long. 116° 41′ E. was discovered by Capt. T. J. Duggan, of the ship *Truro*, in Sept. 1857. He states, "Whilst taking my forenoon observation, distinctly saw the bottom, white coral. Got a cast of the lead instantly at 10 fathoms: again, about half a mile more north, had 19 fathoms; steered north for another half mile, and had 22 fathoms, and the next cast no bottom at 40 fathoms. No shoal patches were visible from the mast head."

shore; the land is of moderate height, and in some parts rather low close to the sea, with several rivers; but the country inland is high and mountainous.*

The coast from Pata point to Cape Engaño forms a deep bay with a chain of mountains inland, and a considerable space of moderately elevated, or rather low land fronting the sea, interspersed with villages and intersected by rivers. There is a continued beach along this coast with regular soundings, generally 30 or 40 fathoms, and similar depths extend 3 and 4 miles off shore when farther to the eastward. The only known danger is a sand bank, on which the sea breaks in bad weather. This bank lies about N. b. E. 2 miles from the bar of the Abulu river, fronting the point to the westward of the river, the west end of it bearing about South from the middle of Fuga island; it extends E.S.E. and W.N.W. about 2 miles, and about a mile outside of it there are from 35 to 40 fathoms water, fine black sand.

The entrance of the great river Tajo, about $13\frac{1}{2}$ miles to the eastward of the Abulu, has good anchorage in 10 or 11 fathoms, about 2 miles N.N.E. from its mouth. The point on the south-east side is known by the church and convent of the town of Apari built on it; abreast of which or North from the church is the best anchorage, with the volcanic mountain on Camiguin island bearing N.N.E. easterly. The river is about a third of a mile wide at the entrance, with 2 and $2\frac{1}{2}$ fathoms on the bar, deepening to 5 and 6 fathoms, mud, inside. The coast to the eastward of this river is flat, with soundings of 20 to 25 fathoms, black sand, about 6 miles off shore.

PORT SAN VICENTE, about 30 miles to the E.N.E. of Apari, is formed by a small island of the same name lying between the north-east end of Luzon and its adjacent island called Palabi. There is room in this port for three or four ships, sheltered from all winds; but the entrance is narrow and intricate, being formed between shoals on each side, which project from the south-west part of Palabi, and from Vicente island; a vessel is therefore obliged to warp in.

There is good anchorage in 5 fathoms opposite the mouth of the port, and sheltered from all winds but those between W. and S.W. There is also anchorage along the coast between Apari road and this place, in 15 or 20 fathoms water within 2 miles of the shore; the soundings are pretty regular, excepting a hole in the bank about 9 or 10 miles to the S.W. of Vicente, with 70 and 80 fathoms water about $2\frac{1}{2}$ miles off shore, having close to the edge of it 30 fathoms, black sand.

---

* *See* Charts:—Bashí and Balingtang Channels, No. 1,352; scale, $d = 5\cdot 3$ inches; and China Sea, Sheet 4, No. 2,661.

CAPE ENGANO forms the north-east point of the island of Palabi, and is moderately elevated; the south point of this island is a round hill rather higher, and forms the east point* of Port San Vicente. From the point of the Cape a coral reef, with high breakers and several rocks above water, extends E.N.E. about 3 miles; and patches of shoal water project a mile beyond it. This reef fronts the eastern side of the island, at the same distance, extending southward about 4 miles, until abreast of a round hill forming its south point, and joins the north-east end of Luzon. Close to the northward of the Cape are two islets, the outermost of which, called Lava, is a square steep mass of lava about half a mile in extent, and may be seen at the distance of about 27 miles.

The channel between Palabi island and Camiguin island to the N.N.W. is about 20 miles wide, and clear of danger. As the currents set strong to the northward in the South-west monsoon, it will be prudent for vessels proceeding to the eastward from this coast with light winds to keep on the south side of the channel, to prevent being drifted to the northward near the Guinapac and Didicas rocks.

DIRECTIONS.—When proceeding northerly from the Palawan passage or from Manila in the North-east monsoon, it is customary to beat up the west coast of Luzon to Cape Boliñao, and thence direct for Macao or Hong Kong. But if bound to any of the ports northward, much time may be saved by beating up through the Babuyan and Batan islands, and along the eastern coast of Formosa, thereby avoiding the heavy labour, wear, and loss of time by the attempt to work against the monsoon along the coast of China, which even a clipper sometimes fails in effecting.

In working along the Luzon coast, particularly about dawn and sunset, less sea, and much lighter winds, will be experienced by hugging the coast by short boards, and at times even land breezes may very much facilitate progress; but in the attempt to render these available, great caution should be observed, particularly between Capes Boliñao and Bojeador, as several coastline dangers do not find place in the charts. H.M.S. *Samarang* met with a dangerous patch in the bay near Dile point, being at the time $2\frac{1}{2}$ miles off-shore, a church bearing E.S.E.

Working up northerly, the first strong gust of the monsoon will be experienced on clearing Cape Bojeador, but this should not induce the navigator to stand farther westward than will enable him to make his eastern stretch to weather it, when he will at once experience less wind. This generally is the case on all lee shores backed by mountains, either resulting from obstruction, reaction, or the effect probably, after sunset, of counteracting land winds. There are no dangers among the groups

---

* Horsburgh, Vol. 2, seventh edition, page 587.

north of Luzon which are not easily avoided, and no continuous strong breezes will be experienced, at all comparable in force, or attended by high sea, similar to those which prevail between Cape Boliñao and Hong Kong. On the contrary, good working breezes, and at times light winds prevail, enabling a sailing vessel of moderate speed to make the range of 6 degrees northing in 8 days. Typhoons are likely to happen in both monsoons between the north coast of Luzon and Formosa.

BABUYAN ISLANDS.

The Babuyan or Five islands, named Dalupiri, Fuga, Calayan, Babuyan Claro, and Camiguin, form a kind of circular chain fronting the north coast of Luzon. The channels between them are said to be safe, without soundings, and their coasts are generally steep-to; but as these islands have not yet been examined their positions remain doubtful.

DALUPIRI ISLAND, the westernmost of the Babuyan group, lying about 36 miles to the north-eastward of Cavndian point, has a level appearance, extends 6 or 7 miles in a N.W. and S.E. direction, and may be seen from a distance of 30 miles. About 1½ miles from its south point is Rijutan islet, with shoals extending a considerable distance to the southward; but the water is deep in the narrow channel between the islet and the south end of Dalupiri.*

FUGA ISLAND, distant about 12 miles to the south-east of Dalupiri, is lower, and of an even appearance, terminating in low land at the eastern part. It is about 9 miles long, east and west, and there are irregular soundings along its south-west side, where a vessel may occasionally anchor.

The bay of Musa is formed between the west end of Fuga and two small islands adjacent, called Barrete and Mabag. The best channel into the bay is from the southward, between Barrete and the west point of Fuga, the depths being 14 and 16 fathoms outside, and 9 to 12 fathoms in mid-channel. The west channel into the bay between the two islands is narrow, with soundings of 6 to 10 fathoms. The north channel into the bay is rendered intricate by a reef extending half way across from the north-east point of Mabag towards Fuga, and the tail of this reef, joining the north-west point of Fuga, is a bed of rocks with 5 and 6 fathoms water on it; this channel, therefore, ought not to be attempted unless in a case of necessity, and a vessel to enter by it must borrow pretty

---

* The Spanish map of the Islas Filipinas for 1852, shows a reef extending from the islet to the south end of Dalupiri.

close to Fuga. Barrete island has a reef lying off its west side, and another projecting from its south point. Water may be procured, but with difficulty, some distance inland.*

DIRECTIONS.—Musa bay, although sheltered from the sea, is only fit to run for in case of necessity, the bottom everywhere being coral rock, mixed in some places with a little coarse sand or gravel. The depths are 17 to 12 fathoms in the middle, shoaling to 4 or 5 fathoms near the coral reefs that line the shores on either side, and the breadth of the bay is not more than three-quarters of a mile. The best anchorage is near the north-east side of Barrete island, in 14 or 15 fathoms water, where the bottom is rotten coral and coarse sand; near Fuga it is all very rocky.

TIDES.—The tide rises in Musa bay about 5 or 6 feet, but it is irregular in time and direction.

CALAYAN ISLAND, lying about 15 miles to the north-east of Dalupiri, is formed of mountainous and uneven land, highest in the centre, with low gaps in some places; it is steep-to, without any safe anchorage, and may be seen in clear weather at a distance of 45 miles. Some rocks above water extend about a mile from its south point; and about $1\frac{1}{2}$ miles off the north-east point there is an islet called Panuctan, about a mile in extent north and south.

H.M.S. *Cornwallis* experienced a high topping sea (first reported as breakers) in passing between Dalupiri and Calayan; this was almost immediately succeeded by a glassy smoothness. These effects were attributed to a strong north-west current.

WYLLIE ROCKS, consisting of two clusters above water, with high breakers between them, are dangerous to vessels passing through the Babuyan group in the night. The southernmost rock, which is the largest, bears N.N.E., distant 4 or 5 miles from Panuctan islet; the other cluster lies about $1\frac{1}{2}$ miles in a N.N.E. direction from the largest rock.

BABUYAN CLARO, the most northerly and highest of the Babuyan islands, bears E.N.E. about 21 miles from Calayan. On its west end is a volcano; between the volcano and the mountains, on the eastern part, there is a concave curve in the form of a crescent, when viewed from the north or south; but when the island is seen at a great distance from the eastward, it appears as one round mountain with a detached hummock to the northward. A reef projects from the west point of the

---

* See Plan of Musa Bay, No. 986; scale, $m = 3\cdot 9$ inches.

island. The south point is steep and rocky, with a black rocky islet about a mile off, in the form of a sugar loaf.

CAMIGUIN ISLAND, about 10 miles in extent N.N.E. and S.S.W., is high and hilly, and lies about S. by W. 27 miles from Babuyan Claro. Its shore in some places is bordered with coral rocks, having soundings of 30 to 35 fathoms about a mile off; and the land is low close to the sea, along its eastern and northern sides. The southern part of the island is formed of a high mountain, formerly a volcano, visible at a distance of 60 miles. To the westward of this mountain some steep white cliffs front the sea, about 2 miles to the southward of the south point of Port San Pio Quinto.

PORT SAN PIO QUINTO may be considered the only place amongst these islands tolerably safe for a large ship, for the bottom in it is not so rocky as in Musa bay, Fuga island. It is formed in a concavity of the land about 3 miles wide and $1\frac{1}{2}$ miles deep, a little to the southward of the middle of the west side of Camiguin island, and is sheltered from the sea by Pio Quinto islet, which lies in the middle of the entrance. This islet is high, about $1\frac{1}{2}$ miles in circumference, steep-to seaward, and has on each side a safe channel leading to the port. The south channel, $1\frac{1}{2}$ miles wide with 40 fathoms at the entrance decreasing gradually, inside, is between Pio Quinto islet and the south point of the port, which, with an islet near it, has the colour of iron; and a little to the southward there is a boiling spring of salt water. The north channel, between Pio Quinto islet and north point of the port, is about a mile wide, with soundings fronting it of 28 and 30 fathoms, and 17 and 18 fathoms inside; but there is a rocky patch of only 6 and 8 fathoms, lying rather nearer the islet than mid-channel, and a coral reef projects about a quarter of a mile from the north point of the entrance.*

The bottom in the channels and in the port is mostly soft sand, with a little coral in some places, and the soundings decrease gradually to the shore around. The best anchorage is in 15 or 16 fathoms to the eastward of Pio Quinto islet, abreast of rivulet of fresh water, which bears E.N.E. from the islet.

TIDES.—It is high water, full and change, in Port San Pio Quinto at 6h. 0m., and the springs rise about 6 feet.

GUINAPAC ROCKS, bearing about E. by S. 10 miles from the north point of Camiguin, consist of two rocks like towers, one larger than the other, with some smaller rocks contiguous. There are no soundings

---

* *See* Plan of Port San Pio Quinto, No. 984; scale, $m = 1\cdot6$ inches.

within a short distance of their eastern side; between them and the nearest part of Camiguin is a channel 6 miles wide, which is safe on the island side.

DIDICAS ROCKS, about 7 or 8 miles N.E. ½ E. of the Guinapac, are a group of four sharp-pointed rocks, much higher than the latter, and when seen at a considerable distance appear like ships under sail. They are about 2 miles in extent N.E. and S.W., and among them are many rocks of various sizes, which render their approach dangerous in light winds; for the currents run strong to the northward, producing ripplings like breakers in the vicinity of and among these dangerous rocks, and there are no soundings near them where a vessel could anchor in case of necessity.

BATAN OR BASHÍ ISLANDS.

The Batan or Bashí islands, so called by Dampier from the name of an intoxicating liquor which was much drank there, lie to the northward of the Babuyan group, and consist of a chain of islands, mostly high, extending from lat. 19° 58′ to 21° 13′ N., and the channels among them are thought to be safe and free from hidden danger.[*] They are named Balintang, Batan, Sabtan, Ibugos (or Bashí), Dequez (or Goat), Diogo (or High island), Ibayat (or Orange island), Siayan, Mabudis, Y'Ami, and North island. Batan, Sabtan, Ibugos, and Depuez were surveyed by Captain Sir E. Belcher, H.M.S. *Samarang*, in 1843-1844. During the North-east monsoon strong winds prevail amongst these islands, and the currents are occasionally very strong; the flood sets to the S.W., the ebb to the N.E.

Supplies.—The islands of Batan and Sabtan are mountainous, with many broad cultivated spots; the highest peak, apparently an old volcano, is about 5,000 feet above the level of the sea, and thickly covered with trees. The former is, however, richer in soil, and produces abundance of yams, sweet potatoes, maize, onions, garlic, rice, grain, &c.; indeed the only want appears to be variety of seed. Cattle, pigs, poultry, sheep and goats are abundant; deer are found on Sabtan and Ibugos, as well as quail on all of the islands. Wood is reasonable and plentiful, as well as water; but this latter necessary is difficult to procure, as the rivers are barred by reefs, which prevent boats from approaching or rafting off in sufficient quantities for vessels of war; this, however, would soon be remedied if the visits of vessels rendered it advantageous.

BALINTANG ISLANDS, in lat. 19° 58′ N., long. 122° 14′ E., and the southernmost of this group, consist of three small but high-peaked

---

[*] See Plan of Batan Islands, No. 2,408; scale, *d* = 24 inches.

islets or rocks, visible about 27 miles off, and when in one bear E. by S. and W. by N. The westernmost islet is much larger than the others, and a hole is seen through it when bearing N.E.; they are steep-to, and may be passed on either side at 2 or 3 miles distance, but the sea beats violently against them in bad weather.

The Balintang channel, between these islets and Babuyan Claro, is about 25 miles wide, and is frequently used by vessels when proceeding by the eastern passages to China.

BATAN ISLAND is about 9 miles long in a N.N.E. and S.S.W. direction, and Mount Irada, on its northern extremity, is 3,806 feet above the sea. The rest of the island is mountainous, and has several broad and cultivated spots.

ANCHORAGE.—The *Samarang* anchored in the bay of San Domingo on the western side of Batan island, on a fair clear bottom of fine coral sand, the best anchorage being with the convent barely open, when moored off the northern point of the bay in 13 fathoms; this, however, is not very secure with a northerly wind. Although the holding ground is good, this bay can only be resorted to in the North-east monsoon. There is a patch of rocks, which show at low water, lying N.N.E. 4 cables from Chaguie point, the south point of the bay, having 27 fathoms close-to on the west, and 4½ fathoms on the east side; and at a cable's length E.N.E. from the point is a rock awash at low water.

The authorities recommended the anchorage off San Carlos, about 2 miles to the south-west, as the best for obtaining a supply of water; but this position is exposed, and watering could only be effected in fine weather. The passage through the reef is, however, quite safe for the largest boats, which land on a sandy beach. This passage has been cut to admit schooners of 50 tons, which are generally hauled up when they arrive from Manila with the first of the South-west monsoon.

The next anchorage is that of San Vicente, which is the port of Ivana, or landing place for that village; it, however, ought not to be resorted to, as it is very confined, with sandy bottom close to the reefs, and must be quitted the moment a northerly wind threatens. Several vessels have been driven off, and being unable to purchase their anchors have had to cut or slip, owing to the length of cable out. During the Southwest monsoon other shelter must be looked for, and probably will be found under the north-east part of the island of Sabtan; it has not yet been sounded. There are two deep bays which appear to afford shelter on the north-east side of Batan, the northern and best is named Sonson, the other Mañañion; but both contain many rocks, and have not been sounded.

SABTAN ISLAND lies to the south-west of Batan, and is separated from it by a channel 2 miles wide, which appears clear of danger. Off the north end of Sabtan are two ledges of rock, with a passage between them carrying 14 and 10 fathoms water. These rocks are placed on the chart by land stations, from which they were clearly visible by the edges of their breakers. The sea breaks upon them, and according to the accounts given by H.M.S. *Alceste* they have only 3 feet water on them at springs.

IBUGOS ISLAND is small and rather low, excepting a hill on its south end, where there is a village. It is separated from Sabtan by a channel from a mile to half a mile wide, which affords indifferent anchorage, the bottom being rocky with sandy patches between. There are no facilities for watering, the stream from the rivulet inside the south-west point of Sabtan entering at the coral beach at least half a cable from the spot where boats could float. This is the only safe landing place, the shores on both sides of the channel being bordered by a reef, through some of the gaps in which the native boats can pass in fine weather. Dequez island is also small and rather low, and lies nearly half a mile to the westward of the north-west point of Ibugos.

DIRECTIONS.—The current sets strong to the southward between the above islands. It is therefore advisable in the North-east monsoon to work westerly round Dequez, and not to cross the channel between Batan and Sabtan, until the dividing neck of San Carlos is clearly open, E.S.E., as the stream dividing at Mabatui point, sends one current southerly; the other, which is an eddy, is favourable from thence north-easterly to San Domingo.

If bound to this latter anchorage, work up to the north-west angle of the island until the wind is free to run down, when round-to with all aback, and drop the inner anchor in 12 fathoms; then veer and drop the outer anchor in 25 fathoms, which will afford sufficient room to weigh. When moored the vessel will be in 15 fathoms, and the set of the current will keep a fair strain on both cables.

IBAYAT ISLAND, lying 14 miles to the N.N.W. of Batan, is about 8 miles long in a N.N.E. and S.S.W. direction, and the channel between them is free from danger; Mount Sta. Rosa at its north end rises 680 feet, and Mount Riposet, at its south-eastern part, 800 feet above the level of the sea. The exterior of the island as viewed from the sea presents a blank barren outline, defying disembarkation to any but those acquainted with the locality, and is moreover without anchorage; the interior is, however, highly cultivated, and in many spots exhibiting patches of good timber trees; abundance of refreshments can be easily obtained.

SABTAN ISLAND.—Y'AMI AND NORTH ISLANDS.

DIOGO is a small high island, 848 feet above the sea, lying 3½ miles eastward of Ibayat, and the channel between them is clear of danger. This island is steep-to on its western side, but several small islets lie off its eastern side, the outermost being distant a short half mile.

MABUDIS and SIAYAN ISLANDS.—Mabudis island, lying N.N.E. 6 miles from the north end of Ibayat, is about 1½ miles long in a N.E. and S.W. direction, high and steep-to. One mile S.S.W. of it is Siayan island, about 1½ miles in circumference, having off its north-east side several detached rocks. The channel between Mabudis and Siayan is rendered unsafe by detached rocks; that between Ibayat and Siayan is about 4 miles wide, and free from danger.

Y'AMI and NORTH ISLANDS.—Y'Ami, the northern island of the Bashí group, is about a mile in circumference and tolerably high; the position of the islet lying off its south-west point is lat. 21° 4' 56" N., long. 121° 58' 24" E.

North island, lying 2 miles S.S.W. from Y'Ami, is high and steep-to, except on its eastern side, off which, at a cable's length, there are three islets and some detached rocks. The channel between Y'Ami and North island is safe, and carries soundings with rocky bottom, but too deep for anchorage; that between Mabudis and North island is 9 miles wide, and free from danger.

The North Bashí rocks could not be found by Captain Sir E. Belcher, who states "they have no existence in the position assigned to them in the charts, nor in the visual radius from the mast head of the *Samarang*, 108 feet above the level of the sea."

GADD ROCK or Cumbrian Reef.—The position of this dangerous rock was ascertained by Captain Ross, of the Indian Navy, to be in lat. 21° 43' N., long. 121° 41' E.; Little Botel-tobago sima bearing N. ¼ W. It is about half a cable long, and the boat had a depth of 12 feet about the middle of the reef at, probably, the time of high water, as Capt. Gadd, who appears to have first discovered it in the Swedish ship *Oster-Gothland*, January 20th, 1800, perceived some points of rocks amongst the breakers; for there is a considerable rise and fall of tide hereabout on the springs, affording sufficient cause to think that some parts of the rock must be level with the surface of the sea, or visible above the hollow of the waves at low water, when there is much swell on.

VELE-RETE ROCKS lie on nearly the same parallel, and 44 miles westward of Gadd rock. and about S. ½ W., distant 11 miles, from the low South Cape of Formosa. They are a mass of detached rocks, above even with, and below water; the highest may be seen at about 5 miles. The channel between them and the south end of Formosa is safe; but very turbulent ripplings are often experienced in this and the neighbouring channels, and they have been observed to run so high that the breakers resembled the sea breaking furiously over a dangerous shoal.

DIRECTIONS.—Quitting the Batan islands during the North-east monsoon, or merely working up to the northward past them, it is advisable to make short boards to the north-east on the western side of Batan island until Mount Irada bears S.S.E., then make a stretch to the north-west and work up on the western sides of Ibayat, Siayan, and Mabudis; but on reaching the latter, pass through easterly between it and North island, where the current will favour northerly.

Keep well to the eastward, or endeavour to pass well to windward of Botel-tobago sima, as the currents in that neighbourhood press strong to the westward, and the changes from strong breezes to calm, attended with swell, are troublesome, as well as harassing.

Gadd rock lies in the fair way of the Bashí channel, and to avoid it vessels should keep either toward Botel-tobago sima or towards the northern port of the Batan group, taking great care to avoid the mid-channel track. When passing southward of the above danger in thick weather or in the night, keep well towards the latter group, making allowance for a northerly current, which is generally experienced in light winds and during the South-west monsoon. From lat. 21° 15′ N. to 21° 21′ N. is a good track to preserve when passing between the Batan group and Gadd rock in thick weather. Several vessels during light winds have been drifted by the current between Formosa and Botel-tobago sima.

BOTEL-TOBAGO SIMA* is a high island, 7½ miles in length in a N.W. and S.E. direction, appearing in form of a saddle, or with a gap in it when viewed from a S.S.W. or N.N.E. direction, and is visible about 50 miles from the mast head. It is well inhabited, and its highest part is crowned with trees; the north-east peak is 1,850, and the west peak 1,820 feet above the sea.

There are several large villages on the southern part of the island, and on the north-west side are several rocky points. Detached rocks, remarkable for their spire-like form, lie off the northern extremity The coast is rocky in almost every part, and probably dangerous to

---

* See Charts:—Formosa island, No. 1,968, scale, $d = 6\frac{1}{2}$ inches; and China General, from Hong Kong to Liau-tung. No. 1,262, scale, $d = 2$ inches. Sima signifies island in Japanese; Jama, san, hill; Take, mine, peak; Saki, cape; Sedo, channel; Kawa, river; Fana, hana, point; Mura, village; Umi, nada, lake.

land upon, as these needle rocks are seen in many parts of the island; with the exception, however, of those off the north extreme, they are attached to the island by low land; but the shore under water often assumes the character of that which is above, in which case a vigilant look-out for rocks would be necessary in rowing along the coast.

LITTLE BOTEL-TOBAGO is a small island of considerable height, lying about 3½ miles to the S.E. of the southern part of Botel-tobago. A reef, steep-to, projects about a cable's length from its south end.

The Alceste shoal, marked on the chart in about lat. 22° 5′ N., long. 121° 18′ E., is supposed to have no existence.

FORMOSA OR TAÏ-WAN ISLAND.

Formosa island is about 210 miles long, N.N.E. and S.S.W., and between 60 and 70 miles wide at its broadest part, which is near the middle. The land is generally high in the interior, but low in some places seaward, with soundings near the shore, particularly on the western side. The southern part has on it a high double-peaked mountain, visible 60 miles in clear weather, from which the land slopes down to the southward, terminating in a low projecting point called the South Cape.

From the south-end of Formosa to Liang-kiau, the coast range, from 250 to 300 feet high and composed of dark rocky cliffs, is backed by an inland range, rising to the height of about 2,000 feet above the sea, and terminating to the southward in a remarkable craggy peak of the same height, from which extends to the south-east the long low promontory of the South Cape. The hills about this part of the island are mostly bare, their summits only being wooded.

The wind blowing a brisk gale from the westward prevented the *Inflexible*[*] from examining the coast about the south end of Formosa, but the sea was observed breaking very heavily over the Vele-Rete rocks, and heavy tide ripples extended nearly the whole distance across to them from the South Cape. A good harbour, affording anchorage in 5 fathoms, was reported to exist at 3 miles East of the South-west point at Formosa.

A reef is said to project from this cape, for high breakers, thought to be upon a sunken reef, projecting to a considerable distance from the cape, were observed from H.M.S. *Alceste*, when that vessel was passing between Botel-tobago sima and Formosa. To the north-east of the cape there is

---

[*] Remarks made during the visit of H.M.S. *Inflexible* to the island of Formosa, in search of missing Europeans, by Mr. William Blackney, R.N., Assistant Surveyor, June 1858.

a village and harbour for small vessels; and there are, it is said, soundings near the shore on the west side of the cape.

LAMBAY ISLAND, lying about 9 miles from the western coast of Formosa, and 29 miles to the north-west of the South-west point, is about 180 feet high, and may be seen at a distance of 18 miles. The island is 3¼ miles in length, north and south, has high yellow cliffs on its western side, a small sandy beach on its eastern, and is inhabited by Chinese fishermen.

Capt. Ross of the Indian Navy examined this island in the *Discovery*, and no bottom was obtained with 70 fathoms line westward of it; but about 3 miles eastward of the island a bank of soft mud commences, which, extending off Formosa, has soundings on it of 15 to 26 fathoms. A reef extends off about a mile between the south-east and north-east points of the island; it is steep-to, there being 25 and 30 fathoms water close to its edge.

In steering from the South-west point of Formosa, along the western coast, the *Discovery* had no soundings off the point until within half a mile of the shore, then had 120 fathoms; and with Lambay island bearing about W.N.W., got 30 and 40 fathoms on the mud-bank when about 1½ miles off Formosa, and passed between the island and the coast. At anchor in 15 fathoms, soft bottom, about 3 miles off the town of Pong-li, with it bearing N.E. by E. ¼ E., Lambay island bore from W. ¼ S. to W. ⅔ N., the north-west extreme of the coast, a small black hummock, N.W. ¼ N., and the south extreme of the coast S.S.E.

From Pong-li the *Discovery* worked to the westward, and anchored in 15 fathoms about 3 miles off the coast of Formosa, and 5 or 6 miles from Lambay island, with the black hummock bearing N. ¾ W., brow of western hill N.N.W. ⅓ W.; a town near which there is a river or inlet and many boats at anchor bore N.E. by E., distant 3 or 4 miles, Lambay island from S. by W. ¼ W. to S. ⅓ E., the south-east extreme of the coast S.E. ¾ S. In working across, as Lambay island was approached, the soundings increased from 35 fathoms into deep water, having 52 fathoms about a mile of Lambay, from the south-east and eastern parts of which a reef projects a short distance. From this last anchorage she steered about 4 miles to the westward, then got off the bank of soundings.

LEANG KIAOU BAY.—*At 11 miles N.N.W. from the South-west point of Formosa is Liang-kiau bay, the south point of which is about 300 feet high, and makes like an island from the northward; the point termi-

---

* Mr. W. Blackney, R.N., H.M.S. *Inflexible*, June 1858.

nates abruptly seaward, and slopes gradually towards the head of the bay, leaving a level between it and the high hills trending from Pong-li.

The north point of the bay, bearing about N.N.E. 2¼ miles from the south point, is a low grassy flat extending from the foot of the hills, and on it stands the village of Liang-kiau, which is the most southern Chinese settlement on the west coast of Formosa. The bay is open to all westerly winds, but it affords good anchorage in the North-east monsoon. It is about 1½ miles deep, and from the depth of 9 fathoms, sand, between the outer points, the water shoals gradually towards the long sandy beach at its head. The approach to the bay is quite clear; outside the soundings deepen rapidly from 10 to 15 and 30 fathoms, and from 45 to 50 fathoms at 5 miles from the coast.

PONG-LI is a small Chinese town at 15 miles N. by W. ½ W. from Liang-kiau bay and a short distance inland, about a mile to the northward of a remarkable square clump of trees on the beach. The *Inflexible* anchored abreast the town in 5 fathoms water, at about 3 cables from the beach. Landing was effected in Chinese catamarans, the surf being too high for the vessel's boats, although there had been but little wind the three previous days.

The shore between Pong-li and Ta-kau-kon is a low sandy beach; but a short distance within are numerous clusters of bamboos and Chinese houses, and a country highly cultivated. The depths are from 8 to 9 fathoms at 2 miles from the beach, but in crossing the bight, in which the river Tollatock is marked on the chart, no soundings were obtained with the hand lead.

PORT TA-KAU-KON.—*Ape hill, called by the natives Ta-kau, bears S. by E ⅓ E., 22½ miles from the old Dutch fort of Zelandia. It appears like a truncated cone, on a North and South bearing, and is 1,110 feet high, sloping towards the land side, and makes at a distance like an island. At 4½ miles N.E. of Ape hill is another remarkable hill, 700 feet high, which from its resemblance to a huge whale sleeping on the water, was named Whaleback; and N.N.E., 12 miles, there is a small triangular shaped hill, and a large detached piece of table-land resembling a quoin, on a North and South bearing. These are the only landmarks on this part of the coast (which is all very low), and of these Ape hill is the most useful, as it stands out on the coast line, and is frequently seen distinctly when all the others are shrouded in mist.

---

* The south-west coast of Formosa from Ta-kau-kon to Kok-si-kon was surveyed by John Richards, Master, Commanding H.M. surveying vessel *Saracen*, in February 1855. See Chart:—West Coast of Formosa, with Plans of Port Kok-si-kon and Port Ta-kau-kon, No. 2,409; scale, *m* = three-quarters of an inch.

This hill is one vast block of coral, and although resembling the crater of a volcano in the peculiar form of its apex, no traces could be discovered of volcanic action. From its summit to the southward it descends in a gradual though somewhat rugged slope, and terminates in a huge nearly level block of a mole-like appearance, which, jutting through the beach to seaward for about 300 yards, forms a sheltered anchorage for small vessels in the strength of the North-east monsoon. This mole is separated from the hill by a deep chasm 50 fathoms wide, and within this is the little port of Ta-kau-kon. The south-west part of the mole is a steep cliff, named Saracen head.

This port has a narrow bar, with 11 feet on it at low water, extending from the south side of the entrance, curving to the N.W. and N.N.W. in the direction of Ape Hill; but directly this is passed the water deepens to 4, 6, and 9 fathoms just within the port. The entrance, though narrow, is steep-to and quite safe of approach, but unfortunately the anchorage within is so very confined that there is no room for a vessel to swing; it is therefore necessary to moor head and stern. The tides are also rather strong when near the springs; but this anchorage is susceptible of great improvement at small expense, and as Formosa is opened to commercial enterprise this place must advance in importance.

Supplies.—Good water can be obtained at Port Ta-kau-kon, but not in any large quantity, from the difficulty of transport; also bullocks, pigs, fowls, ducks, eggs, rice, sugar, fish, and vegetables.

VUYLOY SHOAL.—* This dangerous sand-bank, about half a mile in extent, and with only from 8 to 12 feet on it at low water, lies a good mile off shore, S.S.E. 4½ miles from the entrance of Port Kok-si-kon, W. by N. ⅗ N. 4¼ miles from fort Zelandia, and S.W. by W ¼ W. 2¾ miles from Joss islet. With southerly winds the sea breaks heavily on it, but with off shore or north-east winds there is but little break. The soundings are 4½ to 5 fathoms at 1½ miles westward of the bank, and 3 fathoms between it and the shore.

Vessels, when bound from Port Kok-si-kon to Port Ta-kau-kon, will pass well to the westward of this shoal, by not approaching the coast within 3 miles, or not coming into a less depth than 4½ or 5 fathoms, until fort Zelandia bears East.

TIDES.—The flood stream sets in a N.N.W. direction from 1½ to 2 knots an hour along this part of the coast. The ebb runs S.S.E. except near Kok-he-mung, where its direction is S. by W. out of the harbour.

---

* Mr. W. Blackney, H.M.S. *Inflexible*, June 1858.

VUYLOY SHOAL.—PORT KOK-SI-KON.

PORT KOK-SI-KON, the north point of entrance to which, Gull point, is 32 miles to the N.N.W. of Saracen head, can only be recognized by the number of large junks generally at anchor inside, and by three larger clumps of huts than can be found on any of the outer sand-banks which front all this part of Formosa, and which are elevated only 2 or 3 feet above high water. These banks run in lines, generally parallel to the coast, from 2 cables to half a mile broad, and are pierced at every mile or so by narrow channels, having depths varying from 7 feet and under. There is no vegetation in sight from the western sand bar; the main land of Formosa can only be seen in very clear weather from it, and the whole intermediate space seems to be an intricate mass of sand and mud banks and shallows, with occasional patches of sedge.

These sand-banks are occupied by a few poor fishermen, whose miserable huts and bamboo rafts are the only relieving features of this dreary scene. Ape hill to the southward, and the southern islands of the Pescadores to the westward, will be found useful marks to run in for Kok-si-kon, which bears N.N.W. 30 miles from the former, and E. b. S. ¾ S. 26 miles from East island, Pescadores. The old Dutch fort of Zelandia, built in 1634, is just in sight from the anchorage, from which it bears S.E. ¼ S. and is distant 7½ miles.

This port is the outlet of several small shallow streams which here unite and form a channel through the mass of sand-banks fronting the coast. This channel or port runs N.E. and S.W., and, taking the 3-fathoms line as its boundary inside, is three-quarters of a mile long and only 2 cables broad, with 4½ fathoms in the middle; it is, therefore, necessary to moor N.W. and S.E. The bar has 12 feet on it at low water springs. The deepest part is generally marked by the natives with bamboos; but, as the channel is both wide and straight and the bottom remarkably even, it is by no means difficult of access for vessels of 12 or 13 feet draught at high tide. The *Saracen* sailed in drawing 13 feet 2 inches, but then the sea was remarkably smooth; vessels, therefore, drawing over 13 feet should not attempt to enter, particularly with any swell on.

Fresh water is procured from Taï-wan fu, and if a vessel should only require this article, she will do better by anchoring at once off that town, about three-quarters of a mile off shore; where in 5½ fathoms, with the old Dutch fort bearing N.E., she will find capital anchorage and good shelter from December to March. During the rest of the year the chances of south-west winds would render this position unsafe, and anchorage should of course be sought farther out. At the distance of 1¾ miles N.W. of the old fort there is a large clump of trees on the outer sand bar.

The ruins of the old Dutch fort are about two-thirds of a mile inside the sand, about 60 feet above the sea level, and the only conspicuous landmark in this neighbourhood; they can be seen 8 or 9 miles from a vessel's deck. The principal town of the island, Taï-wan fu, stands S.E. 2 miles from the fort, and large junks trading to the place in the North-east monsoon generally anchor off the fort, and send their cargoes by this route to the town. Here the main land of Formosa approaches within a mile of the sand bars fronting the coast, and although it is generally marshy and flat, it is cultivated with rice, &c. The sand bars are also occasionally clothed with bushes and grass, and are densely populated by fishermen, who appear to be well fed and clothed and a happy and contented people, and pursue their vocation generally in divisions under the direction of particular chiefs; and their rafts hauled upon the beach, placed in tiers on their sides, form a feature in the appearance of the coast. Whenever the officers of the *Saracen* landed they were treated with the greatest civility and deference, and the surveying marks, although sometimes made of an article most tempting to them (white calico), were never in one case interfered with.

There is no remarkable feature in the coast until within 8 miles of Ape hill, where commence some low mud cliffs, and there is also a small piece of table land about a mile inland. The coast between the old Dutch fort and Ape hill is nearly a straight line of beach, pierced by four small streams, navigable only for boats.

TIDES.—It is high water, full and change, at Port Kok-si-kon, at 11h. 30m., and the rise is about 3 feet. The tide from the bar inside sets fairly through the channel; its greatest strength being about a knot. Outside the bar the flood sets northward, along the coast, the ebb southward; its strength varies in different positions, running with much greater velocity off the west sand bar or the edge of the deep water than in the shoal water bight off Taï-wan, where it is occasionally variable in strength and direction.

DIRECTIONS.*—The high land of Formosa, immediately over Port Kok-si-kon, may be distinctly seen in very clear weather from the Pescadores, but as it is generally obscured, and the coast low and sandy, it will be prudent at all times, when bound to that port from the westward, to be certain of the vessel's position before losing sight of East island, or one of the southern islands of that group.

The mast heads of a large fleet of junks, usually at anchor in the small harbour of Kok-he-mung, at 5 miles S.E. by E. of Kok-si-kon, will serve as a guide on approaching the coast, and when at a distance of 3 or 4 miles from the shore, three clumps of huts and trees, (the southernmost clump abreast of West point being the largest and most conspicuous,)

---

* Mr. W. Blackney, H.M.S. *Inflexible*, June 1858.

Joss islet, and fort Zelandia, are objects sufficiently well defined to mark the locality. Joss islet has a clump of dark trees on its southern end, and the Joss house on it a white front to seaward. Ung-lo and So-co are remarkable hills, and may generally be seen when the mountains in the interior are hidden. The clouds sometimes rest upon them, when they appear as the highest land in the vicinity. Ung-lo, 1,080 feet high, is the southern termination of a long table range, which falls steeply for a few hundred feet, and rises again to the round hill of So-co, 880 feet high.

The *Inflexible* in 1858 anchored in 6 fathoms off Port Kok-si-kon, Observatory point bearing N.E. by E. 1½ miles. The wind being light from the S.S.E. and South, an attempt was made to enter the port in the vessel's boat, but it was unsuccessful, as the sea broke the whole way across the entrance. The Chinese fishermen stated that the channel and sand-banks have altered considerably since surveyed by Richards in 1855. There were no junks at anchor in the port, but the harbour of Kok-he-mung was crowded with them. A party from the vessel landed inside the latter harbour, and visited Tai-wan fu, the capital of the island. They were civilly received by the authorities, who sent off presents of pigs, goats, fowls, sweet potatoes, &c.

NORTH COAST of FORMOSA.*—The north end of Formosa is high and mountainous, except the north and north-west points, which are low, and have reefs extending a considerable distance off; from it the Tam-sui range rises to the height of 2,800 feet above the sea, and is generally covered with clouds. In the neighbourhood of, and eastward of Ke-lung harbour, the coast hills are wedged shaped, nearly all perpendicular to the north-west and sloping to the eastward.

Petou point, the north-east extreme of Formosa, is a peninsula, 400 feet high, and from a distance appears like an island; the small boat harbour and fishing village of Petou is close to the westward of it. The coast from thence to Ke-lung harbour is steep-to, all the off-lying rocks, which are composed of sandstone, showing above water. The most remarkable feature on this coast is Dome peak, which makes in that form from the N.E. At 5 miles westward of Petou point is the entrance to Chimmo bay, in which a vessel might anchor if in distress, or forced in by a northerly wind. The depths are from 4 to 10 fathoms at the entrance, and 5 and 4 fathoms at the head of the bay, under the lee of the point on its eastern side. This point is foul, and should be given a berth in entering.

At 7½ miles westward of Petou point, and 2 miles eastward of Ke-

---

* This coast, with Tam-sui and Ke-lung harbours, is from the survey by the late Lieutenant Gordon, Commanding H.M.S. *Royalist*, 1847.

lung harbour, is Pe-ta-ou bay, open to the N.W., surrounded with reefs and rocks, and shoal at the head; it might, however, be available to a vessel embayed to windward of it, and in distress. Immediately westward of Ke-lung harbour is a bay 4½ miles across, with numerous reefs running off the points within it; its north-west point is formed by a remarkable sandstone peninsula, named Masou, 250 feet high, and quite perpendicular to the north-west.

To the westward of this peninsula is the deep bay and valley of Masou, in the middle of which is an islet with three rocks lying to the S.E. of it, two of which are covered at low tide, the other always shows. From thence the coast appears bold until Foki point, the north end of Formosa, is approached. This point is very low, and has a dangerous shoal extending off about a mile from it, and then trends round to the westward. It will be prudent to give this point a wide berth, as H.M.S. *Royalist* shoaled suddenly from 30 to 60 fathoms, and there were breakers close to leeward of her.

From Foki point the coast trends to the south-west 8½ miles, to Tam-sui harbour, with a reef fronting it for the first 5 miles, and in some places extending nearly half a mile off shore. The coast rises gradually, and is very flat for several miles inland to the Tam-sui range. From Tam-sui harbour the coast line bends in a W.S.W. direction for 20 miles to an elevated sand hill named Paksa point. The first 5 or 6 miles is table land, about 600 feet high, and is steep-to until within 2 miles of the harbour. This table land is succeeded by low land with sand hills; along this part reefs extend about half a mile off shore, with soundings of 7 fathoms well clear of them, and there are numerous creeks, in some of which junks were seen lying. Here the melancholy loss of the brig *Ann* took place on the 10th March 1842.

From Paksa point the coast runs S.S.W. ½ W. 10 miles to a table hill 360 feet high, with low land on either side of it. From a position off this hill the elevated land about Tam-sui shows over the low land as three hills, and in fine weather the high ranges, rising to the height of about 12,000 feet in the centre of the island, will be seen to the S.E. Along this part of the coast, as to the northward, are numerous creeks and reefs extending a short distance off; the bottom is dark sand and not fit for anchoring on. South of the table hill reefs extend nearly half a mile off shore, and there is a barred river in which several junks were seen lying.

At 9 miles to the southward of the table hill is the port and village of Teukcham, in which were several large junks. This port is only available for vessels of light draught, and is formed by a sand spit running to the southward; with the end of this spit bearing North, and half way between it and a sand bluff to the southward, a vessel may anchor in 4 fathoms at low water; smaller craft can go farther in and anchor in 1¾ fathoms, but the holding ground is bad and open to the N.W. The sand bluff forms the south-west point of the port, and the point, which is steep-to, has some fishing huts on its extreme.

The bay to the southward of Teukcham has a reef extending from its centre, and at the south corner of the bay is a barred creek in which were numerous junks. A low serrated sandy ridge with a low sandy point, on which is a large fishing village, terminates this bay; off the point and off the village, the water is deep well in, but it appears to shoal off the coast to the southward. With the exception of the ridges, the land in this neighbourhood is low, the hills are all sandy, and show little vegetation. A range of hills having a low but remarkable peak at the point forms the southern point of the bay, off which it is shoal; from thence the coast range extends about 5 miles. The coast appeared to be steep-to, but, as in other places along it, the holding ground is bad. Single peak, a round isolated hill 200 feet high, is a remarkable object in this locality. A river disembogues to the southward, with a large bamboo plantation on its north side. The survey was not continued farther than Single peak, but the coast to the southward was observed to be very low.

DIRECTIONS.—As far as Lieutenant Gordon was enabled to examine the north-west coast of Formosa, it was his opinion that a great advantage would be obtained, were sailing vessels, instead of hugging the coast of China or beating up in the middle of the Formosa channel during the North-east monsoon, to reach well over, and at all events during the day to stand close in to the Formosa shore (page 11), particularly on the ebb tide. The latter stream was always found setting strong to the N.E., whereas the flood ran very weak to the S.S.W., the former having the advantage over the latter of at least 8 miles every 24 hours. It was also observed that until the middle of the Formosa channel was passed, there was no southerly set, and though the *Royalist* experienced a continuation of severe weather for several days, during which time she generally stood off and on under easy sail, she generally weathered and was always to windward of her reckoning.

The time of high water, full and change, on the north-west coast of Formosa is at noon, and the bottom is dark sand, with, occasionally, shells and broken stones; soundings of 30 to 40 fathoms near the shore, and 25 to 17 fathoms at 10 to 20 miles off. The water commences shoaling about 1½ or 2 miles from the shore, and the depths decrease rapidly. The sea near the coast in moderate weather is smooth, the wind blowing along the land.

TAM-SUI HARBOUR, the south point of the entrance to which is in lat. 25° 10′ N., long. 121° 26′ E., is formed between a high range (2,800 feet above the sea) to the north-east, and a remarkable double hill (the north peak of which is 1,720 feet, and the other 1,240 feet above the same level) to the south-west. There is a bar across the entrance with only 10 feet on it, but having entered, the depths increase to 4 fathoms, and a large river runs to the south-east, with two smaller branches leading through the valley to the southward. The principal town, named Min-ka, in the north part of Formosa, is said to be about 13 miles up the main branch of this river.*

The following information is from Her Majesty's Consul at Amoy, 11th May 1857 :—"Owing to the prohibition by the authorities at Formosa against the export of rice from Tam-sui harbour, vessels that now arrive load almost entirely with coal at about 1¼ dollars per ton ; and it is said that arrangements may be made for the formation of a stock for the supply of Her Majesty's steamers on very favourable terms. The harbour is quite safe from all storms ; and, although there are only 10 feet on the bar at low water, yet, the rise being from 7 to 12 feet, vessels of moderate draught may enter or leave daily."

Supplies.—There appears to be a large trade carried on between Tam-sui and the Fu-kyen province. The principal exports being coal, oil made from ground-nut, sulphur, camphor, and camphor wood ; rice was also exported before the above prohibition. There is good water running down in large streams from the south hill. Bullocks, pigs, goats, poultry, vegetables and fruit abound.

TIDES.—It is high water, full and change, in Tam-sui harbour at 11h. 45m., and the rise is from 7 to 12 feet. From observations made in the *Royalist* when at anchor off the harbour, October 1847, five days before full and change, the flood set S.S.W. 1¼ knots per hour at its strength, and 3¼ knots the whole tide ; and the ebb ran to the N.E. 3 knots per hour at its strength, and 7 miles the whole tide. The flood appeared to run about 4 hours, the ebb 8 hours. As the ebb sets along shore to the north, and has the advantage of 4 miles in the twelve hours and more during springs, it will greatly assist a vessel when beating to windward during the North-east monsoon.

DIRECTIONS.—The anchorage off this harbour is unsafe, as the holding ground is not good, being a loose sand, and a vessel, though with a good scope of cable out, is likely to drive even in moderate weather. When the wind freshens from the N.E. a heavy sea rolls in, breaking even

---

* See Plan of Tam-sui harbour, No. 2,376 ; scale, m = 3½ inches.

in 3 fathoms, and a sailing vessel must immediately go to sea, for should the wind veer to the N.W. that might not be accomplished.

The former mark for entering the harbour, viz., the small round red fort upon a little hill on the north side of the river, in line with the highest peak, 2,800 feet high, behind the town bearing E. ¼ S. can no longer be used, as the bar is constantly changing its position. The present mark over the bar, in 1857,* is the red fort in line with the lower fort's centre, which will lead in nothing less than 3¼ fathoms at high water; but no stranger should attempt to enter without a pilot.

KE-LUNG HARBOUR.—The north end of Formosa forms, between Fo-ki and Petou points, an extensive bay, 22 miles wide, into which the North-east monsoon rolls a heavy sea; the current during the ebb sets strong to the eastward, and only occasionally changes its direction to the N.W. during the flood. In the middle of this bay is Ke-lung island, a remarkable black rocky island rising precipitously on all sides to the height of 580 feet above the sea, with rather a flat summit. This excellent landmark guides to the entrance of Ke-lung harbour, the entrance to which bears from it S.W. by S., and is distant 2 miles. At a cable's length from the north-west side of the island is a conical rock 100 feet high, and broken water extends 2 cables from its south-west extreme; the other parts of the island are steep-to, though the strong tide ripples around it frequently resemble breakers. No danger exists between it and the shore, the soundings being 30 to 35 fathoms, sand, but the heavy tide ripples must be guarded against.

The country in the vicinity of the harbour is richly wooded to the water's edge, and the land rises in a succession of picturesque knolls and undulating hills, fantastically piled one above the other, the distant prospect being closed by a range of lofty mountains. Few dwellings, or signs of cultivation, are visible, but the brilliancy of the verdure and luxuriance of the vegetation render the *coup d'œil* most striking, a perfect contrast to the sterile-looking mainland of China.

Ke-lung harbour, but for Ke-lung island, would not be easy to find, as a sand spit projecting from the low island or cay (Bush island) on the eastern side of the entrance almost conceals it. The entrance is formed between Palm island on the east and Image point on the west. Off-lying the former, West 3 cables, is Bush island, a low rocky islet about 10 feet high, and covered with shrubs, thus narrowing the entrance between it and Image point to 4 cables. The passage between Palm island and Bush island is used only by boats.

---

* Thomas R. Collingwood, Master of H.M.S. *Comus*, 1857.

Image point is remarkable from the number of detached pieces of rock which the action of the sea has worn into grotesque figures; the summit over it, 390 feet high, has several patches of stratified cliff on the seaward slope. Palm island is three-quarters of a mile long, east and west. The land over its northern coast is 200 feet above the sea. Close to its north-west extreme, and almost connected with it, is Macedonian mound, 140 feet high. Both the island and the mound are fringed with steep shelving rocks, having 7 and 10 fathoms close to; in approaching them from the north-east, Image point should not be brought to the westward of S.W. ½ S.

On the western side, a little within the entrance, is a small bight, named Merope bay, in which a vessel of that name remained 10 days and procured good water and refreshments. The anchorage in it is in 9 fathoms nearest to the coral bank which extends from the northern shore.

The town of Ke-lung is nothing but a beggarly collection of wooden huts, more filthy than the suburbs of Amoy, and the inhabitants, of Chinese extraction, apparently very poor and wretched. There are several coal mines about a mile E.S.E. of the town, on the southern banks of the small shallow stream which branches off in that direction.

Supplies.—The trade of Ke-lung is extensive, principally with the river Min, Chin-chu, Amoy, and Tongsang. For the latter place quantities of coal are shipped, and for the former rice, ground-nut oil, camphor, and camphor wood.

Good water is easily obtained on the western shore of the harbour, in the second small bay within Crag peak. There are several streams on either shore. Pigs, poultry, and sweet potatoes may be purchased, but at high prices, the Chinese having only sufficient for their own consumption.

Coal, from the mines in the vicinity of Ke-lung, was brought off to H.M.S. *Inflexible* in 1858 for four dollars a ton. It was reported good for domestic purposes and for steamers making short passages; but it consumes rapidly and makes much smoke.

Tides.—It is high water, full and change, in Ke-lung harbour at 10h. 30m., and the rise, when uninfluenced by the weather, is about 3 feet. The flood at the entrance sets fairly into the harbour about a knot an hour; the ebb towards the eastern shore and rocks off Bush island. In the narrows of Junk passage the streams run with great velocity. Outside the harbour the flood sets into the bight towards Masou peninsula; the ebb to the south-eastward along the coast.

Directions.—When nearing the entrance of Ke-lung a remarkable hill, named Crag peak, will be seen at the head of the harbour, and by steering for it on a S. ¾ W. bearing, it will lead in mid-channel between Bush island and Image point. The soundings in the middle of the entrance are 14 to 12 fathoms, decreasing a little towards the coral banks which border the shores on either side.

After passing Image point steer S.E. ½ S. for the Sandy bay on the eastern shore, anchoring in 5½ or 6 fathoms, mud, good holding ground, when the points of Junk passage (the channel on the south side of Palm island) are open, Image point bearing N.W. ½ W. The vessel will then be 2 cables to the southward of Inflexible reef which is a sunken ledge of rocks, 1½ cables in extent, with 6 to 12 feet on it at low water; from the western and shoalest part of this reef Crag peak bears S.W. by S. southerly, and Image point W. by N. ½ N. A sunken ledge, with 3 to 9 feet on it, also extends 1½ cables East of Crag peak. The junks anchor in 2 and 2½ fathoms at a quarter of a mile to the southward of the peak, and about a mile from the town of Ke-lung, which can only be approached by boats at high water.

A sailing vessel must use much caution in leaving this harbour during the North-east monsoon, in consequence of the heavy sea rolling in, and there being no anchorage outside. With a light wind short tacks should be made, and the entrance of the harbour kept open until an offing is gained.

COAL HARBOUR, the next inlet to the eastward of Ke-lung, is so called from its proximity to the coal mines opened by the Chinese on the hill sides of the southern shore of Quar-see-kau bay. It offers anchorage and shelter for one or two vessels only, and should the mines ever be worked by Europeans, the coal, which is of good quality, could be conveyed to Harbour rock by means of a railroad along the west shore of the bay, at the base of the hills. A short pier from the north side of the rock would enable a vessel to lie alongside in 3 or 4 fathoms water, and receive or discharge her cargo.

EAST COAST of FORMOSA.—With the exception of Sau-o bay this coast is without harbours, and deep water will be found close in to the land. The mountains rise almost immediately from the sea; their sides in some places are cultivated and scattered houses seen.

This coast is not visited by the full strength of the North-east monsoon. This probably results from the mountainous character of the country, which prevents the breeze blowing home. Sailing vessels, however, experiencing strong gales at 20 miles to the eastward, might feel cautious in venturing in-shore. Nor is there any necessity to run to leeward; but if they should experience the breeze declining in strength, with less sea on the western board, particularly between 9h. a.m. and 3h. p.m., or up to sunset, they will find it advantageous to hug the coast as far as the depth of 20 fathoms; but within this limit caution is requisite, as sudden loss of wind, attended by inconvenient swell, might be attended, if followed by calm, with imminent danger.

STEEP ISLAND,* lying S.S.W. 11 miles from the north-east extreme of Formosa, is inhabited by Chinese, and cultivated in terraces to its summit, which is a sharp conical peak about 1,200 feet above the sea. At the east end of the island there is another peak, 800 feet high, which falls abruptly and overhangs the sea. The *Inflexible* passed between this island and the coast, but had no soundings with 40 fathoms line.

KALEEWAN RIVER.—At 10 miles S.W. ½ S. of Steep island is the entrance to the Kaleewan river, the waters of which irrigate a fertile plain, about 13 miles long and 6 broad. At the time of the *Inflexible's* visit there were only 3 feet on the bar at low water, the rise of tide being from 2 to 3 feet. The surf broke heavily on the beach, and although there was an occasional break across the entrance, the vessel's gig entered in the wake of a junk without inconvenience; in going out, however, with the wind blowing on the shore, two seas broke into the boat, and nearly swamped her. The junks, with their high bulwarks and great buoyancy, enter with comparative ease, the crews poling them across with bamboos.

The general direction of the river is S.W. The entrance is about a quarter of a mile wide, but just within it narrows to 200 yards. At 4 miles up it is only 50 yards wide, and thus far it has a general depth of 5 to 6 feet, clear fresh water. At 7 miles from the entrance the depth is 3 to 4 feet, but the river was scarcely broad enough to allow the use of the boat's oars.

The banks and country on either side of the river were everywhere under cultivation, principally with rice, Indian corn, and millet; sugar cane also in small quantities. The inhabitants, composed of domesticated aborigines and Chinese, of the different villages scattered along the banks, behaved with great civility. The aborigines are of a clear olive complexion, and in feature resemble the Malay. They are a much finer looking race than the Chinese, who have largely intermarried with them. They live in harmony with each other, both having the same dread of the Chiukwan or "raw" savage of the mountains. The population of the plain is about 10,000.

SAU-O BAY, the south point of which is in lat. 24° 36′ N., long. 121° 53′ E., will be found an excellent place of shelter for vessels

---

* The description from Steep island to Black Rock bay is by Mr. W. Blackney, H.M.S. *Inflexible*, 1858.

working up on the east coast of Formosa against the North-east monsoon. The bay is about three-quarters of a mile wide at entrance, and a mile deep, and in it are two small inlets; that in the south-west corner is a sheltered nook called Lam-hong-ho bay, which is said to have 5 fathoms water in it, and where two or three ships might lie moored, secured from all winds; that in the north-east corner is named Pak-hong-ho bay, in which there is good anchorage in 5 fathoms sheltered from all winds, except those from South to S.E., which seldom blow.*

The westernmost and largest rock of the Sau-o reef is 70 feet above low water, and lies N.E. 1½ miles from the south point of Sau-o bay, and E. by S. two-thirds of a mile from the north point. Two other smaller rocks bear E.N.E. 3 cables from it: the space between having rocks awash and others just above water, generally breaking.

To the north-west of the Sau-o reef and N.E. from the north point of the bay are rocks awash, generally breaking, a quarter of a mile in extent N.W. and S.E. There was no opportunity of examining these dangers, but to all appearance the ground was foul from the Sau-o reef to the north point.

The Breakwater reef lies nearly in the centre of Sau-o bay, and parts of it are uncovered and others are awash. The reef is 1¼ cables in extent, N.E. and S.W., and there is a conical rock 15 feet high rising from its north-east extreme.

Supplies.—The natives in this bay are mostly Chinese fishermen, several domesticated aborigines living with them. Fresh supplies could not be obtained except in small quantities; but if the bay is occasionally resorted to, it would no doubt cause a little traffic between the villagers on the Kalcewan river and those of Sau-o.

ANCHORAGE.—The *Inflexible* first anchored in the outer part of Sau-o bay in 13 fathoms, with the south point bearing South, and the conical rock on Breakwater reef W.N.W.; this, however, would be an unsafe position in easterly winds, although the holding ground is good, black sand and mud. She then shifted her berth to a safe anchorage in from 5½ to 6 fathoms water, inside Breakwater reef, with the conical rock bearing East, distant about a quarter of a mile.

TIDES.—The streams are weak in Sau-o bay. The flood sets in a N.N.E. direction along the coast, the ebb S.S.W., about a knot an hour. It is high water, full and change, at about 10h., and the rise is 3 to 4 feet.

DIRECTIONS.—When approaching Sau-o bay from the northward, pass

---

* *See* Plan of Sau-o Bay; scale, m = 1·3 inches, on Chart of Formosa Island, No. 1,968.

half a mile to the eastward of the Sau-o reef, the highest rocks on which may be seen 8 or 10 miles off in clear weather, then haul up West for the anchorage. From the south-east the approach is quite clear, and the points of the bay may be passed at a cable distance. The passage between Breakwater reef and Rugged point, which lies S. by E. 3 cables from it, is clear, and the depth 5½ fathoms in mid-channel.

The soundings in the outer part of the bay increase quickly seaward to 17 and 20 fathoms, and decrease gradually towards the beach. The north-west corner of the bay is rocky.

CHOCK-E-DAY.—Dome point, 650 feet high, is 3 miles south of Sau-o bay, and from thence to Chock-e-day, in lat. 24° 6¼′ N., the coast is the boldest and most precipitous that can be conceived, the mountains rising almost perpendicularly from the water's edge to the height of 7,000 feet. No soundings with 70 fathoms at from 1 to 1½ miles off shore. The inhabitants of Chock-e-day village were communicated with, but the high surf prevented landing. The aborigines were nearly naked, and used threatening gestures, brandishing their long knives and spears. A few Chinese were among them, and appeared much afraid that the natives would be injured, in which case they said their lives would be taken in revenge. The river marked on the chart in this latitude was not seen. At a mile off shore there was no bottom with 115 fathoms of line.

BLACK ROCK BAY.—H.M.S. *Plover* anchored in this bay in lat. 23° 8′ N., long. 121° 24′ E., and rode out a S.W. gale; but the bottom is uneven and rocky, the vessel swinging from 13 to 22 fathoms, and the anchorage by no means to be recommended.

With the centre of the group of rocks (120 feet high) in this bay bearing S.W. by S., 2 miles, the depth was 29 fathoms, black sand; the next cast to seaward, no bottom with 70 fathoms.

The east coast of Formosa, north of Black Rock bay, is rugged and rocky. The lower slopes of the hills are covered with grass; behind the hills the mountains rise to the height of 5,000 and 6,000 feet above the sea, and are clothed with dense forests.

SAMASANA ISLAND by Collinson is in lat. 22° 41′ N., long. 121° 28′ E., and lies N. by W. ¼ W. about 34 miles from Botel-tobago sima, page 278. Its north extreme is described as being a long, low point, with a double hillock on it, and a pinnacle rock lying off it; the south point falls abruptly. The island, when visited by Capt. Belcher in H.M.S. *Samarang*, June 1845, had a population of about 150 persons, living in a village concealed within a bamboo hedge skirting the sea. It appeared to be under cultivation, chiefly rice; and the village valley was laid out in gardens, producing maize, cucumbers, cabbage, and customary Chinese produce. It will be prudent to avoid the lee side of the island, as calms, eddies, and variables are likely to cause inconvenient delay.

HARP ISLAND.—The supposed position of this island is lat. 23° 45′ N., long. 122° 4′ E. Lieutenant Boyle, of the U.S. Navy, in 1853, reports having seen a volcano in a violent state of eruption, distant about 10 miles from the land, in lat. 24° N., long. 121° 50′ E.

PINNACLE, CRAIG, and AGINCOURT ISLANDS are three islets lying to the north-eastward of the north end of Formosa. They have often been sighted by passing vessels, but as yet no description has been given of them; their positions are as follows;—Pinnacle, lat. 25° 27′ N., long. 121° 58′ E.; Craig, 25° 29′ N., 122° 9′ E.; and Agincourt, 25° 38′ N., 122° 8′ E.

HOA-PIN-SU, PINNACLE, and TI-A-USU ISLANDS.—This group forms a triangle, of which the hypothenuse, or distance between Hoa-pin-su and Tia-a-usu, extends about 15 miles, and that between Hoa-pin-su and the southern Pinnacle island about 2 miles. Within this space are several reefs; and although a safe channel exists between Hoa-pin-su and the Pinnacle islands, it ought not (on account of the strength of the tides destroying the steerage) to be attempted by sailing vessels if it can be avoided.*

The extreme height of Hoa-pin-su is 1,181 feet, the island apparently being cut away vertically at this elevation, on the southern side, in a W.N.W. direction; the remaining portion sloping to the eastward, where the inclination furnished copious rills of excellent water. That this supply is not casual is proved by the existence of fresh-water fish found in most of the natural cisterns, which are connected almost to the sea, and abound in weeds which shelter them. The north face of the island is in lat. 25° 47′ 7″ N., long. 123° 30½′ E. There are no traces of inhabitants, indeed the soil is insufficient for the maintenance of half a dozen persons.

The Pinnacle group, which is connected by a reef and bank of soundings with Hoa-pin-su, allowing a channel of about 12 fathoms between it and the channel rock, presents the appearance of an upheaved and subsequently ruptured mass of compact gray columnar basalt, rising suddenly into needle-shaped pinnacles, which are apparently ready for disintegration

---

* *See* Chart of the Islands between Formosa and Japan, with the adjacent Coast of China, No. 2,412; scale, $d = 3$ inches.

by the first disturbing cause, either gales of wind or earthquake. On the summits of some of the flat rocks long grass was found, but no shrubs or trees. The rocks were everywhere whitened by the dung of marine birds.

Ti-a-usu, bearing N.E. northerly 15 miles from Hoa-pin-su, appears to be composed of huge boulders of a greenish porphyritic stone. The capping of this island, from about 60 feet to its summit, which is about 600 feet above the level of the sea, is covered with a loose brushwood, but no trees of any size.

RALEIGH ROCK, in about lat. 25° 57′ N., long. 124° 11′ E., rises abruptly from a reef to a height of 90 feet above the sea, and is perpendicular on all sides, covering an area of probably 60 feet in diameter, and appearing in the distance as a junk under sail. Captain Belcher states that the weather would not allow him to fix its position, but that as he found it lying upon the computed bearing, as given in the charts, from Ti-a-usu, its position cannot be much, if at all, in error.

MEIACO-SIMA GROUP.

This group forms the westernmost portion of a chain of islands extending in an easterly and a north-easterly direction from Formosa to the southern extremity of the Japan islands.

It lies between the parallels of 24° 0′ and 25° 6′ N., and the meridians of 122° 55′ and 125° 30′ E., and is divided into two divisions. The Pa-chung-san or western division, consists of ten distinct islands, of which five only are at all mountainous; the remainder are flat, like the coral islands in the Pacific, and similarly belted with reefs, which connect them into a distinct group. Chung-chi is a high uninhabited mass of rocks; and to the W.N.W. of it is Kumi island, which is conspicuous by the peculiar sharpness of its lofty peak, 770 feet high, and table base.*

KUMI ISLAND is composed of coralline limestone, all its ranges are capped with trees and brushwood, but excepting the pine fir, which contains a great portion of resin, none attain any size. There are four villages on the island, one on the west, and two on the north side, one of which is inland, in a basin-shaped valley. The principal town and port is on the north, in which were several junks of about 50 tons riding at anchor; but the entrance from the sea is so narrow and shallow, that

---

* See Plan of the Meiaco-sima group, No. 2,105; scale, m = half an inch.

ingress and egress can only be effected at spring tides and with very smooth water. Temporary anchorage, in fine weather, may be found on a sandy ledge to the northward of the town.*

KU-KIEN-SAN and PA-CHUNG-SAN ISLANDS afford several commodious harbours, and are, with good charts, quite safe of approach. Port Haddington, on the western side of the latter island, would shelter a large fleet, but it abounds with coral patches, rising suddenly from 10 or 15 fathoms almost to the surface. Except on the northern side of Ku-kien-san and the latter port, watering would be found very difficult, as the reefs extend a great distance from the mouths of the streams. Seymour bay, at the south-west angle of Ku-kien-san, must also be excepted, for there a fine stream enters the sea in deep water, and a vessel might be moored sufficiently close to lead the hoses from Hearle's pumps into her, without the intervention of boats and casks.

With respect to the various harbours of Ku-kien-san, there are two or three adapted for shelter for small vessels, or even those drawing 18 feet, where a refit might be accomplished in still water in any monsoon, or where steam vessels might lie safely for the purpose of obtaining wood; and there are two other open bays, well sheltered in the North-east monsoon, admirably adapted for watering; but there is not any other inducement to visit this island; all the dangers are well marked by the coral fringe which extends about a cable's length from the outline.

Of the dangers on the northern side of this group, it would not be prudent that any vessel should run the risk of being hampered by the shoals, and therefore should not come farther eastward, when beating up for Chusan, than to sight Chung-chi island. The currents as these islands are approached press more southerly and easterly than those that are experienced on the coast of Formosa, and stronger breezes prevail as a vessel advances easterly. Indeed it blows incessantly at this western group.

TAI-PIN-SAN ISLAND.—The islands composing the Taï-pin-san or eastern division, are Taï-pin-san, Yer-ra-bu, Ku-ree-mah, Y-ki-mah, and Ohotake; the two islets, Mitsuna and Tarara, between Taï-pin-san and Pa-chung-san, are said to be a continuation of the reefs which extend

---

* A dangerous shoal is reported as extending E. b. N. and W. b. S. 3 miles, and bearing N.W. b. W., distant about 10 miles from Kumi.—Horsburgh, vol. 2, seventh edition, page 605.

On the evening of the 16th November, with Kumi island bearing E. b. S. 3 leagues, saw heavy breakers ahead and on the lee bow, apparently on a dangerous shoal, extending E. b. S. and W. b. N., and bearing from Kumi, S.W. b. W. 3½ leagues distant. Having dark cloudy weather with rain, and a heavy sea running, it was too late to send a boat to sound; but the breakers were seen continually from 4.30 p.m. until 6 p.m.—*Nautical Magazine* for 1844, p. 244.

to the N.E., North, and N.W. of Taï-pin-san, and on which H.M.S. *Providence* was lost in 1797. Captain Belcher, in H.M.S. *Samarang*, looked in vain for Ykima island.

Taï-pin-san island is surrounded by an extensive chain of coral reefs, upon which the islands of Ku-ree-mah, Yer-ra-bu, Y-ki-mah, and Ohotake respectively are situated to the West, N.W., North, and N.E. The reefs do not extend far to the westward from Ku-ree-mah, unless in patches unconnected with the main belt. Off Yer-ra-bu they extend 3 or 4 miles, but close towards its north-western angle, where a deep water channel admits vessels within the belt up to Ohotake island and into the main harbour of Taï-pin-san. The reefs again spit out on the south-west angle of Y-ki-mah, and sweep northerly, as far as the eye can reach (from 100 feet elevation), round to east in continuous lines of breakers, edging in towards the south-east extremity of Ohotake. A high patch of rocks lies on the north-east angle of this outer belt, probably 10 miles from the northern point of Taï-pin-san.

Safe anchorage during the south-west monsoon might be found inside the reefs of Ohotake island, and also safe in the other monsoon; but the passage in or out at that season would be attended with risk, as sudden squalls, gales, and numerous patches beset the whole eastern side of Taï-pin-san. The southern coast line, from the south-east breaker patch to the south-west anchorage, does not offer many dangers if a tolerable look-out be observed. The reefs do not extend more than half a cable from the shore, and generally less.

There can be no inducements for any vessel to visit Taï-pin-san; neither wood, water, nor any other necessaries could be procured. A few pigs, fowls, and sweet potatoes might be obtained for cabin use, but this would hardly warrant the risk and detention on such a dangerous coast.

DIRECTIONS.—*Great caution is requisite in approaching the Meiaco-sima group from the north-east, east, or south, particularly with fresh breezes, and in the absence of the sun, by the aid of which reefs below water can be detected. They are from their greenish hue, being covered by seaweed, less distinct than at other places, and therefore, where they are not marked on the chart, it must not be presumed that the space is free from danger; the lead will not afford timely warning.

Approaching the group from the south-west, the island of Ku-kien-san from its great height will be first distinguished, presenting a round backed summit closely clad with trees; knolls occur, elevated 2,000 feet above the sea, but as they seldom present the same appearance owing to those

---

* Capt. E. Belcher, H.M.S. *Samarang*, December 1844.

nearer the coast eclipsing them, their accurate measurement could not be obtained; Adam peak, which may be noticed on the south-eastern outline, was determined to be 1,200 feet. As the island is neared, the high rocky basaltic islet of Chung-chi will show out when the western limit of Kukien-san bears N.E. b. N., and working for this islet no danger can be feared, and should night befall, all the space on the north-west of Kukien-san up to the island of Kumi is safe.

The *Samarang* entered the group from the westward, passing within 2 miles of the southern reefs or breakers of Hasyokan or Sandy island, and standing on close hauled to the eastward, intending to make Ykima, and beat up from it to Taï-pin-san. On the morning following, not seeing Ykima (which is supposed not to exist), and the weather very boisterous, she stood on to the westward to get under the lee of Pa-chung-san, and endeavour to reach some place of shelter. On nearing the latter island, she ran down the eastern and southern side, reaching the south-western extremity of its reef about 4 p.m.

Here was a barrier of breakers as far as the eye could reach from the mast-head, and apparently connecting Hasyokan island with the group of larger islands. An opening, however, was found into the reef, and after due examination the vessel was shot up into 13 fathoms, into Broughton bay, and warped into a snug position, where she was moored with just sufficient room to swing, the depths up to the coral ledges varying from 13 to 7 fathoms.

BROUGHTON BAY.—The only directions which will assist the seaman in finding this snug little anchorage (safe only, however, during the North-east monsoon) are as follows:—

Approaching from the westward, as Chungchi is neared, Hasyokan or Sandy island will soon be seen, and avoiding the space included northerly of a line between Chungchi and it, a vessel may safely stand on, passing within one mile of the southern limit of Hasyokan, and work for the south-west angle of Pa-chung-san, avoiding the reefs, which extend from it in a direct line N.E. and S.W. to Hasyokan. A high rock, named South rock, will point out the outer reefs of Pa-chung-san. The dangers between it and Pa-chung-san must be avoided by eye, the shoals being visible in 5 or 6 fathoms, and breaking upon those of 2 and 3 fathoms. The opening of the reef is in the heart of a deep indentation, just to the northward of the low south-west point of the island, and it has apparently a centre bar. The right-hand opening is the proper one.

From the eastward there are no dangers which are not clearly visible. After making the land, edge along the southern and eastern breakers until the abrupt turn of the breaker line is seen, at which moment the extreme south-west point of the bay will open. The breakers have regular soundings off them, but the course in will probably lead in 7, 8, or 9 fathoms, deepening to 14 or 15 off the inlet. As the breeze generally blows out, it will be advisable to send a boat to find clear ground off the opening, and shoot up and anchor. The vessel may then be warped in. But if merely intending a cursory visit, the outer anchorage appears good.

At Broughton bay, neither wood nor water can conveniently be procured; and the only reason for noticing it is, that a port of refuge with still water, in case of disaster, may be found on this side of the island; when a disabled vessel could not beat round to the more secure harbour of Port Haddington, on the northern side.

PORT HADDINGTON.—No safe anchorage is to be met with between Broughton bay and Port Haddington, which is on the west side of Pa-chung-san; although during the south-west monsoon there are several good bays on the northern side of the island, where anchorage might be found, but certainly not adapted for refit.*

When rounding the north-eastern extremity of Pa-chung-san the two low coral islets of Mitsuna and Tarrara ought to be avoided at night, but the dangers by day are clearly denoted by breakers. To the northward of these islets the ground is foul, and the *Samarang* was compelled to tack to the westward in 7 fathoms, at least 10 miles north of them.

A vessel bound from Broughton bay to Port Haddington, after rounding the north-east end of the Pa-chung-san breakers, and running to the westward the length of the island, should haul close round the north-west angle, and edge along southerly within about a mile of the breakers. The port will then open out, into which, with the prevailing breeze of the North-east monsoon, it will be necessary to beat. Off Hamilton point, the north point of the port, will be seen a remarkable little rocky hummock upon which was left a very large pile of stones. The bottom, for more than a mile off the point, is rocky and dangerous; but as all the dangers of this port are visible from aloft, there is no risk with a proper look-out. The inner parts of the port have numerous shoals, but there is still abundance of excellent anchorage without, and where the vessel will be land-locked. The *Samarang* anchored about a mile or less within Hamilton point, in 10 fathoms, clear bottom.

From the westward Port Haddington may be sought and reached more expeditiously by working up on the north-west side of Ku-kien-san, round-

---

* There is a passage from Port Haddington into Broughton bay, which was used by H.M. Ships *Lilly* and *Contest*, but abounding with coral reefs.—*Commander J. W. Spencer, H. M. Sloop Contest*, 1852.

ing Isaac island and running down off the danger line from Melros point round the reef which extends off Hamilton point one mile, and shoot into 15 fathoms. The chart exhibits several awkward patches, but a vessel which works decently can thread her way between them, if the sun be bright, as all the shoals may easily be traced from aloft.

Supplies.—A convenient watering-place was established by sinking a cask and suspending the suction hose of Hearle's pump over it, so as to prevent the sand from being sucked in. The stream from above was regulated by dams to ensure not more than a sufficient supply, by which means the water obtained was beautifully clear. Here wood is abundant, and the position is farther preferable by being so far from the villages as to prevent the authorities from feeling alarmed. During the North-east monsoon this is a most convenient port; it is land-locked it is true, but there is a long fetch for the sea with a S.W. gale, and in that season Typhoons are said to be very violent about this region. Sufficient firewood was cut at the beach (of Tamanu) to fill the ship, and trees were obtained of pine and other woods adapted for plank.

PORT HADDINGTON to TAI-PIN-SAN.—After quitting Port Haddington, the *Samarang* beat to the northward, and endeavoured to weather Mitsuna and Tarara. She had passed the breakers leaving them about 5 miles under her lee, when finding the depths decrease to 7 fathoms the vessel was immediately tacked. Capt. Belcher strongly suspects that extensive banks or ledges of coral connect these islands (northerly) with Tai-pin-san; and a good reason for this offers in the fact of their being included by the natives in the Tai-pin-san group when they are much closer, by half the distance to Pa-chung-san.

Upon nearing Tai-pin-san, and having tacked twice, rather close to two off lying patches, and obtaining soundings with 15 fathoms, a boat was sent ahead. Upon a given signal, for "danger discovered," the anchor was let go, and the vessel found to be in a secure berth in 12 fathoms, the boat being on the reefs. This turned out to be the only anchorage at Tai-pin-san. It is merely an indentation formed by the reefs connecting the western island Ku-ree-mah with Tai-pin-san, and is very unsafe, a very heavy sea tumbling in with a southerly wind. The observatory at Tai-pin-san (at the most convenient landing-place, within the reefs, and the last rocky point towards the long sandy bay) is in lat. 24° 43′ 35″ N., long. 125° 17′ 49″ E.

LU-CHU OR LIU-KIU ISLANDS.

The Lu-chu or Liu-kiu islands, to the north-east of the Meiaco-sima group, consist of one large island, Okinawa sima, surrounded by smaller ones. Okinawa sima, or Great Lu-chu, is of considerable size and well-inhabited; it is about 56 miles long in a N.E. and S.W. direction, and preserves a tolerably uniform breadth of about 10 or 12 miles. The north end is high and bold, with wood on the top of the hills. The north-east coast is also abrupt but quite barren, and the north-west side is rugged and bare. The south-east side is low, with very little appearance of cultivation. The south, south-west, and western coasts, particularly the two former, are of moderate height, and present a scene of great fertility and high cultivation, and here the mass of the population reside.

KERAMA ISLANDS.—To the westward of the south end of Okinawa sima are the Kerama islands, the Amakirrima of Basil Hall in 1816, and Kern sima of Siebold. The Kerama group consists of four islands, Zamami, Accar or Yakai of Siebold, Ghirum, and Twkaschi, of which all but the last are very small.

Captain Mathison, of H.M.S. *Mariner*, in 1849, states, "that in the chart of the Kerama islands there are six islands marked with apparently clear passages between them; whereas, as well as could be judged, there must be a greater number, and all the spaces between them appear filled with reefs and breakers. There is a shoal lying between the east Kerama island and the south-east end of Okinawa sima, the breakers on which were clearly visible. Reefs also extend to the eastward about 5 or 6 miles from the north-east point of Kume sima, the island lying to the W.N.W., on one of which the ship *Elizabeth and Henry* was lost."

The small coral islands lying off Napha are called Tzee (Kei of Siebold), and by Captain Basil Hall, Reef islands.

NAPHA-KIANG ROAD.—Napha, on the south-west side of Okinawa sima, is the principal sea-port of the island, and perhaps the only one possessing the privileges of a port of entry. The inner, or Junk harbour, carries a depth of 2 to 3 fathoms, and though small, is sufficiently large to accommodate with ease the fifteen or twenty moderate sized junks which are usually found moored in it. These are mostly Japanese, with a few Chinese and some small coasting craft, which seem to carry on a sluggish trade with the neighbouring islands. The outer harbour, or Napha-kiang road, is protected to the eastward and southward by the main land, whilst in other directions it is surrounded by merely a chain of coral reefs, which answer as a tolerable breakwater against a swell from the northward or westward, but afford, of course, no shelter from the wind. The holding ground, is so good, however, that a well-found vessel could here ride out almost any gale in safety.*

---

* *See* Plan of Napha-kiang Road, with views, No. 990; scale, *m* = 3 inches.

Abbey point, forming the south extremity of the road, may be known by its ragged outline, and by a small wooded eminence, called Wood hill, about 1½ miles south of it. The mainland here falls back and forms a bay, which is sheltered by coral reefs extending to the northward from Abbey point; they are, however, disconnected, and between them and the point there is a channel sufficiently deep for the largest ship.

Nearly in the centre of this channel, outside withal, there is a coral bank named Blossom reef, having a good passage on either side of it. The south channel, between it and Abbey point, should be adopted with southerly winds and flood tides, and the Oar channel, between Blossom and Oar reefs, with the reverse. A reef extends from Abbey point to the south-west, and also to the northward. When off Abbey point, Kumi head, a rocky headland, will be seen about 1½ miles north of the town; and upon the ridge of high land beyond it are three hummocks to the left of a cluster of trees. In the distance, a little to the left of these, is Mount Onnodake, in lat. 26° 27' N. A remarkable rock, which from its form has been named Capstan head, will next appear; and then to the northward of the town a rocky head, with a house upon its summit, called False Capstan head. At the back of Capstan head is Sheudi hill, upon which the upper town, the capital of Lu-chu, is built.

Water.—An abundance of water can always be obtained at the fountains in Junk river, where there is excellent landing for boats. There is a good spring near the tombs, at Kumi bluff; but unless the water is quite smooth the landing is impracticable, and under any circumstances it is inconvenient from the want of sufficient depth, except at high tide.

Buoys.[*]—A *black* spar-buoy is moored on Blossom reef half way between its eastern and western extreme; a *red* spar-buoy on the point of reef to the W.N.W. of Abbey point; and a *white* spar-buoy on the south-east extremity of Oar reef. Flags of corresponding colours are attached to all these buoys, and they afford good guides for the South and Oar channels. There are two large stakes on the reefs to the eastward and westward of the North channel, planted there by the natives, this being the channel mostly used by junks trading to the northward.

Directions.—Vessels bound from Hong Kong to Great Lu-chu island during the South-west monsoon, should pass through the Formosa channel, giving Pinnacle, Craig, and Agincourt islands, off the north end of Formosa, a safe berth, as there are said to be reefs among them, and the currents are strong and variable in their vicinity. From thence a course should be shaped to pass to the northward of Hoa-pin-su, Ti-a-usu,

---

[*] The spar-buoys may be displaced, or entirely removed by the heave of the sea, and should, therefore, not be implicitly relied on.

and the Raleigh rock, after which haul to the eastward to sight Kume sima, and pass either to the northward or southward of it, Kurama, and the small islet near the latter, but *not* between them, as reefs are said to have been seen there. If to the northward give a good berth* to Tu sima, a small rocky islet a quarter of a mile in extent, with a reef projecting $1\frac{1}{2}$ miles to the northward and about 4 cables in other directions; it is about 80 feet high, much broken, and lies N. by E. $\frac{1}{3}$ E. $13\frac{1}{4}$ miles from the northernmost peak of Kume sima, and W. $\frac{1}{2}$ N. from the centre of Agenhu. Pass to the southward of Agenhu, which will be readily recognized by its bold south point and wedge-shaped appearance. The Kerama group will be seen to the S.S.E., Lu-chu visible on the eastern horizon, and in a short time the Reef islets will heave in sight to the southward and eastward; these latter are low and sandy, slightly covered with vegetation, and surrounded by coral reefs.

During the North-east monsoon, round the south end of Formosa, and with the strong current setting to the northward, beat along its eastern shore to the northward and eastward. Pass between Hoa-pin-su and the Meiaco-sima group, and either to the northward or southward of Kume sima; if to the southward, a vessel may hug the northern shores of the Kerama islands, as it is believed there are no hidden dangers near them.

During the Typhoon season, however, it is advisable to pass to the southward of Formosa and the Meiaco-sima group, in order to have plenty of sea room, in the event of encountering one of these storms. The passage to the southward of the Kerama islands is clear with the exception of the Heber reef and Sandy island; the former is said to be a rock 6 feet out of water surrounded by reefs; the latter to be just above water; and lying respectively W. by S. $\frac{3}{4}$ S. and W. by N. $\frac{1}{2}$ N. 7 miles from the south point of Great Lu-chu.

Vessels bound into the road from the southward may pass close round Cape Yakimu, the south extreme of Great Lu-chu, and sail along the western coast at the distance of 3 or $3\frac{1}{2}$ miles, leaving Heber reef and Sandy island to the westward.

Through SOUTH CHANNEL.—There are three passages leading into Napha-kiang road, named the North, the Oar, and the South channel. To sail into the road by the South channel, between Blossom and Abbey reefs, having well opened Capstan head, haul towards Abbey reef, and bring the right-hand hummock about half a point eastward of Kumi head; this mark will lead through the South channel, in about 7 fathoms, over the tail of the Blossom reef. A vessel may now round Abbey reef tolerably close, and steer for the anchorage in 7 fathoms, about half a mile to

---

* Lieutenant H. K. Stevens, U.S. Surveying Expedition, 1857.

the N.N.W. of False Capstan head.* Should the wind veer to the eastward in the passage between Blossom reef and Abbey point, with the above mark on, do not stand to the northward, unless the outer cluster of trees near the extremity of Wood hill is in line with, or open to the westward of Table hill, a square rocky headland to the southward of it. This mark clears also the tongue of Oar reef.

The best anchorage is in Barnpool, at the north-east part of the road, in 7 fathoms water, where a vessel may ride with great security. The outer anchorage would be dangerous with strong westerly gales. H.M.S. *Blossom* anchored there in 14 fathoms, muddy bottom, Abbey bluff bearing S.W. ¼ S., and Capstan head E. by S. ¼ S.

The entrance to Barnpool is between Barn head and the reef off Capstan head. In entering, do not approach Barn head nearer than to bring the north edge of Hole rock in one with the before-mentioned flat clump of trees on the hill south of Sheudi, until the point of the burying ground (Cemetery point) is seen just clear of False Capstan head. Anchorage may be taken in any part of Barnpool.

The following directions for the South channel are by Lieutenant S. Bent of the United States American expedition to Japan, to accompany a plan of Napha-kiang road surveyed by that officer in 1853, and on which are marked two patches of only 2½ and 1½ fathoms water; the former named Lexington reef lying W. ½ S., 1¼ miles from Abbey point; and the latter of 1½ fathoms, W.S.W. 1¾ miles from the point :—

The clearest approach to Napha-kiang road from the westward is by passing northward of the Kerama islands and sighting Agenhu island, which will be recognized by its wedge-shaped appearance; from thence steer a S.E. course for the road, passing on either side of the Reef islands; being careful, however, not to approach them too near on the western and southern sides, as the reefs below water in these directions are said to be more extensive than is shown on the chart.

After clearing the Reef islands, steer for Wood hill on a S.S.E. bearing until getting upon the line of bearing for the South channel. This will lead well clear of the Blossom reef, yet not so far off but

---

* Care must be taken to avoid the Ingersoll patches, on which there is only a fathom water. They are inserted in the Admiralty plan of Napha-Kiang road as discovered in 1837, and bear from Capstan head W. ¼ S., and from South fort N. b. W. ¾ W. The French survey of 1846 by the officers of *La Sabine* does not show these rocks, but three patches having over them respectively 2, 4, and 4½ fathoms. From the 2-fathoms patch Abbey point bears S.b.W.¾W., and False Capstan head E. by S. ¼ S.; from the 4-fathoms patch Capstan head bears S.E. by E. ¾ E., and Abbey point S.W. ¾ W. ; and from the 4½-fathoms patch Abbey point bears S.S.W.¼W., and False Capstan head S.E. by E. ¾ E.

that the white tomb and clump of trees or bushes to the southward of Kumi head can be easily distinguished. An E. by N. ¼ N. course now until Abbey point is in one with outer trees will clear S.W. rock, when haul up for Kumi head, and select a berth about half a mile to the northward and westward of False Capstan head. This channel, being quite straight, is more desirable for a stranger entering the harbour than the Oar channel, which, though wider, has the disadvantage of its being necessary for a vessel to alter her course some four or five points, just when she is in the midst of reefs which are nearly all below the surface of the water.

Through OAR CHANNEL.—If the wind be to the north-eastward it will be advisable to beat through the Oar channel, in preference to the South channel. To do this, bring False Capstan head in line with a flat cluster of trees on the ridge to the right of the first gap south of Sheudi. This will clear the north tongue of Blossom reef; but unless Table hill be open eastward of Wood hill, do not stand to the southward, but tack directly the water shoals to less than 12 fathoms, and endeavour to enter with the marks on. Having passed to the N.E. of Blossom reef, which will be known by Wood hill being seen to the right of Table hill, stand towards Abbey point as close as convenient, and on nearing Oar reef take care of a tongue which extends to the eastward of it and of the S.W. rock, and be careful to tack immediately the outer trees of Wood point open with Abbey point. In entering at either of the western channels, remember that the flood-tide sets to the northward, over Blossom reef, and the ebb to the southward.

A good mark to run through this channel is to bring the centre of the island in Junk harbour (known by the deep verdure of its vegetation) to fill the gap between the forts at the entrance of that harbour, and steer a S.E. ½ E. course, until Capstan head bears East, when haul up E.N.E. and anchor as before directed.

Through NORTH CHANNEL.—The North channel into Naphakiang road is much contracted by a range of detached rocks extending out from the reef on the west side, and should not under ordinary circumstances be attempted by a stranger; as at high water the reefs are almost entirely covered, and it is difficult to judge of the vessel's exact position unless familiar with the various localities and landmarks. To enter by this channel, bring a remarkable notch in the southern range of hills in line with a small hillock just eastward of False Capstan head, and stand in with this mark bearing S. by E. ½ E. until Kumi head bears E. ¼ N., when open a little to the southward, so as to give the reef to the eastward a berth, and select an anchorage.

Sailing from Napha during the North-east monsoon, it will be better to pass round the south end of Great Lu-chu, in order to avoid beating through the Montgomery group, of which there is no reliable survey; they are said to consist of five islands, surrounded by reefs. But with a southerly wind and fine weather it will be to the advantage of a vessel bound to the Bonin islands to pass round the north end of Great Lu-chu, where she will feel the influence of the current, which will assist her to the eastward.

TIDES.—It is high water, full and change, in Napha-kiang road, at 6h. 30m., and the rise is from 5 to 7½ feet; but this was very irregular during the *Blossom's* stay at this anchorage.

DEEP BAY*—the observatory spot at the head of which is in lat. 26° 35′ 35″ N., long. 127° 59′ 42″ E.—is formed on the western side of Great Lu-chu island, and although open to the West and S.W. affords good anchorage off the town of Naguh at about half a mile from its head; for winds from these quarters rarely blow home, and if they do they never raise a sea, as the latter is broken by the great depth of the bay.

The country around the head of this bay is fertile and populous, Motubu and Naguh being the largest towns. At the town of Oon-sah there is a good ship and timber yard where junks are built; here also the natives were found more affable and sociable than on any other part of the coast. This part of Great Lu-chu (extending to beyond Nacosi on the opposite coast) appears to be in a high state of cultivation; rice and sweet potatoes are the principal productions; but on the northern side of the peninsula, north of Deep bay, extensive fields of wheat were seen extending uninterruptedly for several miles. Cotton was observed also in many places, but the growth was small and the yield poor. Peas, beans, radishes, turnips, and sugar-cane were growing in considerable quantities, also mustard and ginger. On the Natchijen mountains, 1,488 feet high, cinnamon was growing wild, and there was also a fine growth of timber, which furnishes most of the spars for the native junks; Nakazuni cove, on the north side of the peninsula, being the principal depôt, whence they are transported to the other parts of the island.

SUCO or SETEI ISLAND, lying about a quarter of a mile from the north-west coast of Great Lu-chu, to the northward of Deep bay, affords excellent anchorage on its eastern side, between it and the coast, protected from all winds, and wood, water, and fresh provisions can be easily procured. There is free egress to the northward and southward, and although

---

* Deep and Shah bays are from the surveys of Lieutenants W. B. Whiting, and A. Barbot, of the U.S. Ship *Vandalia*, 1854.

the anchorage appears open to the southward, yet it is well sheltered in that direction by the reef extending S.S.E. nearly half a mile from the south end of Suco, and by the southern shore of Deep bay.

PORT ONTING, or Melville, is on the north-west part of Great Lu-chu, and its entrance is between the eastern side of Kuī or Herbert island and the western side of the reef which fronts the peninsula, and which projects 5 or 6 miles to the westward, having a small islet near its extremity. Iye sima or Sugar-loaf island, lying about 12 miles to the westward of the entrance, is a good guide for it, the island being low and flat, with the exception of a sharp conical peak rising 561 feet above the sea, at its eastern end.*

Water.—Good water is to be obtained at the village of Onting.

DIRECTIONS.—When bound to Port Onting from the westward, passing to the northward of Iye sima, an E.S.E. course will lead to the entrance. It will be advisable to heave-to here, or anchor in 20 or 25 fathoms water, until boats or buoys can be placed along the edges of the reefs bordering the channel; for without some such guides, it will be difficult for a vessel of large draught to find her way in between the reefs, which contract, in places, to within a cable's length of each other, and are at all times covered.

In entering, steer for the western shore of Kuī island until Hele rock is in line with Double-topped mountain (a distant double-topped hill the second highest of the range,) bearing S.E. ½ S. Steer in on this mark, until Chimney rock bears S. ¼ E.; then for Chimney rock until Rankin point bears S.W. ½ W.; then for that point until the port is entered, when anchor, giving the vessel room to swing clear of the reef extending northward of Rankin point, and she will be as snug as if lying in dock, with good holding ground, completely land-locked, and sheltered almost entirely from every wind.

TIDES.—In Port Onting it is high water, full and change, at 6h. 35m., and the rise is about 8 feet.

SHAH BAY, at about 8 miles to the E.S.E. of Port Onting, is a beautiful land-locked sheet of water, but the reef fronting the entrance prevents its being accessible to vessels of larger size than the junks which frequent it; within the entrance the water deepens to 12 and 8 fathoms, the bottom being soft mud. On the southern shore of the bay was found iron ore, mineral coal, and sulphur. The coal appeared of poor

---

* See Plan of Port Onting, with views, No. 2,436 ; scale, $m = 3\cdot4$ inches.

quality and mixed with earth, but good coal might perhaps be found by digging.

BARROW BAY is a deep inlet, bounded by shoals, near the middle of the eastern side of Great Lu-chu island. The following description* is by Lieutenant G. B. Balch, of the U.S. ship *Plymouth*:—

" A reef, of coral formation and bold to approach, commences 5 miles from the south point of Great Lu-chu, and extends in an unbroken chain, outside all the small islands, as far as the north-east point of Ichey island, with the exception of a narrow channel between the islet off the north-east end of Kyoko or Kudaka island, and the island of Taking. Ichey island forms the south-eastern point of Barrow bay, which is useless for all purposes of navigation, being exposed to the east winds and ocean swell. There is, however, secure anchorage in about 15 fathoms water on the western sides of Ichey, and of Hanadi, the next islet to the southward; this anchorage is the only place of shelter on the eastern coast of Great Lu-chu."

---

* From Nautical Remarks by the officers of the United States Expedition to Japan, 1854.

CHAPTER IX.

ISLANDS SOUTH-EAST, EAST, AND NORTH OF THE LU-CHU GROUP; AND OFF THE SOUTH-EAST COAST OF NIPON.

VARIATION in 1861. Borodino islands, 1° 50′ W. Ladrones islands, 2° 40′ E. Bonin islands, 0° 30′ W. Islands off S.E. coast of Nipon, 2° 30′ W.

CAUTION.—The mariner in navigating the space about to be described, between Formosa and the Mariana and Bonin islands, should use extreme caution in approaching the locality of many of the islands and shoals marked on the chart, especially at night, or during dark and hazy weather, for the existence, or at least the positions, assigned to some of them appear to be very doubtful. They would seem to have been inserted on the charts from the uncertain reckoning kept on board whaling vessels, or others, which, from the very nature of their pursuits, cannot be entitled to much confidence. Those only will be noticed whose existence has been determined.

BORODINO ISLANDS.—These two islands lie in a N.N.E. and S.S.E. direction from each other, distant about 4 miles. The southernmost,—the centre of which is in lat. 25° 52¾′ N., long. 131° 12¼′ E.,—is the largest, being about 3 miles in extent east and west; it is low, of coral formation, and covered with vegetation. A reef extends along its southern shore, affording no visible harbour. The northernmost island is also flat, but has a rise near the north end gradually sloping towards the sea.

BISHOP ROCKS, in lat. 25° 20′ N., long. 131° 15′ E., were discovered by Captain Bishop, in the ship *Nautilus*, in 1796, but they do not appear to have been seen since.

RASA ISLAND.—KENDRICK ISLAND.—Rasa island, seen by the Spanish frigate *Magellan* in 1815, and the French frigate *La Cannonière* in 1807, is in about lat. 24° 27′ N., long. 130° 40′ E., about 4 or 5 miles long N.W. and S.E., low, covered with bushes, and surrounded with rocks.

Kendrick island is said to be in lat. 24° 35′ N., long. 134° 00′ E., and to be low, and about 6 miles in length. It is very probable that Kendrick may be identical with Rasa, but this cannot be decided without further examination.

PARECE VELA or DOUGLAS REEF, discovered Sept. 1789, by Captain Douglas, is in about lat. 20° 31′ N., long. 136° 6′ E. It is described by Mr. Sproule, of the ship *Maria,** who partially examined it in his boat, March 1848, as being a narrow perpendicular belt of coral, enclosing an oblong lagoon of water, with heavy breakers rolling over its north and north-east sides. The boat rowed under the lee of the reef, its whole length, which is only 2 miles, and three-quarters of a mile wide at one-third from the eastern point. Sharp heads of coral appeared frequently through the surf, and one isolated rock, about 12 feet high and 15 feet long, rose from the smooth water lagoon near its western extreme. The rock, when first seen from the vessel, about 3 miles off, appeared like a boat's tanned lug-sail. The reef, from its being visible only a short distance off, should be carefully approached, especially in dark and blowing weather.

LINDSAY ISLAND, discovered by Mr. Lindsay, of the British schooner *Amelia,* December 1848, is in lat. 19° 20′ N., long. 141° 15½′ E. It appeared about 40 feet high, 4 miles long, and very barren.

MARIANA or LADRONES ISLANDS.—This archipelago is composed of a chain of volcanic islands, which extend in a north and south direction for a space of 420 miles. Magalhaens, the first circumnavigator, discovered them on March 6th, 1521, but he only saw Scypan, Tinian, and Guam. The Spaniards named them Ladrones, from the great propensity to thieving evinced by the natives. In 1668 they received the name of Mariana, in honour of the widow of the King of Spain, Philip IV., Maria Anna of Austria.

GUAM or Guajan island,† the southernmost and principal of the Mariana group, is about 29 miles long N.E. by N. and S.W. by S., and 3 miles broad at its narrowest part, which is near the centre. It is bordered throughout a greater part of its circuit with a chain of reefs which are uncovered at times. It contained, in 1858,‡ a population of about 4,000 persons, and there was a plentiful supply of water and fresh provisions.

At a distance this island will appear flat and even. There is no anchorage on its east side, which is bordered with steep rocks, against which the sea dashes with great violence. The west side of the island is rather low, and full of small sandy bays, divided with as many rocky points. The soil is reddish, dry, and indifferently fruitful. The produce is chiefly rice,

---

* Nautical Magazine, 1848, p. 242.

† Voyage of the French corvette *Uranie,* by M. Louis de Freycinet, 1819.

‡ Captain N. Vansittart, H.M.S. *Magicienne,* July 1858.

pine-apples, water melons, oranges and limes, cocoa-nuts, and bread-fruit. The cocoa-nut trees grow on the western side in great groves 3 or 4 miles long and 1 or 2 miles broad.

PAGO HARBOUR.—The eastern coast of Guam, as far southward as Tarofofo harbour, in lat. 13° 18' 12" N., affords no shelter to the navigator, and therefore ought to be avoided during the North-east monsoon. The only openings are Pago harbour, in lat. 13° 24½' N., accessible only for boats, and Ylic bay, 2 miles to the southward, and equally unimportant.

TAROFOFO HARBOUR is formed of two inlets, the northern of which, Tarofofo, is open to the east, in which direction it is 1½ miles long, and it is about a third of a mile wide; the other is smaller, and called Paiepoue bay. The Tarofofo river, the most considerable in Guam, enters the head of the harbour. The Madreporic hills, very steep, rise abruptly from both sides of the harbour; the Mahiloue, on the north side, are celebrated in the history of the country. There is no village near this locality.

This harbour is the only port, next to San Luis, that will receive a vessel at all seasons of the year. There are no rocks in it, nor is there any danger.

HOUNLODGNA, YNARAHAN, and AGFAYAN BAYS.—From Tarofofo to Hounlodgna bay, 1¼ miles to the south-west, the land is low, with sandy beaches and rocky points; this bay is only fit for boats.

Ynarahan bay, at a mile farther to the south-west, is a quarter of a mile wide at the entrance, half a mile deep, and open from East to South. A vessel would be quite safe in it during westerly winds, but not with the opposite. The town of Ynarahan stands on its south side.

Agfayan bay is three-quarters of a mile to the south-west and much smaller than the latter bay; it may have good anchorage for vessels of less than 15 feet draught. It is open to the E.N.E. At its head is a small brook, where boats can easily obtain water.

SOUTH END of GUAM.—Ahayan point, the south-east extreme of Guam, forms the eastern side of Ahayan bay, which is so obstructed by reefs that it is dangerous of approach if there is much sea on. The south end of Guam is an uninterrupted sandy beach fronted by a reef, with two or three small islands on it. This reef, after encircling Dancono or Cocos island, lying 2½ miles from the south-west extreme of Guam, trends the same distance to the northward to the south-west point of Guam, where is the small harbour of Marizo, only fit for boats.

CHAP. IX.] MARIANA OR LADRONES ISLANDS. 313

SANTA ROSA SHOAL, of which no description has ever been given, is in about lat. 12° 30′ N., long. 144° 15′ E. A Spanish galleon is said to have struck on it, and to have lost her rudder. Dampier, when approaching the south end of Guam, in 1686, sailed over a rocky shoal with 4 fathoms on it. Galvez also, in 1740, discovered a bank lying about 20 miles from the south-west end of Guam in lat. 13° N.

UMATA BAY, at about 1¾ miles to the northward of the south-west point of Guam, is a mile deep, east and west, and two-thirds of a mile wide at entrance. The bay is sheltered from North, round east, to South, but in the season of westerly winds, or from June to September, it would be imprudent, or perhaps impossible, to remain in it, on account of the heavy sea sent in.

Tougouéne point, the narrow, low, south point of entrance to this bay, has a reef projecting nearly 2 cables from it to the westward. The north point of entrance is an isolated and picturesque rock with a fort on it, which is approached by steps cut in the rock.

The south shore of the bay is mountainous to the head of the bay, where the river Umata or Saloupa enters the sea, and where a church is built at the foot of the mountains. The town stands on the north shore, which is low. Behind the town the hills rise in an amphitheatre, and are neither high nor remarkable. Excellent water can be obtained from a rivulet which flows into the sea on the south shore, between Mount Inago and another mount with a fort on it farther west.

PORT SAN LUIS D'APRA.—From Umata bay the west coast of Guam trends N.N.W. 3 miles to Facpi point, and has several sinuosities, the deepest of which is Cetti bay, as large as that of Umata. Facpi point is remarkable for being pointed, projecting, and terminating in an isolated rock joined to the shore by a reef uncovered at low water. From hence to Oroté point, 6 miles to the N.N.W., the coast falls back 2 miles, and forms a bay, in which are several coves and islets.*

Oroté point is the extremity of a narrow peninsula, projecting 3½ miles in a N.W. by W. direction from the coast, and which cannot be traversed on account of the number of large rocks and precipices which cover it. A small islet lies close to the north side of the point, and from it the north coast of the peninsula trends E. by S. and S.E. by S. to the village of Apra, which is built on the isthmus connecting the peninsula to the main land. From thence it bends round to the East and North, forming a deep indentation named Port San Luis d'Apra, which is nearly in the

---

* *See* Plan of Puerto de Apra, No. 1,102 ; scale, $m = 0·75$ of an inch.

shape of the letter V, the opening to which is nearly closed by a long narrow island, named Apapa or Cabras, the Luminan reefs, and the Calalan bank.

This port is extensive and safe, but it is much encumbered with numerous banks, rocks, and islets. The usual entrance is to the northward of the islet lying close to Oroté point, by a deep channel a third of a mile wide, between that islet and the west end of the Calalan bank. There is a channel, with $5\frac{1}{2}$ fathoms in it, between the east end of this bank and the Luminan reefs, but it should not be used without buoys. In the centre of the port is a rock level with the water, on which fort Santa Cruz is built. The general anchorage is about 2 cables north of this fort, in a space carrying 5 to 15 fathoms, muddy bottom, surrounded by coral patches 2 or 3 feet below water. Some of the channels leading to this anchorage are narrow; the last before entering is not more than 120 yards wide; but the patches are steep-to, and may be approached almost to touching.

The watering place is at a small river which falls into the port at three quarters of a mile from Santa Cruz fort. The casks are usually filled at low tide, and the boats sent at high water to bring them off.

The coast from the east point of Port San Luis d'Apra trends E.N.E. $1\frac{1}{2}$ miles to a perpendicular rocky point, named Acahi-Fanahi, near to which is Gapan islet. At $1\frac{1}{4}$ miles eastward of this islet is Adeloup point, better known to the inhabitants as Punta del Diablo, on account of the rapidity of the currents, which make it difficult to be doubled. From hence a sandy beach commences, which trends to the east and north, forming Agagna bay, in the middle of which is the harbour of Agagna, only fit for small craft.

From Agagna bay the north-west coast of Guam is of steep rocks. At $1\frac{1}{2}$ miles to the north-east of Agagna is Toumon bay, which appears filled with reefs, but there are several passages through them, where boats can reach the shore without difficulty. Near the middle of the coast of this bay is the village of Gnaton. From hence the coast is barren and uninhabited.

A short distance inland from Ritidian point, the north extreme of Guam, the perpendicular hills form, scarcely without interruption, the circuit of the island. In the middle of the north face of the island there is a bay, but it is fronted with coral reefs.

rota or Sarpan island, bearing N.N.E. $\frac{1}{2}$ E. 30 miles from the north end of Guam, is 12 miles long, N.E. by E. and S.W. by W., $5\frac{1}{2}$ miles broad, and its highest part is about 600 feet above the sea. Its east and centre portions are hilly, but the land becomes lower to the south-west, to a low sandy isthmus on which the villages of Sossan Hagno and Sossan Haya are built. To the south-west of this isthmus is the south-west point, a hill, terminating in a level and regular plateau. Some cattle, pigs, cocoa-nuts, bread-fruit, bananas, and a few other vegetables, constitute the entire riches of the island. Three wells furnish the inhabitants with water; two of them are artificial and the water bad; the third, which is natural, affords better, though it is brackish. On the east coast of the island, at 5 miles from the villages, there is a rivulet of very good water.'

The south-east end of the island is tolerably high and perpendicular on the sea shore, presenting thus a straight wall, and at its angles vertical fissures like the embrasures of a fort. In other parts the land descends gradually to the sea, terminating in long and low points. The portion of the island not inhabited is so encumbered with bushes that it is difficult to penetrate; on the north side there are some cocoa-nut trees.

Rota is nearly surrounded by reefs. Its north-east coast and the south-east side of the isthmus are bordered with numerous rocks, on which the sea breaks more or less, according to the direction of the wind. A large bay, 4 miles wide and 2 miles deep, is formed between the south and south-west points of the island, in which vessels can find good shelter from winds between East, round north, and west; but the bottom is foul. On the west side of the isthmus a sort of stage or jetty makes an easy access for boats.

AGUIJAN ISLAND, at 44 miles N.N.E. from Rota, is small, being not more than 3 miles long and 2 miles wide. At its north part are high, perpendicular, and nearly naked rocks, except at their summits, which are crowned with thick wood. Goats are numerous. The only points fit for landing are on the west and north-west sides, where there are some very small creeks lined with sandy beaches. The coasts of the island appear clear of danger, with the exception of a rock lying a mile off its south-west point.

TINIAN or Buena Vista island, separated from Aguijan by a channel 5 miles wide, is 10 miles long, north and south, and $4\frac{1}{2}$ miles broad. It was celebrated, in Anson's voyage in 1742, for its healthy and dry soil, the beauty of its meadows, and its diversified woods, lawns, valleys, and hills, abounding with herds of thousands of cattle, guanacoes, wild hogs, wild fowl, guavas, cocoa-nuts, limes, oranges, and bread-fruit. Later navigators have not given its description in such glowing colours. Instead of valleys and lawns, the trees were thick and almost impenetrable, the land overgrown with a stubborn reed or rush, the climate insufferably hot, the water scarce and bad, the plague of fleas intolerable, and swarms of scorpions, centipedes, large black ants, and other venomous insects without number; all the refreshments, however, mentioned above could be obtained. Not a long time previous to Anson's visit the island had been depopulated by the Spaniards. In 1819 there was not more than twenty inhabitants on it.

The only anchorage for large ships is in Anson road, off the south-west shore of Tinian, abreast the village of Sunharom. This road, however, being open to the westward, and the bottom full of pointed coral rocks, cannot be recommended as an anchorage, particularly between the months of June and October, during the season of the western monsoon. From the middle of October to the middle of June the weather is settled and the road secure. H.M.S. *Centurion* anchored here, however, from 27th August till the end of October 1742, about 1½ miles off shore, abreast a sandy bay, in 22 fathoms, on a bottom of hard sand and coral, with the two extremes of the island bearing N.W. ¼ N. and S.E. ¼ E., the centre of Aguijan S.S.W., the peak of Seypan seen over the northern part of Tinian N.N.E. ½ E., and a reef of rocks lying between the vessel and the shore E. by S. ¾ S. The anchoring bank is shelving, and free from danger, except a reef of rocks lying about half a mile off shore, and affording a narrow passage into the small sandy bay, which is the only place the boats can land.*

SEYPAN ISLAND, at 3 miles to the north-east of Tinian, may be recognized by its lofty peak, 2,000 feet above the sea. The island † is generally visited annually by three or four whalers, which anchor off its north-west side, outside a long reef extending in a southerly direction from the north-west part of the island towards the north part of Tinian; there is, however, a passage through this reef off the north-west part of the island, near a small island, with anchorage inside for several vessels. It was stated by the inhabitants that there is no passage up inside the reef, although it extends a long distance off shore from the south-west end of Seypan; but that there is a ship channel between the north end of Tinian and the south end of Seypan and reef.

The only village on Seypan is on its north-west side. It is ruled by a man appointed from Guam island, and inhabited by old and young settlers, and natives, in all about 300 persons. The natives for the most part are fine young men from the Caroline islands; they are quiet, and in a state of nudity, with the exception of a narrow slip of cotton round the lower part of their bodies.

---

* See Plan of Tinian Bay, No. 1,102; scale, $m = 0·75$ of an inch.

† The description of this island is by Captain N. Vansittart, C.B., and John W. H. Harvey, Master R.N., H.M.S. *Magicienne*, July 1858.

Supplies.—There is no water at Seypan, except what is caught during the rains, and the rainy season in August, September, and October. Cocoa-nuts, bread-fruit, and limes are plentiful; there are also many wild pigs and bullocks—the latter belong to the Spanish government. Pigs, poultry, and fruit can be obtained at the village.

MAGICIENNE BAY.—H.M. steam frigate *Magicienne*, on her passage from the Sandwich islands to Hong Kong, July 1858, having lost the North-east trades in long. 156° 38′ E., and being short of fuel, steered for Seypan island, and anchored in Magicienne bay on its south-east side, in lat. 15° 8½′ N., long. 145° 44′ E.*

This bay cannot be recommended to a sailing vessel, as the water in it is deep, and the anchorage so close to a coral reef bordering its shore, that with a southerly wind there would be no room to weigh. The depth is 30 fathoms, over coral with sandy patches, at only a third of a mile from the bluff at the head of the bay, decreasing rapidly to 3 fathoms close alongside the coral reef, which nearly dries at low water. The *Magicienne* anchored in 18 fathoms, with the south-west point of the bay bearing S. ¼ E. about 2½ miles; the south-east point, which is a bluff, E.S.E. 1¼ miles: and a wooded bluff at the head of the bay, N.N.W. ½ W. nearly a third of a mile. When the vessel swung to the shore, there were 9 fathoms, coral patches, under her stern, and she was distant only a cable's length from the reef; at a cable to the southward of her anchor there was no bottom at 70 fathoms. The bay is well protected, being open only from E.S.E. to South. During the few days the vessel remained in it, the wind was light from the S.E. during the day, and at night a light air from the land between N.E. and N.W.

Supplies.—There is a plentiful supply of wood growing on the shores of Magicienne bay, sufficient to supply any number of vessels, being for the most part the thickness of a man's body, white when cut, and in substance something between a bad ash and a poplar. The best place for landing is on the sandy beach to the eastward of the wooded bluff at the head of the bay. The crew of the *Magicienne* cut down and brought on board 172 fathoms in six days, the wood growing close to the beach, and easily carried to the boats, which could lie afloat close to the reef. No water could be obtained; wells were dug, but the water from them was brackish.

The wood soon dried, but being freshly cut, it was necessary to split and bark it on board before using it for steaming purposes, also to use a small quantity of coal with it; the bark was very sappy. The wood being free of resinous substances, it did not give out so much heat as might have

---

* *See* Plan of Magicienne Bay, No. 1,102; scale, *m* = 2½ inches.

been expected ; three and a half fathoms of it being only equal to one ton of good Welsh coal.

TIDES.—It is high water, full and change, in Magicienne bay, at 6h. 45m., and the rise is about 2½ feet.

FARALLON de MEDINILLA or Bird island, flat, barren, and with perpendicular sides, and distant about 41 miles to the N.N.E. of the north end of Seypan, is 2 miles long N.E. and S.W., and its breadth is much less. On the south and west sides are some deep caverns or grottos. The south point is terminated by a small hill, which appeared to be joined to the island by a narrow neck of land.

ANATAJAN ISLAND, at 27 miles to the north-west of the latter, is about 5 miles long east and west, and has two high and steep peaks, which are on the same parallel. The island has every appearance of being volcanic ; in clear weather it is visible at a distance of about 21 miles.

SARIGUAN ISLAND, to the N.N.E. and distant 18 miles from Anatajan, appears to be merely a high hill, of the form of an upright cone, with nearly a circular base, 1½ miles in diameter. Its summit is rounded. It is almost without vegetation, and seems to be of volcanic origin.

ZEALANDIA BREAKERS.[*]—The ship *Zealandia*, at 4h. p.m. 3rd December 1858, when steering W.N.W. to pass between Sariguan and Farallon de Torres, two patches of breakers were seen right ahead, distant about three-quarters of a mile. The course was altered to pass northward of them, and a good look-out kept from the top-sail yard. At 4h. 20m. p.m. Sariguan bore S. by W. ½ W. about 11 or 12 miles, when the breakers were in line with the island, and distant half a mile from the ship; this would place them in about lat. 16° 50′ N., long. 145° 54′ E. The two patches, which at times broke heavily, bore N. by E. and S. by W. about a quarter of a mile apart, with dark water between and around them. The wind was light and easterly, but the lateness of the day and the unsettled state of the weather did not permit a close examination.

FARALLON de TORRES, lying about 36 miles to the northward of Sariguan, is a small island, 2½ miles long N.N.E and S.S.W., and about a mile broad. It is of moderate height, much resembling the Farallon de Medinilla. Its north point is the lowest; it has a most barren aspect, is perpendicular, and unapproachable on all sides.

---

[*] Mercantile Marine Magazine, September 1859, p. 284,

MARIANA OR LADRONES ISLANDS.

GUGUAN ISLAND, one of the highest islands in this archipelago, bears North, and is distant 16½ miles from Farallon de Torres. It is 2¼ miles long north and south, and on it are two peaks, the highest of which is estimated to be about 2,000 feet above the sea. To the south and east the slope of the hills is extremely rapid, and the rock, which descends to the sea, is composed of lava. At its south end there are some white and red spots, and to the west a point covered with trees. The north side is not so steep as the south. The highest point on the north side is a vast crater, from whence smoke has been seen to issue.

ALAMAGUAN ISLAND, about 25 miles North of Guguan, was only seen from the *Uranie* at a distance of 18 miles. It seemed to be divided into two portions, joined by low land, and to be about 8 miles long N.E. and S.W. Its highest parts are angular; that to the north-east was like a volcano.

PAGON ISLAND, of which but little is known, lies North 9 miles from Alamaguan. Several peaks were seen in passing it, and a small island near its south end. It is said that there is anchorage close to the south part of Pagon, but it must be exposed to the southward.

GRIGAN ISLAND, at 30 miles N. by W. of Pagon, is about 6 miles long N.W. and S.E., and on it are two high peaks, estimated to be 2,026 feet above the sea. It is stated that on the south-west side of the island there is a small plain, in front of which there is bad anchorage on account of the violence of the currents.

This island is larger than Asuncion to the northward of it, and like the latter is volcanic, having a few trees on its north and south sides, which descend gradually from what appears to be a crater, having at some period deposited streams of lava, or black ashes, a considerable distance down its sides.

MANGS ISLANDS are some small islets which the Spanish charts place in the middle of numerous reefs. They were seen from the mast-head of the *Uranie* in 1819, and the bearings then taken serve to point out their approximate position. They lie about N.N.W., 43 miles from Grigan.

ASUNCION ISLAND, called also the Great Volcano, which is expressive of its character, is 54 miles N. by W. from Grigan. It is a remarkable object, being a perfect volcanic cone rising abruptly from the sea to the height of about 1,450 feet. Its whole circumference at the base is not more than 3 miles. In clear weather both this island and Grigan can be seen about 45 miles off.

URACCAS or Mangs islands, according to Lapérouse, are a small rocky group lying N.N.W. ½ W. 15 miles from Asuncion, in lat. 19° 57′ N., long. 145° 20′ E.

GUY ROCK, in 20° 30′ N., 145° 32′ E., and the northernmost of the Mariana group, was discovered by Captain Douglas, Sept. 1789.

Los JARDINES or Marshall islands.—Two small islands, in lat. 21° 40′ N., long. 151° 35′ E., were discovered by Captain Marshall of the ship *Scarborough* in 1788. They are said to be the same as Los Jardines of Alvaro de Saavedra, in 1529.

SEBASTIAN LOBOS or Grampus islands.—The Grampus group, discovered by Captain Meares in April 1788, consists of three islands, two of which are close together, and the third to the south-west of them. The south-west island is assumed to be in 25° 10′ N., 146° 40′ E. The Sebastian Lobos, with which these are supposed to be identical, are nearly in this latitude. Meares does not give their position, but Krusenstern places them from his track in 25° 40′ N.

FORFANA ISLAND, which has not been seen since its discovery by the ship *San Juan* in 1543, is said to lie 90 miles E. ¼ N. of the Volcano islands, in lat. 25° 35′ N., long. 143° 00′ E.

VOLCANO ISLANDS, three in number, were discovered in 1543 by Bernard de Torres, and received their name from the volcano on the central one. There is no doubt of their being the same as the Sulphur islands of Captain King in 1779. They were also seen by Krusenstern in 1805. The southern island is named San Augustino, the centre Sulphur island, and the northern San Alessandro.

San Augustino, the summit of which is in lat. 24° 14′ N., long. 141° 20′ E., is a single mountain of a square form, flat at the top, and 396 feet high.

Sulphur is about 5 miles long N.N.E. and S.S.W., and its summit is in 24° 48′ N., 141° 13′ E. The south point of the island is a high barren hill, flattish at the top, and, when seen from the W.S.W., presents an evident volcanic crater. A low narrow neck of land connects this hill with the south end of the island, which spreads out into a circumference of 9 or 12 miles, and is of moderate height. The part near the isthmus has some bushes on it, and a green appearance, but that to the north-east is barren and full of detached rocks, many of which are white. Dangerous breakers extend 2½ miles to the east and 2 miles to the west from the middle part of the island.

San Alessandro, the northern island, is a single mountain of considerable height, like Volcano island, and its peak, which is of a conical shape, is in 25° 14′ N., 141° 11′ E.

The **MAL ABRIGO** or Margaret islands, in lat. 27° 20′ N., long. 145° 45′ E., are a group of three islands seen in 1773 by Captain Magee. They have been considered to be the Mal Abrigo islets (bad shelter) of Bernard de Torres in 1543.

ARZOBISPO or BONIN ISLANDS were re-discovered by an English whaler in 1825, and were formally taken possession of by Captain F. W. Beechey, H.M.S. *Blossom*, in 1827. The group consists of three clusters of islands extending in a N. by E. direction from the parallel of 27° 44½′ to 26° 30′ N. and beyond, but that was the limit of their view to the southward. They had no signs of ever having been inhabited. The climate is excellent, and the soil productive. In 1853 there were a few whites and Sandwich islanders, 36 in all, settled on Peel island.

The northern cluster, named Parry group, after Captain Sir E. Parry, is composed of small islands, and pointed rocks, and has much broken ground about it, which renders caution necessary in approaching it.

The middle group consists of three islands, of which Peel island, the largest and most southern, is 4¼ miles in length. The northern island is named Stapleton, and the centre one Buckland. This group is 9¼ miles in length, and is divided by two channels so narrow that they can only be seen when abreast of them. Neither of them is navigable; the northern on account of rocks, which render it impassable even for boats, and the other on account of rapid tides and currents, which, as there is no anchoring ground, would drift a ship on to the rocks. Vessels will find good anchorage at the south-west angle of Buckland island, in a sandy bight named Walker bay, but they must be careful, when bringing up, to avoid being carried out of soundings by the current.

The Bailey or Coffin islands, the southern cluster, were visited by a whale ship, commanded by Mr. Coffin, in 1823 ; they were named Bailey by Captain Beechey in 1827. A reconnoissance of these islands was made in 1857, by the officers of the U.S. ship *Plymouth*, and Lieutenant Balch, in his report, says :—" With the exception of New port, in lat. 26° 36′ N., long. 142° 9′ E., on the south-west side of Hillsborough island (formerly known as Fisher island), and a small cove just to the northward of it, there is no place on the shores of any of the islands suitable for a coal depôt, nor can New port or the cove be recommended as places suitable for such a purpose. They are both open from S.W. to N.W., and the holding ground is not good, being sand and rocks; vessels could, however, always get to sea on the approach of

[c.]

a gale. The *Plymouth* anchored in 14 fathoms about half a mile from the head of the port.

"Hillsborough (the largest of the group) is 7½ miles long N.N.W. and S.S.E., 1¼ miles wide, and the greater part is hilly and rocky. There are some wild hogs upon it, fish is abundant, and turtle plenty in the season. Wood and water can be obtained; the former from the head of the above cove. From May to December easterly winds prevail; after that the westerly winds blow till May with the regularity of a monsoon."

FITTON BAY, at the south-east angle of Peel island, affords good anchorage, but it is open to the S.E.; as the winds from this quarter blow generally during summer, it will be prudent not to anchor there during that season. A small islet lies off the south-west end of Peel.

PORT LLOYD.—The entrance to this harbour, on the western side of Peel island, is well defined, so that it can scarcely be mistaken. Before entering it would be well to place a boat on the shoal which extends South fully 2 cables from the eastern point of Square rock, lying off the northern point of entrance. The shoal can, however, be easily seen from aloft even when there is no swell on, and its centre is awash with a smooth sea. A coral rock, with 8 feet water on it, lies about a cable's length to the northward of the Southern head.*

Supplies.—Peel island, as well as those surrounding it, is chiefly visited by whale ships, and its products, therefore, are such as to suit their wants. Potatoes, yams, and other vegetables, fruits of various kinds, together with wild hogs and goats, can be procured from the few whites and Sandwich islanders settled at Port Lloyd. Wood is good, and plentiful; and water can be had, though in limited quantities, and slightly tainted by the coral rocks from which it springs. The best watering place is in Ten-fathom Hole; but it will be necessary to be cautious of the sharks, which are very numerous.

DIRECTIONS.—Having ascertained the position of Port Lloyd, steer boldly in for the Southern head, taking care, when approaching from the southward, not to bring it to the northward of N.E. ¼ E., or to shut it in with two paps on the north-east side of the harbour, and which will be

---

* See Plan of Port Lloyd, No. 1,100; scale, m = 4 inches. In this plan the longitude of Ten-fathom Hole is given as 142° 11′ 30″ E. The Master of the U.S. ship *Susquehanna* gives the longitude 5′ farther East. In a meridian distance measured from Hong Kong with 5 chronometers, interval 33 days, Collinson, in 1852, made the longitude of Port Lloyd, 142° 15′ E., assuming Hong Kong as 114° 10′ 48″ E.

seen nearly in one with it on this bearing. In this position they are a safe leading mark ; to the southward of this line there is broken ground.

In a sailing vessel, if the wind be from the southward, which is generally the case in the summer season, round the Southern head, at the distance of a long cable's length, close to the sunken rock, which will be distinctly seen in clear weather. Keep fresh way upon the vessel, in order that she may shoot through the eddy winds, which baffle under the lee of the head, and to prevent her coming round against her helm, which would be dangerous. She will at first break off, but will presently come up again ; if not, be ready to go about, as the vessel will be close upon the reefs to the northward, and the helm must be put down before the south end of the island off the port to the north-westward comes on with the west side of Square rock.

If the vessel comes up, steer for a high Castle rock, at the eastern part of the port, until a Pointed rock (which is white on the top with bird's dung, and looks like an island) on the sandy neck to the eastward of the Southern head comes in one with a high sugar-loaf-shaped grassy hill to the southward of it ; after which bear away for the anchorage, taking care not to open the sugar-loaf again to the westward of the Pointed rock. The best anchorage, Ten-fathom Hole excepted, which it will be necessary to warp into, is at the northern part of the port, where the anchor is marked in the Admiralty plan.

In anchoring, take care to avoid a spit extending off the south end of the small island near Ten-fathom Hole, and not to shoot so far over to the western reef as to bring a rock at the outer foot of the Southern head, in one with some *black* rocks, which will be seen a short distance to the south-westward. The depth will be 18 to 20 fathoms, clay and sand. The anchorage is fair, though open to the south-west.

If the wind be from the northward, turn to windward between the line of the before-mentioned sugar-loaf and Pointed rock, and a north and south line from the Castle rock. This rock, on the western side, as well as the bluff to the northward of it, may be passed close to, if necessary. The hand lead will be of little use in beating in, as the general depth is 20 or 24 fathoms.

TIDES.—It is high water, full and change, at Port Lloyd at 6h. 8m., and springs rise 3 feet. At New port, Hillsborough island (the largest of the Bailey islands), it is high water at 11h. 32m., and the rise $3\frac{1}{2}$ feet.

ROSARIO or Disappointment island, in lat. 27° 16′ N., long. 140° 51′ E., and about 70 miles W. by N. from Peel island, one of the Bonin group, is small, being only about a mile long, east and west, and its highest part 148 feet above the level of the sea. It is rugged, and apparently unsusceptible of cultivation, and is surrounded by numerous isolated rocks.

ISLANDS NORTH OF THE LU-CHU GROUP.

To the northward of the Lu-chu group inhabited islands extend in a northerly direction, with many safe channels between them, as far as lat. 30° 51′ N. The mariner is, however, cautioned not to place too much dependence either on their configurations or positions, as shown on the chart of this part of the ocean, for they are by no means correct; they are from the Japanese as collated by Siebold, and from detached surveys and corrections by English, French, and American navigators. The French corvette *La Sabine* examined them in the year 1846; their positions, to which we have given the native names and restored those of former explorers,* appear on the chart of her track to be as follows :†

YORI SIMA, or Julo of Basil Hall in 1816, centre, lat. 27° 2′ N., long. 128° 25′ 24″ E.

YEIRABU SIMA of Siebold, or Wukido of Basil Hall, south peak, lat. 27° 21′ N., long. 128° 31′ 34″ E., height 889 feet, (lat. 27° 14′ N., long. 128° 33′ by Collinson in 1845).

TOK SIMA of Siebold, or Crown island of Broughton, in 1797, highest peak lat. 27° 44′ N., long. 128° 59′ E.; height, 2,461 feet. The northern peak is 2,034 feet above the sea; a village is built on its north-west face. This island is 14 miles long north and south, and 7 miles east and west.

All three of the above islands are well wooded, and appear to be inhabited.

IWO SIMA, or Sulphur island, in lat. 27° 51′ N., long. 128° 14′ E. (128° 19′ E. by Collinson); height, 541 feet; is a volcanic mountain still in action.

---

* It is greatly to be regretted that navigators will not endeavour to ascertain the names of places as given by the natives; or, failing these, that they will not retain the names affixed to islands by the first discoverers. In the present case there are three and, occasionally four names for each of the islands in this archipelago. So long as this practice is pursued our charts will remain a maze of confusion.

† *See* Chart of Islands between Formosa and Japan, No. 2,412.

OHO SIMA, or Harbour island, is the largest of the chain of islands lying between Great Lu-chu and Japan. It is about 30 miles in length, in a N.E. and S.W. direction, is high, well cultivated, and, from the number of villages seen along the coast, must contain a large population. There are two peaks upon its south end, 1,674 and 1,420 feet respectively above the sea.

This island was partially surveyed by the American squadron in 1856, and by their chart the outline of the coasts appears much broken and deeply indented with numerous bights, most of which are very bold. Wood and water are good and plentiful; but refreshments scarce. The inhabitants are timid and harmless. The north end of the island is high, and being connected with the main part of the island by a narrow low isthmus, it has the appearance, on some bearings, of being isolated. Foul ground appears to extend about 2½ miles N.E. by E. from the north end, and two rocks to rise from it, the northern of which is about 80 feet high. North extreme of the island, lat. 28° 31' 40" N., long. 129° 40' 12" E.; south extreme, lat. 28° 6' 30" N., long. 129° 22' E.

The south end of the island is separated from Katona sima, by a narrow channel, in some places not more than half a mile wide. The *Vincennes* anchored within the eastern entrance of this channel, in a small bay formed at the north end of Katona sima. In entering the channel an anchor should be ready to let go, in case of being set too near danger, for the entrance is narrow and the current strong.

TIDES.—By three days' observations in Vincennes bay, at the north end of Katona sima, it was high water, full and change, at 7h. 30m.; and the rise and fall 5½ feet.

KIKAI SIMA, lying about 15 miles to the E.S.E. of the north end of Oho sima, is moderately high, about 7 miles in length, N.N.E. and S.S.W., and inhabited. The summit (867 feet high) is in lat. 28° 18' N., long. 129° 57½' E.*

GERMANTOWN REEF.†—The U.S. ship *Germantown,* on the 23rd March 1859, when beating along the south-east side of Oho sima, struck on a coral reef said to lie in lat. 28° 16' N., long. 129° 58' E. From the shoalest spot found, 6 feet, the highest terrace on Kikai sima bore N.E. ¼ E. 6 or 7 miles. The reef is about a mile long in a N.N.E. and S.S.W. direction, and half a mile wide.

Another shoal spot was found lying North 2 miles from the centre of this reef, with apparently a clear passage between. Reefs were also

---

* From American chart, 1855.   † Hong Kong Register, 26th April 1859.

seen from aloft, extending from one to two miles from the south-west and south-east points of Kikai sima.

ANCHORAGE.—The *Germantown* anchored at 1¼ miles from the shore in 25 fathoms, coral and shell, with the south-east point of Kikai sima bearing S.E. by E. ½ E., and the south-west point N. ½ E. The tides here set strong ; the ebb from E.N.E. to N.E., and the flood from West to W.N.W. The strength was about 2 knots per hour, with an undertow of at least double that velocity.

SANDON ROCK (Constantine of the French charts), about 20 feet high, resembling a small haycock, lies N. by E. ½ E. 12 miles from the north point of Oho sima.

The LINSCHOTEN ISLANDS, or Cecille archipelago (so called in the French charts after Admiral Cecille, by whose directions the islands were examined in 1846), extend from lat. 28° 49' N. to 30° 6' N., and from long. 129° to 130° 3' E.

YOKO SIMA, or Ogle island of Belcher in 1845, rising to the height of 1,623 feet above the sea, is an extinct volcano, the highest part of which is in lat. 28° 49' N., long. 128° 59' E.; there is a small islet about a mile to the northward of it.

TOKARA SIMA, 885 feet above the sea, is in lat. 29° 8' N., long. 129° 11' E.

SIMAGO, or Cooper group of Belcher, are four small islets, the highest of which, 738 feet above the sea, is in lat. 29° 13' N., long. 129° 19' E. The easternmost islet bears from it about E. ½ N. 3 miles.

AKUISI SIMA, or Samarang island of Belcher, 2,184 feet above the sea, is in lat. 29° 27' N., long. 129° 35' E.; a small islet lies off its north-west face.

SUWA SIMA, or Volcano island of Belcher, is an active volcano, 2,805 feet high, in lat. 29° 38' N., long. 129° 42' E.

FIRA SIMA, or Disaster island of Belcher, lying to the W.N.W. of Suwa sima, is 879 feet high, and in lat. 29° 41' N., long. 129° 31' E.

NAKA SIMA, or Pinnacle island of Belcher, is 3,287 feet above the sea ; its peak is in lat. 29° 53' N., long. 129° 50' E.

HEBI SIMA, or St. François Xavier island on French chart, rises to the height of 1,820 feet. The peak is in lat. 29° 55' N., long. 129° 32' E. There is a small islet off its north-west face.

LINSCHOTEN ISLANDS.

KOZEBI SIMA, or Forcade rock on French chart, 984 feet above the sea, is in lat. 29° 53′ N., long. 129° 36′ E.

KUTSINO SIMA, or Alemène island on French chart, is 2,116 feet above the sea, and its highest part is in lat. 29° 59′ N., long 129° 55′ E.

BLAKE REEF, or Lapelin rocks on French chart, the highest islet on which is 90 feet above the sea, and in lat. 30° 5′ N., long. 130° 3′ E., consist of several distinct islets and rocks, extending about 3 miles in a N.E. and S.W. direction.

YAKUNO SIMA.—To the north-eastward of Blake reef is the island of Yakuno, the highest peak of which, Mount Motomi, 5,848 feet high, is in lat. 30° 21′ N., long. 130° 29′ E. The island, which has not been examined, is about 12½ miles long north and south.

TANEGA SIMA has never been examined, but it is said to be level and covered with trees. It appears to extend from lat. 30° 22′ to 30° 43′ N., and from long. 130° 54′ to 131° 5′ E. According to the French chart of 1846 there is the outline of a good harbour on its western side, and the sites of some towns are noticed.

SERIPHOS or Omuru rock is marked in the French chart as under water; it is in lat. 30° 44′ N., long. 130° 45′ E.

YERABU SIMA, or Julie island on French chart, is an active volcano, rising to the height of 2,067 feet above the sea. Its highest peak is in lat. 30° 27′ N., long. 130° 11′ E. The island is about 6 miles long, in an E.S.E. and W.N.W. direction; its greatest breadth is 3 miles.

TAKE SIMA, or Apollos island on French chart, is high, and about 2 miles in circumference; its centre is in lat. 30° 48′ N., long. 130° 24′ E.

IWOGA SIMA, or Volcano island on French chart, is an active volcano; its highest peak, which rises 2,345 feet above the sea, is in lat. 30° 42′ N., long. 130° 17′ E.

POWHATTAN REEF.—This dangerous reef, in lat 30° 41′ N., long. 130° 19′ E., was discovered by the U.S. frigate *Powhattan* in January 1860. From the centre rock, which is about 18 feet above the sea, the south-west point of Iwoga sima bore N.W., the east point N. ½ W., and the east point of Take sima N.E. ¼ N.

Other rocks were seen awash, or a few feet above water, stretching out about three-quarters of a mile from the centre rock. Some reefs were also observed extending about three-quarters of a mile from the eastern points of Iwoga sima; and a rocky spit about a quarter of a mile from the east point of Take sima.

TRIO ROCKS are three distinct islets, of about an equal height; the centre islet, which is 223 feet above the sea, is in lat. 30° 45′ N., long. 130° 5′ E.

KURO SIMA, or St. Clair island on French chart, rises to the height of 2,132 feet; its centre is in lat. 30° 50′ N., long. 129° 55′ E.

INGERSOLL ROCKS.—The Ingersoll, Morrison, or Larne rocks, eight in number, extend in a N.E. and S.W. direction about $5\frac{1}{2}$ miles; the highest, 446 feet above the sea, is in lat. 30° 51′ N., long. 129° 26′ E.

ISLANDS OFF SOUTH-EAST COAST OF NIPON.

LOTS WIFE or Black rock, in lat. 29° 47′ N., long. 140° $22\frac{1}{2}$′ E., lying off the south-east coast of Nipon, is a tall pinnacle rising about 300 feet above the sea; in clear weather it can be seen at a distance of 25 miles, and bears a remarkable resemblance to a ship under all sail. A cast of the lead within 8 miles of this rock gave no soundings with 160 fathoms of line.

PONAFIDIN or St. Peter island was discovered in 1820 by Lieutenant Ponafidin of the Russian Navy, and named by him Three Hills island, from its having apparently three hummocks. It was seen by one of the vessels of the United States squadron in 1853, and its position is given as lat. 30° 33′ N., long. 140° 15′ E.

SMITH ISLAND.—H.M.S. *Tribune*, 18th January 1859, passed a high pinnacle-looking rock, in lat. 31° 18′ N., long. 139° 50′ E., about three-quarters of a mile in circumference, with heavy breakers extending apparently a quarter of a mile from it, and a small rock close to its north side. As this rock is nearly in the same position as that assigned to Smith island on the chart, there is every reason to believe they are identical.

BAYONNAISE, or King William isle, in lat. 32° 0′ 40″ N., long. 140° 0′ E., forms a curve a cable long north and south, its northern summit rising 20 feet; several rocks extend off it a quarter of a mile to the north-west.*

ONANGA SIMA, or South isle, in lat. 32° 30′ N., long. 139° 50′ E., is 3 miles long, and is visible 36 miles off in clear weather. Its coasts are steep, and the only landing place is on the east side, where there is a rock level with the water at a little distance from the land. Onanga is

---

* *See* Chart of Nipon Island, Kiusiu and Sikok, and part of the Coast of Korea, No. 2,347, scale, $d = 2\cdot3$ inches; corrected to 1861.

inhabited and cultivated on the north and north-west sides. Cheyne, who saw the isle in July 1853, places it 7' farther north and 12' farther west.

FATSIZIU ISLAND, 12 miles long, N.W. and S.E., has the appearance of two islands joined by a low plain, while a high islet near its north end makes it look like three islands on making it from seaward, whence it may be seen from 40 to 50 miles. It is inhabited and well cultivated. Lat. (of centre) 33° 6' N., long. 139° 43' E.

BROUGHTON ROCK, in lat. 33° 42' N., long. 139° 17' E., is about 50 feet high, flat at the top with steep slopes, except on one side, where it presents a broken and precipitous face.

MEAC SIMA and MECOURA are two high and bold islands; the first in lat. 34° 6' N., long. 139° 29' E., and Mecoura in 33° 54' N., 139° 35' E.; the latter lies about 17 miles to the north-east of Broughton rock. Mecoura is called Prince island and Meac sima Volcano island in former charts.

There is a cluster of rocks about 2½ miles to the south-west of Meac sima, and Broughton says, "there are, in addition, some black rocks 2 or 3 miles from the eastern point of the island."

REDFIELD ROCKS.—This dangerous cluster of small, sharp pointed rocks, from 15 to 20 feet above water, lies about 20 miles to the south-west of Kozu sima, in lat. 33° 56' 50" N., long. 138° 49' E. They are said* to rise from the north-east extremity of a reef which extends from them 2 miles in a south-westerly direction, and only breaks in bad weather. A vessel, therefore, in passing westward of them in fine weather should give the tail of the reef a good berth, as it might not then be marked by breakers.

KOZU SIMA, in lat. 34° 13¼' N., long. 139° 8' E. (centre), is the most south-western of the chain of islands fronting the Gulf of Yedo. It is 3½ miles long N.E. and S.W., and may be recognized by a remarkable white cliff on its western side, and a white patch on its summit, to the northward of the cliff. It has an elevation of 2,000 feet above the sea. There is a safe channel 15 miles wide between Kozu sima and Meac sima.

About 2 miles to the southward of the south-west point of Kozu sima are the Brood rocks, which should be given a safe berth, as their jagged appearance would lead to the belief that there are many hidden dangers in their immediate neighbourhood.

SIKINI SIMA is low, with a small islet off its north end. It is 1½ miles long N.N.E. and S.S.W., and lies 5 miles to the north-east of Kozu

---

* Edward H. Hills, Master of H.M.S. *Highflyer,* 1859.

sima. One of the vessels of the American squadron passed between these islands and saw no danger; there is therefore reason to believe that the channel is safe.

NEE SIMA is about 1½ miles to the north-east of Sikini, and from its broken outline appears from a distance as several islands. Its extent is 5 miles, north and south, and its most elevated part is 1,490 feet above the sea. There is a small low islet a short distance to the south-east.

UTOMA is a conical islet, 660 feet high, lying N. ½ E., about 2½ miles from Nee sima; it has detached rocks lying near its shores.

TO SIMA, bearing N. ½ W. 2 miles from Utoma, is one mile in diameter, pyramidal shaped, and its summit 1,730 feet above the sea.

OHO SIMA, or Vries island, the largest and most northern of the chain fronting the gulf of Yedo, is 10 miles to the N.N.E. of To sima, its south-eastern point being in lat. 34° 39½' N., long. 139° 28' E. It has an active volcano near its summit, which rises 2,530 feet above the sea. The slopes of the mountain are extensively cultivated and dotted with towns and villages. The vapour which sometimes ascends from the volcano, and condenses in masses on the mountain, renders the island a good land-fall for vessels approaching this part of the coast of Nipon. The passage between it and To sima is believed to be free from hidden danger.

CAUTION.—The current sets strongly to the north-east through the various passages between the above islands, and this should be remembered, particularly in bad weather. In their vicinity, in fact throughout the path of the current from the south end of Formosa to Behring strait, constant heavy tide rips will be encountered, which, in light winds, frequently render a vessel for a time unmanageable.

PORTSMOUTH BREAKERS.—Captain Foote of the U.S. frigate *Portsmouth* reports* that he nearly lost his vessel on a reef of rocks (not laid down in the charts) about 35 miles in a south-westerly direction from Simoda, and 13 miles from the nearest land.

In steering† for the Gulf of Yedo the *Furious* passed inside the above position assigned to these breakers, and although the water changed its colour very decidedly, no indication of danger was seen, nor was any bottom obtained with 13 fathoms. From the mast-head the line of discoloration could be traced from the shore to as far as could be seen seaward. It was from 4 to 6 miles wide in an E.N.E and W.S.W.

---

* Shipping Gazette, 30th March 1858.
† Stephen Court, Master of H.M.S. *Furious*, 1858.

direction where the *Furious* crossed, and on emerging from it into blue water, the boundary line was as plainly marked as on entering it from the westward.

The direction in which this discoloured water extended, S.E. and N.W. and its extent, 6 miles between the vessel and the shore, and about the same distance seaward, in all 13 miles, would lead to the supposition that it terminated in the Portsmouth breakers. Vessels, therefore, bound either way should be well assured of their reckoning when passing this locality; a good look-out should also be kept, the lead constantly hove, and every precaution taken when nearing the position given for these breakers.*

WINDS and WEATHER.—The South-west moonsoon sweeps over the Lu-chu group, and reaches the southern shores of the Japan and Bonin islands. At Napha, Lu-chu, during the visit of the United States squadron, in 1854, it prevailed steadily in May and June, and veered to the southward and eastward in July. In August the wind was changeable, and blew at times quite strong, with squally, rainy weather.

The North-east monsoon set in about the 1st September, and continued until the departure of the squadron, on the 7th February, being, however, interrupted during the winter months by fresh gales from the northward and westward, which were generally accompanied with heavy rain.

At the Bonin islands, in April, the wind was variable; in June it was from the southward and westward; and in October, from the northward and eastward. The passage from Lu-chu to these islands, in October, was exceedingly boisterous.

The Mariana islands lie in the region of the northern tropic, and consequently in that of the north-east trades. But this is not the prevalent wind. The N.E. and S.W. monsoons, which are met with in the China Sea and on the coasts of China, extend as far as the Mariana islands, and sometimes even beyond them; so that the limits between the monsoons and the trade winds must be found near this Archipelago. The months of July to November are the season of bad weather, storms, thunder, and rain.† In December, January, and February the weather is variable. March, April, May, and June are the finest months; the wind is then from East and N.E. The winds blow strongest in August, September, October, and November; their direction at these periods is from N.W. to S.W. by W., sometimes from South and S.E., but in general between

---

* In page 336, it is stated that where whirls and eddies are produced by the inequalities of the bed of the Japan current, strong tide rips are encountered, often resembling heavy breakers on reefs or shoals.

† Voyage of the French corvette *Uranie* in 1819, by M. Louis de Freycinet.

North and West than from North itself. Hurricanes are rare, but are not unknown; there had not been one of these scourges for seven years prior to the *Uranie's* visit. Earthquakes are frequent.

TYPHOONS.—The whole region from Formosa to the Bonin islands is within the track of these storms; though it is believed they seldom reach the coast of Japan. The season during which they may be expected is from May to November, inclusive; but in the neighbourhood of the Bonin islands they seem to occur more frequently in October.

CHAPTER X.

JAPAN AND KURIL ISLANDS, AND SOUTH-EAST COAST OF KAMCHATKA.

VARIATION in 1861. Japan islands, 2° 30′ W. Kuril islands, 1° 40′ W. to 2° 30′ E. Avatcha bay, 3° 40′ E.

JAPAN ISLANDS.—The empire of Japan is composed of three large islands, Kiusiu, Sikok, and Nipon, and numerous smaller islands. Nipon, the largest and most important of the group, and that which gives name to the whole empire, is more than 700 miles in length in a N.E. and S.W. direction, and its breadth varies from 50 to 150 miles. South of Nipon, and separated from it by a narrow channel, is the island of Kiusiu, about 180 miles in length, north and south, and about 80 in average breadth.

Lying north-east of Kiusiu, and eastward of the south extreme of Nipon, is the island of Sikok, about 130 miles in length, N.E. and S.W., and 60 in breadth. It is separated from Nipon by a long strait named the Misima Nada, and from Kiusiu by the Boungo channel. The islands Kiusiu, Sikok, and the western end of Nipon form a basin or interior sea named Suwo Nada or the sea of Suwo. This sea, through which one European ship only (H.M.S. *Cruizer* in 1859) has ventured, has many islands, and according to the Japanese the largest ships may navigate it. It is separated from the Pacific ocean by the island of Sikok and communicated with by the Kino channel to the east, and the Boungo channel to the west, and with the Japan sea by the Fiki channel between the islands of Notch and Wilson. North of Nipon, and separated from it by the strait of Tsugar, is the large island of Yezo, a conquest and colony of Japan. Its form is that of an irregular triangle, and its area is computed at 30,000 square miles. The southern portion of the island of Saghalin, which is separated from Yezo by Lapérouse strait and the three southernmost of the Kuril islands—Kunashir, Iturup, and Urup—belong to Japan.

The Japanese islands are exceedingly broken and mountainous, with numerous peaks rising to a considerable height. Mount Fuzi or Fuziyama is the highest; it is in about lat. 35° 36¾′ N., long. 138° 48¼′ E., about 12,450 feet above the level of the sea, and an excellent landmark for vessels approaching the Gulf of Yedo.

CLIMATE.—The climate of Japan must vary considerably between its southern and northern extremities; but except at a few points we possess very little information on the subject.

At Nagasaki, in the island of Kiusiu, the average temperature in the month of January was 35°, and in August 98° of Fahr. At this place the weather is very changeable. Rain is frequent at all seasons, but especially in July and August. In December and January the ground is covered with hoar frost, and occasionally with snow, except in very mild winters. In summer the land is cooled by the sea breeze, which blows from the south during the day and from the east at night.

At Simoda, on the south-east coast of Nipon, the climate is more or less variable in the winter and spring. The presence of snow upon the lofty peaks, although there is seldom frost or snow at Simoda itself, and the not unfrequent rains, with the fogs, give an occasional humidity and rareness to the atmosphere which must be productive of occasional inflammatory diseases. The change of wind alternates often between the warm sea breezes from the south, and the cold blasts from the snow-capped mountains inland. In summer it is occasionally very hot in the daytime, but the nights are refreshed by the sea breezes.

Of the climate of the still more northern part of the empire we have no precise account. It is stated that at Hakodadi, on the southern coast of Yezo, that severe frosts are uncommon, yet the temperature is often two degrees below freezing point. In summer the rain pours in torrents at the least twice a week, the horizon is obscured by dark clouds, the winds are violent, and the fogs frequent. Apples, pears, and peaches hardly attain ripeness, and the orange and lemon trees will not bear fruit.

Fogs are very prevalent on the coasts of Japan, and thunderstorms and earthquakes are frequent. On the 10th November 1855 an earthquake at Yedo is said to have caused the destruction of 100,000 dwellings and 54 temples, and the death of 30,000 persons.

WINDS.—During the stay of the American squadron in Japan, from February to July 1854, the weather was generally fine, but occasionally interrupted by strong winds and heavy rain. Northerly winds were prevalent in February, March, and April, south-westerly winds in May and July, and variable winds in June. The gales came on suddenly from the south-west, with a low barometer, and continuing for a short time, hauled round to the northward and westward and moderated. There were no easterly gales; in fact the wind was rarely from that quarter, except when veering round from the northward (as it invariably did) by the east, to the southward and westward. In the bay of Yedo the mean temperature for the month of February was 46° Fahr., and the apricot and camelia japonica were in full bloom. There were but few fogs; they commenced at Hakodadi about the 1st June, but did not extend as far southward as Simoda.

In the months of August and September 1858—the period H.M.S. *Furious* remained in Yedo bay—heavy gales from the E.N.E., shifting round to the S.W. and increasing in force, were frequent. Winds from West, round northerly, to E.N.E. generally brought fine weather, and rain when between S.E. and S.W.

From information obtained at Yedo, it appears the prevailing winds throughout the year are to the northward of East and West, and that those to the southward generally bring bad weather; always, however, causing the barometer to fall in sufficient time to enable a vessel to obtain a 60 or 80 miles offing, should she be near the coast.

Strong winds from the S.E. are generally accompanied with thick weather and rain. At such a period it is recommended that a vessel bound to the westward from Yedo should run through the chain of islands to the southward of Van Diemen strait, instead of passing through the strait; for by taking this latter route, and not making sufficient allowance for the north-easterly current, she would in all probability find herself embayed on a lee shore to the northward of Cape Chichakoff and possibly of Cape D'Anville.

During H.M.S. *Saracen's* survey of the Strait of Tsugar, May, June, July, and August 1855, the prevailing winds were from the South, with much fine clear weather. The wind was less frequent from the N.W. than any other quarter. Dense fogs prevailed in May and June; after that period they were comparatively rare.

The wind in shifting usually followed the course of the sun. After a few days of light southerly wind and fine weather it freshened, and veered to the westward, accompanied by fine clear and cold weather. At N.W. it usually died away, or flew round suddenly to the eastward; in the latter case it was always followed by a dense fog or a gale; the weather getting fine again as the wind veered to the southward.

The KURO-SIWO or JAPAN STREAM is an immense oceanic current, which from observations appears to have its origin in the great equatorial current of the Pacific, from which it is separated by the south end of Formosa. The larger portion of this current, when it reaches the point just named, passes off into the China Sea; while the other part is deflected to the northward along the eastern coast of Formosa, until reaching the parallel of 26° N., when it bears off to the northward and eastward, washing the whole south-east coasts of Japan, as far as the Strait of Tsugar, and increasing in strength as it advances.

Near its origin the stream is contracted, and is usually confined between Formosa and the Meiaco-sima group, with a width of nearly 100 miles;

but to the northward of the latter it rapidly expands on its southern limit and reaches the Lu-chu and Bonin islands, attaining a width to the northward of the latter of about 400 miles. Its average maximum temperature is 86°, which differs about 12° from that of the ocean, due to the latitude. The north-western edge of the stream is strongly marked by a sudden thermal change in the water of from 10° to 20°; but the southern and eastern limit is less distinctly defined, there being a gradual thermal approximation of the air and water.

Along the borders of the stream, where it chafes against the counter currents and torpid waters of the ocean, as also in its midst, where whirls and eddies are produced by islands and the inequalities in its bed, strong tide rips are encountered, often resembling heavy breakers on reefs or shoals. Its average velocity between the south end of Formosa and Tsugar strait has been found to be from 35 to 40 miles in 24 hours.

This current is, however, much influenced, both in direction and velocity, by local causes. It is sometimes entirely checked for a day by a north-east wind; when it may be again expected to resume its former course, and possibly run with greater rapidity than usual for one or two days. On one occasion, off the Gulf of Yedo, its maximum strength is recorded as high as 72, 74, and 80 miles respectively, on three successive days.

To the northward of lat. 40° N., in long. 143° E., there is a cold counter current intervening between this stream and the south coast of Yezo, as shown by the sudden thermal change in the water from 16° to 20°, which it is believed sets to the westward through the Strait of Tsugar.

VAN DIEMEN STRAIT.—The islands (described in page 327) on the south side of this strait, and the south end of Kiusiu on the north side are high, and apparently of safe approach. On the north side is a large and deep bay, of which Cape Chichakoff forms the south-east point, and a remarkable mountain, named Horner peak, the north-west point. This mountain, with a similar mountain on Iwoga sima on the southern side of the strait, both of great height, form two conspicuous land-marks when approaching the strait from the westward. The whole of the above bay, excepting to the north, is surrounded by high mountains, covered with verdure. At its head is the town of Kagosima and the island of Sakara.

Cape Chichakoff is about 500 feet high, and has three remarkable rocks lying close off it, one of which, bearing West from the extreme of the cape, is of a sugar-loaf shape, and perforated at its base. These, with a small island lying South about 2 cables, and another East about a quarter of a mile from the pitch of the cape, will enable the seaman at all times to recognize this headland with certainty.

Soundings from 25 to 40 fathoms, were obtained by H.M.S. *Roebuck*, in 1859, between the parallels of 31° and 32° N., and the meridians of 124° and 126° E.; and 8 fathoms in Van Diemen strait, about $2\frac{1}{2}$ miles off Cape Chichakoff, the perforated rock of that cape bearing N. $\frac{1}{4}$ E.

ANCHORAGE.—In Van Diemen strait, H.M.S. *Furious*, in 1858, experienced a strong breeze from the N.E., and being accompanied with a falling barometer, an anchorage was sought for under and to the westward of Cape Chichakoff, where the chart by Siebold appears to point out a small harbour. In searching for this anchorage, soundings were first obtained in 30 fathoms, with the cape bearing S.S.E., and the vessel eventually came to in 13 fathoms, abreast a small village near the centre of the first bay westward of the cape, with the extreme of the cape bearing S.S.W., Horner peak N.W., and north-west extreme of the same bay N.N.W.; here she was well sheltered for thirty-six hours, with the wind steady between E.N.E. and E.S.E., although it was blowing heavily outside, as evinced by the heavy gusts off the land and low barometer; but on its veering to the southward of S.S.E. the bay became too exposed from that quarter, which rendered it advisable to weigh and proceed to sea.

There was no opportunity of verifying the existence of the above harbour; but there was every appearance of a small but well-sheltered inlet in the north-east corner of the bay in which the *Furious* anchored, which, if examined, might prove of great service to vessels meeting with adverse winds, when bound eastward through Van Diemen strait.

OHOSAKA BAY, formed at the south-west part of Nipon, is 35 miles deep, N.E. and S.W., and is bounded to the south by a peninsula, and to the west by the large, mountainous, and wooded island of Awadji. The shores of the bay are in general high and thickly wooded; in some places, however, they are low and sandy. A very considerable coasting trade is carried on. Ohosaka is one of the five Imperial cities and has a large population. It stands on the north-east shore of the bay, on the left bank of the Sedogawa, a small stream which has its source in the lake Oity, at a day and a half journey into the interior; large boats can pass up to Miako, and the river is crossed by several bridges. At the river's mouth are two large towers; the position of the tower before the city is lat. 34° 41′ 38″ N., long. 135° 29′ 27″ E. The depth in the mouth of the river is sufficient to allow boats to enter to procure water.

Two small islands, lying between Awadji and the peninsula, divide the entrance of the bay into three straits, of which the widest is the western. The centre strait, between the islands, is narrow and rocky. The eastern or Dzinosetto strait, is between Dzino island and the peninsula. In approaching this latter strait from the southward, through the Kino channel, a mountainous and wooded headland, with a small summer-house on its summit, will be first seen on the eastern shore. Steer for this headland on a North bearing, and pass it at half a mile distant; a reef extends 3 cables from it. Farther, in a small bay running into the peninsula, is the town of Kada, the position of which is lat. 34° 14′ 28″ N., long. 135° 4′ 20″ E. After passing this reef a N. by E. course will lead into the middle of the strait, from the points on both sides of which rocky reefs extend 1½ cables. The depth obtained in the strait was 16 fathoms, over a bottom of sand and small shells.

From this strait the bay is quite clear to the city of Ohosaka, the course to which is N.E. There is anchorage on either side of the bay; the depths are less on the eastern than on the western shore, where there is anchorage only in a few small bays. Within 2 miles of the city the water shoals, and vessels should not go into less than 4 fathoms, for there is sometimes a heavy swell. The rise of tide is about 2½ feet.

PORT FIOGO* is in the north-west part of Ohosaka bay, abreast of Ohosaka, from which it is distant 10 miles. This port by treaty is open for trade, and it probably will be the principal trading port with the empire of Japan. The Japanese have expended a large sum of money to improve the anchorage, and they have built a breakwater. The town is said to be as large as Nagasaki.

ENORA BAY, on the eastern shore of Tutomi gulf and in lat. 35° 10′ N., long. 138° 53′ E., is 9 cables long, north and south, and 6 cables deep. Ara sima, a wooded island affording shelter from westerly winds, lies off the south point of entrance, and there is a small monument on the north point. The soundings in the bay are deep, 20 to 30 fathoms, and it is open to the west, but there is good shelter from all winds in a small bay in its northern part, where the depth is 13 fathoms over a bottom of fine sand.†

In steering for this small bay, keep midway between its western shore and the cliffs on the eastern. A town stands on the western shore. There is a river in Enora bay, but as water is obtained from it with difficulty on account of its shallowness, it is best to procure it from the

---

* Port Hiogo of the French charts.

† The description of Enora, Heda, Arari, and Tago bays is by Lieut. Elkin of the Russian frigate *Diana*, 1853-55. See Plans of these Bays, scale, $m = 1\frac{3}{4}$ inches, on Chart of Nipon Island, No. 2,347, corrected to 1861.

town wells. There is abundance of fish and vegetables. The rise and fall of tide is about 4 feet.

HEDA BAY, in lat 34° 58′ 11″ N., long. 138° 48′ E., is 6 cables in extent north and south, and 5¾ cables east and west, and carries a depth of 8 to 22 fathoms over a bottom of fine sand. It is sheltered on all sides by high mountains. There is a village in a valley. Six rivers empty themselves into the bay, but it is preferable to obtain water from the wells. Fish and vegetables are abundant.

The entrance into this bay is a quarter of a mile wide and open to the N.W. It is to the northward of a low and sandy spit extending half a mile in a northerly direction from the southern shore and covered with wood. The rise of tide is 5½ feet.

ARARI BAY, sheltered also from all winds, is in lat 34° 50′ N., long. 138° 46′ E. Its extent is 4 cables north and south, and 2 cables east and west, and the depths are 6 to 12 fathoms over fine sand. The shores of the bay are mountainous. Water may be conveniently obtained from the village on the eastern shore ; fish is plentiful.

The entrance is open to the N.W. ; in entering keep in mid-channel, and when a small island opens, steer between it and the sandy point to the S.W. After rounding this point the course is South for the middle of the bay, where the depth is 7 fathoms.

TAGO BAY, in lat. 34° 47′ 3″ N., long. 138° 46′ E., is 4 cables in extent north and south, and half a mile wide, is sheltered from all winds, and carries a depth of 12 to 20 fathoms, soft mud bottom. There is a small town here, and water can be obtained from the wells. Fish and vegetables can be procured.

In steering for the entrance of this bay, which is also open to the N.W. two islands (lying half a mile W.N.W. from the south point of entrance) will be seen, with rocks and breakers extending to the southward from them. Pass to the northward of these islands, between them and the mainland. After passing about a cable to the northward of another island, lying off the south point of entrance, steer S.E. for the middle of the bay, where there is anchorage in 13 fathoms.

DIRECTIONS.—These four bays just described will serve as a refuge from S.W. winds, which cause a great swell in Tutomi gulf. Their coasts are wooded and mountainous, attaining the height of 1,000 feet. The entrances must be approached fearlessly, for the high coast conceals them, and the bays only open when within a mile.

The whole of the western coast of the Idsu peninsula is shelving, and may be safely approached to 2 miles ; islands lie off it, but not beyond the distance of a mile. The current is stronger along the shores than in the middle of the gulf.

SIMODA HARBOUR is near the south-eastern extremity of the peninsula of Idsu, which terminates at the cape of that name in lat. 34° 36′ N., long. 138° 50′ 35″ E. Cape Idsu will be recognized by a conspicuous white cliff a short distance to the north-west of it, and a conical rocky peak a few miles farther to the northward and westward, forming the south-western extreme of the peninsula. To the northward of the harbour a high ridge intersects the peninsula; and south of this, all the way to the cape, it is broken by innumerable peaks of less elevation. To the E.S.E. of the cape, distant half a mile, is a rock about 20 feet above water; and a similar rock lies a third of a mile off Nacane point.*

Vandalia bluff, on the east side of the entrance to the harbour, will be known by a grove of pine trees on the summit of the bluff, and the village of Susaki, which is about a third of the way between it and Cape Diamond. This cape is a rocky islet, lying immediately off a point at 1½ miles eastward of the entrance, and to the northward of it is the bay of Sirahama, which is deep, and, as it has several sand beaches, it may be mistaken for Simoda harbour; but as it is approached, Cape Diamond will shut in the Ucona rocks and Rock island to the southward, while in Simoda road they are visible from all points. The town of Simoda stands on the west shore of the harbour, and Kakisaki village on the west. There is good landing for boats in Simoda creek, and also at the village.

Supplies.—Wood, water, fowls, and eggs, also sweet potatoes and other vegetables, may be procured from the authorities at Simoda. It will be necessary to supply them with casks to bring the water off.

ROCK ISLAND, about 120 feet high and a third of a mile in length, with precipitous shores and an uneven outline, bears E. by S. ¾ S. about 5 miles from Cape Idsu; it has a thick matting of grass, weeds, moss, &c., on its summit. Between this rock and the main land are a number of rocks awash and above water, among which the junks freely pass; but a vessel should not attempt to run inside Rock island unless in case of urgent necessity, particularly as the north-easterly current which sweeps along this coast, seems to be, at this point, capricious both in direction and velocity.

From the summit of this island overfalls were seen, bearing N. ½ W., distant a mile or a mile and a half. These may have been caused by a

---

* See Plan of Simoda Harbour, with views, No. 2,655; scale, $m = 3 \cdot 8$ inches.

rock or reef. An attempt was made to find it, but the strong current and fresh wind prevented a satisfactory examination. The Japanese fishermen, however, deny the existence of any such danger.

UCONA ROCKS, two in number, though they generally appear as one, bear N. by W., distant 2 miles from Rock island; the largest is about 70 feet high. Between these and the island the current was found setting about E.N.E. fully 4 miles an hour.

CENTRE ISLAND lies nearly in the middle of Simoda harbour, and bears N. ½ E. 5½ miles from Rock island, and N. by E. ½ E. 3½ miles from the Ucona rocks. It is high, conical, covered with trees, and a cave passes entirely through it.

Buisaco islet, a quarter of a mile N.N.E. from Centre island, is about 40 feet high, and covered with trees and shrubs.

SOUTHAMPTON and SUPPLY ROCKS.—There are but two hidden dangers in Simoda harbour; the first is Southampton rock, which is in mid-channel S. by E. ¾ E. 2 cables from south point of Centre island, and N. ½ W. from Vandalia bluff, about three-fourths of the way between it and Centre; it is about 25 feet in diameter, has 2 fathoms water on it, and is marked by a *white* spar-buoy. The other is the Supply rock, bearing S. by W. a short distance from Buisaco or Misana islet; it is a sharp rock, with 11 feet water on it, and is marked by a *red* spar-buoy.

Both of these buoys are securely moored, and the authorities of Simoda have promised to replace them should they by any cause be removed. Should the buoy on the Southampton rock be removed, the east end of Centre island in line with the west end of Buisaco will lead to the westward.

Off the village of Susaki, at a third of a mile from the shore, is a ledge of rocks upon which the surf is always breaking; give them a berth of 2 cables in passing.

TIDES.—It is high water, full and change, in Simoda harbour at 5h. 0m.; extreme rise of tide, 5¾ feet; mean rise, 3 feet.

DIRECTIONS.—In navigating the south-eastern coast of Japan, after passing Cape Chichakoff, in Van Diemen strait, if the weather be thick, the vessel's position should be well ascertained before she is hauled to the north-eastward, as the land appears to trend in an E. by N. direction for about 10 miles from the pitch of the cape, instead of N.E., as shown in the present charts. It should also be borne in mind that, as far as our knowledge extends, the current on this coast generally runs to the E.N.E. at the rate of 40 miles a day; it may, however, be entirely checked for twenty-four hours by a north-east wind, when it may be again expected to resume its former course, and possibly run with greater rapidity than usual for one or two days.

Vessels therefore bound to the eastward must allow for this current, and should keep not more than 30 miles off shore, so as to be enabled, if necessary, to verify their reckoning by sighting the land. The positions of the headlands appear to be sufficiently correct for navigation, but until this coast has been more correctly examined, the mariner is warned not to place implicit reliance on any chart, and to be prepared to meet with unknown dangers. In approaching the Gulf of Yedo, the remarkable high mountain of Fuzi-yama (page 334), so different in form from any other land in its vicinity, cannot fail to be of great service in directing vessels either to Simoda or Yedo; Cape Idsu is in line with it when bearing N. ½ E.

Vessels bound to Simoda harbour from the southward and westward should make Cape Idsu, from which Rock island bears E. by S. ¾ S., distant about 5 miles; and if the weather is at all clear, the chain of islands off the Gulf of Yedo will at the same time be plainly visible. Omae saki, the west point of entrance to Tutomi gulf, cannot be mistaken for Cape Idsu, the former being low, with a sandy beach and low sand hills, with occasional patches of trees, extending 30 or 40 miles to the westward; whereas the cape is high and rocky, and its summit generally hidden in the clouds. Rock island being low, unless the weather is clear, will not be seen until long after the cape and Volcano island.

Giving Rock island a berth of a mile, the harbour will be in full view, bearing N. ½ W. distant 5 miles. Standing in from this island, a vessel will probably pass through a number of tide rips, but no soundings will be obtained with the hand lead until near the entrance of the harbour, when the depth will be 14 to 27 fathoms. Should the wind be from the northward and fresh, she should anchor at the mouth of the harbour until it lulls or shifts, or until she can conveniently warp in, as it is usually flawy and always baffling.

Approaching from the northward and eastward, a vessel can pass on either side of Oho sima or Vries island, from the centre of which Cape Diamond bears W. by S. ¼ S., distant about 20 miles. Between Oho sima and Simoda, no dangers are known to exist; but the north-easterly current must be borne constantly in mind, particularly at night and in thick weather. Its general strength is from 2 to 3 knots per hour; but as this, as well as its direction, is much influenced by the local winds, headlands, islands, &c., neither can be relied upon.

Should Oho sima be obscured by thick weather, before reaching Cape Diamond endeavour to sight Rock island, for there are no conspicuous objects on the main land by which a stranger can recognize the harbour at a distance, and the shore appears as one unbroken line. To the westward of the harbour there are several sand beaches, and three or four sand banks; these can be plainly discerned when within 6 or 8 miles, and are good landmarks.

Approaching from the southward and eastward, pass westward of Kozu sima, from which the harbour bears N. by W. ½ W. distant about 26 miles.

In the outer road, or mouth of the harbour, a disagreeable swell is sometimes experienced; but inside the Southampton rock and Centre island vessels are well sheltered, and the water comparatively smooth. Moor with open hawse to the south-west. The bottom throughout is mud.

YEDO BAY.—Cape Sagami, in lat. 35° 6½' N., long. 139° 42¾' E., bears N.E. ¾ E. 44 miles from Cape Diamond, and forms the south-west extreme of the Uraga channel, leading into the bay of Yedo, which is 12 miles wide, 30 miles deep, with excellent holding ground, and capable of sheltering the fleets of the world. The survey of this bay, by the U.S. Naval Expedition to Japan in 1854, embraced the western shore only, from Cape Kamisaki to Beacon point, there being no opportunity of examining the eastern side. The soundings from Treaty point, across in an E.S.E. direction, are regular, and 3 fathoms were found about 1½ miles from the opposite shore. A reconnoissance was made of the western shore only of the Uraga channel.*

Supplies.—At Yoku-hama, on the western shore of Yedo bay, the Japanese authorities supplied the vessels of the squadron with wood, water, a few vegetables, fowls, eggs, oysters, and clams.

DIRECTIONS.—When bound to Yedo bay from the southward, pass westward of the chain of islands lying off the Gulf of Yedo, but beware not to mistake the deep bay of Wodawara or Kawatsu for the entrance of the Uraga channel, for on the north-east side of this bay there is a ledge of rocks extending several miles from the shore, and bearing about N.W. by W., distant 5 miles from Sakura point, and upon which one of the American squadron grounded. A stranger without a correct chart is liable to make this mistake, as the opening of the channel is not seen at a distance from this quarter, the shore appearing as an unbroken line.

The entrance of the Uraga channel bears N.E. ¾ N. distant about 25 miles from Oho sima. Steer in upon this line, and the saddle hill to the northward of Cape Sagami will be readily recognized, as well as the round black knob on the eastern side of the channel. When nearing

---

* *See* Plan of Yedo Bay and Harbour, with views, No. 2,657; scale, m = one inch.

the channel the Plymouth rocks will be plainly seen on its western side; pass half a mile to the eastward of these to clear the Ingersoll patch, a sunken rock with but 6 feet water on it, and the only known danger in the channel. Between these rocks and Cape Kamisaki the ground is clear and the anchorage good, if care be taken to get pretty well in, so as to avoid the strong tides which sweep around the latter with great rapidity. A spit extends a short distance to the southward from this cape, but to the northward of the cape the shore is bold and the water deep.

After rounding Cape Kamisaki, if bound to the city of Yedo, steer N.W. ¼ N. until Perry island bears South, when Webster island will bear West; this will clear the Saratoga spit, which extends well out from the eastern shore. Then haul up N. by E., and run on this course until Treaty point bears S.W. by W., which keep on this bearing by steering N.E. by E. until the beacon on Beacon point bears N.W., when good anchorage will be found in 15 fathoms water. At this point the survey terminated; a clear channel, however, was found with plenty of water for the largest vessels several miles farther northward, and within a few miles of the city.

* "As far as Beacon point the American chart appears to be correct, and the directions given sufficient for navigating the bay. The Saratoga spit, however, is dangerous, and great caution is necessary in passing it. By rounding Cape Kamisaki at not more than a mile distant, and hauling up N.W. when abreast it, or even more westerly, keeping Perry island half a point on the port bow, a vessel will ensure clearing the spit.

From abreast Treaty point, and 2 miles from it, a N.E. course is recommended, instead of a more easterly one, on account of the eastern point of Yedo harbour, which bears N.E. ⅛ E. 9 miles from Beacon house, having also a projecting and very steep spit off it. Beacon point is well marked by this house, which is very distinct; the opposite point has nothing to distinguish it. A bank extends a good 2 miles from the house when it bears North. In approaching the point, therefore, from the southward, do not come within 2½ miles of it, nor stand into less than 10 fathoms, while the beacon on the point is between the bearings of N. by E. and W. by S.

Having passed Beacon house at not within 2½ miles, continue steering N.E. until the house bears W. ½ N., when if that distance off the point the soundings will be 12 fathoms, and no vessel should come within that depth. A N.N.E. course may then be steered, taking care not to shoal the water under 10 fathoms until the house bears S.W., when the Japanese ships and forts will be seen bearing about N.W. The course

---

* Stephen Court, Master of H.M.S. *Furious*, 1858.

will then be N.W. and N.N.W. to the anchorage, the water shoaling so regular and gradual, that a berth may be taken in any convenient depth. The *Furious* anchored in 15 feet, at low water, over a bottom of soft mud, good holding ground, with five well constructed and armed forts immediately in front of the western part of the city bearing from N. ½ W. to N.W., Beacon house S. ½ W., and the peak of Fuzi-zama W. ½ N. (N. 87° 42' W. true), distant about 50 miles."

American Anchorage.—If bound to American anchorage, from Cape Kamisaki, steer N.W., and anchor in 8 or 10 fathoms water with Perry island bearing S.S.E., and Webster island S.W. by S.

Powhattan Bay.—There is good anchorage in this bay in 6 or 7 fathoms water. Near the anchorage there are two snug coves, in which vessels may conveniently repair and refit.

Susquehanna Bay, at 3 miles W.N.W. from Cape Kamisaki, is well sheltered, but it contains a number of reefs and rocks, and is, therefore, not recommended as an anchorage.

Mississippi Bay, at 4 miles to the northward of American anchorage, is well sheltered from the prevailing winds. Upon anchoring it is necessary to give the shore a good berth, to avoid a shoal which extends out from half to three-quarters of a mile. The conspicuous headland or yellow bluff on the north side of this bay is called Treaty point; a shoal surrounds the point from two-thirds of a mile to a mile distant.

Between the American anchorage and Treaty point the soundings are irregular, shoaling suddenly from 12 to 5 fathoms on banks of hard sand.

Yoku-hama Bay is immediately to the northward of Treaty point, and N.N.W. 14 miles from Cape Kamisaki. To reach this anchorage steer for the wooded bluff, which terminates the high land on the north side of the bay, on a N. by W. ½ W. bearing, until Treaty point bears S.W. by S.; this clears the spit off the point; then haul up about N.W. by N. for the bluff over the town of Kanagawa, and anchor in 5½ or 6 fathoms, with the Haycock just open eastward of Mandarin bluff, which is the steep bluff a mile northward of Treaty point.

A flat extends one to two miles from the northern shore of this bay, between Kanagawa and Beacon point. There is also a shoal projecting a mile northward from Mandarin bluff.

TIDES.—It is high water, full and change, in Yoku-hama bay, in the bay of Yedo, at 6h. 0m., and the greatest rise is 6 feet.

The tidal streams run strong in the middle of Yedo bay, and off the tail of the Saratago spit, Perry island, and Cape Kamisaki, their velocity is much increased; but in Yoku-hama bay they are scarcely felt.

DIRECTIONS from YEDO to STRAIT of TSUGAR.—Vessels bound along the eastern coast of Nipon, from the bay of Yedo to the eastern entrance of the strait of Tsugar, may pass within a few miles of Capes Susaki* and Sirofama, after which they will experience the full force of the current setting them to the E.N.E.

Caution is requisite in doubling Cape Blanco, a bold chalky bluff, as the American squadron passed over the edge of a reef in 22 fathoms water S.S.E. from this cape, distant about 5 miles, and from the heavy overfalls, in which fishing boats were anchored, there must be much less water upon its shoalest part. As it was near nightfall it was impossible to examine this reef, but its position is about lat. 35° 8′ N., long. 140° 34′ E., and Cape Blanco in lat. 35° 13′ N., long. 140° 32½′ E.†

From this cape to Tsugar strait no dangers were seen,‡ nor did the squadron approach the coast sufficiently near to test the accuracy of the charts, until arriving off Cape Nambu, the north-east point of Nipon. From the northward and eastward, at the distance of 6 or 8 miles, the outline of this point resembles the back of a sperm whale, with its head to the southward, the Dodo rocks, off the point, forming the flukes. On nearing the entrance of the strait, the water thermometer suddenly falls 15° or 20°, as the vessel runs from the north-easterly current into the cold current setting through the strait. From Cape Nambu a N.W. by W. course made good will lead to Hakodadi head, (page 360).

WEST COAST OF NIPON.§

August 15th.—At 11 p.m. the *Saracen* weighed from Hakodadi, and cleared the strait of Tsugar at noon on the 16th. On the 17th at noon, in lat. 40° 44′ N., long. 139° 14′ E., the left extreme of the land near Cape Gamaley bore E. ¼ N., and the right extreme S.E. by E. ¼ E. ; off shore about 20 miles. Since yesterday the current set the vessel 20 miles to the N.N.E., and a heavy southerly swell.

---

* On the north side of Cape Susaki there is an excellent harbour, named Susaki bay, which affords excellent shelter from north, round easterly, to W.N.W. Mr. McDonald, Commander of the ship *Medina*, 1860. *See* Plan of Susaki Bay, on Chart of Nipon, No. 2,347.

† These positions are doubtful, as the unfavourable weather prevented observations near them. They are probably 6 or 8 miles too far eastward.

‡ The Russian frigate *Arkold* reports the existence of a reef in lat. 36° 15′ N., long. 141° 28′ E.; but this position must be considered as doubtful, as the vessel had no observations for two days previous to seeing the reef. The *Highflyer* passed within a mile of it on an unusually bright moonlight night, and saw nothing.—Edward H. Hills, Master H.M.S. *Highflyer*, 1859.

§ The description of this coast is by John Richards, Master Commanding H.M. surveying vessel *Saracen*, 1855.

BITTERN ROCKS.—At 4 p.m. on the 17th the Bittern rocks bore S. by E. about 3½ miles; altered course to S.W. by S., and brought them abeam at the distance of 1½ miles. This group of three small rocks, two above water and one awash, was discovered by H.M.S. *Bittern*, 8th July 1855. They lie close together, within the space of 2 cables' lengths, steep-to on their western side, having 15 and 17 fathoms at the distance of 2 cables, and no bottom with 140 fathoms at 1½ miles. The face of the rocks were covered with seals, which were with difficulty dislodged, the larger ones barking and snarling in a vicious manner, and carrying off musket balls from within 10 yards' distance without any apparent inconvenience.

The south-western or largest rock, in lat. 40° 31′ N., long. 139° 31′ E., and lying W. by S. about 15 or 17 miles from Cape Gamaley, is about 15 feet high, and in size and appearance resembles the hull of a vessel of about 200 tons. The smaller rock, lying rather more than a cable E.N.E. from the larger one, is about 7 feet high, and has broken water extending from it about a third of a cable to the northward. The third rock, awash, lies to the S.E. of these two, forming nearly an equilateral triangle with them.*

OGA SIMA PENINSULA.—On the 18th, at noon, in lat. 40° 10′ N., long. 139° 6′ E., current East 4 miles, the right extreme of the land of Oga sima bore S.E. by E. From this it would appear that the whole peninsula of Oga sima is placed about 12 miles too far south on Krusenstern's chart. From noon the *Saracen* ran S.W. by S. 10 miles, when Oga sima (centre) was abeam, and bore S.E. by E. 15 or 20 miles.

TABU SIMA.—At daylight on the 19th saw Tabu sima.† At 5h. 10m. a.m. it bore S.E. by E. ¾ E. five or six leagues. At noon, in lat. 39° 13′ N., long. 138° 21′ E., current North 10 miles, a very high peak (the only land in sight) bore E. ¾ S. The position of Tabu sima was fixed from the above bearing at 5h. 10m. a.m., and another taken at 10 a.m. At the latter position the island was very indistinct (it had been covered with mist in the interval). This would place Tabu sima in 39° 31′ N., 138° 53′ E., and give it an elevation of 610 feet. The weather being fine, the vessel pressed on, hoping to fix some of the points of Sado island before sunset, but owing to the great error in its position (it appears on Krusenstern's chart 43 miles too far north), the *Saracen* did not get up to the north point of the island before late, when she rounded to for the night.

---

* *See* Chart of Nipon, Kiusiu, and Sikok, and part of the Coast of Korea, No. 2,347; scale, $d = 2 \cdot 3$ inches; corrected to 1861.

† Owing to hazy weather, Mr. Richards appears to have mistaken the high land on the main for Tabu sima, the position of which has been subsequently found by H.M.S. *Actæon* to be in lat. 39° 12′ N., long. 139° 35′ E.

SADO ISLAND.—At daylight on the 29th bore up for the north point of Sado island, and at 9 a.m. shaped a course for the south-west point, and commenced observations for a track survey of the western side. The island is high, bold, and safe of approach; but no place like an anchorage was seen.

There are two flat rocks about 40 feet high, and one remarkable conical rock, 700 feet high, off the northern point of the island; the Cone rock is in lat. 38° 18′ N., long. 138° 32′ E. The west point of the island, in 38° 01′ N., 138° 17′ E., has a bold and clean shore, and from thence the coast recedes to the south-eastward, and is low; two large gaps in the coast line were observed, probably the entrance to harbours, as some Japanese fishermen, who visited the ship, pointed in that direction, and appeared to wish her to proceed thither. The north point of the island rises somewhat abruptly to an elevation of 3,500 feet, from whence the land, descending slightly, runs in rocky ridges to the south-west for 9 miles, and terminates in a sharp nipple of 4,500 feet elevation. From thence the land descends in a gentle uniform slope until it meets the low lands at the southern part of the island.

In sailing along its western shore numerous villages were seen, and the ground appeared highly cultivated. The fishermen came on board without hesitation when invited, and evinced great curiosity at everything they saw. Their behaviour was respectful, and they received with gratitude the trifling presents made them.

PORT NIEGATA,[*] on the west coast of Nipon, was opened for trade in 1860. The coast in the environs is formed of low sand hills; the entrance from the offing is only to be made out by the masts of junks which are at anchor. In fresh breezes from seaward the sea breaks across the entrance, and at that time not even a boat could take the pass without risk. The best anchorage, at Sado island, is abreast the village of Oda, where a vessel may be anchored so as to be sheltered from all winds during the winter. The shores in the environs of Sawa-umi bay at the south-west part of this island, are well peopled and the land well cultivated. The Japanese inhabiting Sado are tolerably civilized and well disposed to trade with Europeans.

YÚTSI SIMA.—At daylight on the 24th, Cape Noto bore from the *Saracen*, South about 20 miles; steered West for Yútsi sima. At 10 a.m. saw the Astrolabe rock from the mast head, bearing about S.W., and shortly afterwards Yútsi sima about W. $\frac{1}{4}$ N. At noon, in lat. 37° 49′ N., long. 137° 9′ E., current W. by N. $\frac{1}{2}$ N., 17 miles, Yútsi sima was found

---

[*] Renseignements Hydrographiques sur les Iles du Japon, 2nd edition, 1860, page 191.

to be in lat. 37° 50½' N., 136° 55' E.; Astrolabe rock in 37° 35' N. 136° 54' E.; Cape Noto in 37° 28' N., 137° 22' E.

Yútsi sima is 40 feet high, about two-thirds of a mile in diameter, level, and cultivated; there are a few stunted trees on it, and a small village on its southern side. With the island bearing N.N.W. 5 miles, the depth was 46 fathoms, fine sand, and regular soundings both on approaching and leaving it.

The Astrolabe rock, 200 feet high, and about a quarter or a third of a mile in diameter, is the largest and highest of a rocky group of five, which extend nearly 1½ miles in a N.E. and S.W. direction, and vary from 200 to 70 feet in height.

CAPE NOTO is elevated about 700 or 800 feet, the land rising to the westward of it to 1,200 feet. At the distance of 12 miles W. by S. from the cape there is a remarkable white cliff that shows well to the westward; from this cliff the coast bends in to the southward and forms a bay. The opposite point of the bay is about 8 miles distant, and immediately above it is a sharp peak elevated 2,000 feet, which is the highest point in the neighbourhood, the land being generally level and of an uniform height of about 600 or 700 feet. At 10 miles W. by S. from this point, or 30 miles W. by S. from Cape Noto, the coast trends away sharp to the southward. None of those rocks and islands were seen which so thickly stud the coast in Krusenstern's chart, although, as the *Saracen* passed at the distance of 19 miles, it is possible that small rocks may exist close in shore.

OKI ISLANDS.—On the 25th, at 6h. 15m. p.m., observed the Oki islands bearing S.W. about 20 miles. At 7h. 15m. p.m. the land bore S.W. about 15 miles, which would make it agree very nearly with the reckoning.

At daylight on the 26th, squally with rain; weather very threatening. Saw the Oki group, consisting of two large islands and a number of small islets and rocks; they lie N.E. and S.W. of each other, and occupy an extent of 28 miles in that direction. There is an open and apparently safe channel 4 miles wide between the two large islands.

The north-east or largest island, about 9 miles in diameter, has a number of detached high rocks close to its northern point, which is steep and cliffy. The south-west extreme of this island is remarkable from its terminating in a high steep bluff; the south-east point is comparatively low. The highest point of the island was estimated at 3,000 feet elevation, but no indication of a port, nor even the prospect of tolerable anchorage, was observed along its western shore.

The south-east island is about 5 or 6 miles in diameter, and being deeply indented by bays, it is probable good anchorage may be found. The north point of this island rises to a sharp peak of 1,700 feet elevation; all the rest of the island is about one-half that height. It appears to be thickly populated, and the hills are cultivated to their very summits. There are three or four small islands lying eastward of it, separated by channels apparently not exceeding half a mile wide, but they appear rocky and precipitous.

A track survey was made of the west side of this group, and the sun peeping out for a minute enabled us to get the time, but bad weather setting in soon after prevented our completing what was begun, and for the latitude of the north point we are indebted to Mr. Collingwood, who passed it near noon on two several occasions in the hired steamer *Tartar*. Taking this latitude 36° 30' N., our chronometers place the north point of the north-east island in long. 133° 23' E. At sunset the *Saracen* was scudding before a fresh gale from N.N.E. with sharp squalls from N.W. and much rain.

MINO SIMA.—At daylight on the 27th, saw Mino sima bearing S.S.E. about 15 miles. The centre of this island is in lat. 34° 48' N., long. 131° 9' E., and rises to an elevation of 484 feet. The island is nearly 2 miles in diameter, with cliffy sides, except to the north-east, where there appeared to be a sandy bay. It appeared steep-to, and quite safe of approach on all sides. The south-east point has a large square rock lying off it at the distance of half a cable. With the island bearing South about 12 miles, sounded in 45 fathoms, sand and coral, and from this position an even bottom was carried towards the land with very little decrease in the soundings.

CAPE LOUISA.—At noon, in lat. 34° 44' N., long. 131° 22' E., (Mino sima was found to be placed on Krusenstern's chart 28 miles N.E. b. E. of its true position,) a remarkable sharp peak on the eastern extreme of the mainland bore E.S.E. about 9 miles; this was named Cape Louisa. The western extreme bore S.W. by W. about 25 miles, and appeared about 700 feet high, tapering towards the sea, where it is about 500 feet high; it may be easily known by a remarkable Square rock or hill on the point, about 200 feet high, which being entirely detached from the high land at the point, gives it the appearance of an island. At 7 or 8 miles E. by N. from this point there appeared to be a fine port; the eastern point of its entrance is marked by a sharp cone of about 700 feet elevation. Between Square rock point and Cape Louisa there is a deep bay, with many islands in it, generally elevated about 200 feet, flat-topped, with cliffy inaccessible sides. The outer island, which is about midway between Cape Louisa and Square rock point, and just within the chord of the bay, was named Richards island.

Cape Louisa is in 34° 40′ N., 131° 36′ E., and the sharp peak immediately above it is elevated 1,800 feet. Richards island is in 34° 32′ N., 131° 18′ E., and elevated 400 feet. Square rock point is in 34° 27′ N., 130° 59′ E.

COLNET and OBREE ISLANDS.—At daylight on the 29th, Colnet island bore N.N.W. ¼ W. about 14 miles. A new island bearing W.S.W. 12 miles was named Obree island; it is half a mile in diameter, and elevated 500 feet, and is in lat. 33° 51′ N., long 130° 02′ E. The north point of Kiusiu island was not seen, but there was a mass of islands in that direction, the outer one of which was named Wilson, and the point inside, Winchester point.

WILSON ISLAND is about 1½ miles in diameter, and remarkable from its prominent position as the outer and northern island of the Kiusiu group, as well as from its sharp peak, in lat. 33° 54′ N., long. 130° 25′ E., forming a conspicuous landmark visible nearly 30 miles. At 10h. 30m. p.m. the *Saracen* anchored with kedge in 22 fathoms, Cone rock bearing W. by N. and Obree island N. by W.; but at daylight a reef of rocks was observed about 2 cables to the southward of her. The western stream or flood made here at 5h. 30m., which, supposing a regular tide, would give the turn of stream, on full and change days, at about 3h. 0m.; no tidal stream was found at a greater distance than 5 miles from the land. After weighing, a number of Japanese government boats hove in sight from between the islands to the eastward, and after passing disappeared among the islands to the westward; they were supposed to be going to Nagasaki, through the Firando channel.

At noon on the 30th, in 33° 40′ N., and 130° 2′ E., Cone rock bore N. 60° W. 1½ miles. Wind light, with frequent calms. At 8 p.m. drifted to the westward within 1½ miles of the north point of an island, and anchored when the western stream stopped, which took place about 10h. 20m. p.m. In the morning weighed and made sail, working to the westward against light airs and frequent calms.

Not a single point on Krusenstern's chart could be recognized, and the *Saracen* was surrounded by innumerable islands in every variety of form and size; but as they are generally safe of approach, and the adjacent seas seem singularly free from danger, they will offer no impediment to navigation when properly surveyed. At present no vessel should attempt to pass inside Yki island and Wilson island in the night, or during thick weather, although during the light airs and calms which frequently prevail in the Japan sea in the month of August, a well-manned ship proceeding to the southward would benefit considerably by keeping in shore, out of the north-easterly current, and taking advantage of the tides by dropping a kedge when unable to make way. The soundings obtained along the coast showed a remarkably even bottom; generally about 50 fathoms at 10 miles from the outer islands, decreasing to 20 fathoms close to and along them.

Firando island was expected to have been sighted, and it was then intended to have passed between it and Kiusiu, but not seeing it, the *Saracen* still coasted along, carefully tracing the outline of the outer islands and narrowly examining every opening. On rounding the north-east point of a large island (Harbour island) bearing S.W. by S. 13 miles from Yki, a fine bay was observed about three-quarters of a mile wide, which appeared a promising anchorage. After rounding the north part of this island the *Saracen* hauled in for a bay to the S.E. distant about 7 miles, and anchored at midnight in 32 fathoms.

YKITSK ISLAND.—On the 1st September, at daylight, the *Saracen* weighed and worked up between two islands for shelter, and the weather having set in thick and stormy from the southward, she anchored inside the outer island in a snug bay off a Japanese village about a quarter of a mile from the shore in 7 fathoms. Soon after anchoring she was boarded by a Japanese boat, containing several respectably dressed officials, who seemed by their manner desirous of knowing what was wanted. After comprehending the purport of her visit, and satisfying their curiosity by a minute inspection of the ship, and taking numerous notes, they departed.

Early the next morning an official visit was paid by the authorities. They readily pointed out the direction of Nagasaki but they could not be prevailed on to give a pilot. The island the *Saracen* anchored under they called Ykitsk, and the large one to the eastward Firado (Firando they did not appear to know). They stated at first, that there was a channel east of Firado, but afterwards when they saw the vessel proceeding in that direction, they came off in haste, and intimated by signs that there was not sufficient depth of water in it. Whilst thus engaged on board, Mr. Obree (my assistant) landed on the western part of Firado, at the narrowest part of the channel between it and Ykitsk, and was fortunate enough to secure equal altitudes for time, as well as the sun at noon; he also sounded the channel across, and ascertained that it was quite clear, and had a depth of 14 fathoms in the middle.

A breeze springing up just as the boat returned the *Saracen* weighed, and ran for the north point of Firado island, with the intention of passing through, or proving the impracticability of the channel to the eastward of that island; but after proceeding about 4 miles in that direction, she was becalmed, and came to with stream anchor in 30 fathoms off the entrance to an apparent harbour.

HARBOUR ISLAND.—On the 3rd, at 8 a.m., with a light breeze from the northward, and occasional rain squalls, weighed, and worked to windward against a freshening breeze from the north, and passed between the two outer islands north of Firado. The outer and largest island is about 5 miles long in a N.N.E. and S.S.W. direction, and 2 miles broad, and has a fine bay or harbour, Port Lindsey, on its south side, as well as the deep bay before noticed on the north-west; it was therefore named Harbour island. After passing between the islands, bore up for the channel (Spek strait)* eastward of Firado, but on nearing it, it appearing so narrow and unpromising, hauled to the wind again, and passing inside all the islands west of Firado, rounded its south point at 4 p.m., and shaped a course for Nagasaki.

None of the rocks or islands laid down on this line in Krusenstern's chart could be recognized, but a broad deep channel was found to Nagasaki, and not more islands than just sufficient to make good landmarks in cloudy weather. At midnight passed between the two outer islands of a group marked in the chart as 22 miles N.N.W. of Nagasaki (but which the reckoning placed only 12 miles N.W.), and fearing to run past the place, anchored for the night in 32 fathoms. Weighed at daylight and proceeded towards Nagasaki with a light air from the northward, and anchored at noon in the outer harbour.

The *Saracen* remained in this port until the 1st October, when she weighed and proceeded towards Cape Gotto for observations, but finding a fresh gale blowing outside, hauled up for Kabe sima, and entering the third channel of the old Dutch chart, worked into a snug harbour on its eastern side, and anchored within a cable's length of a small islet admirably adapted for an observation spot.

FIRADO ISLAND, lying off the west side of Kiusiu, was visited in

---

* Lieutenant H.O. Wichers, of the Dutch Royal Navy, in the year 1857–59 examined the south and west shores of Kiusiu, and corrected the coast line of Siebold's chart, which he found 4¾ miles too far West. He also verified Richard's positions in this locality, in 1855, and states that the channel (Spek strait) on the east side of Firado, is navigable. *See* Chart of Nipon, Kiusiu, and Sikok, and part of the coast of Korea, No. 2,437, corrected to 1861.

1859 by H.M.S. *Roebuck*, on her passage from Hong Kong to Japan, and a spacious harbour found on its north-west side.

The harbour runs east and west, is 2½ or 3 miles deep, and sheltered from all winds. A good guide to recognize its position is a small island lying off the north point of entrance. There are three islets in the harbour, and good anchorage in 4 fathoms was found between the southern islet and the shore. There is a deep bay running in a southerly direction, inside the southern point of entrance, but there was no time for its examination.

MEAC-SIMA GROUP.—The relative positions of the Meac-sima or Asses Ears group, and Pallas rocks, not having been accurately determined, the *Saracen* steered for the first-named group, but the wind being light, did not come up with them until the evening of the following day, when she lay to for the night, and kept her position pretty well by the lead, finding bottom in from 80 to 85 fathoms.

The Meac-sima group are two small islands, extending nearly 4 miles in a N.E. and S.W. direction, but not more than a mile broad.

Taka sima, the north-east island, is nearly 1½ miles long, and three-quarters of a mile wide. It is elevated 610 feet above the sea, and nearly level at the top, with cliffy precipitous sides, excepting to the southward, where there appeared to be some shelter for fishing junks, as several were observed at anchor; in clear weather it may be seen upwards of 30 miles.

Kusa-saki, the south-west island, is less than half the size of the north-east island, but 100 feet higher and very craggy; its remarkable peak, which is in lat. 32° 2′ 47″ N., long. 128° 30′ 42″ E., probably suggested the name of the group to its first discoverers.

Me sima and Wo sima, the intermediate small islands and rocks, are high and cliffy, the latter partaking generally of the sugar-loaf form.

The only outlying rocks noticed extend South about a third of a mile from the south-west island, and may be almost considered part of the main group.

The approach * to these islands from the northward is quite clear. Between the Asses Ears and Pallas rocks the ground is pretty even, and the general depth is about 81 fathoms.

---

* The *Furious* passed about 3 miles from these islands; but until the ground in their vicinity has been more minutely examined they should be approached with caution, as the sea was seen breaking heavily nearly a mile from the shore.—Stephen Court, Master, H.M.S. *Furious*, 1858.

PALLAS ROCKS are three in number, two of which lie close together, and one N.E. 1½ cables from the largest, which is the south-western of the group.

The largest rock does not exceed a third of a cable in diameter, and is about 60 feet high; the other two are about one half that elevation. They are steep-to, and soundings were obtained at the distance of a mile South from them, in 95 fathoms, sand and shells. The largest rock is in lat. 32° 14′ 17″ N., long. 128° 13′ 30″ E.

UDSI SIMA, Parker, or Roche Poncié islands,* are four in number, the largest of which is about 2 miles in circumference, 1,049 feet above the sea, and in lat. 31° 12′ N., long. 129° 23′ E. Two and a half and one mile respectively to the south-west of the larger island are two small islets; and to the eastward of the peak of the large island, about a mile, is the fourth islet.†

RETRIBUTION ROCKS, three in number, were discovered by H.M.S. *Retribution*,‡ 6th August 1858. They lie north and south of each other, about 2 cables apart, in lat. 31° 23′ N., long. 129° 37½′ E. The two southernmost rocks rise in a needle form about 60 feet above the sea; the northern rock is flat, and only 10 feet high. The vessel passed about a quarter of a mile to the northward of this group, which appeared to be steep-to; no discoloured water was seen, nor bottom obtained with the hand lead.

KOSIKI ISLANDS (Meac sima of the French chart), consisting of two large and several small islands, extend in a N.N.E. and S.S.W. direction from lat. 31° 35′ to 31° 52′ N., and from long. 129° 36′ to 129° 51′ E.; they are very little known, neither are the Nadiejda rocks, which are supposed to lie in lat. 31° 48′ N., long. 129° 36′ E.

TSUKURASE ISLANDS.—Off the south end of the Kosiki group, at the distance of 8 miles, are the Tsukurase or Symplegades islands, three in number, also very imperfectly known. H.M.S. *Highflyer* passed about 2 cables to the eastward of them in September 1859, and they are described as three islets forming a triangle, the sides of which are each

---

* Thus named (and deservedly so, if an European name is admissible), after the Ingénieur-hydrographe of the French Expedition under Admiral Cecille in 1846.

† The Udsi sima are high, and of considerable extent. The easternmost island appeared larger and its position to be farther north than marked in the chart; no foul ground was visible near them. Stephen Court, Master of H.M.S. *Furious*, 1858.

‡ These rocks were subsequently seen by the United States steamer *Mississippi*, and named the Mississippi rocks. Captain G. S. Hand, H.M.S. *Sampson*, 1859.

not more than a cable in extent; a rock awash was observed lying a quarter of a mile to the southward of the group.

NAGASAKI HARBOUR is formed at the head of a bay running in a north-east direction on a peninsula at the western extremity of Kiusiu island. Cape Nomo, the southern point of the bay, is the extreme of a promontory, which consists of a hill with a split or double summit, and at a distance has the appearance of an island; it may also be recognized by the islets in its vicinity, and in clear weather can scarcely be mistaken when within the distance of 6 or 7 miles. The city of Nagasaki covers a plain at the head of the harbour, but it has overgrown its area, the houses cluster up the spurs of the hills, and the streets are, in places, so steep as to render steps necessary. The population, in 1858, amounted to about 70,000.*

The harbour of Nagasaki may be divided into three parts, for it contains three distinct anchorages; but the outer ones cannot be recommended at all seasons, the water being too deep, and the swell too heavy to render them safe. The first or outer road is westward of Papenberg island; the second or middle is to the eastward; and the third is at the head of the harbour in front of the city.

The entrance to the outer road is between the north end of Iwo sima and Fakuda saki. The anchorage is in from 22 to 25 fathoms, over a bottom of thick green ooze, with fine sand, good holding ground. A vessel will lie here sheltered, except from the N.W.; but as the wind blows seldom from this direction during the North-east monsoon, and never very strong, it is quite safe at this season.

The only known danger in this road is the Barracouta rock, lying off the southern side of Kamino sima; but the ship *Templeman*, 23rd July 1860, is reported to have struck on a rock, from which the bluff south end of Papenberg bore E. by S., and Tree rock N.E. The vessel was leaving the harbour, the wind was light, and she was passing a rock which was thought to be the Barracouta.

The anchorage in the middle road, eastward of Papenberg, is in about 17 fathoms, more towards Papenberg than the eastern shore. To the N.N.E. of Papenberg, and distant a third of a mile, is a small flat wooded island, named Nezumi; and about the same distance farther in the same direction is the small bay of Kibats, in which there are 6 to 10 fathoms water. This, in all the harbour of Nagasaki, is said to be the best place to refit a ship, for in the inner road the shore is everywhere so muddy that no ship can approach it.

---

* *See* Plan of Nagasaki Bay, No. 2,415; scale, $m = 1$ inch.

From the middle to the inner road, abreast the city, the direction of the harbour is N.E. ½ N., and the distance 2⅓ miles, the depth decreasing gradually from 18 to 5 fathoms. The harbour is not more than half a mile wide, and in some places only 3 cables. The anchorage near the city, which stands on the eastern shore, is in about 6 fathoms over a bottom of thin clay.

Supplies.—Coal is abundant at Nagasaki, and will be more so when the mines, which are reported to be numerous, are worked. Water and wood, and all supplies, except beef and mutton, are plentiful, cheap, and easily obtained.

There is a native steam factory, at which small repairs to steamers may be effected.

TIDES.—The time of high water, full and change, in this harbour is 6h. 28m., and the greatest rise 6¼ feet.

DIRECTIONS.*—When leaving the Yang-tse kiang for the Japan islands, if bound to Nagasaki, a direct course may at once be steered for Meac sima, or Asses Ears group, the highest island of which is visible in clear weather at about 37 miles.† With Meac sima bearing South from 3 to 6 miles, a N.E. by E. course for 62 miles will place a vessel off the Mitsuse rocks, which lie 3 miles to the north-west of Cape Nomo. The Mitsuse are in line with this cape when bearing S.E. by S.; and two rocks lie off it, one S.E. by S. half a mile, and the other S.W. by W. not quite that distance; Kaba sima will also be seen just shutting in with the cape when bearing S.E. by E.

From the Mitsuse steer for the north-west extreme of Iwo sima, passing about half a mile to the north-west of two rocky patches named Kutsnose and Sotonohirase, which are always visible, being just awash at high water. After rounding Iwo at any convenient distance, as no apparent danger extends off it, steer about E. ¾ S. for the entrance of the channel leading to Nagasaki, midway between Papenberg island and the Hungry rock, passing northward of the Hirase Rock, which, like the Hungry, just covers at high tide.

When the channel leading to the city comes fairly open haul up to the north-east for it, keeping the western shore aboard while the Dutch consul's flag-staff bears between E.N.E. and E. by S., to avoid the Desima shoal, which extends about the third of the way over from the eastern shore, and narrows the channel considerably. Vessels can anchor in mid-channel either above this shoal, with the flag-staff to the southward of

---

* Stephen Court, Master of H.M.S. *Furious*, 1858.

† Captain Fleetwood Pellew, R.N., recommends that a sailing vessel should make the land in about lat. 32° 48′ N., as N.E. winds blow here the greatest part of the year.

E. ⅔ S., or below it, with the flag-staff to the northward of E.N.E. The outer anchorages cannot be recommended, the water being too deep and the swell too heavy to render them safe. A rocky ledge, which uncovers at low water, extends about half a cable's length off the western shore and is marked by a pole beacon, from which the above flag-staff bears about East.

TSUGAR ISLAND.

The Strait of Tsugar, separating Nipon from Yezo island, is about 40 miles in length in an E.N.E. and W.S.W. direction, 9½ miles wide at its narrowest part. The following description is by John Richards, Master Commanding H.M.S. *Saracen*, who surveyed the strait in 1855 :*

CAPE GAMALEY.—In approaching the western entrance of Tsugar strait from the south-west, the Bittern rocks will be seen lying W. by S. about 16 miles from Cape Gamaley, page 349. The land about this cape is moderately elevated and level. The coast between it and Oho saki, to the northward, is low and sandy. Sasagota bay (of Krusenstern's chart), at 6 miles to the southward of Oho saki, appears to be nothing more than a large shallow lagoon; its entrance is narrow and barred right across, with only sufficient depth to admit junks at high water. Between the bay and Oho saki the coast is safe of approach, having regular soundings, and fair anchorage in N.E. winds.

OHO SAKI, or Cape Greig, is remarkable from its peculiar form, and being the commencement of the high land extending to Tatsupi saki, which bears from it N.N.W. ¾ W., 8½ miles. The bay between these points, although containing much foul ground, may be useful to a vessel not able to get through the strait during an easterly gale. The best anchorage is in 12 fathoms, and three-quarters of a mile off shore, at one-third the distance from Oho saki to Tatsupi saki. The bottom of the bay is very foul.

TATSUPI SAKI, or Cape Tsugar, the south point of the eastern entrance to Tsugar strait, is a bluff, 362 feet high, from whence the land rises to the height of 2,200 feet, at the distance of 4 miles inland. A large rock, 300 feet high, lies 2 cables to the N.E. of the cape, and is connected to it by a low neck of sand and stones. On a N.W. and S.E. bearing this rock makes like an island. The cape is steep-to, but the strong eddies near it make it prudent not to approach it nearer than a mile.

---

* See Chart of Strait of Tsugar, No. 2,441 ; scale, $m = 0\cdot3$ of an inch.

TSUGAR STRAIT.

Gun cliff, at 9¼ miles E. by S. ¼ S. from Tatsupi saki, is steep-to, and has a battery of six guns on its apex, which is 200 feet high. There is a remarkable black rocky cliff three-quarters of a mile to the westward.

In the bay between these points, off the town of Memoyah, about half a mile from the shore in 8 fathoms, there is capital anchorage, indeed the best in the strait next to Hakodadi. A little to the southward of the town there is a fine stream of good water, which may easily be obtained. Wood is also abundant.

After passing Gun cliff the shore is less steep, and bottom will be found in 30 to 40 fathoms right across to the opposite coast of the peninsula of Nambu. From the south point of this coast a steep cliffy shore, with deep water close to, trends to the northward, nearly in a straight line to Toriwi saki. The cliffs are coloured with the most brilliant and varied tints, and, like the entire coasts of the strait, are of basaltic formation. Among the most remarkable are the Red cliffs, rising to the height of 1,600 feet, at 17 miles to the southward of Toriwi saki. At 9 miles farther to the northward are two remarkable pointed cliffs, named Double head. Nearly 2 miles to the S.W. of this head is a rock 42 feet high; and North about 3 cables from this is a rock awash at low water.

TORIWI SAKI is a low tapering point, off which, at the distance of a cable, is Low islet or Omaski sima, 40 feet high. The ground all around this cape and islet is very foul, except to the N.E., where a vessel may anchor to wait tide in 13 fathoms, with the centre of the islet bearing S.W. by S. distant about a mile. This is a useful anchorage for vessels approaching Hakodadi from the eastward, particularly during the light south-westerly winds, common to the strait during the summer months. There is a tide race, near the full and change of the moon, 3 miles North of Low islet, and heavy overfalls with a north-east swell. On such occasions care must be taken to avoid this locality. There is a clear channel between the race and the islet.

SIRIYA SAKI.—From Low islet the coast to the eastward is foul for about 3 miles, after which it may be approached without fear. At 10½ miles from Low islet is a remarkable red cliff, which shows well to the westward; and at 2 miles to the westward of this cliff there is a high sharp bluff, and a high round bluff 2 miles to the eastward.

From the latter bluff the coast is low to within 4 miles of Siriya saki, or Cape Nambu, where it rises to 1,265 feet, and descends again towards the cape in a gentle slope, making at a distance like an island. There is good anchorage in the deep bay formed between this cape and the red cliff, but the best is on its western side, abreast the coast line where the high and low land meet, in 15 fathoms, with the above round bluff bearing W.N.W. 2 miles. Off the cape, at a distance of 3 cables, is a small white rock 70 feet high. There is also another rock, rather larger, lying a cable off shore, at 2 miles to the south-west of the cape. The coast within 4 miles of Cape Nambu is studded with rocks, and altogether foul.

CAPE YESAN, or Esamu, the north point of the eastern entrance to Tsugar strait, is the east extreme of a bold promontory, with several remarkable dome-shaped mountains in the rear. The cape itself is a steep cliff about 600 feet high; the volcano immediately above it is 1,935 feet high, and frequently capped with a light cloud of steam, but not otherwise active. The west side of this mountain is covered with patches of sulphur, having the appearance of snow at a distance. There is anchorage in the bay about 2 miles westward of the cape; a kedge or stream anchor is recommended to be used when unable to make way against the tide.

At $8\frac{3}{4}$ miles S.W. by W. of Cape Yesan is Conical islet, 200 feet high, lying close to the coast, which in its immmediate neighbourhood is high and cliffy, and the approach steep and safe: there is, however, a dangerous low point one mile to the westward; and at $2\frac{3}{4}$ miles to the eastward is Foul point, which is low, and has a dangerous reef extending 2 cables from it.

CAPE SIWOKUBI, or Cape Blunt, bears W. $\frac{2}{3}$ S. $2\frac{1}{2}$ miles from Conical islet, E. by S. 12 miles from Hakodadi head, and N. by E. $\frac{3}{4}$ E. $9\frac{1}{2}$ miles from Low islet, which is the narrowest part of the strait. This cape is steep-to, and the north-east current frequently runs with greater strength close to the rocks than out in the stream. The summit of the bluff immediately above the cape is 1,022 feet high; from thence the high land ranges in towards the Saddle mountain. The coast for about 7 miles to the westward is a level plain of an average elevation of 200 or 300 feet; beyond 7 miles, it descends to the low beach connecting the high land of Hakodadi head with the main.

HAKODADI HEAD is the south extreme of a bold peaked promontory, 1,136 feet high, standing well out from the high land of the main, with which it is connected by a low sandy isthmus. The head is steep and precipitous, and safe of approach. At $4\frac{2}{3}$ miles west of the head is Mussell point, off which a reef extends 2 cables and is steep-to. The coast from thence to Cape Saraki, at $4\frac{1}{4}$ miles to the S.W., is level but fringed with rocks, and requires caution in approaching. To the westward of the cape the shore is low, with a sandy beach safe of approach, and clean ground for anchorage to within 3 miles of Cape Tsiuka.

HAKODADI HARBOUR.—Hakodadi bay, on the northern side of Tsugar strait, is 4 miles wide and 5 miles deep, and for accessibility and safety is one of the finest in the world. Its entrance is betweeen Hakodadi head and Mussell point, which bear East and West of each other, distant 4¾ miles. The harbour is in the south-eastern arm of the bay, and is completely sheltered, with regular soundings and excellent holding ground. The town of Hakodadi stands on the north-east slope of the promontory, facing the harbour, and in 1854 contained about 6,000 inhabitants.*

Supplies.—Excellent wood and water may be procured from the authorities at Hakodadi; or, if preferred, water can be easily obtained from Kamida creek, which enters the harbour to the northward and eastward of the town.

The season, at the time of the visit of the American squadron to this harbour, in 1854, was unfavourable for procuring supplies; a few sweet and Irish potatoes, eggs, and fowls, however, were obtained, and these articles, at a more favourable period of the year, will no doubt be furnished in sufficient quantity to supply any vessel that may in future visit the port.

The seine supplied the squadron with fine salmon and a quantity of other fish, and the shores of the bay abound with excellent shell-fish.

TIDES.—It is high water, full and change, in this harbour at 5h. 0m., and the extreme rise and fall of tide is 3 feet.

DIRECTIONS.—In entering Hakodadi harbour, after rounding Hakodadi head, and giving it a berth of a mile to avoid the calms under the high land, steer for the sharp peak of Komaga-daki, bearing about North, until the eastern peak of the Saddle mountain, bearing about N.E. b. N., opens to the westward of the round knob on the side of the mountain, then haul up to the northward and eastward, keeping them well open until the centre of the sand hills on the isthmus bears S.E. b. E. ¾ E. (these may be recognized by the dark knolls upon them). This will clear a spit which runs in a N.N.W. direction two-thirds of a mile from the north-western point of the town;† then bring the sand hills a

---

* See Plan of Hakodadi Harbour, No. 2,672; scale, $m = 2$ inches.

† The north-east end of the fir trees in line with the Joss house clears the eastern side in 5 fathoms water; and the foot of the hill in line with the middle of the sandy point leads along the northern side of the shoal.—Captain K. Stewart, H.M.S. *Nankin*, 1855.

If these leading marks should be in the clouds, as they generally are, keep the western extreme of the promontory of Hakodadi bearing South or S. ½ W., which will clear the spit, and haul to the eastward when the centre of the sand hills on the isthmus bears S.E. b. E. ¾ E.—Commodore the Hon. C. Elliot, H.M.S. *Sybille*, 1855.

point on the port-bow, and stand in until the north-western point of the town bears S.W. ½ W., when the vessel will be in the best berth, in 5½ or 6 fathoms water.

If it is desirable to get nearer in, haul up a little eastward of South, for the low rocky peak which will be just visible over the sloping ridge to the southward and eastward of the town. A vessel of moderate draught may approach within a quarter of a mile of Tsuki point, where there is a building-yard for junks. This portion of the harbour, however, is generally crowded with vessels of this description; and unless the want of repairs or some other cause renders a close berth necessary, it is better to remain outside.

If the peak or Saddle is obscured by clouds or fog, after doubling the promontory, steer N.N.E., until the sand hills are brought upon the bearing above given, when proceed as before directed.

A short distance from the tail of the spit is a detached sand-bank, with 3½ fathoms on it. The outer edge of this is marked by a *white* spar-buoy. Between this and the spit there is a narrow channel with 4½ fathoms water. Vessels of moderate draught may pass on either side of the buoy, but it will be prudent to go to the northward of it.

Should the wind fall before reaching the harbour, there is good anchorage in the bay, in 25 to 10 fathoms water.

CAPE TSIUKA, at 11 miles S.S.W. of Cape Saraki, is a high cliffy point, and may be further known by three rocks which extend a quarter of a mile from a point one mile eastward of it; the outer rock of the three is of a conical form and 70 feet high. The land to the westward for 4 miles is high and cliffy; about half way between the cape and the end of the cliffs there are two waterfalls.

Vessels can anchor in the bight of the bay, between Capes Tsiuka and Sirakami; but as a southerly wind on the western tide sends in a cross swell, it would not be prudent to anchor far in. The best position is in 15 to 20 fathoms, with the southern white cliff bearing West about a mile.

CAPE SIRAKAMI, or Nadiejda, the north point of western entrance to Tsugar strait, is a high bluff similar to Cape Siwokubi, but not so safe of approach. The coast, for more than a mile on each side of the cape, is bordered with numerous rocks, generally above water, some of which run off nearly 2 cables. As it is not known whether the dangers extend under water beyond this distance, it will be prudent to give the cape a good berth in passing.

From Cape Sirakami the coast trends W. by N. ¼ N. 5 miles to Cape Matsumae, which is low, and off it is a conical islet with a small temple or building on it. The bay between is very rocky, excepting off the east end of the city of Matsumae, where there is good anchorage in 12 fathoms at half a mile off shore; but this anchorage would, of course, be unsafe in southerly winds. H.M.S. *Sybille* in 1855 coasted along this bay at about 2 miles off shore, and had irregular soundings, 16 to 25 fathoms, shoaling suddenly when near the city from 13 to 5 and 7 fathoms.

CURRENTS and TIDES.—During the survey of Tsugar strait by the *Saracen*, in June, July, and August 1855, a constant N.E. current set through the middle of the strait, the breadth of which varied considerably according to the state of the wind and weather. Before and during a N.E. wind its strength was much diminished; but with the wind from the opposite direction, it would expand and fill up two-thirds of the channel against the strength of the western tide.

The tide in the stream runs about 12 hours each way near the full and change of the moon, and there are only two regular tides by the shore in 24 hours. At full and change, the flood or eastern stream makes at Tatsupi saki at 6h. 30m. a.m., at 7h. 0m. at Cape Tsiuka, and at 7h. 30m. at Toriwi saki. The western stream begins about 12 hours later. The turn of the stream takes place $1\frac{1}{2}$ hours later every day.

KO SIMA, which lies W. $\frac{1}{2}$ S. $16\frac{1}{2}$ miles from Cape Sirakami, the north-west point of entrance to the strait of Tsugar, has a high round peak, 974 feet high, and there are two remarkable sugar-loaf islets or rocks lying close off its west end.*

U SIMA, lying about 24 miles to the W.N.W. of Ko sima, is higher and longer.

OKOSIRI ISLAND lies off the south-west coast of Yezo, about 30 miles N. by E. of U sima. On two occasions H.M.S. *Sybille* passed through the channel between Okosiri and Yezo. The south extreme of Okosiri is low, and detached rocks lie about 2 miles south of it. Some of these rocks are 10 to 15 feet above water, and apparently a reef connects them with the island. There may be anchorage off the south point, but the eastern side of the island is steep-to, and no bottom could be obtained when sailing through the channel, which is about 10 miles wide at its northern entrance. The north-east point of the island appeared from a distance to have a rocky ledge running out a short distance from it. On the Yezo side of the channel the land is high, and the coast apparently bold-to.

---

* Captain K. Stewart, R.N., 1855.

DIRECTIONS.— Sailing vessels approaching Tsugar strait from the westward during foggy weather should guard against being carried by the current to the northward past the entrance. Should the weather be clear when nearing Cape Gamaley, it may be as well to sight it; but if doubtful, shape a course (allowing for the probable current) direct for Cape Greig. Should a fog come on suddenly when nearing this cape, recollect that the coast is clear and sandy, and the soundings are regular to the southward, but rocky with irregular soundings to the northward of it. The cape is steep-to, and standing out prominently from the coast line, forms a good landmark.

No particular directions are required in passing through this strait to the eastward, as there are no hidden dangers, and the north-easterly current will always be found strongest in the middle of the stream. After passing Cape Tsugar, if the weather is thick, and the vessel bound to Hakodadi, endeavour to make Cape Tsiuka and proceed from thence to Mussell point; or giving Cape Tsiuka a berth, feel the way up into the bay, between it and Cape Saraki, by the lead, and anchor till the weather clears.

Approaching the strait from the eastward, steer for Cape Nambu, and endeavour to make it on a N.W. bearing. Pass the cape at about a mile distant, then haul in to avoid the current and to anchor should it fall calm. In this case, by keeping this shore close aboard, the vessel may probably be drifted up to Low islet, off Toriwi saki, by the western stream, when the north-east current is running like a mill stream in mid-channel.

At the anchorage off Low islet the vessel must wait a favourable opportunity for crossing the strait. During the summer months the winds are generally light from the south-west for a considerable period; the wind, however, generally freshens a little when the western stream makes, and this is the right time to weigh. Pass about half a mile from Low island, and in crossing the current take care not to be set to leeward of Hakodadi.

Proceeding from Hakodadi to the westward against S.W. winds, keep well inside Cape Tsiuka, and if unable to round it, anchor with the stream or kedge about 2 miles to the north-east, weighing again when the next western tide makes. Should the wind be very light, a vessel may not clear the strait in one tide; in this case it will be better to wait a tide to the eastward of Cape Sirakami, and take the whole of the following tide to clear the strait, than run any risk of being swept into the strait again by the current. Vessels passing through the strait, particularly to the westward, should have a good kedge and 150 fathoms of hawser ready for immediate use, and must keep the land close aboard.

VOLCANO BAY and ENDERMO HARBOUR.—The United States frigate *Southampton* visited both this bay and harbour in 1854, and verified the accuracy of Captain Broughton's survey made in September 1796. Cape Yetomo, at the entrance of the harbour, is in lat. 42° 21′ N., long. 140° 56½′ E.*

The following description of this bay and harbour is from Broughton's voyages, pp. 102, 104 :—He there states, "I have seen few lands that bear a finer aspect than the northern side of Volcano bay. It presents an agreeable diversity of rising ground, and a most pleasing variety of deciduous trees, shedding at this time their summer foliage.

The entrance into this extensive bay is formed by the land marking the harbour, which the natives call Endermo, and the south point, which they call Esarmi; they bear from each other N. by W. ½ W., and S. by E. ½ E., distant 33 miles. There are no less than three volcanoes in the bay, which induced me to call it by that name. The depth is 50 fathoms in the centre, and the soundings gradually decrease on the approach to either shore. During our stay at the period of the equinoxes we experienced generally very fine weather, with gentle land and sea winds from N.E. and S.E., and no swell to prevent a vessel riding in safety, even in the bay; and the harbour of Endermo is quite sheltered from all but bad weather, by bringing the bluff on the extreme part of the isthmus, which forms the starboard point in coming in, to bear N.W.; in this position, in 4 or 5 fathoms water, the port entry point on the north shore was in one with the bluff."

In running for the harbour, the island must be kept open with the starboard entry point till within half a mile of a small islet (which is only so at half tide), and then steer in to the S.W., when the water will shoal, and any convenient berth taken. The soundings gradually decrease from 10 to 2 fathoms, soft bottom. A few houses are scattered on the south side of the harbour, and towards the head the shores are low and flat, so much so as to prevent boats landing within one hundred yards; in all other parts wood and water are procured with the utmost convenience. High water, full and change, at 5h. 30m.; rise and fall 6 feet.

KURIL ISLANDS.

This extensive chain of islands extends nearly in a uniform N.E. and S.W. line of direction from the north-east point of Yezo island to the south extreme of Kamchatka. The Boussole channel separates the

---

* *See* Plan of Endermo Harbour; No. 2,674; scale, m = 5¼ inches.

chain into two portions ; that to the northward belonging to Russia, and the southern portion to Japan. The fog in which these islands are constantly enveloped, the violent currents experienced in all the channels separating them, the steepness of their coasts, and the impossibility of anchoring, are such formidable obstacles, that it tries to the utmost the patience and perseverance of the mariner to acquire much knowledge respecting them.*

KUNASHIR ISLAND, the south-eastern of the Kuril islands, is separated from the north-east end of Yezo by a channel, named by Kruseustern the strait of Yezo, which is about 8 miles wide in the narrowest part. This end of Yezo has not been yet explored ; all that is known of it is, that it forms a deep bay, and that the south-east part of Kunashir advances very far into this bay, so that Cape Nossyam or Broughton, the eastern extreme of Yezo, entirely hides the strait, from which cause Kunashir and Yezo are often taken for one island.

St. Anthony peak, rising near the north-east end of Kunashir, is in lat. 44° 31′ N., long. 145° 46′ E. The island is stated to be surrounded with rocks and dangers. Its south-west part forms a bay, the two points of entrance to which bear N.W. by W. ½ W. and S.E. by E. ½ E. from each other, distant 11½ miles. The south-east side of the bay is a low sandy tongue of land, nearly 7 miles in length ; the opposite side, also a tongue of land, is only 5 miles. The Japanese have an establishment in the bay. The tide rises about 4½ feet.

CHIKOTAN or Spanberg island, at 30 miles N.E. by E. ½ E. from the east extreme of Yezo, is about 5 miles long, east and west, and the same north and south. In the centre of the island there rises a mount, even and uniform to the summit. It is said that there is a good harbour in the south-west part of the island.

The space between Yezo and Chikotan is much contracted by rocks and breakers, which occupy an extent of 20 miles. At 9 or 10 miles to the south-west of Chikotan are several rocky islets, named Walvis islands, one of which is about a mile in length, flat, narrow, and covered with verdure. The only safe channel is between the islets and Chikotan.

ITURUP or Staten island is separated from Kunashir by the Pico or Catherine channel, about 18 miles wide. Very little is known of this island. It appears to be about 135 miles in extent, N.E. by E. and S.W. by W., and 25 miles at its broadest part. Its north-east point, Cape Okchets, is said to be high and perpendicular, and to be in

---

* *See* Charts :—Kuril Islands, No. 2,405 ; scale, *d* = 2·1 inches : and Pacific Ocean, Sheet 1, No. 2,459 ; scale, *d* = 0·8 of an inch.

lat. 45° 38½' N., long. 149° 14' E.; its south point, Cape Rickord, in 44° 29' N., 146° 34' E.

URUP, or Companys island, lying on the 46th parallel of latitude, is separated from the north-east end of Iturup by Vries strait, 13 miles wide, and free from danger. The island is 58 miles long, N.E. and S.W., and its greatest breadth about 15 miles. Near its centre there is a remarkable peak, in shape like a haycock, which can be seen in clear weather at a distance of 50 miles, and is often visible when the other portion of the island is obscured by fog. The south-west point of the island is low and steep, and continues so for about 15 miles in a northerly direction, when it rises to a lofty mountain range; a high and almost perpendicular rock, appearing like a sail when seen at a distance, lies S.E. about one mile from the point.*

The north-west side of the island is for the most part rugged, steep, and without the least appearance of vegetation, sign of an anchorage, or inlet sufficient to shelter a boat. The north-east point is of a dark colour, moderately high, and slopes towards the sea. A chain of rock runs off about 5 miles in an E.N.E. direction from this point; and at a distance of a mile from the shore there is a large rock of pyramidal form, with two others smaller.

PORT TAVANO.—The small port of Tavano lies on the eastern side of Urup and is quite open to the eastward; with the wind from that quarter a heavy swell rolls in, which, with the shallow water and rocks it contains, do not recommend it as a safe anchorage. The entrance has a depth of 8 and 10 fathoms, and is 120 yards across; nearly in the centre of the port there are some rocks just above water, with 4½ and 5 fathoms close to.

Water can easily be procured from two rivers at the head of Port Tavano. Salmon and rock fish are plentiful.

DIRECTIONS.—The only good mark for Port Tavano is a small island lying S.E. by E. one quarter of a mile from the south point of the bay; it is steep-to (except for a few yards on its north side), high, conical-shaped, and of a blackish colour. N. by E. 1¼ miles from this island there is a remarkable point with a hole through it near the water, which can be seen at some distance when bearing North. After passing the small island just noticed, the port will be seen open bearing about West, and when the islet bears South the soundings will be 27 fathoms (sand and mud), gradually shoaling as the harbour is approached.

---

* The description of some of the islands which follow is by George L. Carr, Master of H.M.S. *Pique*, 1858, and William H. Drysdale, Master of H.M.S. *Spartan*, 1855.

BROUGHTON, REBUNTSIRIBOI, and BRAT CHIRNOEF ISLANDS.
—Between Urup and Simusir, the next large island to the north-east, are three small islands. The northernmost, Broughton or Makanruru island (Round island of Broughton), is of good height, bold, and abrupt, sloping a little to the southward, near which end are some rocks, and apparently the only place where a landing could be effected.

The two other islands, named Rebuntsiriboi or Chirnoi, and Brat Chirnoef or Chirnoi Brothers, lie N.N.E. and S.S.W. from each other, distant 1½ miles. Rebuntsiriboi, the northernmost of the two, is remarkable from its having two conspicuous peaks of sugar-loaf form. A reef, which much resembles an artificial breakwater, extends a mile east from its north point, and at its extremity there is a high rock.

SIMUSIR ISLAND is separated from the last-mentioned group by the Boussole channel, which, as before stated, divides the chain into two portions. The island is 27 miles long N.E. and S.W., and 5 miles in breadth. The coast appears safe to approach, no dangers being visible. It is destitute of any anchorage except in the northern part, where there is said to be a good harbour, named Broughton bay, which, although spacious, is only navigable for small vessels, on account of a reef lying in the middle of the entrance.

About 10 miles to the south-west of the north-east point of the island is Prevost peak, which can be seen at a great distance. The south end of the island is low, but rises suddenly to a lofty mountain, having a crater on its north-west side.

KETOY ISLAND is separated from the north-east end of Simusir by the Diane strait, about 15 miles wide. The island is high and mountainous, and about 8 miles in circumference. Rocks and islets are said to extend some distance off its north-east and eastern sides.

USHISHIR, 12½ miles to the N.E. by N. of Ketoy, is composed of two islands, each about 1½ miles long N.N.E. and S.S.W., which are connected by a reef 2 cables long. A reef of rocks terminated by a small islet extends to the northward from the northernmost of these islands. The channel between Ushishir and Ketoy is safe.

RASHUA, the next island to the north-east, is about 15 miles in circumference, high, uneven, and barren.

MATUA ISLAND is separated from Rashúa by Nadejda strait, 18 miles wide, and free from danger; the currents in it, however, are strong, and must be particularly guarded against, and unless the wind blows right through, a sailing vessel should not attempt to take it.

SIMUSIR ISLAND.—PARAMUSHIR ISLAND.

The island is 7 miles in length, north and south, and may be easily known by its lofty peak, which rises near its south-west side.

RAIKOKE ISLAND, lying 5 miles to the N.N.E. of Matúa, from which it is separated by Golovnin strait, is small but hilly, and has a high peak. In using this strait vessels should be prepared for strong gusts of wind.

MUSIR ISLANDS, at 22 miles to the north-east of Raikoké, are four small islets, or rather rocks, one of which is awash. They were discovered by Krusenstern in 1805, and named by him the Snares, on account of the great hazard he ran of being drifted on them by the strong current.

SHIASH-KOTAN ISLAND, at 12 miles to the north-east of the Musir, is 12 miles in extent, N.N.E. and S.S.W., and 5 miles in breadth.

The two islands Ekarma and Chirin-kotan lie respectively N.W. by W. 5 miles, and W. by N. 24 miles from the north-west point of Shiash-kotan.

KHARIM-KOTAN, the next island to the north-east of Shiash-kotan, is of a round form, 7 miles in diameter, and a peak rises near its centre.

ONEKOTAN ISLAND, 24 miles long N.N.E. and S.S.W., and 5 to 10 miles wide, is separated from Kharim-kotan by a channel 8 miles wide, called Shestoi strait, which is free from danger, but the currents in it being strong would make it dangerous to a sailing vessel if overtaken by a calm or light winds. Onekotan makes in several lofty peaks two of which are dome-shaped.

MAKANRUSHI ISLAND lies 15 miles to the W.N.W. of the north-west of Onekotan; and at the distance of 7 miles to the S.W. of it is Avos island or rock, which is surrounded by a dangerous reef awash.

PARAMUSHIR ISLAND is among the largest of this archipelago, being 60 miles long N.E. and S.W., and of an average breadth of 14 miles. It is separated from Onekotan by Amphitrite strait, which is 22 miles wide, safe, and generally used by vessels going from **Okhotsk** to **Kamchatka**, or to or from the American coast.

Paramushir, from its size and appearance, could not well be mistaken; near its south end is a mountain rising to a high and conspicuous peak, very dark, and unlike the land in its neighbourhood; and to the southward of this are three remarkable peaks. The south-east point of the island is long and low, with apparently a reef extending off it to the southward. A small island about 2 miles in diameter, named Shirinki, lies 9 miles from its south-west point; and Alaid, the most northern island of the chain, 11 miles from its north-west point.

SHUMSHU ISLAND, the last island of this chain, is separated from Cape Lopatka, the south extreme of Kamchatka, by Kuril strait, and from the north-east end of Paramushir by another channel, called Little Kuril strait. The south extreme of Shumshu terminates in a tongue of low land; the north end is the same, and is distant 8 miles from Cape Lopatka, which is also low.

Kuril strait is about 3 miles wide, but very dangerous, on account of the strong currents and sunken rocks bounding it on either side. Little Kuril strait is also dangerous.

WINDS.—During the summer months the winds are said to be variable near the Kuril islands, and it 'is then difficult to determine which are the prevalent quarters. With S.S.E. and S.S.W. winds the weather is always foggy, but the haze being dry the sun is often visible through the fog. East and N.E. winds bring rain and bad weather. With N.N.E and N.N.W. winds the weather clears, and the temperature is cold; it is also cold with winds between N.N.W. and W.S.W., and the air is dry and hazy. With the wind between S.S.W. and W.S.W. the sky is clear, and it is rarely foggy.

SOUTH-EAST COAST OF KAMCHATKA.

AVATCHA BAY is formed at the bottom of the large outer bay of the same name on the south-east coast of the peninsula of Kamchatka. It is the principal port of the peninsula, and is so extensive and excellent, that it would afford secure shelter for all the fleets in the world in its capacious basin. The channel leading to this bay is 4 miles long and a mile broad, and there is anchorage throughout its whole extent. It trends in a N. by W. direction, and when the wind is fair it may be sailed through by keeping in mid-channel; but it frequently happens that vessels have to beat in, and as the narrowness of the channel renders it necessary to stand as close to the dangers as possible, in order to lessen the number of tacks, it will be requisite to attend strictly to the leading marks.*

The extensive outer bay of Avatcha is between Capes Povorotnoï or Gavarea, and Chipounskoï, which are the best landfalls for making the port; for should a vessel, when off either of these capes, be over-

---

* See Chart: Sea of Okhotsk, No. 2,388; scale, $d = 2$ inches.

taken by thick fogs or strong winds from East or S.E., it is always possible to keep the sea; whereas should the port be made on its parallel and the vessel surprised by any contrariety, her situation would be precarious, there being no soundings on the coast, nor any anchorage which could be taken under such circumstances.*

The harbour of Petropaulski, on the eastern side of Avatcha bay, is small, deep, well shut in, and a convenient place for a refit of any kind. The town of Petropaulski stands at the head of the harbour, in an amphitheatre on the slopes of two hills, which form the valley, and is composed of a group of small wooden houses, covered with reeds and dry grass, and surrounded by courts and gardens, with palisades. At the lower part of the town, in the valley, is the church which is remarkable for its fantastic construction, and for its roof, which painted green, seems to add considerably to the effect of the picture, surrounded as it is by lofty mountains. Tareinski harbour, in the south-west part of the bay, is extensive and excellent, but as there is neither population nor commerce in it, it has been of no utility. Rakovya harbour also forms, to the southward of Petropaulski, an equally excellent port, but it is of less easy access than the foregoing, on account of the Rakovya shoal, lying in the middle of its entrance.

The weather is frequently fine at Petropaulski up to the 15th October; but after this period it is wet, and the land begins to be covered with snow, which becomes permanent, and does not disappear until May or June in the ensuing year. In the months of November, December, and January, violent storms are experienced. In fine weather, the morning breeze is from North to N.N.W. lasting until eight or ten o'clock, and sometimes even until eleven o'clock; then shifting to the West and South it sinks altogether; in the afternoon about one or two o'clock, the breeze from the offing sets in, varying from South to East.

During winter the cold is severe; the snow falls in abundance far from common, and frequently rises as high as the houses, which thus become buried until the return of spring; but whatever may be the intensity of the cold, it is rare that the roadstead is entirely frozen over.; the ice does not generally extend more than a cable's length off shore; and after bad weather occasioned by winds from seaward, as well as those from West and North, it becomes detached from the shore, and is carried out into the road. One of the most severe winters remembered was that of 1814, when the road was almost entirely blocked up. In ordinary winters the coves, the bays, and the rivers are only covered, and the ice is not always too thick to hinder a passage by breaking or cutting.

---

* *See* Plans:—Avatcha Bay, with Plan of Petropaulski Harbour, No. 1,040; scale, $m = 1$ inch; and Avatcha Outer Bay, No. 1,041, scale, $m = 0.2$ of an inch.

Supplies.—A vessel in need of repair will only find safe anchorage in Avatcha bay, and must depend on her own resources both for provisions and workmen; for there is no certainty in obtaining wood or water, still less any refitments for the ship. It is, however, possible to procure in urgent cases some slight aid from the Government stores, and some workmen of the port; but these assistances, besides being very limited, are very precarious. A supply of fresh beef may be procured, and a little fresh butter, but it is difficult to get poultry or eggs. There are no sheep nor pigs. Fish is abundant in the bay in the season; it begins with cod and herrings, and is followed by salmon and salmon trout.

Lights.—There are three *fixed white* lights exhibited at the entrance of Avatcha bay, at 449, 294, and 378 feet respectively above the level of the sea. The first is on the eastern point of entrance, and is visible at 24 miles. The second is on the inner point on the west side of entrance, and visible at 19 miles; and the other is on the signal station, about half a mile south of the entrance to Rakovya harbour, visible at 22 miles.

Tides.—It is high water, full and change, in Petropaulski harbour at 3h. 30m.; springs rise 6¾ feet; neaps 2¼ feet.

Directions.—When bound to Avatcha bay from the southward, a sailing vessel should make the coast well south of Cape Povorotnoi or Gavarea, and round it as close as possible, as the wind will in all probability veer to the northward on passing it. If the weather be clear, two mountains will be seen to the west and north-west of the cape, and one far off to the northward and eastward. The eastern one of the two former, Mount Villenchinski, is 7,372 feet high, in lat. 52° 42′ N., long. 158° 20′ E., and is peaked like a sugar loaf. The highest and most northern of the three is Mount Avatcha, which is 11,554 feet high, and may be seen in clear weather a very considerable distance. To the eastward of this peak, and close to it is the Roselskoi volcano, but it emits but little smoke.

These peaks are the best guides to Avatcha bay, until near enough to distinguish the entrance, which will then appear to lie between two perpendicular cliffs. Upon the eastern one of these, the lighthouse bluff, there are a hut and signal staff, and when any vessel is expected a light is sometimes shown. If the entrance of the bay be open, a large rock, called the Baboushka, will be seen on the western side of the channel and three others, named the Brothers, on the eastern side of the lighthouse bluff.

The outer dangers are, a reef of rocks lying S.E., about 2 miles from the lighthouse bluff, and a reef lying off a bank which connects Stanitski point with the point to the southward of it. To avoid the lighthouse reef, do not shut in the land to the northward of the lighthouse bluff, unless certain of being $2\frac{1}{2}$ miles off shore, and when within three-quarters of a mile only, tack when the lighthouse bluff bears North, or N. $\frac{1}{2}$ E.; the Brothers rock in line with the lighthouse, leads close to the edge of the reef.

The first western danger has a rock above water upon it, and may be avoided by not opening Baboushka rock with the point beyond, with a flagstaff upon it. In standing towards this rock, take care that the ebb tide in particular does not set the vessel upon it. A good working mark for all this western shore is the Baboushka open with Direction bluff, the *last* hill on the *left* upon the low land at the head of Avatcha bay. The bay south of Stanitski point is filled with rocks and foul ground. The lighthouse reef is connected with the Brothers, and the light-house bluff must not be approached in any part within half a mile, nor the Brothers within a full cable's length. There are no good marks for the exact limit of this reef off the Brothers, and vessels must estimate that short distance ; they must also, in beating through this channel, allow for shooting in stays, and for the tides, which, ebb and flood, sweep over towards these rocks, running S.E. and N.E. Good way should also be kept on the vessel, as the eddy tides may otherwise prevent her staying.

To the northward of the Brothers, two-thirds of the distance between them and a ragged cape, named Pinnacle point, there are some rocks nearly awash, and off the point, there is a small reef, one of the outer rocks of which dries at half-tide. These dangers can always be seen ; their outer edges lie nearly in line, and they may be approached within a cable's length. If they are not seen, do not shut in the Rakovya signal bluff. There is deeper water off Pinnacle point than in mid-channel, and the soundings are irregular.

To the northward of Stanitski point, the Baboushka rock may be opened to the eastward a little with the signal staff bluff, but be careful of a shoal which extends about 3 cables south of the Baboushka. The Baboushka has no danger to the eastward, at a greater distance than a cable's length ; and when it is passed, there is nothing to fear on the western shore, until N.N.W. of the signal staff bluff, off which there is a long shoal with only 2 and $2\frac{1}{2}$ fathoms on it. The soundings shoal gradually towards it, and the helm may be safely put down in $4\frac{3}{4}$ fathoms ; but a certain guide is not to open the western tangent of the Baboushka with Stanitski point.

When the vessel is a cable's length north of Pinnacle reef, she can stand into Ismennai bay, guided by the soundings which are regular, taking care to avoid a 3-fathoms knoll, lying half-way between Pinnacle point and the second cliff north of it. This bay affords good anchorage, and it may be convenient to anchor in for a tide; there is no danger but the 3-fathoms knoll, and the Ismennai rock may be passed at a cable's length. This rock is connected with Ismennai point by a reef, and in standing towards it Pinnacle point must be kept open with the lighthouse bluff—when in line the depth is only $3\frac{1}{2}$ fathoms—Rakovya signal staff in line with the bluff south of it will place a vessel in 5 fathoms water close to the rocks.

A small reef extends a full cable's length from Ismennai point, and until this is passed do not shut in Pinnacle point with the lighthouse bluff; but to the northward of it a vessel may stand to about a cable's length of the bluffs.

Northward of Rakovya signal staff the only danger is the Rakovya shoal, upon the west part of which there is a buoy in summer, and to clear this keep the Brothers in sight. There is no good mark for determining when the vessel is to the northward of this rocky shoal, which has only $4\frac{1}{4}$ feet least water on it, and as the tides in their course up to Rakovya harbour will be apt to set her towards it, it will be prudent to keep the Brothers open, until it is certain, by the distance, of having passed it (its northern edge is seven-eighths of a mile from Rakovya bluff), particularly as the vessel may now stretch well to the westward, and as there is nothing to obstruct her up to the anchorage. Rakovya harbour will afford good security to a vessel when running in from sea with a southerly gale, at which time she might find difficulty in bringing up at the usual anchorage. In this case the Rakovya shoal must be left to the northward; 5 and $5\frac{1}{2}$ fathoms will be close upon the edge of it, but the water must not be shoaled under 9 fathoms.

In entering Petropaulski harbour it is only necessary to guard against a near approach to the signal staff on the end of the peninsula on the western side of the entrance, named Shackoff point. In approaching this point, on which there is a battery, a white buoy will be seen, marking the extremity of a bank, extending S.S.E. from it;* this may be passed close to, leaving to the westward, and from thence steer for the end of a low

---

* A shoal extends from Shackoff point, the western point of the entrance to Petropaulski harbour, in a south-easterly direction nearly a quarter of a mile from high-water mark, having 3 fathoms only on its extremity, although in some places there are 4 fathoms between it and the point. By keeping the south end of the cliff under the cemetery bearing N.E., until within a cable's length of the beach, when the church of Petropaulski will appear in the centre of the valley, leads clear to the southward of this shoal, after which the direct channel to the inner harbour of Petropaulski has nothing less than 6 fathoms.—Valentine G. Roberts, Master of H.M.S. *President*, 1855.

sandy point which projects at an angle of 45° from the direction of the coast and nearly closes the bottom of the bay, forming it into an excellent natural harbour. This tongue of land, like an artificial causeway, is but little above the level of the sea, and is covered with huts raised on piles above the ground serving to dry fish. This point may be passed close to.

A sailing vessel leaving this harbour with light winds and with the beginning of the ebb, must guard against being swept down upon the Rakovya shoal, and when past it upon the signal bluff. There are strong eddies all over this bay, and when the winds are light, vessels often become unmanageable. It will be better to weigh with the last drain of flood.

CHAPTER XI.

SEA OF JAPAN; GULF OF TARTARY; GULF AND RIVER AMÚR; SAGHALIN ISLAND; LAPÉROUSE STRAIT; AND SEA OF OKHOTSK.

VARIATION in 1861. East coast of Korea, 3° 30′ to 4° 30′ W. Coast of Tartary, 3° 30′ W. Lapérouse strait, 2° 45′ W. Sea of Okhotsk, 4° 45′ W. to 6° 15′ E.

The SEA of JAPAN is bound on the east and south by the Japan islands, and on the west and north-west by the coasts of Korea and Tartary. It is about 900 miles long, N.N.E. and S.S.W., and 600 miles wide, East and West at its broadest part. Surrounded by land on all sides, this sea is only accessible by the following narrow passages:—To the south by the Korea strait, which connects it with the China sea; to the east by Lapérouse and Tsugar straits, by which it communicates with the Pacific; and to the north by the Gulf of Tartary, through which it communicates with the Sea of Okhotsk by the Gulf of Amúr. The Boungo and Kino channels which lead into the Suwo Nada at the south-west end of Nipon, and the connexion of the Suwo Nada with Korea strait north of Kiusiu island, are as yet little known.*

WINDS and CURRENTS.—There exists at present but little information regarding the navigation of the sea of Japan. The winds there appear very variable, and the currents, depending on special causes, are at times insignificant, whilst at others they run with great strength. There is no doubt † that the winds, in the chief principles of their change, coincide with those of the China sea, and that all irregularities in the latter influence in a greater or lesser degree the winds in these parts. It may, however, be remarked with greater certainty that hazy weather is accompanied by fresh and extremely cold north-east winds, or by light south-westerly breezes. The latter, not unfrequently, bring on a dense fog, particularly in the northern half of the eastern coast of Korea. South-westerly and north-westerly winds are certain precursors of fine weather.

In the middle, and towards the eastern part of the sea, the currents appear to set nearly always to the eastward, but occasionally from the

---

* See Chart of the Kuril Islands, No. 2,405; scale, $d = 2\cdot2$ inches.
† Voyage of the Russian frigate *Pallas*, in 1854.

north and south. On the 4th June 1855 the *Constantine*, proceeding from Korea strait to the strait of Tsugar, experienced a current of 36 miles a day, setting to the E.N.E., in lat. 36° 48' N., long. 131° 52' E., while on the following days, the rate varied from 6 to 9 miles, and the direction was from N. by E. to E. by N., until near the parallel of Tsugar strait, when its rate was 11 miles a day to the N. ½ E. During the whole of this passage the weather was fine and the winds light from West to South ; two days only they were from South to East, with cloudy showery weather.

In November 1855, the same vessel proceeding from Tsugar strait to that of Korea experienced a current setting E. by N. at a rate of 28 miles on the parallel of the strait ; to the S. by E. 7 miles in 40° N. and 138° E. ; to the S. ½ E. in 39° N. ; and the day following E. by S. 41 miles in 38° N. During all this time, the winds were from N.W. and N.N.W. and the weather fine ; but on the 16th the wind changing to East and E.N.E. accompanied by rain, the current set N.N.E. ½ E. at the rate of 26 miles.

Variable winds were experienced in this sea by H.M.S. *Hornet* in May and June 1856 ; in July and August they were mostly from the southward and eastward. There were fogs, more or less, in all these months. In August there was much heavy rain. From May to July little or no current was felt, either in this sea or in Korea strait ; but in the latter end of July and the month of August it ran strong to the N.E., on one occasion as much as 48 miles in twenty-four hours ; it, however, gradually decreased as the vessel proceeded to the northward. These currents were evidently much influenced by the wind, for on one occasion, in the vicinity of the Liancourt group, she was set 1½ miles per hour to the S.E.

On the 12th October H.M.S. *Barracouta*, in 36½° N. and 130° 22' E., after a series of variable winds from the N.N.E., the barometer going down, and heavy rain having fallen all day, was struck by a gale of wind from the North which obliged her to bear up for Shanghai, where she arrived on the 21st.

There seems to be no regularity in the currents along the western coast of this sea; they are in general very feeble, increasing only off the abrupt points of the coast. Broughton, who in October 1797 passed close along this coast, notices a current setting to the S.S.W. at the rate of one mile an hour. On the 18th of July 1856, the French frigate *Virginie*, with a good breeze from the northward, coasted it at a distance of 2 miles from Barracouta harbour to Cape Clonard, and had fine weather. From Barracouta harbour to Yung-hing bay, as in the preceding year at the same time, she had light winds from the North, varying to the N.E. and very fine weather, with the exception of some hours of fog in crossing D'Anville bay. Weak currents set to the northward at a mean rate of 10 miles in twenty-four hours.

According to Krusenstern, North and N.E. winds blow for 10 months of the year along the east coast of Korea, and if sometimes they change to the South, which is rare, it is only for a short time.

SENTINEL ISLAND, the south-easternmost of the group forming a belt round the south coast of Korea (respecting which we have but little information), may be considered as the western limits of the Broughton channel, which lies between the Korea and the double island Tsus sima. It is isolated, conical, and rugged, about 328 feet high, and barren. Its position, by M. de Montravel, is 34° 30′ 11″ N., 128° 38′ E. In the French chart it is placed in 34° 34′ N., 128° 53′ E., according to M. Roquemaurel.

TSUS SIMA, extending about 37 miles in a N.N.E. and S.S.W. direction, separates the Korea strait into two channels, the western of which is named Broughton and the eastern Krusenstern. The island is divided into two parts by a large inlet named Tsus sima sound.* The southern portion is high and mountainous,† with two sharp peaks on its northern part, forming like asses' ears when bearing S.E. The northern portion is comparatively low and level, except about one-third from its north-east extreme, where it rises into an high mountain.

M. de Montravel places the south-east point of this island in lat. 34° 5½′ N., long. 129° 16′ E.; and Kiku-saki, the south-west point, in 34° 10′ N., 129° 10′ E.

H.M.S. *Sybille*, in April 1855, passed through Broughton channel, between Tsus sima and the Korean coast, and it appeared to be clear of danger on the island side. A dangerous reef was seen extending a mile or more from the south-west point of Tsus sima; and the north-east point appeared to terminate in a rocky reef. Captain Forsyth, who passed through Korea strait in H.M.S. *Hornet*, in 1856, states that Krusenstern channel on the east side of Tsus sima is to be preferred, as it is the widest and less current was experienced in it.

CAUTION.—When navigating Korea strait, the barometer should be carefully watched, especially at night, as sudden shifts of wind frequently occur with heavy gusts, which give but little warning of their approach.

---

* *See* Plan of Tsus sima Sound, surveyed by Commander J. Ward, R.N., in 1859, No. 2,710; scale, $m = 3$ inches.
† F. G. L. Street, Master of H.M.S. *Nimrod*, June 1860.

MATU SIMA (Dagelet island of the French and Dajette of the Russian charts) is by the Russian frigate *Pallas* in lat. 37° 22′ N., long. 130° 56′ E.* It is of circular form, about 20 miles in circumference, and its peak, rising from the centre of the island, is 2,100* feet above the sea level. Its shores are cliffy and almost inaccessible.

LIANCOURT ROCKS are named after the French ship *Liancourt*, which discovered them in 1849; they were also named Menalai and Olivutsa rocks by the Russian frigate *Pallas* in 1854, and Hornet islands by H.M.S. *Hornet* in 1855. Captain Forsyth, of the latter vessel, gives their position as lat. 37° 14′ N. long. 131° 55′ E., and describes them as being two barren rocky islets, about a mile in extent N.W. by W. and S.E. by E., and a quarter of a mile apart, and apparently joined together by a reef. The western islet, elevated about 410 feet above the sea, has a sugar-loaf form; the easternmost is much lower and flat-topped. The water appeared deep close-to, but they are dangerous from their position, being directly in the track of vessels steering up the sea of Japan for Hakodadi.

TAKO SIMA or Argonaut island, marked doubtful on the charts, does not exist in the position assigned to it, in 37° 52′ N., and 129° 53′ E. In the year 1852, the French corvette *Capricieuse* twice crossed this position without perceiving any land.

WÁYWODA ROCK, so named after a Russian corvette, which on her way from Hakodadi to D'Anville gulf is said † to have discovered a rock which appeared to be 12 feet high, and 70 feet broad. This danger lies approximately in lat. 42° 14½′ N., long. 137° 17′ E.

The EAST COAST of KOREA, unlike the western coast, is steep-to at a short distance from the shore. A running survey was made by the Russian frigate *Pallas* in 1854 of its whole extent for about 600 miles, from the high detached rocks at the entrance of Chosan harbour to lat. 42° 31′ N., long. 131° 10′ E. Port Lazaref, Napoléon road or Posiette harbour, and the Tumen river were surveyed by her boats.

The frigate had steady fair winds, and kept at a distance of from one to 4 miles off shore, approaching even nearer at some parts, and following its windings. The coast has a uniform appearance; sometimes, however,

---

* M. Roquemaurel places it in lat. 37° 30′ N., long. 130° 53′ E., and gives its height as 4,000 feet.

† Renseignements Hydrographiques sur la Mer du Japon, &c., 2nd edition, page 124, 1860.

it changes suddenly; from being mountainous and rocky it becomes low and sandy, assuming its former appearance after a short interval.

TSAU-LIANG-HAI, or Chosan harbour of Broughton, who discovered it in October 1797, is formed on the south-east coast of Korea in lat. 35° 6′ N. The entrance is between Capes Young and Vashon, which bear N.N.E. $\frac{1}{3}$ E. and S.S.W. $\frac{1}{2}$ W., and are 3 miles apart; from thence the harbour trends in a N.W. direction $4\frac{1}{2}$ miles, and is from one to $1\frac{1}{2}$ miles wide. At 3 miles within the entrance there are some rocks always above water, one of which, 6 feet high, lies nearly in mid-channel, and the passage on either side of it is much narrowed by foul ground extending from both shores; otherwise the harbour is clear of danger, and the soundings gradually decrease from 12 fathoms at entrance to 3 fathoms at its head.*

Chosan is one of the most important harbours on the Korean coast, on account of its being the entrepôt of trade between that country and Japan. It affords safe anchorage. All around the shores are populous villages; the land is well cultivated, and numerous streams falling into the sea, afford easy watering places. Cattle, pigs, fish, poultry, wood, and vegetables can be easily obtained; also wild fowl in the cold season.

TIDES.—It is high water, full and change, in Chosan harbour at 7h. 45m.; springs rise 7 feet, and neaps 5 feet; neaps range 3 feet.

DIRECTIONS.—The entrance to this harbour may be easily recognized from the offing by the Black rocks, 90 feet high, lying off Cape Young, the north point of entrance. These rocks are bold-to and may be approached to a cable's length on their south side. At $1\frac{1}{2}$ miles westward of Cape Young is an abrupt headland, 565 feet high, named by Broughton Magnetic head, from its affecting the compass needles. His vessel anchored to the westward of this headland in a fine sandy bay, with good anchorage. H.M.S. *Nankin* also anchored here in September 1855, and it is described as being a convenient place to water at, with good holding ground, but a considerable swell must run in during a S.E. gale.

In steering to the westward for the best anchorage in the harbour, pass about a cable's length to the southward of the rock, 6 feet high, lying in mid-channel. West of the most southern of the rocks above water, and nearly abreast a wooded hill, on which a white house stands out among the trees, is a Japanese factory. The anchorage is about half a mile eastward of the factory jetty, in a depth of 7 fathoms, black mud.

---

* *See* Plan of Tsau-liang-hai or Chosan Harbour, surveyed by Commander J. Ward, R.N., in June 1859, No. 1,259; scale, m = 2 inches.

EAST COAST OF KOREA.

The COAST from Cape Young, at the entrance of Chosan harbour, trends first in a N.E. by N. direction 28 miles to Tikmenef point, then N. by E. 35 miles to Cape Clonard. Over the whole of this extent it preserves a uniformly mountainous and desert character. The space between the high mountain range, running in a parallel direction with the shore, is occupied by rows of sand hills which are detached from the range, and projecting towards the sea in abrupt points, form an unbroken series of little bays which give no shelter. Close to the southward of Tikmenef point there is an island of considerable size, which appears to hide the mouth of a river or large bay.

UNKOFSKY BAY.—Cape Clonard, it lat. $36° 53\frac{3}{4}'$ N., is the extremity of a rather low peninsula, which advances boldly towards the north-east, and is visible 18 or 20 miles off. It forms the easternmost point of this part of the Korea, and the south side of the bay of Unkofsky, the extent of which has not been determined; but which is 4 miles wide at its mouth, comprised between the peninsula and Crown point to the north. The trend of the bay is to the south-west; it has 17 to 12 fathoms at the entrance, and 7 fathoms with good anchoring ground at 2 miles from its coast, but it is open to the prevalent autumnal winds, and those from the north-east in winter. It appears clear of danger throughout, but the north part of Cape Clonard must be avoided, as foul ground extends a little distance from it.

The northern and southern shores of the bay have a marked difference in character. Those on the south are elevated, the hills of the main range here approaching the sea; they then retire at the centre of the bay, resuming their original direction parallel with the coast. The northern and western shores are of moderate height; the northern consists chiefly of abrupt or steep sandy cliffs of snowy whiteness, with hills of the same description. Several large settlements spread along the shores, the inhabitants being probably employed in fishing.

PING-HAI HARBOUR, in lat. $36° 36'$ N., is a small creek in the sandy coast surrounded by high hills. A conical shaped island lies at its entrance and shelters the whole roadstead. On the side of the island is a large group of houses standing near the sea. The *Pallas* had 23 and 32 fathoms water within $1\frac{1}{2}$ miles of the land; and the *Virginie*, which coasted it for 3 miles, had 27 fathoms. Moreover, the steep and rocky shores seem to show that they are steep-to; there was no apparent danger between the frigate and the shore, and the current towards the north averaged 10 miles in twenty-four hours.

Three mountains rise to the S.W. of Ping-hai, and three rounded summits of one sloping mountain at a little distance to the north. The mountains resemble each other; they are however so situate, N.E. and S.W., that their summits seem to rise one above the other.

The **COAST** from Unkofsky bay trends in a northerly direction to a large round headland slightly advanced, named by the *Virginie* Cape Pélissier. Its less mountainous character does not, however, dispel the appearance of sterility, rows of round sand-hills occurring for a considerable distance inland. To the W. by N. of the cape is a high mountain, Mount Popof, of regular pyramidal form, and the highest of those that occur near the shore.

From Cape Pélissier the coast changes its direction, and runs nearly in a straight line N.N.W. for 120 miles to Cape Duroch. Between these two capes there is not a bay worth mentioning; the coast appears steep, and there are a few houses near the water's edge.

Beyond the parallel of 38° N., the land becomes gradually depressed. Groves of cedar trees cover the sides of the sloping mountains. The Sedlovaya, or Saddle-shaped mountain, is the most remarkable elevation along the coast. Its two blunt summits, with an arched slope between them, rise about 5 miles inland in lat. 38° 10½' N.; the western summit is the highest. Several considerable mountains, having more or less pointed summits, surround this mountain on all sides, particularly on the north.

Several small islands occur near the coast north of Sedlovaya mountain. They are mostly desolate and rocky, intermingled with high reefs. The bay, in lat. 38° 42' N., is a small indentation of the coast, about a mile wide, and less than a mile deep. Two islets lie at half a mile to the west of its entrance, and are connected with its north extreme, which is high and rocky, by a bed of rocks some of which are uncovered. The islets at the south extreme of the bay are nothing but bare rocks, and serve as a refuge for innumerable seals and penguins.

CAPE DUROCH, or Peschurof, forming the south point of Broughton bay, is a massive cliff presenting to the north-east a front 8 miles long and bordered by rocks level with the water and close to the land. The depth was 104 fathoms at 5 miles to the east, and 23 fathoms at less than a mile north of the cape.

BROUGHTON BAY (Korea gulf of the French charts), is 93 miles wide between Cape Duroch on the south and Cape Petit Thouars on the north, and 55 miles deep, and Yung-hing bay and Port Lazaref at its head offer excellent shelter. The shores of the bay are winding and mostly low, and vessels can anchor in a moderate depth off them with safety.

The north and south shores of Broughton bay are commanded by lofty heights near the sea; the Belavenz mountains, about 15 miles to the south west of Cape Duroch, are respectively 6,092 and 5,884 feet above the sea; and to the north at 24 miles in the interior, W.N.W. of Cape Petit Thouars, Mount Hienfung reaches the height of 8,113 feet. The shores, although wooded and verdant, are varied occasionally by waste lands and rocky cliffs.

After passing Cape Duroch, a moderate sized bay will open out with a low sandy shore, and it probably affords good anchoring ground during southerly winds, as in general all the bays examined on this coast invariably do. A group of small islands and sunken rocks lie in the middle and in north-west parts of this bay; several of the islands have a few cedar trees. From the north extreme of the bay the coast again becomes winding and rocky, and gradually falls towards Felény point.

FELENY POINT, with the islands adjoining it, is the southern boundary of a shelving bay of moderate depth, 18 fathoms, at a distance of 3 or 4 miles from the coast, and 5 fathoms within a few cables of it. Mian-tsin-liang point, the north extreme of this bay, is high and abrupt, and at a distance appears as an island, being separated from the shores of the bay by low land. A rocky islet, named Observation island, lies in the centre of the bay, at $1\frac{1}{2}$ miles from the shore.

MOUCHEZ ISLAND (or Khalczov of the Russian chart), lies 8 miles to the N.E. of Observation island, and $7\frac{1}{4}$ miles off shore, the channel between being clear and safe. It is 475 feet high, and visible 24 miles off. Coming from the southward, it is an excellent land-mark when bound for Yung-hing bay. A shoal extends about half a cable from its south-west point.

The COAST from Mian-tsin-liang point trends in a north-east direction 16 miles to Codrika point. On the north side of the low isthmus which connects Mian-tsin-liang point with the mainland, is a small circular-shaped bay, into which a clear rivulet empties itself, and on the shore of which is the large village of Lon. The shores of the bay contain much basalt; the rocks of its northern point consisting almost exclusively of basalt columns. From Lon bay, northward, the coast is indented by small inlets with high points.

Codrika point, on which is a large village, is 328 feet high, and steep-to. At three-quarters of a mile N.W. $\frac{1}{2}$ W. of the point is Basalt or Michaud islet, near which the depth is 6 to 8 fathoms. This islet has two remarkable grottos, in each of which boats can freely enter; their sides consist of perpendicular basalt columns about 40 feet high.

On the west side of Codrika point is a small bay, which is sheltered from almost all winds except those from the N.E. The average soundings are 8 fathoms in the entrance, between the point and Basalt islet, and 11 fathoms in the bay. Two small rivulets flow to the west of a high point in the centre of the bay; some isolated huts appear on their banks, forming together the village of Anbian.

YUNG-HING BAY,[*] comprised between Ilary or Périer point to the south, and Desfossés point to the north, has an area of about 10 square miles. The depths are moderate, 11 fathoms in the middle, and 7 or 8 fathoms at 2 or 3 cables from the shores of the bay, over good holding ground; but the best shelter will be found to the northward in Port Lazaref, which is formed on the western side of Nakhimof peninsula, and which has an area of 4 square miles. The entrance to the bay is fronted by a group of islands and rocks, which protect the anchorages on their western sides from easterly winds.

NIKOLSKI, the largest and southern outer island of this group, is of oblong form and 3 miles in circumference. It is high, and covered with hills or mounds overgrown with cedars; a small islet lies near its north point, and another at half a mile from its south extreme. There is a village on its west coast.

The French corvette *Capricieuse* anchored in 7 fathoms about a mile to the south-west of this island; but this outer anchorage is of little value, being 8 miles from the chief villages on the coast. The safest channel into the bay is between the south-west point of Nikolski and Ilary point; it is about 3 miles wide, and carries a depth of 8 to 10 fathoms.

KUPRIANOF ISLAND, lying 1¼ miles north of Nikolski, is somewhat smaller than the latter, also rounder and more desolate, forming, as it were, one sloping mountain. The channel between these two islands is apparently deep and safe, and convenient for vessels entering the bay.

ANNENKOF ISLAND, at 3 miles west of Nikolski, is one mile long, east and west, and three-quarters of a mile wide. It consists of a series of hills crowned with cedars. The north side is steep and straight; the remaining shores though precipitous form little bays, at the head of one of which is a village.

Two small but high islands lie off the south-west face of Annenkof,

---

[*] This bay and Port Lazaref were partially surveyed by the Russian frigate *Pallas* in 1854, and the description is chiefly from the report of her voyage.

and form with it a snug anchorage, carrying a depth of 10 fathoms. The channel between these islands and Muravief point to the south-west is three-quarters of a mile wide, and has 6 and 7 fathoms water in it; this point is the north extreme of a peninsula, which being connected with the main by low land, makes at a distance like an island.

VISHNEVSKI ISLAND, lying 2 miles north of Annenkof, is high and narrow, and about half a mile in extent N.E. and S.W. With the exception of its southern side, it is surrounded by a reef of sunken and dry rocks. Within the reef is another small islet or rock.

The channel between this island and Annenkof carries a depth of 16 fathoms in the middle, decreasing gradually on both sides to 10 fathoms, at a distance of one or two cables from the shore.

In addition to the foregoing chief islands of the group, there are, as already stated, several islets, which, with the exception of one, lie eastward of Vishnevski, between it and Kuprianof island. Of these, Cone islet is remarkable from its conical form; all the others are more or less alike. There are several rocks above water surrounded by reefs, between these islets, as also north of them off the south face of Nakhimof peninsula, of which Desfossés point is the south-east extreme: some of these rocks from their whiteness are visible from a considerable distance.

These dangers would offer some difficulties if attempting the channel between the islets and Desfossés point. The channel is half a mile wide between this point and the most northerly white rock, but a sailing vessel might have baffling or light winds when passing the south end of the peninsula.

PORT LAZAREF (Virginie bay of the French charts), the entrance to which is in the northern part of Yung-hing bay, is formed on the west side of Nakhimof or Bosquet peninsula, which shelters it from the eastward. It is 2 miles wide at entrance, from thence it trends about 6 miles in a northerly direction, and its breadth varies from 1¾ to 3½ miles. It carries a depth of 7 to 10 fathoms, over a mud bottom. The peninsula is covered with high hills, from the centre of which rises a sloping mountain with two sharp summits W. by N. and E. by S. of each other. It is connected with the mainland by a narrow sand spit, on which are several elevations having the appearance of islands. Its western side is indented with small bays, on the shores of which are villages. A sand spit extends a quarter of a mile from Aliman point, its south-west extreme.

[C.]

In the northern part of the port are two islands, and in its north-west angle the mouth of the Dun-gan river. Butenef, the westernmost and the largest of these islands, is 3½ miles in circumference, and all its shores are steep except its eastern, from which a long spit extends in that direction. The best anchorages are at 1½ miles, and at 3 miles within the entrance. The first or outer anchorage is in 10 fathoms, mud, in the middle of the port, the water shoaling gradually to both shores. The other or inner anchorage is in 8 fathoms, mud, in the channel a mile wide between the south end of Butenef island and the shore to the southward; the latter is a convenient position for merchant vessels receiving cargo.

From the inner anchorage towards the mouth of the river, the soundings gradually decrease to 7 fathoms; they then suddenly shoal to 5 and 3 fathoms to the banks which front the entrance. These banks approach very near the north-west end of Butenef island, and of Trifanof point to the south-west, leaving narrow channels between. The channel between the banks and the island is silted up at its northern part; that between the banks and the point leads into a wide though shallow bay.

This shallow bay extends westward 2 miles from Trifanof point. Its breadth at the entrance is about half a mile, becoming wider farther in. The shores are winding and varied in character. There is an island at the north-west angle, and two islets in the middle of the bay. The whole of its northern part is occupied by a continuous reef, carrying only a few feet water. This reef surrounds the two islets and extends to both the north and south shores of the bay. The bay is thus divided into two parts; one west, and the other east, of the meridian of the islets. The first or inner part is inaccessible from all sides; the other or outer part is connected with the north-west part of Port Lazaref by the channel between the banks fronting the mouth of the river and Trifanof point, and may serve as an excellent shelter for small vessels during winter.

The river Dun-gan* (Eastern river) falls into the north-west part of Port Lazaref, and with its collateral branches, occupies the whole plain extending between two ranges of low hills. Winding along the plain, it approaches first one and then the other range which border it. At its mouth, the river divides into several shallow channels, formed by a group of low marshy islands covered with villages. The breadth of the principal channel is about 3 cables, and the entrance to it is difficult, from its shallowness, and from its winding between the steep part of the banks

---

* River Aube of the French charts.

before approaching the islands. The current of the river is weak, and the water for a considerable distance from its mouth is of a bad colour and taste.

Admiral Guérin, who visited this port in the *Virginie* in the year 1855, states that the surrounding land is well cultivated and populous, particularly at the river's mouth. On the marshy island at the termination of the left bank there is a large village, apparently the commercial emporium of a considerable town in the interior, and which the natives said was the chief city of the province; the information gathered from them only making the distance from Yung-hing to Séoul (the capital of Korea) 30 miles, and establishing the possibility of going there by water; the river probably extends as far as Séoul.

The bar of the river had 10 feet over it at low water, a depth which seems to be the limit of the lowness of the river at this season (the height of summer). From thence the course of the river was to the north, the soundings gradually deepening to 5 fathoms to the distance of 6 miles from its mouth, when it divided into several branches and appeared to run to the westward through a large plain terminating in a chain of high parallel mountains, covered with houses, hamlets, and villages.

Supplies.—Fresh provisions, oxen, pigs, poultry, and vegetables, are plentiful both in Yung-hing bay and Port Lazaref. Instead of procuring water from the Dun-gan river, it will be preferable to obtain it from the river Giffard, a clear and rapid stream in the south-west corner of the bay.

TIDES.—The tide streams are weak in Yung-hing bay; their influence being only felt in Port Lazaref. High water, full and change, at 5h. 20m., and springs rise 2½ feet.

DIRECTIONS.—The approach to Yung-hing bay and Port Lazaref is not accompanied with difficulty. Besides the mountain with the sharp summits on Nakhimof peninsula, the proximity of the bay may be recognized by the islands fronting it. As before stated, the channel into the bay between the south end of the peninsula and the rocks and islets fronting it, is narrow and intricate, and a sailing vessel might have light or baffling winds when passing Desfossés point.

The middle channel between the two outer large islands is apparently safe and convenient for vessels entering the bay. If entering by the southern channel between Nikolski island and Ilary point, and between Nikolski and Annenkof islands, do not stand towards the shore to a less depth than 7 fathoms. The channel between the two small islands lying off the south-west face of Annenkof and Muravief point, the southern inner point of the bay, although narrow, is clear and deep. The only danger to avoid in entering Port Lazaref is the reef extending from Aliman point, the east point of entrance.

The COAST after passing Nakhimhof peninsula trends to the North and N.E., 54 miles to Cape Rouge (Cape Goncharof of the Russian charts). As far as Trompeuse or Losef point, which is the eastern extreme of a high headland in lat. 39° 35′ N., the shore is low and sandy; several sloping hills rise at a little distance inland. Between this headland and Muséef point, 13 miles to the N. by E., the coast forms an extensive bay, the shores of which are low and sandy, and the soundings regular, 8 and 9 fathoms, over a bottom of fine sand. A round hill rises on the parallel of the centre of the bay, and may serve as a good land-mark for vessels entering, as it stands detached on a plain at a distance from the adjacent hills and mountains.

In the northern part of this bay is an island about 2 miles in circumference named Hodo (Flowery), with two islets off its eastern extreme, separated from it by a narrow rocky channel. The northern part of the island, as well as a portion of the western part, are sandy; the remainder is rocky and high. The channel between its north-west end and the main is about a mile wide, but it is partly blocked up by a shoal which extends nearly half way across it from the island towards the main. The other part of the channel is probably shallow; large fishing boats were, however, seen to pass through it.

At 3 miles to the S.E. of Hodo are two small isolated rocks. Leaving the bay with an easterly wind, the *Pallas* had 9 fathoms between these rocks and the coast, but in attempting to pass between them and the island, the depth decreased to 6 fathoms, which compelled her to go south of them. A river disembogues in the northern part of the bay.

The coast from Muséef point trends to the north-east towards Cape Petit Thouars, and is moderately high, and indented by bays with cliffy shores and points. Along its whole extent it is bordered by scattered rocks and some islands, the largest of which are, an island 3 miles in circumference, lying off the south side of Anjou point, and an island lying off Cape Rouge. This latter has rocky shores, and is divided into two parts by a narrow creek, and forms with the coast a channel about half a mile wide, with an islet at its eastern end. This channel, providing there are no dangers in it, presents a good and apparently a safe harbour.

On the parallel of Muséef and Anjou points two high peaks show themselves inland. Anjou point is perpendicular and high. Cape Rouge from its peculiar aspect, and having a high peak about 5 miles to the north-west of it, may be easily recognized. There are several villages hereabouts at the bottom of the creeks and bays, and traces of cultivation on the heights. Weirich point to the north-east of this cape, is a long narrow spit with three hills rising on it. There is a village on this point.

CAPE PETIT THOUARS, or Schwartz, seen from the south, makes in the form of a mamelon or small eminence, commanded by a conical but flat-topped hill a little west of it. The face of the cape is formed by a great reddish cliff, 5 miles long, off which is a small pointed islet. To the W.N.W. of the cape, at 20 miles inland, the range of the Hienfung mountains, 8,113 feet high, is a conspicuous object. Between this cape and Cape Rouge the soundings were 27 to 37 fathoms at 2 miles outside the points.

CAPE BRUAT or Boltin, 60 miles farther to the north-east, forms the eastern extreme of a chain of mountains, which presents two pointed summits 50 miles apart, the northern or Mount Tao Kwang reach 6,309 feet; the range slopes by a gradual declivity towards the east, where it terminates in a low cliff. Cape Bruat rises 1,541 feet above the mean level of the sea, and forms with Cape Schlippenback, 27 miles to the south-west, a deep bay in which the lead gave 23 to 24 fathoms at above 2 miles off shore.

COCKS COMB ROCKS.— The approach to the above bay is well marked by some jagged white rocks in the form of a cock's comb. They lie 10 miles from the coast in lat. 40° 41¼' N. There is also a group of three islets, named Arevief isles, connected together by shoals, in the northern part of this bay, in front of and at 1¾ miles from a projecting hill. The *Pallas* passed inside these islets, between them and the coast, and found the channel deep, 16 fathoms, and clear of shoals.

The COAST from Cape Bruat turns abruptly to the northward and takes a N. by E. trend for 31 miles to Cape Kozakof, or D'Après of the French charts. For the first 10 miles it is an inaccessible and perpendicular wall. The vegetation along its extent is very poor, not a tree or hovel being visible until near the cape, when villages again appear near the shore.

Cape Kozakof is a bluff rocky projection, about 2,615 feet high, and the limit of the high coast line; beyond it the coast turns to the N.W., and forms an extensive bay, 24 miles wide, north and south, and 6 miles deep, but open to the east. The depth was 40 fathoms in its northern part, and no danger. Mount Stolovaya, a conspicuous object on this part of the coast, rises 5 miles inland West of the cape. To the north of this elevation several detached hillocks and hills present themselves along the coast.

Daussy or Kolokozev point, elevated 1,791 feet, and in lat. 41° 47′ 40″ N., is the north-east termination of a small bay formed between it and Urusof point to the south. From hence the coast curves to the north-east, assumes a more hilly appearance, and small islands and rocks lie off. To the north-east of Bougarel or Linden point the shore is indented by a series of small bays as far as the large bay of Goshkevich. At 7 miles to the N.E. of Bougarel point, and near the shore, is a group of high, rocky, and desolate islets, named Avvakum. The group consists of two large islets, about the same size, with some smaller ones; the soundings were 20 to 26 fathoms near them.

GOSHKEVICH BAY is formed between a steep point to the north-east of the Avvakum group and Casy or Sisdro point, the south extreme of the Sisuro peninsula. In some of its nooks it has convenient anchoring ground, protected nearly from all sides. A round island, rather remarkable for the regularity of its form, and a smaller one near it, lie at the head of the bay. An island with white cliffs, and surrounded by rocks, named by the Russians Bielaya Skala, or White Cliffs, lies in the middle of the entrance, at 3 miles S.W. of Casy point.

The northern shores of the bay are sloping, and the bay is connected to the north-east with a deep basin. This connexion is not visible from seaward. The basin was accidentally discovered during the survey of the shores of the Tumen river, and was not examined.

Sisuro peninsula, the south-eastern extremity of this bay, is connected with the mainland by a very narrow isthmus; its sides, with the exception of the one facing the isthmus, are rocky. Sisuro village stands on the east side of the peninsula, on the shores of a small bay, convenient of access for fishing boats; the village is large and surrounded by low stone walls. This little bay is formed by the peninsula, the mainland, and a small promontory, which is also united with the mainland by a narrow spit. This promontory, and the inconsiderable precipitous mountains to the N.E., are the principal landmarks for the mouth of the Tumen, without which there would be difficulty in finding it on the sandy coast. The limit of muddy water extends too far off shore to serve as a guide to the position of the river's mouth.

M. Rocquemaurel, captain of the French corvette *Capricieuse*, in his report says, "Casy point is only 153 feet high. At 10 or 12 miles off it resembles a large rock scarped and jagged like the profile of a Venetian fort. Completely isolated from the mountains which rise above it to the north, it appears severed from the shore, to which it is joined by a long low beach trending N.N.E. About 18 miles inland, to the west, Mount Chienlong, rising 4,215 feet above the sea level, appears to form the north-eastern range of heights that extends some 30 miles to the south-west, at about 10 miles inland. A large river enters the sea at 4 miles to the east of Casy point."

"To the eastward of the river the range of mountains suddenly change their direction to the N.W., leaving a large valley between them. The remarkable structure of the ground consists generally in beds of gneiss or schist thrown up so to form cliffs or pointed hills, which rise abruptly from a low swampy shore; everything bears the stamp of a complete change in soil, climate, and people."

The **TUMEN KIANG**, or Tsing-hing, the position of which is vaguely marked on the charts of Lapérouse and Broughton, separates the Korea from the vast territory occupied by the Manchu Tartars; and by the Russo-Chinese treaty of 14th November 1860, it now forms the southern limit of the Russian territory on the shores of the sea of Japan. Its embouchure, which opens in the middle of a low plain, is at 4 miles eastward of Casy point, in lat. 42° 19′ N.

This river, which was examined for the first 10 miles from its entrance by the boats of the *Pallas* in 1854, not only serves as the northern boundary of the Korea, but is the limit of the mountainous coast line. It is bounded on the west from its fall to the sea, a distance of 10 miles, by a high projection of the coast, on the west side of which is Goshkevich bay, and by a sloping depression of the land on the east.

From sea the entrance is only observable on account of the elevation of its right bank. A long sand-spit, extending from the left bank, narrows the entrance, and forms a bar carrying 9 feet water. The channel is winding, the soundings in it are irregular, from 3 to 1 fathoms, and even less; and it runs between numerous shoals and banks, several of which appear above water in the form of low islands. This renders the navigation of the river difficult, and higher up from its mouth it becomes so shallow as to be fordable.

The right bank of the Tumen is hilly, and terminates in steep sandy slopes; the left bank presents an extensive plain, covered with lakes, or rather marshes, filled with brackish water. The banks preserve this character for about 10 miles (the extent examined). As regards its course higher up, no positive information could be procured from the Koreans, who are in general taciturn; judging, however, from the first 10 miles, it was supposed that the banks gradually became more sandy.

D'ANVILLE GULF is comprised between Casy point to the south, and Cape Hugon or Gamova to the north, about 33 miles apart. This latter cape, rising 1,800 feet above the sea, is the end of a long peninsula extending to the south, and is formed by a conical mountain sloping gently to the north, where, as in the south, it ends in steep cliffs. It is visible 30 miles off, and coming from the north-east is a good landmark for this gulf, as Casy point does from the south.

The land at the bottom of the gulf is low, marshy, and broken by a range of steep bluffs, resembling at a little distance a group of islands. At the head of the gulf are the anchorages, named by the French Capricieuse bay, Port Louis, and Rade Napoléon; the latter is called by the Russians Posiette harbour.*

CASSINI ISLAND, or Furngelm of the Russian chart, lying in the south-west part of D'Anville gulf, is 3 miles in circumference and 413 feet high. It forms a good mark for entering the gulf, when distinguishable from the high lands to the north-west, with which it is generally blended when approaching it from the southward. There are two villages on its west face. A reef of rocks extends more than half a mile from its south-west point, and another reef projects from its north-west point. At half and two-thirds of a mile N.N.E. $\frac{1}{4}$ E. and E. by N. $\frac{3}{4}$ N. respectively from its north-eastern point lie the Pilier and Buoy rocks, near which no soundings have been made; the latter rock hardly shows at high water.

The channel between Cassini and the coast is $3\frac{1}{2}$ miles wide, and carries 12 to 18 fathoms water, but in the middle lies the Baleine or Whale rock above water.

ANSE aux MOULES, or Mussel cove, is partly sheltered to the south-east by Cassini island, and to the north-east by Bodisco peninsula, the summit of which, named Mount Direction, is 820 feet high; the peninsula is joined to the coast by a low isthmus, on which is a village. The cove is $1\frac{3}{4}$ miles deep, north and south, and one mile wide at its entrance, which is narrowed by a reef of rocks, awash, stretching nearly 2 cables off the west shore.

The depth, 8 fathoms, at the entrance of this cove, decreases gradually up to the sandy beach, near which there are 16 feet. The holding ground is not good, being composed of fine sand and gravel with but little clay. It is open from S.S.E. and E.S.E., but the force of the sea from the latter direction is expended on Patau point, the south-east end of the peninsula. With the wind from S.E. to S.S.E. a heavy swells sets in, and makes the

---

* The French corvette *Capricieuse* examined this Gulf and the anchorages at its head in 1852. It was still farther examined by the Russian frigate *Pallas* in 1854, and it was here that her survey terminated. In 1855 a survey was made of D'Anville Gulf, and the gulfs of Guérin and Napoléon by F. H. May, Master of H.M.S. *Winchester*, and D. H. Wilder, Master of H.M.S. *Nankin*. *See* Plan of Victoria Bay and D'Anville Gulf, No. 2,432; scale, $d = 15$ inches: and Plan of Napoléon Road and Port Louis, No. 2,507; scale, $m = 2$ inches.

anchorage unsafe. The highest tides observed in this cove hardly rose 3½ feet.

CAPRICIEUSE BAY, comprised between Mount Direction to the south and Klaproth point to the north, is 5 miles deep N.W. and S.E., with a mean width of 3 miles. It affords fair anchorage within Balbi and Malte-Brun points, the inner points of the bay, in from 12 to 5 fathoms, over a bottom of mud. The best position is in 7 fathoms, mud, a long mile south of Porzic point. But as this bay is open to S.E. winds, which throw in a heavy swell, it can only be considered as a temporary anchorage for vessels waiting to pass the entrance into Napoléon road, which, with Port Louis, form secure basins at its head.

DIRECTIONS.—When wishing to enter Capricieuse bay it will be necessary for a vessel either to sight Casy point or Cape Hugon, to make sure of her position; sighting Cassini island must not be depended on, although it is in the offing, because it is often blended with the high lands which tower over it to the north-west.

If beating up against north-west winds, which are often fresh, a vessel should, when standing to the south-west towards the Pilier and Buoy rocks, go about when Malte-Brun and Bergasse points come in line bearing N.W. ¼ N.; the latter rock is nearly covered at high water.

The *Winchester** left Tsugar strait on the 16th August, and about noon on the 20th, in lat. 42° 41′ N., long. 134° 27′ E., made the coast of Tartary. A thick fog coming on, no distinct view of the coast could be obtained until the morning of the 23rd, when high land was observed to the northward, extending from N.E. by N. to N.N.W. ½ W.; that to the westward was supposed to be Cape Hugon. Stood towards it, and after making the Pelées islands shaped a course for Capricieuse bay, carrying regular soundings from 25 to 13 fathoms; but this bay not affording shelter from the S.E., ran into Napoléon road, the plan of which being found very incorrect, a fresh survey was made.

NAPOLEON ROAD is a good and secure harbour, and affords a safe retreat for vessels not wishing to ride out a south-easterly gale in Capricieuse bay. The entrance to it—between the Musoir rock, which is the northern termination of the sandy spit at the head of this bay, and Porzic point, the west extreme of Cuvier peninsula—is nearly half a mile wide, but it is divided into two channels by a large bare rock named the Mingan. The eastern of these channels should on no account be taken, as the ground is rocky and uneven, with a depth of only 4½ fathoms

---

* The remarks on Victoria bay are by Francis H. May, Master of H.M.S. *Winchester*, 1855.

in its centre. The channel on the west side of the Mingan must be entered with caution, for it is narrow, and a shoal extends nearly half way across from the Musoir, so that it is advisable rather to close the Mingan, as it is almost steep-to on its western side.

On entering the road, take care to avoid a small knoll with 3½ fathoms water on it, and 5 or 6 fathoms close to, which can be done by borrowing on the northern side of the fairway, as the knoll is directly or very nearly in the centre.

The observation spot at the entrance on the Musoir rock is in lat. 42° 37' 22" N., long. 130° 44' 10"; the meridian distance from Hakodadi being 0h. 39m. 43s. 8, differing 7 miles from the chart of the French corvette *Capricieuse*.

Water can be procured from the river in the north-east part of Napoléon road, but on account of the shallowness of the entrance none but small boats can go in except at high tide. The *Winchester* was a whole day procuring one turn of water with the boom boats, as the casks had to be rolled from the bar to about half a mile up the river to the fresh water, it being brackish at the entrance.

TIDES.—It is high water, full and change, at the entrance to Napoléon road, at 2h. 30 m., and the rise is about 2½ feet.

PORT LOUIS.—After passing through the channel into Napoléon road between the Musoir and Mingan rocks, Port Louis will open out to the eastward, and Napoléon road to the westward. There are no dangers after the port is open, and the spits off some of the points can be avoided by attention to the lead. There is anchorage any where in mid-channel, with good holding ground, mud and sand, and land-locked.*

A hard sandy spit runs across the upper part of the port, where several very fine bullocks and horses were observed to cross at low water. Above this spit is a large lagoon, about 2½ miles across each way, with irregular soundings in it, from 2½ fathoms to 4 feet. The port abounds with fish.

Water.—There are several small runs of good water, which are marked on the plan of Port Louis, and one spring, from which about 10 tons per day might be collected, all close to the shore. There are also various beds of oysters, and one of large mussels. The hills are covered with long grass, and abound with pheasants, partridges, and foxes; and the low ground, which is swampy, with woodcock and snipe.

---

* The description of Port Louis is by David H. Wilder, Master of H.M.S. *Nankin*, 1854.

GULF OF GUERIN.—The *Winchester* left Napoléon road September 3rd, and rounding Cape Hugon, passed between it and the Pelées islands; from thence she proceeded in a north-easterly direction into the Gulf of Guérin, 40 miles in extent, in which there are several good anchorages. The only danger seen was a rock awash, lying about N.W. ½ W. three-quarters of a mile from the northernmost of the three small islets west of the Pelées islands. To the north-west of these islets is a bay in which the *Styx* anchored in 8 fathoms sandy bottom, but it only affords shelter in northerly and westerly winds.

WHITE CLIFF BAY.—Passing between Red Cliff island and the main, a bay, named White Cliff, was observed, which apparently affords good shelter, but there was no opportunity of examining it. The south point, with rocks off it, should not be approached too near, as irregular soundings were obtained when in its vicinity.

NINEPIN ROCK.—The *Winchester* passed eastward of the two islands fronting White Cliff bay, (there may be a channel between them and the main,) near the northernmost of which there is a remarkable rock called the Ninepin. About 3 miles northward of these islands is a conspicuous cliffy point, which forms the south-east point of a secure and deep bay, named Port Bruce.

PORT BRUCE is an excellent harbour and well protected; the Eugénie archipelago, which fronts it, forming a good breakwater in south-easterly winds. There is no difficulty in working into this port, as there are 10 fathoms close to Column point, the south point of entrance, and the same depth near the islands on its northern side. The soundings are regular, and a vessel may anchor near the head of the port, where there is a small river, in 4 fathoms water, good holding ground. Fish and potatoes were procured from the natives, and they appear to be amply provided with those necessaries.

SANDY POINT.—From Column point, a high table hill will be seen to the north-eastward, with a tuft of trees on its summit, called Mount Virginie, and another farther in the same direction named Mount Winchester, the eastern slope of which terminates in a low sandy point, in lat. 43° 09′ N., long. 131° 50′ E. The *Winchester* anchored near this point in 6½ fathoms, mud, with Sandy point bearing S. by W. ½ W.; River islet, N.N.E. ¼ E.; East islet, E. ⅓ S., and Aube bluff, E. by N. ¼ N. This anchorage is good and quite land-locked.

There is an extensive bay in the north-east part of Guérin gulf, but the water in it is shallow. The eastern side of the gulf is formed by Albert peninsula, and an extensive archipelago.

Water.—At the head of Guérin gulf there is a river apparently of some extent, where good water can be procured about a mile from the entrance; but the same difficulty would be experienced in watering as at the one already alluded to in Napoléon road. The distance from Sandy point to the entrance of the river is 7 miles. A vessel might anchor within 3 miles of the entrance, as the depths gradually shoal towards it.

Wood can be obtained on any part of the coast, the land being covered with trees and the beach with drift-wood.

NAPOLEON GULF.—The *Winchester* anchored near the western entrance of Hamelin strait in 14 fathoms, mud, and on the 10th of September the *Styx* proceeded through the strait, and steamed round an extensive inlet called Napoléon gulf. The boats only landed at its head, where there is a river, but nothing could be seen of the least indication of a good harbour, although it was thought there was one on its eastern side. The gulf is about 30 miles long and about 16 broad, abreast of Hamelin strait, and the land everywhere is bold, and resembles that of the Gulf of Guérin.

HAMELIN STRAIT, which is the passage between Albert peninsula and the islands forming the Eugénie archipelago, is 6 miles long, and about half a mile broad in its narrowest part. At its western entrance there is a sand spit stretching half-way across the passage from the northern point (Knob point); the spit is steep-to, having 28 fathoms water within a ship's length of it. When past the spit there is no danger, and a vessel may run or work through, by paying common attention to the lead.

PORT MAY, on the northern side of Hamelin strait, is an excellent harbour, where a vessel may lie quite land-locked in 8 or 9 fathoms water over a muddy bottom. The land is high, and for the most part thickly wooded.

EUGENIE ARCHIPELAGO extends in a south-west direction from Albert peninsula, and separates Guérin and Napoléon gulfs. There are some good harbours in it, named **Port Deans Dundas,*** Port Stewart, and Wilder bay, affording shelter for ships of the largest draught, and no doubt good water can be procured at these as well at every other

---

* See Plan of Port Deans Dundas, No. 2,407; scale, $m = 3\frac{1}{2}$ inches.

part of this coast by digging wells; but no running streams of any size were seen.

A stranger might be led to suppose that there is a passage through Wilder bay into Napoléon gulf; but a bar runs across, with only 2½ fathoms water on it.

There are two other excellent outlets from Guérin gulf into Victoria bay, called Fellowes passage and Currie channel; by the latter the *Winchester* proceeded through to sea.

HORNET BAY is formed at the eastern extreme of Victoria bay, and Captain C. C. Forsyth, who discovered it in H.M.S. *Hornet* in July 1856, describes it as a spacious inlet, carrying a moderate depth up to its head, but open to southerly winds. There is snug anchorage on its western side, between the mainland and an islet, named Fox island, in lat. 42° 41′ N., long. 132° 56′ E. At the head of the bay, a river, apparently of some magnitude, was observed flowing round the foot of a peculiar conical hill; it was named Lyons river after Lord Lyons. Many villages and several herds of cattle were seen. The country is thickly populated, and the natives freely bartered their vegetables for old clothes, bottles, &c.

ISLET POINT. — The *Hornet* in running along this coast, observed several small islands on either side of a cape, named Islet point, which is in lat. 42° 49′ N., and 37 miles eastward of Hornet bay. An anchorage might be found inside these islands, for there was no apparent danger near them.*

PORT MICHAEL SEYMOUR (Olga bay of the Russian charts), in lat. 43° 46′ N., was discovered by the *Hornet*, July 1856. It is open to the south-east, but is protected by high land from north-west and south-westerly winds. There is shelter for a few vessels in all winds in its northern part, in 10 fathoms, over a muddy bottom. At the north-east part of the port there is a narrow passage, named Brown channel, leading into another harbour or estuary, which is well adapted for careening purposes, as there is but little rise and fall, the water at all times smooth, and deep close to its southern bank; a fresh water river empties itself at its head. Brown channel has from 3 to 4 fathoms water in it, deepening to 6 and 7 fathoms towards the Careening harbour.†

---

* This anchorage, named Siau Wuhu bay, was surveyed in 1859 by Commander Ward, R.N.

† *See* Plan of Port Michael Seymour, No. 2,511; scale, $m = 2$ inches.

A large river (Gilbert river) empties itself into the north-west angle of the port. The mouth is broad and shallow, but soon deepens when over the bar. The river flows in the bed of a deep valley, with high mountains on either side of it, and few places are void of vegetation, save some abrupt and precipitous crags; the valley itself consisting of marshy and turfy land, with flat islands in the course of the river. A few miles up the river divides into branches, and becomes shallow and narrow.

Supplies.—In this port the seine will always procure an abundance of capital fish, such as salmon and trout, in any spot where it is possible to haul it, and particularly near a stream; a few fowls may also be procured, and a small supply of fresh beef. The watering stream runs through a valley on the eastern side of the port, near the ordinary anchorage. A thick growth of oaks, hazel, and willows completely hides it from view, and fine ferns, though limited in variety, grow on the banks and under every cliff. Wood may be obtained in any quantity; also good oak knees for boats.

Tides.—At Port Michael Seymour it is high water, full and change, at 5h. 30m., and the rise is about 3 feet.

Directions.—The position of Port Michael Seymour may be easily known when approaching it from the northward by Brydone island, a high island on the eastern side of its entrance, and the only one on this part of the coast. This island is steep on its eastern side, and sloping on its western, and appears to be joined to the mainland by a reef of rocks; no bottom was obtained with 20 fathoms line within a quarter of a mile from its southern side, which appears to be free from danger. When within the entrance keep the starboard shore aboard, but not nearer than a good cable's length, as it is fronted by a steep bank.

When approaching from the southward the entrance will be plainly distinguished by the opening in the land. A few miles to the southward of it is a low flat cape, named Low point, which should not be approached nearer than half a mile, as heavy ripplings were seen off it.

From Low point the coast trends more to the westward and its features become changed; the country is less wooded, and several sandy beaches, many villages, cattle, and boats were observed.

St. Vladimir Bay, the entrance to which is in lat. 43° 54′ N., was discovered by the Russian frigate *America*, in 1857. Its entrance, formed by the peninsula Baliuzeka to the north, and Vachofski to the south, which are each joined to the mainland by a low isthmus, is $1\frac{1}{4}$ miles wide, and open to the east. From thence the bay extends westward, and forms three inlets, one to the north, one to the south, and the other, which is the smallest, to the west. Those to the north and south are vast circular basins separated from each other by a hilly and wooded peninsula, and both are surrounded and sheltered by adjacent high hills covered with wood. The depths are 5 to 15 fathoms in the northern, and the soundings are about the same in the southern inlet, where the *America* anchored.*

In this latter inlet, which is more completely sheltered, would be found every facility for careening and repairing a ship. To the south it communicates by a stream with a natural basin, in which are 16 to 10 feet water, and with little trouble it might be formed into a graving dock. The northern inlet forms an excellent outer harbour, where a ship might anchor when ready for sea.

In entering St. Vladimir bay, and having passed between the two elevated peninsulas which form the entrance, steer to the southward for the southern inlet. Be careful to keep midway between the two points of entrance of this inlet, as they are fronted by rocks and foul ground, which extend upwards of 2 cables off shore. The best anchorage is in 10 fathoms, about half a mile south of Low point, the east point of entrance, and a quarter of a mile from the eastern shore.

There is no fixed population in this bay, and from the report of other natives, it appears that these inlets are only frozen over from the middle of December to the middle of February ; the entrance of the bay remains in a great measure free from ice by the effect of the offing swell.

TIDES.—The tides are irregular in this bay. It is high water, full and change, at 1h. 0m., and the range is about 2 feet.

The COAST from St. Vladimir bay trends in a N.E. by N. direction to Barracouta harbour, in 49° 2′ N., and then N. by E. to Castries bay, and is free from apparent danger. Its outline was but imperfectly known until 1855, when H.M. ships *Sybille* and *Barracouta* sailed along it, and determined the following points and anchorages.

SHELTER BAY, in 44° 28′ N., affords shelter from N.E. and E.N.E. winds. Good fresh water can be obtained in a river a cable wide, with a bar at its mouth, within which there is a depth of 9 feet, over a fine sandy bottom. The land is high to the northward of this bay, and its aspect is that of bluff and barren headlands.

SYBILLE BAY, in 44° 44′ N., affords shelter against S.S.E. winds ; its entrance is remarkable from having on either side some prominent pinnacled rocks, high and isolated. The south side of the bay is high and rocky, with small deep valleys opening towards the sea ; many large rocks lie along the beach. The north side consists of a series of hills,

---

* *See* Plan of St. Vladimir bay, surveyed by Commander J. Ward, R.N., in 1859, No. 2,773 ; scale, m = 3 inches.

clay and sand. The bay is closed by a broad valley, through which some streams run and form a small river, which empties itself into the bay. The bay becomes gradually shallow towards the valley; the bottom is rocky, and covered by a thick growth of seaweed.

PIQUE BAY, in lat. 44° 46′ N., has good shelter from N.E. and easterly winds. The best anchorage is in 5 fathoms, with the point bearing S.E. by E. There is good water in a river, separated from the bay by a spit of sand, at the mouth of which is a bar. Cattle may be obtained at this anchorage.

BULLOCK BAY, in 45° 2′ N., affords but bad anchorage. Bullocks and fowls can be procured, but with difficulty, although they may be frequently seen near the beach; but it is necessary to search for the natives in the interior.

Luké point, in 45° 19½′ N., is high, bluff, and woody. Cape Disappointment, 22 miles to the north-east of this point, has rocks extending a cable from it.

GULF OF TARTARY.

From Cape Disappointment, in lat. 45° 40½′ N., the coast of Tartary trends to the north-east, and forms with the west coast of Saghalin island a long channel named by Lapérouse, the Manche or strait of Tartary. Strictly speaking, however, it should still be the Gulf of Tartary, as the strait can hardly be considered to begin until the parallel of about 51° N. On the parallel of Castries bay the coasts of Tartary and Saghalin converge rapidly; and abreast of Cape Catherine, in 51° 57′ N., they are only 7 miles apart; 17 miles farther north, between Capes Lazaref and Pogobi, they approach to within 3½ miles; and this is the gorge of the strait, and the entrance into the Gulf of Amúr.

The only ports at present known in the gulf are Barracouta harbour and Castries bay on its western side, and Jonquière bay on its eastern; there are, however, on both coasts many indentations, which may afford occasional shelter. The anchorage along the eastern shore is safe during the summer months, when easterly winds prevail; but a vessel must be prepared to weigh, should the wind veer to the west.

The coast on both sides of this gulf is high. On the eastern shore the country is everywhere thickly wooded, but in no way cultivated; towards the end of May or beginning of June, when the snow disappears, it has a most inviting appearance, and presents an agreeable variety of hill and dale, with streams and torrents. It appears to be thinly inhabited by a simple-minded race of people of small stature and dark complexion, with features partaking of, but more developed than the Chinese character ; they subsist chiefly upon fish, consequently locate themselves on the sea shore.

No dangers have as yet been observed off either coast of the gulf. The soundings decrease gradually towards its shores, though near the headlands, and where the coast looks bold, as may be expected, deep water is carried closer in, and the lead gives shorter warning of approach to land ; a peculiarity, however, in the soundings has been noticed off some parts of the west coast of Saghalin, the water there being found to deepen when approaching within 3 or 4 miles of the land.

The soundings in the northern part of the gulf, from 51° N. northward, vary from 46 to 26 fathoms, which latter is on the parallel of Castries bay, whence they rapidly decrease to 16, 10, 7, and 4 fathoms, in 51° 42′ N., abreast Cape Chihachef, and which may be considered the bar of the strait, unless, indeed, there be deep water close under this cape or over towards Saghalin, which seems doubtful.* To the northward of this spot the water deepens to 8 and 10 fathoms, which depth, with few exceptions, is carried up to the gorge.

The navigation of the gulf would be simple enough, but the fogs render it dangerous, requiring the greatest caution to be observed. It has been remarked by former voyagers that on nearing the land in these seas a vessel will suddenly emerge from the fog or find it lift ; this was found frequently to be the case by H.M. ships in 1855–56, but of course it would not be prudent to run for the land in full reliance on such an occurrence ; the best method appears to be to stand in cautiously for the land in the daytime, when a clearance from fog, as just described, may be met with.

SUPPLIES.—Wood and water can be procured in abundance, and with facility on all parts of the coast of the Gulf of Tartary, and coal of fair quality in any quantity at Jonquière bay.

Fish and wild fowl are plentiful ; very fine codfish have been caught in soundings from 73 to 30 fathoms.

WINDS.—During May, June, and July, the winds in this gulf prevailed from the southward and eastward, sometimes blowing a double and even treble reefed topsail breeze ; occasionally in May a furious south-easterly gale and snow storm, lasting 10 or 12 hours, has been experienced, and they may be expected with a previously freshening breeze from East and S.E., and sometimes a rapidly falling barometer ; the

---

* It does not appear clearly whether Admiral Kusnetsof with his squadron in 1858 passed to the eastward or westward of this bar.

wind in these storms rises suddenly and falls equally so, and will most probably veer to South, S.W., West, and perhaps N.W.

In the end of August and during October the winds were chiefly from S.W. to N.W and North; they are probably the same in September.

In October foul weather appears to come on as the wind draws to the eastward of North, but by all accounts the heaviest gales may be expected from N.W., and these prevail through the winter.

FOGS.—From March till August fogs are almost continuous in the gulf, with scarcely any clear interval for more than a day or two at a time; they are most prevalent and dense in June, and are immediately dispersed in S.W. winds; the mercury is little affected by them.

In August, September, and part of October, fair clear weather comes in agreeable contrast, although in the latter month snow occasionally falls. The change of weather about the middle of October is sudden, winter generally usurping the warmth of summer in a day and setting in with all its vigour; at this period ice begins to form in the northern part of the gulf, and the season for sailing vessels to be in the gulf on ordinary occasions must be considered as having terminated.

The barometer in May, June, and July, 1853, ranged between 29·65 in. and 30·28 in., once, however, falling to .29 in. in a heavy south-easterly gale. The temperature in April and May was registered by thermometer from 30° to 50°, in June and July 45° to 60°; in the latter half of October, in Castries bay, it was seldom above freezing point, and sometimes as low as 18°.

TIDES.—The tides in the Gulf of Tartary are regular close in shore, the flood stream setting to the northward, the ebb to the southward. The currents are uncertain and irregular.

SUFFREN BAY, formed by Cape Suffren to the south, in lat. 47° 20′ N., is but an exposed anchorage. Its shores are fringed by a low shingle beach on which the sea constantly breaks; it is surrounded by vast forests, which extend out of sight. The cape is fronted by rocks, which stretch half a mile out to the offing.

FISH RIVER, in 47° 55′ N., has anchorage off its entrance in 9 fathoms at about a mile from the land sheltered from N.W. and westerly winds.

LOW CAPE, in 48° 28′ N., has high land behind it, and a depth of 8 fathoms was obtained a mile off shore.

From Low cape the coast line northward is irregular for 15 miles; it then trends N. by E. 20 miles to Beachy head; it is steep-to, the lead giving 14 to 17 fathoms at 2 miles from the shore. High snowy mountains are visible inland, and pine forests cover the ground to the coast, which is hilly, and in cliffs.

BARRACOUTA HARBOUR.—The entrance to this harbour, (named also Hadshi bay and Port Imperial), in lat. 49° 2′ N. is between Freeman point,—the north point of entrance,—and Tullo island, which bear N. by W. and S. by E. from each other, distant three-quarters of a mile. The general depths in the bay are 5 to 15 fathoms over a mud bottom. The entrance is open to the eastward, but within the bay are several inlets, which afford shelter for all classes of vessels. The only danger is the Carr bank, with $1\frac{1}{2}$ and 2 fathoms on it, which extends a cable from the shore on the north-west side of the entrance, one third of the distance between Sybille head and Freeman point.*

This bay remains frozen for about the same time as Castries bay. Its shores are covered with wood fit for building purposes, such as the larch, fir, and stone-pine.

DIRECTIONS.—If approaching Barracouta harbour from the northward, and having made Barren bluff, in lat. 49° $18\frac{1}{2}$′ N., the three hills over Beachy head will be seen if the weather is clear, and by bringing the centre one to bear S. by W. it will lead to the entrance. At 3 miles N. by E. $\frac{1}{2}$ E. from Freeman point the shore should not be approached within 2 cables, to avoid a rock on which the sea occasionally breaks ; it lies about a cable off a point bare of trees, at 2 cables from which the depth is 12 fathoms, and thence to Freeman point 13 to 15 fathoms. Pass the latter point at a convenient distance, and avoiding Carr bank proceed into the harbour, where the best anchorage, with the exception of Pallas bay, is on its south side to the S.S.W. of Fortescue island.

Vessels bound into the harbour from the southward or south-east should make the land near Beachy head, which may readily be distinguished by some rocks above water, half a cable's length off its extremity. From Beachy head, which may be passed at a convenient distance, Tronson point, on the south side of the harbour entrance, bears N.N.W. distant about 5 miles. This point is low, rugged, and devoid of trees, and should not be approached within $1\frac{1}{2}$ cables. Between Tronson point and Tullo island the shore is clear of danger. Give Tullo island a berth of a cable's length, at which distance there are 12 or 13 fathoms water, and then proceed to an anchorage.

CAUTION.—Vessels approaching this harbour in foggy weather should not shoal their water within 40 or 35 fathoms, unless well

---

* *See* Plan of Barracouta Harbour, No. 2,508 ; scale, $m = 3·0$ inches.

assured of their position, and in all cases due allowance must be made for currents, which generally set along the coast to the southward in northerly winds, and to the northward in southerly winds. In strong breezes they have been found to run as much as 25 and 30 miles a day ; they are sometimes, however, extremely variable.

TIDES.—In Barracouta harbour the time of high water, full and change, is at 10h. 0m., and the rise is 3 to 4 feet.

The COAST from Beachy head trends in a northerly direction to Castries bay, and is steep-to, with 15 to 18 fathoms at 2 miles off shore ; depths, however, of 25 and 30 fathoms have been obtained at a cable's length off Cape Lesseps, in lat. 49° 33′ N. The coast assumes a bolder aspect in proceeding northwards. From Beachy head to Barren bluff, the land, backed by high mountains, forms two ranges when seen from the offing ; one of these mountains, cone-shaped and covered with snow, is visible 50 miles. The steep coast is undermined in some places by the sea ; large masses of rock, which are detached from time to time, fall down, and form a dangerous rocky shore. The coast-line is irregular, but by keeping outside the line, which joins the promontories of Beachy head and Barren bluff, all the intermediate points are cleared.

Barren bluff is a high perpendicular headland. Vessels will find shelter under it during winds from N.N.E. to N.W., in 5 or 6 fathoms water, with its extreme point bearing N.E.

Before approaching Cape Lesseps, there is a wide and low forest, extending for miles westward, backed by a high range of mountains. This cape is a bluff headland, bleak, and rugged, with many traces of continual disintegration.

Cape Destitution, in 49° 46′ N., is bold, high land, having a bay on its north side, which affords good shelter from S.E. to S.W., in 9 or 10 fathoms, with the extreme of the cape, E.S.E., distant about a mile.

Cape Dent, in 50° 00′ N., declines to the eastward, and has a bay on its north side ; at half a cable from its extremity, is the Gulf rock, of pinnacle shape, and about 20 feet high.

CASTRIES BAY.—The entrance to this bay is between Closter-Camp, or Quoin point, and Castries point, which bear North and South, and are distant about 4 miles ; Quoin point, the south point of entrance, is in lat. 51° 28′ N. Although the greater part of the bay is open to easterly winds, which throw in a heavy sea, yet vessels, if their draught will permit, will find shelter behind the islands in it, particularly on the west side of Observatory island. The bay is covered with ice from the middle of November or December to April, being open to navigation for 7 or 8 months in the year. The isthmus which separates it from the principal branch of the Amúr is not more than 40 miles across, and lake Kyzi is only 15 miles distant.*

A dangerous rock upon which the sea occasionally breaks heavily, but which does not show in smooth water, lies in the middle of the entrance to this bay, with the western part of the peninsula (the east extreme of which is Quoin point) in line with the bluff headland beyond, bearing about South ; the north end of Observatory island, at the head of the bay, bearing West, leads well to the northward of it. A flat rocky bank carrying 11 feet water, but with only 5 feet on its northern part, is said to extend between Oyster and South islands.

The general nature of the bottom of this bay appears to be mud, but from many circumstances this is supposed to be a mere superficial covering to a rocky or otherwise treacherous bottom. The anchorage is exposed to easterly winds, which cause a heavy swell to set in.

In making Castries bay from the southward two small high and barren islets will be seen near the coast, about. 16 miles southward of Quoin point. In entering the bay and passing to the southward of Danger rock, the reef which extends from the north end of Oyster island must be guarded against, but that channel is in other respects clear.

TIDES.—It is high water, full and change, in Castries bay at 10h. 30m., and the rise is about 6 feet.

JONQUIÈRE BAY is on the west coast of Saghalin island, about $1\frac{1}{2}$ miles to the north-east of Cape Otsisi, in lat. 50° 54′ N. Its position may be recognized by three remarkable detached pinnacle rocks, about 50 feet high, off its south point. The coast to the southward is bold, but becomes less so in the bay and to the northward as the high land recedes. This bay should be looked upon only as a fine weather anchorage ; it affords shelter from N.E., round east, to South, but is exposed to all other winds, and the holding ground is mostly bad ; there are some good spots, however, for anchorage in 9 to 7 fathoms water. A small river finds an outlet in the bay, and boats can pass over its bar when the tide is in.†

Coal.—A few huts of the natives will be seen on the south part of the above river entrance, and between these and Pinnacle point are seams of good surface coal, some of which, being close to the water's edge, can be easily worked.

---

* *See* Plan of Castries Bay, scale, $m = 0·7$ of an inch, on Chart of Kuril Islands, No. 2,405.

† *See* Plan of Jonquière Bay, scale, $m = 2$ inches, on Chart of Kuril Islands, No. 2,405.

Supplies.—Large quantities of fish were taken in Jonquière bay by hauling the seine on the beach to the northward of the huts, and good sized flat fish were caught with hook and line about a quarter of a mile off shore, in 3 or 4 fathoms water. Wild fowl and white hares are numerous.

The watering place is inconvenient. Drift wood is plentiful.

TIDES.—The time of high water, full and change, in Jonquière bay, is at 10h. 0m., and the rise is about 6 feet.

GULF AND RIVER AMUR.

The GULF of AMÚR, or Saghalin, is 70 miles long, north and south, and 25 miles in its greatest breadth. The waters of the river Amúr, which empty themselves into this vast basin with great rapidity, have formed banks of sand and mud, which cover almost its whole surface, barely leaving the shallow channels by which the stream flows on one side to the Sea of Okhotsk and on the other to the Strait of Tartary; this renders the entrance of this great river difficult and at times dangerous.*

Immediately north of Cape Lazaref, at the south entrance of the gulf, the channel from the Strait of Tartary divides into two branches. That which goes to the N.N.W., narrow and slightly winding, is called the South Fairway, and keeps close to the Tartary shore, passing the isles of Chomé and Hagemir; it bends abruptly around Cape Prongé, and thence holds a W.N.W. course to Nikolaevsk, which is 65 miles from Cape Lazaref. The channel varies from three-quarters of a mile to 2 miles in width; the depths are generally small, but occasionally are as much as 14 and 19 fathoms. The least water is $2\frac{1}{4}$ fathoms, and a flat with this depth extends for nearly 10 miles between Capes Koisakoi and Prongé, and this may be considered the real bar of the river, and must be crossed to enter it. Beyond this bar the water deepens, and 11 fathoms are found abreast the town of Nikolaevsk, above which the river is said to be navigable for 1,500 miles.

The N.N.E. branch or Saghalin channel is wider and deeper than the other, the least depth being 18 feet at low water, according to the Russian chart made between the years 1849 and 1854. It keeps along the Saghalin shore at about 5 miles distant, for nearly 60 miles, until just north of Cape Halezof, where it almost touches the coast; and 20 miles farther north, between Capes Golovachef and Menshikof, 16 miles apart, it opens out into the Sea of Okhotsk.

---

* *See* Chart of the Strait of Tartary and the entrance of the Amúr river, No. 2,650 ; scale, *d* = 14 inches.

The above chart also shows a narrow gut, dignified with the name of North channel, leading close to Cape Tebakh, the north point of entrance of the Amúr, and then in a N.E. direction 30 miles to the Sea of Okhotsk. The least water in it is 13 feet, until within one mile of the sea, where apparently there is a bar or flat of 6 feet (probably a closer examination would discover a deeper pass) ; but with a 6 feet rise of tide in the Sea of Okhotsk, or with a northerly wind, in fine weather, gun boats or despatch vessels under 12 feet draught might, there is little doubt, pass over it.

The **RIVER AMÚR** or Saghalin Ula is formed of the streams Shilka and Argun, which unite in lat. 53° 30′ N. on the frontiers of Russia and China. The former of these consists of the Ingoda and Onon ; the latter being the main stream which rises south-east of Lake Baikal, in the mountain chain called Khing-khan Ula by the Chinese, and Yablonoi Krebit by the Russians.

The river flows east as far as Nertchinsk, here it is said to be 600 yards wide, and very deep ; then north, then again east, when it receives the Argun which comes from the south near Baksanova. The united streams, under the name of Amúr, continue to the east and south-east, receiving from the south the affluents Songari and Usúri, and reaching its southern limit in 47° 48′ N., whence it turns abruptly to the north-east and east, falling into the Gulf of Amúr between Capes Prongé and Tebakh, which are 8 miles apart. The length of the Amúr, including all its windings, is about 2,500 miles ; it is navigable for large vessels as far as Nertchinsk, 1,500 miles from its mouth, in the summer season ; in the winter it is frozen over.

The fortress of Nikolaevsk is built on the left bank of the river, at 22 miles from the entrance. It is surrounded by a few houses, and defended by batteries and strong advanced works. The channels leading from the gulf to the anchorage abreast it are frequently changing, owing to the great débris sent down by the strong current of the river, and with the constant fogs, frequent squalls, and gales, render the approach both difficult and dangerous.

Owing to the vicinity of the Sea of Okhotsk, with its masses of ice, and the easterly winds prevailing during spring, the seasons at Nikolaevsk are much more inclement than higher up the Amúr. Even at the commencement of September continued rains set in. October brings snow and cold, and at the end of the month or commencement of the next, the mouth of the river is frozen over. November, and the first half of December, are mostly clear ; the temperature is low, the minimum being 39° Fahrenheit. At the latter part of December, and during January, severe snow storms, with westerly winds, render the communication between the different houses difficult, and even dangerous. The temperature during that time rises often above freezing point. The river navigation does not open before May, while snow and ice are to be found in the forests and more sheltered bays as late as June.

DIRECTIONS.—A vessel entering the Gulf of Amúr from the Strait of Tartary, should proceed with great caution, with a boat sounding on each bow, and an anchor ready at a moment's notice.

A Russian steam squadron, under the command of Admiral Kusnetsof, visited this gulf in September 1858, but they found that it could not be navigated safely without being buoyed and beaconed, and also that many alterations had taken place in the South Fairway channel leading to Nikolaevsk since the survey of 1854. The squadron often grounded, and were 13 days in getting through. Merchant vessels often remain aground in it for weeks together, and frequently throw a portion of their cargoes overboard to lighten.

The squadron in steering for the south entrance of the gulf, between Capes Catherine and Liak, passed at the distance of half a mile from the edge of the shoal, on the west side of the entrance, and which was found to extend 6 miles to the southward from Glasenap island, and to break heavily in easterly winds. After passing Cape Nevelskoi one of the vessels grounded on a small bank (not marked on the chart) lying N.E. by N. about $2\frac{1}{2}$ miles from the cape; it had 12 feet on it, and 6 and 7 fathoms close-to.

In proceeding towards the Amúr, and endeavouring to cross the bar between Capes Djaoré and Prongé, some of the vessels of the squadron grounded, there being not more than $13\frac{1}{2}$ feet over its shoalest part; but the water rising 2 feet with a northerly wind they were able to proceed over it in safety. Whilst aground in the southern part of the channel between Capes Nevelskoi and Muravief, the tide rose about 6 feet, and the morning tides were higher than the evening; the water remained stationary at its highest level for an hour, and then commenced to fall rapidly.

At Cape Lazaref the tide flowed twice in the 24 hours, the rise was 5 feet, and the ebb ran $3\frac{1}{2}$ to 4 knots. Abreast of Chomé island, the rise was 4 feet, and the water remained at its highest level about fifty minutes. From Cape Djaoré to Cape Prongé, there appeared to be no regularity in the tidal action, it being greatly influenced by the winds. It was high water only once in 24 hours, and the tide rose one foot with a southerly and 3 feet with a northerly wind.

During the few days H.M. ships were off the northern entrance of the Gulf of Amúr in 1855–56, the greatest rise of tide observed was 5 feet. The current from the Amúr set to the N.N.E. over the banks, sometimes at the rate of 3 knots per hour.

LAPÉROUSE STRAIT.

This strait is formed between Cape Notoro, the southern end of Saghalin, and Cape Soya, the north extreme of Yezo. The soundings in it are mostly 35 to 40 fathoms and upwards, decreasing to 25 and 20 fathoms as the shores are neared; but as these latter depths will be found in the middle of the strait near Opasnost or Dangerous rock, and in other places, the soundings in thick weather cannot always be trusted to ensure safety.* Dangerous rock, lying S.E. by E. 10 miles from Cape Notoro, is well above water, and may be seen 8 miles distant in clear weather; other rocks, on which the sea breaks, surround it for a mile.

There are heavy over-falls, giving the appearance of a reef, between Cape Notoro and Dangerous rock, but deep water was found on passing through. Neither this Cape nor Cape Nossyab should be closed without a commanding breeze, on account of the tide race off them.

TOTOMOSIRI, or Monneron island, lying N.W. by W. ½ W. 32 miles from Cape Notoro, is of moderate height, without the volcanic appearance of Refunsiri or of Risiri, the islands to the southward of it; some rocks lie off its eastern side.

Water.—There is a spring on Totomosiri, from which whalers are in the habit of watering, but with great difficulty and labour.

REFUNSIRI, lying to the southward of Totomosiri, on the south side of the western entrance of Lapérouse strait, is a high and irregular shaped island, having an anchorage on its eastern side. Its western side is bold, but the north-west end is faced with reefs and sunken rocks, and some miles off it a long low rock just above water was seen by H.M.S. *Bittern* in 1855; this part of the island should therefore be avoided. The island is inhabited.

RISIRI is about 7 miles to the south-east of Refunsiri, and there is a clear passage between them. This island has a magnificent volcanic cone, which forms a conspicuous landmark, and is visible in clear weather from a distance of 70 or 80 miles. The cone (The Pic de Langle of Lapérouse) rises to an elevation of 4,500 feet, and terminates in a sharp summit, which often peers out most usefully above the harassing fogs; it is generally covered with snow, and frequently presents a strikingly beautiful appearance.

CAPE NOSSYAB, the north-west point of Yezo island, is the abrupt but rather sloping termination of a remarkable table land, and appears like an

---

* The description of Laperouse Strait, the Gulf of Tartary, and the Sea of Okhotsk is chiefly from the Remark books of Officers of H.M. Ships, 1855-1857.

island at a distance. Extending a mile to the northward from the cape is a flat narrow tongue of land, only a few feet above the sea, having upon it a few huts, and a fishing station, which are conspicuous objects before the low land, on which they stand, rises to view 5 or 6 miles distant. From the extreme point of this low land a shoal rocky spit, partly covered with weed, extends off in a N.N.W. direction for upwards of a mile, with but little water over it in places, and at its extremity a depth of $2\frac{1}{2}$ fathoms, which rapidly deepens to 6 and 7 fathoms. Heavy breakers sometimes disclose this spit, but in smooth water they do not extend to its outer end. No vessel should near it within the depth of 12 or 14 fathoms.

ANCHORAGE.—There is an anchorage in a bay 6 or 7 miles to the southward, on the west side of Cape Nossyab, with the cape bearing N.N.E., south extreme of land S. by W. $\frac{1}{4}$ W., and the Pic de Langle S.W. by W. $\frac{1}{2}$ W., in 12 fathoms, the soundings decreasing gradually; a little farther north the ground is foul, in a less depth than 10 fathoms, at three-quarters of a mile off shore. The shore abreast this anchorage is low and swampy, covered with long rank grass and weeds, and backed by higher ground terminating in Cape Nossyab.

CAPE SOYA, the north extreme of Yezo, may easily be recognized, sometimes even in a fog, by a remarkable white rock lying off it to the westward, and which appears to be surrounded with broken ground.

ROMANZOV BAY.—Between Capes Nossyab and Soya, the coast forms an extensive bay, in which the land, covered with rank verdure, slopes towards the sea margin, and its formation in the south part or bottom leads to the supposition of a large river being in that direction. Several huts are distributed along the shores of this bay; and within it, at about 5 miles from Cape Soya, is a large Japanese village or fishing station, having near it, on an elevated position, an earthwork with embrasures, but no guns were seen.

About 2 miles within the bay, on the Cape, Soya side, the water shoaled suddenly from $12\frac{1}{2}$ to 8 and 2 fathoms on the outer edge of foul ground, extending nearly 2 miles off shore; a vessel, therefore, intending to take shelter here from a south-east wind, or wait the clearance of fog, should be careful to keep 3 or 4 miles off shore, where there is good anchorage for the purpose in 17 or 18 fathoms.

CAPE NOTORO or Crillon, the south extremity of Saghalin island, is low, sloping gradually towards the point, and has a hollow behind it, so that, at a distance of 10 or 12 miles, it appears to be detached, and makes like an island. A reef of straggling rocks extends off it a short distance.

On the western side and 3 miles to the northward of this cape and near a Japanese fishing station, is an extensive patch of dangerous rocks, covered at high water; they lie about a mile off shore, with irregular soundings of 3 fathoms within them, and 7 fathoms near their outside. No vessel should approach these rocks in less than 12 or 13 fathoms water, as then the ground is broken and foul.

WATER.—On the north-west side, and 6 or 7 miles from Cape Notoro, is an excellent watering place, used by H.M. squadron in 1855, at a running stream over a bed of sand and gravel into the sea. Anchorage off it should be taken in from 10 to 9 fathoms, sand; the soundings decrease gradually towards the shore.

CAPE SIRETOKO, or Aniwa Vries, the south-east extreme of Saghalin, is a remarkable promontory, the more so from a chain of high mountains near it, stretching away to the northward, between which and the cape is a hollow that gives it the appearance of a saddle. The headland itself is a steep abrupt mass of rocks, quite barren, and having a deep inlet at its point.

ANIWA BAY is an extensive bight, about 45 miles deep, occupying the southern end of Saghalin between Capes Notoro and Siretoko, which bear W. $\frac{1}{2}$ S. and E. $\frac{1}{2}$ N. of each other, and are distant 65 miles. At its head is Salmon cove, where there is a Japanese settlement, composed of a few houses built of wood, also the huts of the natives, which are of the most wretched description; the coast here is moderately high and level, and when first seen has an appearance of chalk cliffs, which continues for about 12 miles to the south-east, when it becomes higher, irregular, and of a dark colour.

The point to the southward of Salmon cove runs out shoal, and, until better known, should not be approached within 2 miles. To the northward of the north point, seen from the anchorage off the village, a shoal flat fronts a low plain covered with trees, and farther back the land rises in an undulating form to high hills. About midway between Capes Notoro and Siretoko the depth is 58 fathoms, and the soundings decrease gradually to the anchorage. Aniwa bay is open to the southward, but the holding ground is good.

Supplies.—Water is plentiful in Aniwa bay, but it is of inferior quality, and difficult to be procured. Excellent fire-wood, in any quantity, can be obtained from the authorities.

Fish are abundant; a small kind of salmon weighing about 3 lbs. is taken in large numbers during a few weeks in June and July; herrings are earlier in season. Supplies of any other description are not to be obtained.

TIDES and CURRENTS.—In and about Lapérouse strait the tides are very irregular, and they are probably much influenced by prevailing winds. They are felt mostly in shore, particularly round Capes Notoro and Nossyab, where at times they become perfect races. It is high water near these capes, full and change, between 10h. and 11h., and the rise is about 6 feet. The flood stream sets to the northward along the west coasts of Yezo and Saghalin, and to the eastward through Lapérouse strait; the ebb sets in the contrary direction.

The currents in the strait can neither be depended on in strength nor direction; near the shore they are probably regulated by and unite with the tides, but in the middle of the strait they will be found setting generally to the E.S.E. or S.E., sometimes at the rate of 2 or 3 knots per hour.

EAST AND NORTH-WEST COASTS OF SAGHALIN.

Saghalin island, Tarakai of the natives, and Krafto of the Japanese, extends nearly north and south along the coast of Manchuria or Tartary, for a length of 510 miles, by a width varying from 25 to 100 miles, and may have an area of 30,000 square miles. Its northern portion lies opposite the entrance of the Amúr, while its southern extremes, Capes Aniwa and Crillon, are separated from the island of Yezo by Lapérouse strait.

The island is mountainous; two ranges extend respectively N.W. and N.E. from its southern extremes, and meet in Bernizet peak in 47° 33′ N. Its western face is steep, the eastern low and sandy. The middle district of the island is flat and swampy, but to the north hilly and fertile. It is well wooded throughout, and large quantities of timber are exported to Japan for building purposes.

There is coal in several parts of this island, and around Jonquière bay it rises to the surface, and is of fair quality. Whales are found on the east and south coasts, salmon and herrings abound, and in the deep bay of Aniwa on the south, into which two large streams fall, the Japanese have established an extensive salmon fishery. Water is abundant at all parts, and drift wood for fuel is found in large quantities along the western coast. The northern portion of the island is inhabited by Ghiliaks, and the southern by Ainos, aborigines of Yezo, a race of small stature. Jonquière bay on the west coast of Saghalin, and Aniwa bay on the south coast, are described in pages 405, 411.

CAPE LÖWENHÖRN, in lat. 46° 23′ N., long. 143° 40′ E., is a steep projecting rock, easily to be distinguished from the rest of this coast by its yellow colour. North of it the coast assumes rather a westerly direction, and consists of a chain of large lofty mountains, covered with snow in May.

Cape Tonin, the next headland to the northward, is of moderate height, and entirely overgrown with fir trees. A chain of rocks stretches to the northward from it; southward of the cape the bottom is rocky, with small stones; to the northward it is entirely of clay.

MORDVINOF BAY is a large bight in the coast to the westward of Cape Tonin, in which plenty of water was procured and abundance of firewood. On the shores of the bay several dwelling houses were seen by Krusenstern, in 1805, but most of them were empty. The natives appeared to be superior to those in Aniwa bay.

CAPE SENIAVIN is a high point of land in lat. 47° 16½′ N. To the northward of it the coast is low, and trends suddenly to the westward; to the southward are lofty mountains covered with snow in May.

Bernizet peak of Lapérouse, is probably the same as Mount Spenberg of the Dutch. It is a lofty, rounded mountain, in 47° 33′ N., 142° 20′ E., near the north-west end of a lofty chain of mountains running through the valley from N.W. to N.E.

CAPE DALRYMPLE. in lat. 48° 21′ N., is formed by a high mountain rising close to the beach, in a north and south direction, and is the more easily known from being altogether isolated, except that to the northward, 12 or 15 miles, is another, very unlike this, apparently consisting of four separate mountains; the coast between is, with the exception of a peak of moderate hight, quite low.

From Cape Dalrymple, the coast trends in a S. by W. direction, and is bounded by lofty mountains, divided by deep valleys, the shore being steep and rocky. In several places are inlets between the rocks, which might afford anchorage; one in lat. 48° 10′, looked more promising than the rest.

PATIENCE BAY.—Cape Soimonof, in lat. 48° 52½′ N., is the western point of entrance to Patience bay, which is limited to the eastward by Cape Patience. The former cape is a high promontory, projecting to the eastward, and was taken for an island when it bore North. The north coast of the bay is mountainous, and the beach craggy. Far inland are lofty snow-topped mountains, except in one part, where an even country stretches away to the northward as far as the eye can reach. In the north-west angle of the bay is the mouth of the river Neva.

Cape Patience, the most prominent and the easternmost cape of Saghalin, is a low promontory, formed by a double hill, terminating abruptly. From this a flat tongue of land projects some distance to the southward; on the north side of the cape, the land is likewise low, the flat hill near Flat bay being the first high land in that direction. By this hill, Cape Patience, (which, owing to its little elevation, is not easily perceived,) may soon be recognized. The cape is surrounded by a rocky shoal, extending a considerable distance from the land.

ROBBEN ISLAND, the centre of which bears about S.W. ¼ S., 22 miles from Cape Patience, is surrounded by a dangerous reef about 35 miles in circumference. Krusenstern examined this reef in 1805. The waves broke violently over it, and to the northward there appeared, as far as the eye could reach, a large field of ice, under which, in all probability, the reef continued. The channel between the cape and the reef was not examined.

FLAT BAR, in lat. 49° 5′ N., is surrounded on all sides by a low country. It is a deep opening, in which, even from the mast head, no land could be descried. From this circumstance it was thought by Krusenstern, to be the mouth of a large river.

CAPE BELLINGSHAUSEN, at 30 miles to the northward of Flat bay, has at 7 miles to the S.S.W. of it, a point which was thought to offer a good harbour. The shore is abrupt, and entirely white. Between two hills that project considerably, the southernmost apparently insulated, is this apparent harbour, and perhaps a small river; it was, however, unexplored. The country about it is regular in appearance.

CAPE RIMNIK, at 40 miles farther to the northward, has at the back of it, some miles inland, a high flat hill, named Mount Tiara, remarkable for having three points on its summit.

CAPE RATMANOF, in lat. 50° 48′ N., terminates in a flat neck of land, stretching a considerable distance into the sea. The coast hereabouts is invariably craggy, and of a yellow colour.

CAPE DELISLE de la CROYÈRE, in lat. 51° 0′ 30″ N., forms the boundary of the mountainous part of Saghalin, for to the northward of it there is neither high land nor a single mountain, the shore everywhere consisting of sand, of a most dangerous uniformity. This cape is connected with Cape Ratmanof by a flat sandy beach, with mountains in the back ground between them.

DOWNS POINT, in lat. 51° 53′ N., is rendered remarkable by a round hill. It is not the boundary of the sand coast, for this continues to the northward of the same features as that to the southward, only that behind this point there is a bay of considerable depth.

To the northward of Downs point is a chain of five hills, having the appearance of islands in this extended plain. The whole coast here, like that to the southward, is scarcely raised above the water's edge; it is entirely of sand, and a little way inland is covered with a seemingly impenetrable forest of low shrubs.

SHOAL POINT (Cape Otméloi), in lat. 52° 32½′ N., so named from its vicinity to the only dangerous shoal Krusenstern met with off this coast, may be easily known by a hill of tolerable height, which on this flat coast almost merits the name of a mountain, and forms a remarkable object.

This shoal might have proved dangerous if great attention had not been paid to the soundings, the depth falling suddenly from 8 to 4½ fathoms. It is in lat. 52° 30′ N., long. 143° 29′ E., and extends, probably, some miles north and south at a distance of 10 miles off shore. The coast, the direction of which from Downs point is North, projects to the eastward nearly on the parallel of this shoal.

CAPE VIRST is in lat. 52° 57½′ N. A long way inland there are several considerable high lands, the coast being, as far as the eye can reach, composed of flat sand.

CAPE KLOKATCHEFF is in lat. 53° 46′ N., and near it appeared to be the mouth of a considerable river, as the land appeared unconnected. The land about it is flat, gradually increasing in height, the shores being flat and sandy.

CAPE LÖWENSTERN, named after Krusenstern's third lieutenant, in 54° 3¼′ N., is a large promontory, from which the coast takes a more westerly direction. To the southward of this cape, the land, which has been hitherto flat and sandy, is high and mountainous, with narrow spaces between the hills, the shore very steep, and in several places consisting of rocks of a chalk like appearance. In front of the cape there is a large rock.

The land between this cape and Cape Elizabeth presents a lofty, dreary, and barren appearance; no traces of vegetation are apparent, and the whole coast is iron-bound, consisting of one mass of black granite rock, with here and there a white spot; the depth at 2 miles off shore was 20 fathoms, and at 3 miles, 30 fathoms, rocky bottom. There are four other promontories between these headlands.

CAPE ELIZABETH, the north extreme of Saghalin, is a high mass of rock, forming the extremity of an uninterrupted chain of mountains. It is rendered remarkable from a number of high pointed hills, or rather naked rocks, upon which neither tree nor verdure of any kind is perceptible. It descends gradually to the sea, and at the brink of the precipice is a pinnacle or small peak. From the northward it makes in two rugged points, the western one being divided into peaks. Seen from the west it bears an extraordinary appearance to Cape Lopatka, the south end of Kamchatka, except that it is higher. On the west side of the cape a point projects, and between them there is a small bay.

CAPE MARIA, about 18 miles to the W.S.W. of Cape Elizabeth, is lower than the latter cape, and consists of a chain of hills all nearly of the same elevation. It slopes gently down to the sea, and terminates in a steep precipice, from whence a dangerous reef appears to project a considerable distance to the north-east.

TIDES.—It is high water, full and change, at Cape Maria, at 2h.; the rise is about 5 feet.

NORTH BAY.—Between Capes Elizabeth and Maria is a large bay of considerable depth, at the head of which, at the foot of a mountain, is a village in a beautiful valley. The locality is fertile, and the mountains covered with forests of fir trees.

This bay, although open, is said to be safe in the summer when north winds are rare. At 1½ miles from the shore the depth is 9 fathoms, fine sand, decreasing at half a cable gradually to 3 fathoms over excellent anchoring ground.

The north-west coast of Saghalin is infinitely preferable to the southwest. Between the mountains, which are entirely evergrown with the thickest forests, are valleys which appear capable of cultivation. The shores are broken, and almost everywhere of a yellow colour. The confines of the high and low lands are precisely in the same parallel as on the opposite shore; and beyond the limits, to the S.S.W., as far as the eye could reach, nothing could be seen but the low sandy shore, with here and there a few insulated but picturesque sand hills.

NADESHDA BAY, named after Krusenstern's ship, is about 10 miles to the south-east of Cape Maria. A plentiful supply of wood and water may be easily procured here, but the bay being open, the bottom rocky, and consequently not a safe anchorage, will preclude its ever being much visited.

OBMAN BAY, at 44 miles to the southward of Cape Maria, has barely sufficient water at its entrance for boats.

SEA OF OKHOTSK.

The Sea of Okhotsk, surrounded as it is on all its northern and western sides by the continent, and to the south-east by the Kuril islands, may be considered as completely land-locked. A large portion of its shores is comparatively unknown, for with the exception of its single important port, from which it derives its name, we have no accurate description of its details.*

Whaling vessels frequent this sea from the beginning of July to the beginning of October, few, if any, being in it by the 10th October. The shores are covered with ice from November to April, but the main expanse continues open throughout the year, and being generally deep without any apparent danger, its navigation is safe, notwithstanding the fogs and storms with which it is often visited.

The western coast of Kamchatka is uniformly low and sandy to the distance of about 25 to 30 miles inland, when the mountains commence; it produces only willow, alder, and mountain ash, with some scattered patches of stunted trees. The soundings are shallow for a considerable distance off shore; nor is there at the entrance into any of the rivers more than 6 feet at low water, with a considerable surf breaking on the sandy beach. Those vessels that navigate this coast endeavour not to lose sight of it, and judge of their distance from the land in foggy weather by the soundings, allowing a fathom for a mile.

SALUTATION BAY.—Near Cape Nagiba, at the north entrance of the Gulf of Amúr, is a bay of shallow water, where vessels, without entering the gulf, may discharge their cargoes, to be sent by boats or by land to Nikolaevsk on the Amúr. It is known as Salutation bay to whalers, who frequent it for fresh supplies; large vessels must, of necessity, lie outside.

SHANTARSKI ISLANDS.—This group lies off the eastern coast of the Sea of Okhotsk, and although the largest island is 35 miles long, east and west, and about the same distance broad, yet it does not appear to afford any port or shelter; but its south-west point projects to the S.W. so as to form a bay on the eastern side. Between this point and the nearest

---

* See Chart of Sea of Okhotsk, No. 2,388; scale, $d = 2$ inches.

point of the continent 14 miles distant to the south-west, are two islets surrounded by rocks and reefs. Soundings of 30 to 40 fathoms, stones, will be found at 8 to 10 miles to the eastward of the group. The tides are regular in strength, running from 1½ to 2 knots an hour.

To the southward of the south points of Great Shantar island are some small islands which have not been examined. At the distance of 6 miles from its west side is Feklistoff island, 20 miles in extent, N.E. and S.W., and 10 miles wide, but it has no port nor shelter.

ST. JONA ISLAND, in lat. 56° 25½' N., long. 143° 15¾' E. is merely a bare rock about 2 miles in circumference, and 1,200 feet high. It is surrounded on all sides, except the west, by detached rocks, against which the waves beat with great violence, and which probably extend a considerable distance under water. With the island bearing North, distant 12 miles, Krusenstern had 15 fathoms water, but when it bore West about 10 miles, no bottom with 120 fathoms.

PORT AIAN.—The coast in the neighbourhood of Port Aian, on the western Coast of the Sea of Okhotsk, is high and bold, and at 3 miles in the offing the soundings are 35 and 40 fathoms, sand.*

This port may be recognized from the southward by Cape Vneshni or Outer Cape, a high barren promontory with several craggy peaks upon it, at a mile to the eastward of the eastern point of entrance of the port. The inner harbour, affording good shelter for small vessels, is from one quarter to half a mile wide, and three-quarters of a mile deep, and the soundings in it vary from 2 to 4 fathoms, muddy bottom.

The outer harbour is exposed to S.W. and southerly winds, which send in a heavy sea; it has depths of 8 to 12 fathoms over good holding ground. On the west side of the entrance there is a reef of rocks with 4 fathoms close-to, barely covered at high water; the eastern shore is steep, and may be closely approached.

The climate here is abominable, and fogs are uninterrupted; the ice breaks up in June, and snow does not always disappear before August. The port is frozen over in November.

Supplies are scarce in Port Aian and difficult to be obtained. To the southward of the entrance is a sandy bay in which the soundings appear to be regular, and where water may be conveniently procured. Wild rhubarb grows close to the sea in most parts of the harbour, and is of great service as an anti-scorbutic. Scurvy is common and fatal among the inhabitants.

---

* See Plan of Port Aian on **Chart of Sea of Okhotsk.**

DIRECTIONS.—Vessels approaching Port Aian should make the land to the southward of Cape Vneshni, which is remarkably prominent; from that direction only can the entrance be seen, and with northerly winds the fog frequently lifts from 2 or 3 miles off shore to leeward of that headland, when it remains thick elsewhere. The high land of the peninsula should be avoided, on account of the calm it occasions. In the event of falling in with the land to the northward of the cape, Malminsk island, if seen, will guide to the harbour; but to judge from the appearance of the coast in that direction, as seen from the heights of Aian, it is not to be made bold with. The tides at this port have not been as yet correctly ascertained, but their rise appears to be about 12 feet. A current sets strong to the eastward, out of the harbour.

OKHOTSK HARBOUR, on the north-west side of this sea, is its principal port, but the shallowness of the water a long distance from its entrance, and the violence and cross set of the tides at the harbour's mouth, preclude the possibility of its being easily accessible, except for vessels of small draught. The mouth of the Okhota has only 9 feet water, and is only accessible from June to September, being blocked up with snow and ice the remainder of the year.

The town stands on a narrow tongue of land at the mouths of the rivers Okhota and Kuktúi. It possesses a shipbuilder's yard, an hospital, and large storehouses. The population in 1842 amounted to 800. Not a tree, and hardly even a blade of grass, is to be seen within miles of the town, and a more dreary scene can scarcely be conceived. Summer consists of three months of damp and chilly weather, succeeded by nine months of dreary winter, as raw as it is intense. The principal food of the inhabitants is fish, which is also the staple food of the cattle and poultry. The Sea of Okhotsk yields as many as fourteen varieties of salmon alone. Scurvy, in particular, rages here every winter.

SURFACE CURRENTS.—Near Cape Elizabeth, and on approaching the Gulf of Amúr, heavy overfalls and ripples occur, which appear to be produced by shallow surface currents, and they often render a vessel quite unmanageable; on some occasions, in a steady 5-knot breeze, vessels have been for hours with their head in the wrong direction, unable to answer the helm or trim of sails. A strong surface current here may naturally be expected, as the immense body of water from the Amúr, meeting with the obstruction caused by Saghalin island, effects its escape by the largest outlet, rushing over the shallow banks at the mouth of the river, and continuing its course, following the line of coast round Cape Elizabeth, causes, especially with East and S.E. winds, a dangerous race, extending off shore 3 or 4 miles, and setting strong to the southward along the eastern coast of Saghalin, where, for some distance, the sea is discoloured by it.

CAUTION.—It would be prudent not to approach Capes Elizabeth and Maria within 20 miles, to avoid the current (vessels are occasionally swept helplessly round North bay, between these capes, towards Cape Elizabeth), and the winds are generally light and variable in shore when strong in the offing ; by standing over to the Tartary coast the current is less, soundings are regular, and depth of water moderate, affording safe anchorage in case of calm or fog.

Care must also be observed in approaching the coasts of Saghalin and the shores of the Sea of Okhotsk, their longitude in connexion with that of parts better ascertained, being most probably erroneous.

The navigation to the northward of the Gulf of Amúr during the summer months in unsurveyed waters where strong currents exist and thick fogs are so universal, is less dangerous than would at first appear, anchorage being commonly found when near the land, for a vessel to await the clearance of fog to ascertain her position.

WINDS.—The prevailing winds in this sea during June, July, and August, may be considered south-easterly and moderate, accompanied with smooth water ; in the beginning or middle of September, southerly gales are said to be experienced, but that month has, in other respects, finer weather than August.

FOGS are nearly constant in the Sea of Okhotsk ; they are most dense in S.E. and easterly winds, but generally disperse with S.W. winds.

The open season is of less duration in the neighbourhoood of the Shantarski islands than at Port Aian ; the ice about the former in July preventing an approach within some miles.

CURRENTS.—The direction of the currents in this sea is uncertain ; they are found to increase in strength as the land is approached.

TIDE TABLE for the COASTS of CHINA, KOREA, and TARTARY, and off-lying ISLANDS; the SEA of JAPAN, GULF of TARTARY, and SEA of OKHOTSK.

Place.	High Water Full and Change.		Rise.		Place.	High Water Full and Change.		Rise.	
			Springs.	Neaps.				Springs.	Neaps.
	h.	m.	ft.	ft.		h.	m.	ft.	ft.
					River Min, Temple Point	10	45	19	14
East Coast of China.					,, ,, Losing Island	12	0		
Canton River, entrance	10	0	8		Changchi Island	9	30	17	
Broadway River, entrance	11	0	7½		Spider Island	10	0	17	
Typa anchorage	10	0	7		Lishan Bay	10	15	16	
Macao	10	0	6¼		Nam-quan Harbour	10	0	17	
Hong Kong Road	10	15	4¾		Namki Islands	8	30	17	
Lintin Island, Canton River	12	0	7¼		Pih-ki-shan Islands	8	30	17	
Fan-si-ak Channel, do.	1	0	7¼	2⅜	Fong-whang group, Bullock Harbour.	8	30	17	
Chuen-pee Point, do.	2	0	7¾						
Whampoa Docks	1	8	7½		Wan-chu River, entrance	9	0	15½	
Canton	2	40	5¼		,, ,, city	9	30	15½	
Niuepin Group	10	0	5		Chin-ki Island	9	20	13	
Tide Cove, Mirs Bay	10	0	6¼		Tai-chu Islands	9	0	14	
Tuni-ang Island, Bias Bay	8	0			St. George Island, San-mun Bay.	10	20	15	
Tsang-chau Island, do.	8	30			Kwesan Islands	9	30	14	
Hong-hai Bay	10	0	6¼		Nimrod Sound	10	30	20	
Kin-siang Point, Hie-che-chin Bay.	7	0	6¼		Vernon Channel (Chusan Archipelago).	9	40	14	
Cupchi Point	8	0	6¼		Ting-hai Harbour	11	0	12	
Hai-mun Bay	9	0	6¼		Pu-tu Island	8	15	12	
Cape of Good Hope	9	0	6¾		Lansew Bay	10	0	13	
Clipper Road, Namoa Island	11	15	7		Volcano Island	11	30	15	
Chauan Bay	11	0	6¼		East Saddle Island	11	0	14	
Tongsang Harbour	11	30	12		Yung River, Chin-hai	11	20	12½	
Chimney Island, Rees Pass	11	30	12		,, Ning-po fu	1	0	0	
Amoy, inner Harbour	12	0	16		Hang-chu Bay, Seshan Islands.	11	45	14	
Hu-i-tau Bay	12	15	16						
Chimmo Bay	10	20	16		,, Fog Islands	11	45	17	
Chinchu Harbour	12	25	17		,, Chapu Road	12	0	25	
Meichen Sound	12	30	17		Gutzlaff Island	11	30	15	
Makung Harbour (Pescadores)	10	30	9½	7	Yang-tse Kiang, entrance	12	0	15	10
White Dog Islands	9	0	19		Wu-sung River, entrance	1	30	15	10

Tide Table for the Coasts of China, Korea, and Tartary, &c.—cont.

Place.	High Water Full and Change.	Rise. Springs.	Rise. Neaps.	Place.	High Water Full and Change.	Rise. Springs.	Rise. Neaps.
	h. m.	ft.	ft.		h. m.	ft.	ft.
Shanghai	1 40	10	7	*Bonin Islands.*			
Yellow Sea, Wei-hai-wei Harbour.	0 30	9					
" Lung-mun Harbour.	10 0	7		Peel Island, Port Lloyd	6 8	5	
Gulf of Pe-chili, Che-fau Harbour.	10 0	8		Hillsborough Island, New Port.	11 32	3½	
" Pei Ho entrance	10 45	7		*Japan Islands.*			
Gulf of Lian-tung, Hulu Shan Bay.	2 30	9					
" Tai-cho and Yang Rivers, entrance.	0 15	6		Kiusiu Island, Nagasaki Bay	6 23	6¼	
" Lau-mu River	1 30	5		Nipon Island, Simoda Harbour.	5 0	5¾	
" Ching River	1 20	6½		" Yedo Bay	6 0	6	
" Peh-tang River	10 0	9½		Yezo Island, Hakodadi Harbour.	5 0	3	
West Coast of Korea, Chodo Island.	6 20	11½					
" Marjoribanks Harbour.	3 30	29	10	" Endermo Harbour.	5 30	6	
" Basil Bay	4 15	17½					
" Ko-Kun-to group, Camp Islet.	2 25	20		*Japan Sea.*			
South Coast of Korea, Port Hamilton.	8 30	11		East Coast of Korea, Chosan Harbour.	7 45	7	5
				" Port Lazaref	5 20	2½	
Bashi and Balintang Channels.				Coast of Tartary, Napolean Road.	2 30	2½	
Babuyan Islands, Port San Pio Quinto.	6 0	6		" Port Michael Seymour	5 30	3	
				" St. Vladimer Bay	1 0	2	
Formosa Island.				Laperouse Strait	10 30	6	
				S.E. Coast of Kamchatka.			
Port Kok-si-kon	11 30	3					
Tam-sui Harbour	11 45	7 to 12		Petropaulski Harbour	3 30	6¾	2¼
Ke-lung Harbour	10 30	3		*Gulf of Tartary.*			
Lu-chu or Liu-kiu Islands.				Coast of Tartary, Barracouta Harbour.	10 0	3¾	
Napha-kiang Road	6 30	6¼		" Castries Bay	10 30	6	
Oho Sima, Vincennes Bay	7 30	5½		Saghalin Island, Jonquière Bay.	10 0	6	
Mariana or Ladrones Islands.				*Sea of Okhotsk.*			
Seypan Island, Magicienne Bay.	6 45	2¼		Saghalin Island, Cape Maria	2 0	5	

TABLE OF POSITIONS.*

ON THE

COASTS OF CHINA, KOREA, AND TARTARY, AND OFF-LYING ISLANDS; AND IN THE SEA OF JAPAN, GULF OF TARTARY, AND SEA OF OKHOTSK.

Place.	Particular Spot.	Latitude, North.	Longitude, East.	Authorities.
	CHINA, EAST COAST.			
		° ′ ″	° ′ ″	
Hong Kong	Point Albert	22 16 27	114 10 48	Belcher, 1841.
Raleigh rock	-	22 2 0	113 47 0	Bate, 1847.
Ninepin rock	-	22 15 45	114 22 7	Collinson, 1845.
Single island	East summit	22 24 6	114 39 12	,,
Tuni-ang island	Summit	22 27 6	114 36 45	,,
Mendoza island	,,	22 30 42	114 50 0	,,
Pedro Blanco rock	,,	22 18 30	115 6 54	,,
Pauk Piah rock	,,	22 32 54	115 1 0	,,
Chino peak	,,	22 44 24	115 46 50	,,
Cupchi point	Hill on it	22 48 7	116 4 26	,,
Breaker point	-	22 56 0	116 27 45	,,
Cape of Good Hope	-	23 14 0	116 47 0	,,
Brothers islets	South-east islet	23 32 30	117 42 0	,,
Tongsang harbour	Fall peak	23 47 15	117 36 48	,,
Chapel island	Summit	24 10 18	118 13 30	,,
Amoy island	Citadel	24 28 0	118 4 0	,,
Dodd island	Summit	24 26 16	118 29 4	,,
Chin-chu harbour	Pisai island	24 49 13	118 41 0	,,
Pyramid point	-	24 52 12	118 58 0	,,
Sorrel rock	-	25 2 18	119 10 36	,,
Ockseu islands	Western island	24 59 0	119 27 30	,,
Lam-yit island	High Cone peak	25 12 0	119 35 0	,,
Hungwha channel	Sentry island	25 16 30	119 45 0	,,
Hai-tan island	Kiangshan peak	25 36 18	119 50 42	,,
Turnabout island	Summit	25 26 0	119 58 42	,,
Pescadores islands, Makung harbour.	Observatory point, the second point on north side of harbour.	23 32 54	119 30 12	,,
River Min	Temple point	26 8 26	119 37 42	Richards, 1854.
Changchi island	Highest peak	26 14 0	120 1 42	Collinson, 1845.
Alligator island	Summit	26 9 0	120 26 0	,,
Tung-ying island	Peak	26 23 12	120 31 0	,,
Cony island	Summit	26 30 0	120 10 0	,,
Double Peak island	Highest peak	26 36 6	120 11 12	,,
Pih-seang islands	Town island	26 42 30	120 22 42	,,
Dangerous rock	Summit	26 53 0	120 34 18	,,
Tae islands	Easternmost	26 59 12	120 43 48	,,

* The positions by Belcher, Collinson, Bate, Gordon, Richards, and Ward, and by H.M. ships, depend upon Point Albert, on the north shore of Hong Kong, being 114° 10′ 48″ East from Greenwich; those by Basil Hall depend upon the fort at the mouth of the Pei ho being 117° 49′ East.

TABLE OF POSITIONS.

Place.	Particular Spot.	Latitude, North.	Longitude, East.	Authorities.

CHINA, EAST COAST—cont.

Place.	Particular Spot.	Latitude, North.	Longitude, East.	Authorities.
		° ′ ″	° ′ ″	
Ping-fong island	Summit	27 9 42	120 32 42	Collinson, 1845.
Pih-quan peak	„	27 18 48	120 28 45	„
Nam-quam harbour	Bate island	27 9 20	120 25 50	„
Port Namki	Eastern Horn	27 26 18	121 6 36	„
Pih-ki-shan island	Summit	27 37 18	121 12 18	„
Fong-whang group	Coin island	27 50 0	121 15 0	„
Pe-shan island	Summit	28 5 30	121 31 48	„
Sondan islet	„	28 15 54	121 44 36	„
Chikhok island	„	28 22 24	121 44 12	„
Tai-chau group	Hea-chu islet	28 23 18	121 55 12	„
Chuh-seu island	Summit	28 40 30	121 47 24	„
Tungchuh island	„	28 42 12	121 55 6	„
Hieshan island	Southernmost	28 50 48	122 14 24	„
Montagu island	North-east point	29 10 30	122 5 0	„
Kweshan islands	Patahecock	29 21 54	122 13 42	„
Mouse rock	Summit	29 32 42	122 13 36	„
Buffaloes Nose island	High part	29 36 12	122 1 24	„
Nimrod sound	Middle island	29 34 20	121 43 15	„
Chusan Archipelago:				
Tongting islet	Summit	29 51 42	122 35 48	„
Chukea island	Peak	29 54 0	122 25 18	„
Just-in-the-way islet	Summit	29 57 42	121 54 12	„
Chusan island	Observation spot, Ting-hai harbour.	30 0 25	122 5 18	„
Video island	Summit	30 8 0	122 46 0	„
Barren isles	Centre	30 43 0	123 7 14	„
Saddle group	North island	30 50 0	122 41 0	„
Cairnsmore rock	-	30 42 10	122 34 40	Ward, 1858.
Chapu	Battery	30 36 0	121 3 0	Collinson, 1845.
Yung river	Chin-hai citadel	29 57 8	121 43 6	„
Yang-tse-kiang	Shaweishan islet	31 25 12	122 14 0	„
„ Wusung river.	Fort A. at entrance	31 23 30	121 20 11	Ward, 1858.
„ Shanghai	British Consul's flag staff	31 14 42	121 28 55	„
„	Hankau city	30 32 51	114 19 55	„

YELLOW SEA.

Place.	Particular Spot.	Latitude, North.	Longitude, East.	Authorities.
Whang-ho or Yellow river.	Entrance	34 2 0	119 51 0	Horsburgh.
Staunton island	Summit	36 47 0	122 16 0	„
Shan Tung promontory	Extreme	37 25 0	122 45 0	Ross.
Wei-hai-wei harbour	East end of Observatory islet.	37 30 19	122 7 0	Ward and Bullock, 1860.
Lung-mun harbour	Ta-shan	37 27 20	121 32 56	„
Chi-fau or Yen-tai harbour.	Fort in Village bay	37 35 56	121 22 33	„
Mian-tau group	Peak of Northern island	38 23 37	120 52 0	„
„	South-west extreme of Miau-tau island.	37 56 0	120 37 12	„
Ta-lien-hwan bay	Observation spot on isthmus on south San-shan island.	38 52 38	121 49 30	„
Encounter rock	-	38 33 50	121 37 0	„
Blonde island	-	39 2 0	122 49 0	H.M.S. *Pylades*, 1840.
Dangerous shoal	-	38 56 0	124 37 0	„

GULF OF PE-CHILI.

Place.	Particular Spot.	Latitude, North.	Longitude, East.	Authorities.
Pei ho	South Taku fort	38 59 52	117 39 19	Ward and Bullock, 1860.

TABLE OF POSITIONS.

Place.	Particular Spot.	Latitude, North.	Longitude, East.	Authorities.
WEST AND SOUTH COASTS OF KOREA.				
		° ′ ″	° ′ ″	
Chodo island	South point	38 27 0	124 34 40	French frigate *Virginie*, 1856.*
Deception bay	Middle of entrance	37 3 0	126 33 0	,,
Caroline bay	West point of entrance	37 1 30	126 25 0	,,
Joachim harbour	,, ,,	36 53 30	126 17 50	,,
Chassériau bank	South extreme	36 59 20	126 18 0	,,
Daniel island	West side	38 17 0	124 56 0	Horsburgh.
Sir James Hall group	North island	37 56 0	124 44 30	Basil Hall, 1816.
Marjoribanks harbour	-	36 25 0	126 25 0	Horsburgh.
,,	Mauzac islet	36 26 45	126 28 0	French frigate *Virginie*, 1856.
Tas-de-Foin islet	-	36 24 30	126 24 0	,,
Wai-ian-do island	-	36 15 45	126 9 50	,,
Basil bay	-	36 7 38	126 42 20	Basil Hall, 1816.
Guérin island	Summit	36 7 0	126 1 9	French frigate *Virginie*, 1856.
Alceste island	-	34 6 0	125 11 9	,,
Quelpart island	Observation spot on middle of west side of Bullock island.	33 29 40	126 58 25	Belcher, 1845.
Port Hamilton group	West point of Observatory island.	34 1 23	127 20 34	Richards, 1855.
ISLANDS OFF COAST OF CHINA.				
Pratas island	North-east part	20 42 3	116 43 22	Richards, 1858.
Balintang islands	Centre of group	19 58 0	122 14 0	Horsburgh.
Batan group	Islet off south-west point of Y'Ami island.	21 4 56	121 58 24	Belcher, 1843.
Gadd rock	-	21 43 0	121 41 0	Ross, 1817.
Vela Rete rocks	-	21 42 0	120 52 0	,,
Botel-Tobago sima	South extreme	22 1 40	121 39 45	Beechey, 1826.
Little Tobago sima	-	21 57 30	121 40 30	,,
Formosa island	Ape hill	22 38 3	120 16 30	Richards, 1855.
,,	Saracen head	22 36 14	120 16 33	,,
,,	Port Kok-si-kon, Observatory point.	23 6 0	120 5 0	,,
,,	Tam-sui harbour, Sand point.	25 10 6	121 26 6	Gordon, 1847.
,,	Ke-lung harbour, Ruin rock.	25 9 0	121 47 0	,,
,,	Foki point	25 19 0	121 37 0	Collinson, 1845.
,,	Petou point	25 8 0	121 57 0	,,
,,	Sau-o-bay, south point	24 36 0	121 53 0	Mr. Blackney, H.M.S. *Inflexible*, 1858.
Samasana island	-	22 41 0	121 28 0	Collinson, 1845.
Hoa-pin-su island	North face	25 47 7	123 30 31	Belcher, 1845.
Meiaco-sima group	Kumi island, north beach.	24 26 0	122 56 0	,,
,,	Broughton bay, landing place.	24 21 30	124 17 40	,,
,,	Port Haddington, Hamilton point.	24 25 0	124 6 40	,,
,,	Tai-pin-san, south-west bay.	24 43 35	125 17 49	,,
Lu-chu group	Napha-kiang road	26 12 25	127 42 20	Beechey, 1827.
,,	Deep bay, observatory spot at the head.	26 35 35	127 59 42	American Chart, 1854.
,,	Port Melville, Onting village.	26 40 42	128 0 0	Basil Hall, 1816.

* The *Virginie's* positions depend upon Quelpart island (observation spot on middle of west side of Bullock island) being 126° 58′ 25″ East from Greenwich, The position of Chodo is doubtful.

TABLE OF POSITIONS.

Place.	Particular Spot.	Latitude, North.	Longitude, East.	Authorities.
ISLANDS SOUTH-EAST AND EAST OF LU-CHU.				
		° ′ ″	° ′ ″	
Borodino islands	Centre of south island	25 52 45	131 12 17	American Chart, 1854.
Bishop rocks	Centre	25 20 0	131 15 0	Bishop, 1796.
Rasa or Kendrick island.		24 27 0	130 40 0	*La Cannonière*, 1807.
Parece Vela or Douglas reef.		20 31 0	136 6 0	Sproule, 1848.
Lindsay island		19 20 0	141 15 30	Lindsay, 1848.
Santa Rosa shoal	West extreme	12 30 0	144 15 0	Raper.
Guam island	Fort San Luis	13 26 0	144 45 0	French corvette *Uranie*, 1819.
Rota or Sarpan island	North-east point	14 12 0	145 23 0	,,
Aguijan island	Centre	14 54 0	145 38 0	,,
Tinian or Buena Vista island.	Sunharom village	14 59 0	145 43 0	,,
Seypan island	Peak	15 13 0	145 49 0	,,
,,	Magicienne bay	15 8 30	145 44 0	H.M.S. *Magicienne*, 1858.
Farallon de Medinilla or Bird island.	South point	16 0 0	146 7 0	French corvette *Uranie*, 1819.
Anatagan island	East point	16 20 0	145 47 0	,,
Sariguan island	Centre	16 40 0	145 52 0	,,
Zealandia breakers		16 50 0	145 54 0	Foster, 1858.
Farallon de Torres	Centre	17 18 0	145 57 0	French corvette *Uranie*, 1819.
Guguan island	East point	17 36 0	145 57 0	,,
Amalaguan island	North-east point	18 6 0	145 58 0	,,
Pagon island	North point	18 17 0	145 52 0	,,
Grigan island	,,	18 51 0	145 43 0	,,
Asuncion island	Peak	19 41 0	145 27 0	Beechey, 1827.
Uraccas or Mangs islands.	Centre	19 57 0	145 20 0	Lapérouse, 1786.
Guy rock	,,	20 30 0	145 32 0	Douglas, 1789.
Marshall or Los Jardines islands.	,,	21 40 0	151 35 0	Marshall, 1788.
Sebastian Lobos or Grampus islands.	South-west island	25 10 0	146 40 0	Raper.
Forfana island	Centre	25 35 0	143 0 0	,,
San Augustino island	Peak	24 14 0	141 20 0	King, 1805.
Sulphur island	,,	24 48 0	141 13 0	,,
San Alessandro island	,,	25 14 0	141 11 0	,, 1799.
Mal abrigos or Margaret islands.	Centre	27 20 0	145 45 0	Magee, 1773.
Bonin islands	Port Lloyd in Peel island.	27 5 35	142 11 30	Beechey, 1827.
,,	Newport in Hillsborough island.	26 36 0	142 9 0	American Chart, 1854.
Rosario or Disappointment island.		27 16 0	140 51 0	Raper.
ISLANDS NORTH OF LU-CHU.				
Yori sima	Centre	27 2 0	128 25 24	French Chart, 1846.
Yeirabu sima	South peak	27 21 0	124 31 34	,,
,,	,,	27 14 0	128 33 0	Collinson, 1845.
Tok sima	Highest peak	27 44 0	128 59 0	French Chart, 1846.
Iwo sima		27 51 0	128 19 0	Collinson, 1845.
Oho sima	North extreme	28 31 40	129 40 12	American Chart, 1854.
Kikai sima	Summit	28 18 0	129 57 30	,,
Yoko sima	,,	28 49 0	128 59 0	French Chart, 1846.
Tokara sima		29 8 0	129 11 0	,,
Sima-go islands	Highest	29 13 0	129 19 0	,,

TABLE OF POSITIONS.

Place.	Particular Spot.	Latitude, North.	Longitude, East.	Authorities.
ISLANDS NORTH OF LU-CHU—*cont.*				
		° ′ ″	° ′ ″	
Aknisi sima - -	- - - -	29 27 0	129 35 0	French Chart, 1846.
Suwa sima - -	- - - -	29 38 0	129 42 0	,,
Fira sima - - -	- - - -	29 41 0	129 31 0	,,
Naka sima - -	Peak - - -	29 53 0	129 50 0	,,
Hebi sima - -	,, - - -	29 55 0	129 32 0	,,
Kohebi sima -	- - - -	29 53 0	129 36 0	,,
Kutsino sima -	Summit - -	29 59 0	129 55 0	,,
Blake reef - -	Highest rock -	30 5 0	130 3 0	,,
Yakuno sima -	Mount Motomi -	30 21 0	130 29 0	,,
Seriphos rock -	- - - -	30 44 0	130 45 0	,,
Yerabu sima -	Highest peak -	30 27 0	130 11 0	,,
Take sima - -	Centre - - -	30 48 0	130 24 0	,,
Iwoga sima -	Highest peak -	30 42 0	130 17 0	,,
Powhattan reef -	- - - -	30 41 0	130 19 0	U.S. frigate *Powhattan*, 1860.
Trio rocks - -	Centre rock -	30 45 0	130 5 0	French Chart, 1846.
Kuro sima - -	Centre - - -	30 50 0	129 55 0	,,
Ingersoll rocks -	Highest - -	30 51 0	129 26 0	,,
Udsi sima - -	Largest - -	31 12 0	129 23 0	,,
Retribution rocks	- - - -	31 23 0	129 37 30	H.M.S. *Retribution*, 1858.
Nadiejda rocks -	- - - -	31 48 0	129 36 0	French Chart, 1846.
ISLANDS OFF SOUTH-EAST COAST OF NIPON.				
Lots Wife rock -	- - - -	29 47 0	140 22 30	American Chart, 1854.
Ponafin island -	- - - -	30 33 0	140 15 0	,,
Smith island -	- - - -	31 18 0	139 50 0	H.M.S. *Tribune*, 1859.
Bayonnaise island	- - - -	32 0 40	140 0 0	American Chart, 1854.
Onanga sima -	- - - -	32 30 0	139 50 0	,,
Fatziziu island -	Centre - - -	33 6 0	139 43 0	,,
Broughton rock	- - - -	33 42 0	139 17 0	,,
Meac sima -	- - - -	34 6 0	139 29 0	,,
Mecoura island -	- - - -	33 54 0	139 35 0	,,
Redfield rocks -	Centre - - -	33 56 50	138 49 0	,,
Kozu sima -	,, - - -	34 13 15	139 8 0	,,
Oho sima - -	South-east point -	34 39 30	139 28 0	,,
,, - - -	North point - -	34 47 30	139 24 0	,,
JAPAN ISLANDS.				
Kiusiu island -	Nagasaki harbour, Nezumi sima.	32 43 22	129 50 33	Richards, 1855.
Nipon island, south-east coast.	Gulf of Tutomi, Enora bay.	35 10 0	138 53 0	Russian frigate *Diana*, 1853-55.
,, - -	,, Heda bay -	34 58 11	138 48 0	,,
,, - -	,, Arari bay -	34 50 0	138 46 0	,,
,, - -	,, Tago bay -	34 47 3	138 46 0	,,
,, - -	Cape Idsu - -	34 36 0	138 50 35	American Chart, 1854.
,, - -	Rock island - -	34 34 20	138 57 10	,,
,, - -	Simoda harbour, Centre Island.	34 39 45	138 57 30	,,
,, - -	Yedo bay, Cape Sagami	36 6 30	139 42 45	,,
,, - -	,, Webster island	35 18 30	139 40 34	,,
Strait of Tsugar -	Islet off Cape Matsumac.	41 24 54	140 7 20	Richards, 1855.

TABLE OF POSITIONS.

Place.	Particular Spot.	Latitude, North.	Longitude, East.	Authorities.

JAPAN ISLANDS—cont.

Place.	Particular Spot.	Latitude, North.	Longitude, East.	Authorities.
		° ′ ″	° ′ ″	
Strait of Tsugar	Hakodadi harbour, entrance to Kamida creek.	41 47 8	140 45 34	Richards, 1855.
,,	Small islet on west side of Cape Nambu.	41 25 24	141 28 32	,,
,,	Red Cliff point	41 28 7	141 9 0	,,
,,	Centre of Low island off Toriwi saki.	41 33 34	140 56 36	,,
,,	North side of Tatsupi saki.	41 16 17	140 22 37	,,
,,	Small rock off south side of Cape Greig.	41 5 39	140 20 19	,,
Nipon island, west coast	South-west Bittern rock	40 31 0	139 31 0	,,
,,	Tabu sima	39 31 0	138 53 0	,,
,,	West point of Sado island.	38 1 0	138 17 0	,,
,,	Yútsi sima	37 50 30	136 55 0	,,
,,	Astrolabe rock	37 35 0	136 54 0	,,
,,	Cape Noto	37 28 0	137 22 0	,,
,,	North point of Oki islands.	36 30 0	133 23 0	,,
,,	Centre of Mino sima	34 48 0	131 9 0	,,
,,	Cape Louisa	34 40 0	131 36 0	,,
,,	Richards island	34 32 0	131 18 0	,,
,,	Obree island	33 51 0	130 2 0	,,
,,	Peak of Wilson island	33 54 0	130 25 0	,,
,,	Rock in centre of channel on north-west side of Firado island.	33 21 30	129 26 11	,,
,,	South side of Yenoi sima.	32 59 44	129 21 24	,,
,,	North side of island within Hardy harbour.	32 49 0	128 56 33	,,
,,	Peak of Kusa-saki island.	32 2 47	128 30 42	,,
,,	Pallas rocks, largest	32 14 17	128 13 30	,,
Yezo island	Volcano bay, Cape Yetomo.	42 21 0	140 56 30	American Chart, 1854.

KURIL ISLANDS.

Place.	Particular Spot.	Latitude, North.	Longitude, East.	Authorities.
Kunashir	St. Anthony peak	44 31 0	145 46 0	Golownin, 1811.
Chikotan	Centre	43 53 0	146 43 30	,,
Iturup	Cape Okebets	45 38 30	149 14 0	,,
,,	Cape Rickord	44 29 0	146 34 0	,,
Urup	Cape Kastrikum	46 16 0	150 22 0	,,
,,	Cape Vanderlind	45 39 0	149 34 0	,,
Brat Chirnoef	-	46 29 15	150 33 30	,,
Rebuntsiriboi	-	46 32 45	150 37 10	,,
Broughton	-	46 42 30	150 28 30	,,
Sinusir	Prevost peak	47 2 50	151 52 50	,,
Ketoy	South point	47 17 30	152 24 0	,,
Matua	Peak	48 6 0	153 12 30	Krusenstern, 1805.
Raikoke	,,	48 16 20	153 15 0	,,
Musir	-	48 35 0	153 44 0	,,
Shiash-kotan	Centre	48 52 0	154 8 0	,,
Kharim-kotan	Peak	49 8 0	154 39 0	,,
One-kotan	South-west point	49 19 0	154 44 0	,,
Makanrushi	Centre	49 51 0	154 32 0	,,
Shumshu	,,	50 46 0	156 26 0	,,
Alaid	,,	50 54 0	155 82 0	,,

TABLE OF POSITIONS. 429

Place.	Particular Spot.	Latitude, North.	Longitude, East.	Authorities.

KAMCHATKA, SOUTH-EAST COAST.

Place.	Particular Spot.	Latitude, North.	Longitude, East.	Authorities.
Mount Villcuchinski	Peak - - - -	52 42 0	158 20 0	Beechey, 1827.
Petropaulski - -	Church - - - -	53 0 58	158 43 30	,,

SEA OF JAPAN AND GULF OF TARTARY.

Place.	Particular Spot.	Latitude, North.	Longitude, East.	Authorities.
Sentinel Island -	- - - -	34 34 0	128 53 0	French corvette *Capricieuse*, 1852.
Tsus sima - -	Observatory rock, Tsus-sima sound.	34 18 55	129 12 0	Ward, 1859.
Matu sima -	Peak - - - -	37 22 0	130 56 0	Russian frigate *Pallas*, 1854.
Liancourt rocks -	- - - -	37 14 0	131 55 0	H.M.S. *Hornet*, 1855.
Chosan harbour -	Observation spot -	35 6 6	129 1 49	Ward, 1859.
Cape Clonard - -	- - - -	36 5 45	129 33 30	Russian frigate *Pallas*, 1854.
Port Lazaref -	Observation point, South 1½ miles from south end of Butenef island.	39 19 12	127 32 48	,,
Napoléon road -	Musoir rock, west point of entrance.	42 37 22	130 44 10	H.M.S. *Winchester*, 1855.
Guérin Gulf - -	Sandy point - -	43 9 0	131 50 0	,,
Hornet bay -	Fox island - -	42 41 0	132 56 0	H.M.S. *Hornet*, 1856.
Islet point - -	- - - -	42 49 0	133 51 0	,,
Port Michael Seymour	Observation spot at head of port.	43 46 0	135 19 0	,,
St. Vladimir day -	Low point - -	43 53 40	135 27 21	Ward, 1859.
Shelter bay -	- - - -	44 28 0	136 2 0	H.M.S. *Barracouta*, 1856.
Sybille bay - -	- - - -	44 43 45	136 22 50	,,
Pique bay - -	- - - -	44 46 15	136 27 15	,,
Bullock bay -	- - - -	45 2 0	136 44 0	,,
Luké point - -	- - - -	45 19 30	137 10 15	,,
Cape Disappointment	- - - -	45 40 30	137 38 15	,,
Cape Suffren -	- - - -	47 20 0	138 58 0	,,
Fish river - -	- - - -	47 55 0	139 31 0	,,
Low cape - -	- - - -	48 28 0	140 10 0	,,
Beachy head -	- - - -	48 56 0	140 21 0	,,
Barracouta harbour -	Tullo island - -	49 1 50	140 19 0	,,
Castries bay -	Quoin point - -	51 28 0	140 49 30	H.M.S. *Hornet*, 1855.
Jonquière bay -	- - - -	50 54 0	142 7 0	,,

LAPEROUSE STRAIT.

Place.	Particular Spot.	Latitude, North.	Longitude, East.	Authorities.
Risiri - - -	Pic de Langle -	45 11 0	141 12 15	Krusenstern, 1805.
Refunsiri - -	Cape Hieber - -	45 27 45	141 4 0	,,
Cape Notoro -	- - - -	45 54 15	141 57 56	,,
Cape Nossyab -	- - - -	45 25 50	141 34 20	,,
Dangerous rock -	- - - -	45 47 15	142 8 45	Lapérouse.
Cape Siretoko -	- - - -	46 2 20	143 30 20	Krusenstern, 1805.

EAST AND NORTH COAST OF SAGHALIN.

Place.	Particular Spot.	Latitude, North.	Longitude, East.	Authorities.
Cape Löwenörn -	- - - -	46 23 10	143 40 0	,,
Cape Tonin -	- - - -	46 50 0	143 33 0	,,
Cape Seniavin -	- - - -	47 16 30	142 59 30	,,
Bernizet peak -	- - - -	47 33 0	142 20 0	,,

Place.	Particular Spot.	Latitude, North.	Longitude, East.	Authorities.

EAST AND NORTH COAST OF SAGHALIN—cont.

Place.	Particular Spot.	Latitude, North.	Longitude, East.	Authorities.
Cape Moulovskoi	-	47 57 45	142 44 0	Krusenstern, 1805.
Cape Dalrymple	-	48 21 0	142 50 0	,,
Cape Soimonof	-	48 52 30	143 1 30	,,
Cape Patience	-	48 52 0	144 46 15	,,
Robben island	Centre	48 32 15	144 23 0	,,
,,	N.E. edge of reef	48 36 0	144 33 0	,,
,,	S.W. edge of reef	48 28 0	144 10 0	,,
Cape Bellingshausen	-	49 35 0	144 25 45	,,
Cape Rimnik	-	50 12 30	144 5 0	,,
Mount Tiara	-	50 3 0	143 37 0	,,
Cape Ratmanof	-	50 48 0	143 53 15	,,
Cape Delisle de la Croyère.	-	51 0 30	143 43 0	,,
Downs point	-	51 53 0	143 13 30	,,
Cape Ouméloi	-	52 32 30	143 14 30	,,
Cape Virst	-	52 57 30	143 17 30	,,
Cape Klokatcheff	-	53 46 0	143 7 0	,,
Cape Löwenstern	-	54 3 15	143 12 30	,,
Cape Elizabeth	-	54 24 30	142 46 30	,,
Cape Maria	-	54 17 30	142 17 45	,,

SEA OF OKHOTSK.

Place.	Particular Spot.	Latitude, North.	Longitude, East.	Authorities.
Great Shantar island	North point	55 11 0	137 40 0	,,
St. Iona island	-	56 25 30	143 15 45	,,
Port Aian	Cape Vneshni	56 25 28	138 25 50	Russian chart, 1851.

INDEX.

	Page
Abbey point	303
———— reef	304
Aberdeen Harbour	31
————— island	264, 265
Abulu river	269
Acahi—Fanahi point	314
Accar island	302
Achau island	28, 43
Acong rock	70, 71, 75
Actæon shoal	218
Adam peak	299
Adams, port	237
Adeloup point	314
Adolphe islet	252, 253, 254
Agagna bay and harbour	314
Agenhu island	304, 305
Agfayan bay	312
Agincourt island	299, 303
Aguijan island	315
Ahayan point	312
Aian port	418
Ai-chau islands	24, 25, 28, 44
Akuisi sima	326
Alaid island	370
Alamaguan island	319
Albert peak	151
———— peninsula	396
Alceste island	218, 220, 221, 230, 260
———— shoal	279
Aiemène island	327
Alessandro, San, island	320
Alfred peninsula	252
———— islet	254
Algerine point	172
Alligator point	172, 184, 186
———— island	135
Aliman point	385, 388
Amakirrima islands	302
American anchorage	345
———— reach	54, 58
Amherst rocks	181, 192, 196
———— point	57

	Page
Amoy harbour,	6, 7, 11, 97–101, 105, 106, 290
——— city	101
——— inner harbour	103
——— island	3, 9, 10, 101, 102, 104, 123
——— outer harbour	99
——————— entrance	99–103
Amphitrite strait	369
Amúr, gulf of	406
——— river	407
Anatajan island	318
Anbian village	384
Andrew, St., island	159
Angle island	140
Aniwa bay	411
Aniwa Vries, cape	411
Anjou point	388
Annenkof island	384, 387
Anson bay	51
Anthony, St., peak	366
Anung-hoy island	38, 51, 52, 53, 56
——— peak	48, 52
——— point	48, 49, 51, 52
Ao-shan island	167, 177
Apari	269
——— road	269
Apapa island	314
Ape hill	281–284
Apollos island	327
Apomee island	18
Apomi point	15
Apra village	313
Ap-tan-shan island	168
Arari bay	339
Arevief isles	389
Argonaut island	379
Argun river	407
Ariadne rock	192, 197
Arzobispo islands	321–323
Asses ears peaks	28
——— group	354, 357
Astrolabe rock	349

INDEX.

	Page
Asuncion island	319
Aube bluff	395
Auckland mount	262
Augustino, San, island	320
Avatcha bay	370–375
——— mount	372
Avos island	369
Avvakum islets	390
Awadji island	337
Awoota rock	95, 98
Babonshka rock	372
Babuyan Claro island	271, 272, 273, 275
Babuyan islands	270, 271–273
Baikal lake	407
Baily islands	321
Baksanova	407
Balcine rock	392
Balintang channel	275
——— islands	274
Baliuzeka peninsula	398
Ballast island	15, 16
Bangao islands	153
——— rocks	153
Bangui point	268
——— port	268
Barker island	208
Barlow island	263
Barn head	305
Barnpool anchorage	305
Barometer	3, 5
Barracouta harbour	400, 403
——— rock	356
Barren bay	150
——— bluff	403, 404
Barren island	157
——— isles	177, 178, 181, 182, 195
Barrete island	271, 272
Barrow bay	308
———, island	262
Basalt island	65, 66, 383
Bashi islands	3, 4, 7, 274–279
Basil bay	258, 259
Batan islands	270, 274–279
Bate island	71, 73, 74, 75, 76
——— rock	53, 56
Bateman island	157
Bay islet	65, 66

	Page
Bay rock	249
Baylis Bay	87, 88
Bayonnaise isle	328
Beachy head	403
Beacon hill	16, 170, 185, 186
——— house	344
——— point	344
——— rock	170
Beak head	161
——— channel, 12, 156, 158, 161, 162, 184
——— island	161
——— island	156, 161
Bear islet	155
Beaufort island	32
Becher islets	180
——— point	211
Beehive rock	178, 181
Belavenz mountains	383
Bell channel	168, 169, 184, 185, 186
——— island	92, 167, 168, 171, 184
——— rock	168
Bella Vista island	149
Bellingshausen cape	414
Bentinck bank	189
Bergasse point	393
Bernizet peak	412, 413
Biss bay	70–75
——— point	71, 73
Bielaya Skala island	390
Big island	73
Bill islet	85, 86
Bird island	128, 318
——— rock	159
Bishop rocks	310
Bit rock	182, 183
Bittern island	139
——— rock	133
——— rocks	347, 358
Black Cliff head	133
——— head	95, 96, 131, 132
——— Ink city	217
——— mount	82
Blackney reach	211
Black-peaked rock	119, 120
Black point	199
Black rock, 22, 126, 127, 135, 136, 170, 328
——— bay	10, 294
——— point	81
——— rocks	380, 96

INDEX. 433

	Page
Blackwall channel	12, 169, 172, 173, 184
——— island	172, 173
——— pass	172
——— point	173
Blake point	54
——— reef	327
Blanco, cape	346
Blenheim passage	58
Blockhouse island	193, 196
Blonde island	249, 250
——— rock	185, 187
——— shoal	201, 202
Blossom Reef	303, 304
Blue river	60
Bluff head	21, 68
——— island	65, 66
Blundell rock	100
Blunt, cape	360
Boat islet	40
——— rocks	89
Boca Tigris	19, 47, 48, 49, 51, 52, 53, 55, 56
Boddam Cove	23
Bodisco peninsula	392
Bojeador, cape	268, 270
Boliñao, cape	270, 271
Boltin, cape	389
Bonham isles	182
Bonin islands	4, 321–323
———, winds	331
Boot sand	6, 109, 111
Borodino islands	310
Bosquet peninsula	385
Botel-Tobago sima	8, 278, 279, 294
Bouët island	257
Bougarel point	390
Boungo channel	333, 376
Boussole channel	365, 368
Bower point	51, 52, 56, 57
Brass Basin island	143
Brat Chirnoef island	368
Breaker island	87, 88, 91
——— point	9, 10, 76, 83
Breakwater islet	121, 122
——— reef	293
——— rock	131, 134
Bremer channel	51, 52
Brig island	86, 87, 88
——— rock	88
Broadway	14–18
Broken island	173

	Page
Brood rocks	329
Brooke island	182
Brother A islet	132
Brother B islet	132
Brothers islets	36, 89, 92, 97, 103, 130
——— rocks	92, 97, 103, 178, 372
Broughton bay	229, 368, 382–388
——— cape	366
——— channel	378
——— island	368
——— rock	329
Bruat, cape	389
Brown channel	397
——— reach	58
Bruce port	395
Brunswick patches	54, 58
——— rock	54
Brydone island	398
Buckland island	321
Buena Vista island	315
Buffaloes Nose channel	155, 156, 158, 161, 184
——— island	155–157
Buisaco islet	341
Bullock bay	400
——— harbour	6, 144
——— island	262, 263
——— reach	210
Bullocks Head gate	27
Buoy rock	392
Burnet island	262
Bush island	191, 289
——— reef	68
Butenef island	386
Bythesea channel	213
Cabras island	314
Cairn hill	230
Cairnsmore rock	182, 183
Cake islet	71
Calayan island	271, 272
Cambodia, coast of	41
Cambrian pass	163, 184
Cambridge reach	58
——— rock	29, 30, 44
Camiguin island	269, 270, 271, 273, 274
Camille island	256
Camp islet	259, 260
Canpn	190

[C.]   E E

434 INDEX.

	Page
Canton	17, 55, 56, 61
——— river	5, 7, 14, 19, 28, 51–58
———, approaches to	14–50
———, directions	56–58
———, freshes out of	7
Cap island	247
—— rock	170, 172
Capricieuse bay	392, 393
Cap-sing-mun passage	37
Capstan Head rock	303
Cap Yit islet	117
Caravallo point	268
Careening harbour	397
Caroline bay	252, 253
Carr bank	403
Cassini island	392, 393
Castellated rock	143
Castle Peak island	37
Castle point	137
——— rock	35, 157, 323
Castles group	264, 265
Castries bay	400, 404
Casy point	390, 391, 393
Catherine, cape	400, 408
——— channel	366
Cau-chau islands	20, 21, 38
Cavndiao point	268, 271
Cecille archipelago	326–328
Cemetery point	305
Central islands	160
Centre isle	68
——— island	341
Cetti bay	313
Chaguie point	275
Chain islands	149
——— rock	52
Challum bay	88, 91
——— island	91
Changchi island	135, 136, 138
——— peak	134
Chang-kia-kau	211
Chang-kon-yam island	256
Chang-pih island	6, 173, 174, 179, 180
Chang-saug chau	205
Chang-shan island	225, 226, 227, 228, 230, 234
——— tail spit	228, 230
Chang-tan island	179
——— peak	179
——— strait	179
Channel banks	50

	Page
Channel island	219, 220, 221
——— rock	295
Chapeau rock	253
Chapel island	7, 97, 98, 103, 106
Chapu	189
——— bay	189, 190, 191
——— road	190
Charer island	256
Chasseriau bank	252
Chauan bay	91, 92, 94, 98
——— head	92
Chauchat rocks	97, 99, 103
Che-chin point	79
Chelang point	79
Chelsieu rocks	91
Cheng rock	151
Chesney island	182
Chiang ho	245, 246
Chichakoff cape	336
Chi-chau pagoda	209
——— islands	24, 27, 28, 29, 43
Chief bay	227
Chienlong mount	390
Chi-fau cape	221–225
——— harbour	221, 224
——— peak	223, 224
Chibachef, cape	401
Chih-seu island	100, 103
Chi-kau	232
Chi-kau ho	232, 233, 237
Chikhok island	148, 149
Chikotan island	366
Chi-kyan province	142
Childers rock	182
Chim bank	117, 119
——— island	116–119
Chi-ma-tau promontory	232
Chimmo bay	11, 108, 109, 285
——— point	108
——— rocks	108
Chimney island	94, 95, 96, 98
——— point	126
——— rock	308
China sea	1–9
Chin-chu	111, 290
——— bay	112
——— harbour	6, 109, 110
Ching ho	245
Chin Keang harbour	6, 166, 171
Ching rock	95
Chin-ha point	97, 98, 99, 101

INDEX. 435

	Page		Page
Chin hai	86, 185–188	Chu-shan pagoda	204
Chinkeamun harbour	6, 166, 176, 177	Chwang-shan island	254
Chin-kiang fu	206	Citadel hill	186
Chin-ki island	146, 147	Clair, St., island	328
Chino bay	80	Clam islet	115
——— hills	81	Claret rocks	137
——— peak	79, 80	Cleft islet	96
——— reef	80	——— rock	96, 141
Chin-quan island	141	Cliff island	92, 116, 151, 153
Chin san island	153, 181	——— islet	92, 113, 160, 173
Chin-tseao patches	99, 100	——— rocks	145, 146, 175, 178, 179
Chipounskoï cape	370	Clifford islands	251
Chirin-kotan island	369	Cliffy islands	117
Chirnoi island	368	Clio rock	23, 44
Chloe island	159	Clipper point	87, 88, 89
Chock-e-day village	294	——— road	87, 88, 89
Chodo island	250, 251	Clonard, cape	381
Choho pagoda	110, 111	Closter Camp point	404
——— reef	110	Club point	77, 230
Chomé isle	406, 408	Cluster islands	174
Chosan harbour	380	Coal harbour	291
Christmas island	210	Coal, Coal harbour	291
Chukea island	163, 164, 165, 184, 195	———, Fu-chu bay	238
——— peak	164	———, Hankau	215
Chung-chau-si island	26, 27, 46, 47	———, Jonquière bay	401, 405
Chus Peak island	163	———, Ke-lung	290
Chu-kiang	51–58	———, Nagasaki	357
Chu-ying bay	258	———, Niu-chwang	240
Chuck-tu aan island	26	———, Shah bay	308
Chuen-pee, fort	52, 56	———, Tam-sui harbour	288
———, hill	51	———, Teng-chau	225
———, island	49, 51, 52, 56	Coast islet	70, 75, 220
———, point	51, 53	Cochin China	3, 6, 8, 41
Chuen-pi island	168	Cocks Comb	61
Chuh isle	166	——— head	212, 213
Chuhpi island	140	——— rocks	389
——— pass	140	Cocos island	312
Chuh-seu island	149, 150	Cod, cape	223
Chukwan island	24, 28, 44	Codrika point	383
Chu-lu-cock island	36	Coffin islands	15, 88, 321
Chung island	32, 120	Coin island	144
——— point	187, 188	Coker rock	102, 103
Chung-chi island	296	Collinson island	213
——— point	109, 110	Colnet island	351
Chung-chau island	26, 27, 43, 44	Column point	395
Chung-hue island	37	Company's island	367
Chusan archipelago	3, 6, 9, 10, 11, 12, 155, 157, 158–184	Cone islet	385
		——— hill	187
———————, northern part	177–192	——— island	70, 71, 137, 138, 139, 152, 265
——— island	162, 163, 164, 165–177, 179, 184, 186, 195	——— peak	92, 93, 188
		——— rock	157, 348, 351

436 INDEX.

	Page
Conical Hill island	161
—— islet	360
Conic isle	66
Constantine rock	326
Contest patches	188
Conway island	161
Cony island	136, 138, 139
—— islet	136
Cooper group	326
Cordelia rock	30, 31
Corkers patches	155, 156
Cork point	97
Cornorandière rock	252
Cornwallis stone	101, 102, 103
Couding island	133, 134
Court reach	212
Cows Horn peak	119, 120, 121
Cox point	136
Crab islet	89, 90, 137
—— point	96
Crack islet	173
Crag island	138
—— peak	290
Craig island	295, 303
Crate island	149
Creek point	242
Crescent island	68, 69
Crillon cape	410
Crookback island	148
Crooked island	68, 69
Crown island	324
—— point	381
Cumbrian reef	277
Cum-sing-mum harbour	21
Cupchi island	82, 83
—— point	81, 82, 83
Current from Formosa to Japan	10, 335, 341
Currents in China sea	6–8
—— Japan sea	376
Currie channel	397
Cuvier peninsula	393
Cyclones	4, 304, 332
Dagelet island	379
D'Aguilar, cape	34
Dajette island	379
Dalrymple, cape	413
Dalupiri island	271, 272

	Page
Damson islets	160
Daneono island	312
Danes island	54, 55, 57, 58
Dangerous reef	249
—— rock	140, 409
—— shoal	209
Daniels island	254
D'Anville, gulf	391, 392
D'Après, cape	389
Dansborg island	95
Daussy point	390
David island	157
Deadman island	184, 185, 186
Deans Dundas port	396
Deception bay	252
Deep bay	307
Deep water bay	31
Deer island	167, 169, 170, 171, 177
—— channel	170
Delisle de la Croyere, cape	414
Dent, cape	404
Dénudées islands	257
Dequez island	274, 276
Desfossés point	384, 385
Destitution, cape	404
Devils peak	41
Dezima shoal	357
Diablo point	314
Diamond, cape	340
Diane strait	368
Didicas rocks	270, 274
Difficult islet	91
—— point	91
Dike islet	151
Dile point	270
Diogo island	274, 277
Dioyu reef	91
Diplo islet	137
Direction bluff	373
—— mount	392, 393
Directions, Formosa channel	287
——, Gulf of Pe-chili to Hong Kong	12
——, Hong Kong to Yang-tse kiang	9–12
——, Japan islands	341, 343
——, Singapore to Canton	40, 58
——, —— to Hong Kong	40–42
——, Yang-tse kiang	195
Disappointment, cape	400
——, island	323

INDEX. 437

	Page
Disaster island	326
Djaoré, cape	408
Dodd island	103, 106, 107
—— ledge	107
Dodo rocks	346
Dome bay	125, 126
—— hill	123, 126
—— island	90, 125
—— peak	285
—— point	294
Dot rock	90
Double haven	69
—— head	359
—— island	68, 69, 86
Double Peak island	138, 139
—— rocks	222, 223
Double-topped mountain	308
Double Yit islet	117
Doub rocks	175, 178
Douglas reef	311
Doves nest	202
—— point	210
Downs point	415
Druid head	104
—— island	163
Duff rock	53, 56
Duffield pass	10, 158, 159, 160, 183, 184
—— reef	159
Dumb islet	185
Dumbell bay	73
Dundas cape	262
—— rock	170
Dun-gan river	386
Dunsterville group	173, 191
Duroc, cape	382
D'Vissers island	126
Dzino island	338
Dzinosetto strait	338
Ear island	218
East cone peak	69
Eastern river	386
East island	123, 283
—— islet	165
—— peak	106
—— rock	165
—— Saddle island	178
—— White Stone	80, 81
E-chau head	30
Eddy island	223
Eden island	263
—— point	178
Edible Plant city	231
Eighteen Yits	114, 117
Elephant island	160, 162, 163, 167, 168, 169, 170
Elgin reach	207
Elias mount	232
Elizabeth, cape	415, 416, 420
Ellicott isle	179, 180
Elliot passage	58
Elliott islets	182
Ellis island	230
Endermo harbour	365
Encounter rock	247
Engaño, cape	269, 270
English reach	55, 58
Enora bay	338
Entrance head	78
—— hill	78
—— island	91, 168
Esamu, cape	360
Esarmi point	365
Eugénie archipelago	396
Facpi point	313
Fakew island	144
Fakuda saki	356
Fall island	150
—— peak	93, 94, 98
False Capstan Head rock	303
False Saddle island	181
Fan-lo-kong harbour and village	73
Fan-si-ak channel	49, 50
—— islet	38, 39, 50
Fa pew	61
Farallon de Medinilla island	318
Farallon de Torres island	318
Farmer rock	142
Fatsiziu island	329
Fat-shan	60
Feklistoff island	418
Feleny point	383
Fellowes passage	397
Fernande island	252, 253, 254
Ferry, the	132, 133
Fielon island	139
Figure rock	81
Fiki channel	333
Finger rock	223

	Page
Fiogo port	338
Firado island	352, 353
Firando island	352
Fira sima	326
Fir cone rock	149
First bar	54, 61
——— island	54, 57, 58
——— shoal	54
First Cone point	157
Fisher island	6, 122, 124, 125, 126, 127, 128, 173, 321
Fishermans group	158, 165, 175, 178, 181
——— rock	226, 229, 230, 231
Fish river	402
Fitton bay	322
Fitz-Roy island	209
Five Brothers rock	82
Five islands	271, 274
Flag island	119
Flak island	107
Flap island	139
Flare island	147
Flask island	144
Flat bay	414
——— cape	238
——— island	54, 57, 58, 67, 126, 133, 163
——— reef	76, 108
——— rock	79, 83, 134
——— Top isle	238
Flats opening	17
Fog islands	190, 191
Fogs	12, 402, 420
Fokai hills	74, 76
——— point	74, 75, 76, 77
Foki point	286
Folkstone rock	88
Fong-chuen	62
Fong-ho islet	139
Fong-whang group	143, 144
Forcade rock	327
Forester rock	60
Forfana island	320
Formosa banks	129
——— channel	3, 10, 12, 13
——— island	2, 3, 4, 6, 8, 9, 10, 11, 123, 126, 278, 279–296
——— island, W. coast	279–285
———, E. coast	291–296
———, N. coast, directions	285–291
Fort head	87, 91
——— bay	10

	Page
Fort hill	146
——— island	86, 87, 88
Fortescue island	403
Fortaumun pass	63, 64
Foto channels	12
Four-feet rock	26
Four-feet needle rock	44
Four Sisters rocks	178
Fox island	397
Freeman point	403
French island	54
——— river	58
Friendly islands	187, 188
Front island	156
Fu-chau bay	130, 222, 225, 237, 238, 239, 240, 241
Fu-chau fu	9, 121
Fuga island	269, 271, 272, 273
Fuh-ning	139
——— bay	137, 138, 139
Fuh-yan island	140, 141
——— pass	140
Fu-kyen province	288
——— boundary	91, 117, 142
Fu-mun entrance	51
Fung bay	66
——— head	66, 67
Funghwa river	157, 186
Funing island	179
Furious rock	205
——— shoals	210, 212, 215, 216
Furngelm island	392
Fu-shan	203, 204
——— hill	203
Fu-to island	159, 160
Futz-kau	212
Fuynng Quoin island	232
Fuzi, mount	333
Fuzi-yama	333, 342
Gadd rock	277, 278
Gales in China sea	6
Gamalcy, cape	346, 356, 364
Gamova cape	391
Gan-ching rocks	175
Gan-keang harbour	255
Gan-su island	
Gapan islet	314
Gap islet	142
——— rock	28, 29, 44

INDEX. 439

	Page
Gap tree	264
Gantau island	66, 67, 69, 150
Gauze islet	25
Gavarea cape	370, 372
Gay-unc islet	29
Gee-fou island	53, 54, 56, 57
—— rock	54
George island	31
Germantown reef	325
Ghiruma island	302
Giffard island	263
—— river	387
Gilbert river	398
Glasenap island	408
Glue city	217
Gnaton village	314
Goat island	78, 274
Golden island	206
Golovachef, cape	406
Golovnin strait	369
Goncharof, cape	388
Good Hope, cape of	10, 83, 84, 85, 86
Goodridge point	139
Goo rock	95
Gordon bay	141
—— islet	135, 137
Gore island	151, 152
Goshkevich bay	390
Gotto, cape	353
Gough pass	160, 161, 183, 184
Grain islet	174, 175
Grampus islands	320
Grass island	67, 68
Grassy Tongue	57
Grave island	171
Gravener island	213
Great Bamboo island	227
—— Black island	228
—— Ladrone island	6, 14, 21, 22, 24, 28, 41, 44, 45
—— Lu-chu island	302
—— Pass	60, 61
—— San-pwan island	144, 145
—— Volcano island	319
—— Wall of China	237, 241, 242
—— West channel to Canton river	14, 21, 45, 46
Green island	41, 42, 72, 73
—— islet	65, 85
Greig, cape	358, 364
Griffin rock	67

	Page
Grigan island	319
Guajan island	311
Guam island	311
Guardhouse isle	166, 168, 169
Guerin gulf	395
—— island	256
Guguan island	319
Guinapac rocks	270, 273, 274
Gulai	112
Gulf rock	404
Gull island	161
—— point	283
Gun cliff	359
Gutzlaff island	12, 182, 183, 190, 191, 192, 193, 194, 196, 197, 198
Guy rock	320
Haddington, port	297, 300
Hadshi bay	403
Hae head	118, 121
Hagemir isle	406
Hai-tan bay	118
—— island	117–122
—— point	118
—— strait	6, 9, 11, 109, 115, 118–120, 122
——, pilots	109
Hailing island	41
Hai-mun	83, 149
—— bay	83, 84
—— point	84, 85
—— river	84
Hainan head	45
—— island	5, 6, 8, 41
Hai-ning	189
Hai-ye-tse	245
Hak-chau island	25
Hakodadi bay	361
——, weather	334
—— harbour	262, 361, 362
—— head	346, 360
Halezof cape	406
Half-tide reef	90
—— rock	133, 154
Hall island	178
Hamelin strait	396
Hamilton point	300
—— port, group	261, 262, 264, 265
Hand bay	249
Han river	10, 85, 86, 87, 214, 215
Hanadi islet	309

INDEX.

	Page
Hang-chu bay	12, 177, 178, 188, 189, 190, 191
Hang-chu fu	190
Han kau	191, 214, 215
—— reach	214
—— to Wusung	215
Han-yang	214
Harbour group	72
—— island	68, 325, 352, 353
—— rock	291
—— Rouse	12, 156, 161
Harlem bay	70, 74, 75
—— peak	75, 76, 77
Harp island	295
Harty island	175
Harvey point	202
Hasyokan island	299
Hat islet	76, 77
—— rock	253
Ha-tse island	167
Hauseu island	103, 104
Haycock islet	251
Hea-che-mun channel	162
Hea-chu islet	148
Heang-shan	17
Hea-ta island	148
Hebe island	159
—— islet	74, 75
—— lock	146, 147
—— reef	134
Heber reef	304
Hebi-sima	326
Heda bay	339
Helens island	258
Hele rock	308
Hen and Chicks rocks	183, 190
Hen point	209
Heong-kong bay	31
Herbert island	308
Hesper rock	226, 229, 230
Hewen rocks	109, 110
Hie-che-chin bay	77, 79, 80, 83
Hienfung mount	383
—— mountains	389
Hieshan group	151, 154, 155
High Cone peak	114, 117
—— Fair Head hill	93
—— islands	65, 66, 92, 123, 124, 125, 129, 274
—— Shar peak	131
Hill islet	25, 29

	Page
Hills passage	59
Hillsborough island	321
Hiogo port	338
Hirase rock	357
Hoa-ock islet	22, 24
Hoa-piu-su island	295
Hoc-kang fort	17
Hodo island	388
Hokeang bank	132
—— island	130, 131
Hokeen island	145
Hok-tau point	17
Holderness rock	154
Hole island	66
—— rock	305
Hong-hai bay	77, 78, 79
—— island	77, 78
Hong-hau	18
Hong-Kong island	9, 12, 13, 14, 31, 32, 33, 34, 35, 40, 42, 55, 63, 270
—— road	33, 34, 37, 41, 63
Hong-shan river	17
Hook islands	171
Hooper island	262, 263
Hope bay	85
—— sound	227, 228, 230
Horns, N.W. & S.W., Rugged island	183
Horner peak	336
Hornet bay	397
—— islands	379
Hou point	125
Houbland islets	174
Hou-ki island	229
Hounlodgna bay	312
House hill	97
—— islet	189
Hsin-shai-kau	244
Hueung river	14–18
Hugon cape	391, 393
Hu-i-tau bay	7, 11, 101, 106, 107, 109
—— point	101, 106, 107, 108
Hulu sban bay	237, 238, 241
Hunchback hills	64, 68
Hung rocks	116, 117
Hungry rock	357
Hungwha channel	115, 117
—— river	115, 116
—— sound	6, 109, 113, 114, 115, 116, 117
—— pilots	109
Hunter islands	157, 212
Hut islet	96

INDEX.   441

	Page
Hut island	163
Hutau bay	145
Hutau island	145, 146
Hu-tau-shan head	95
—— river	95
Hutung point	81
—— river	81
Hwang bay	225, 230
Hwang-chau	215
—— pagoda	213
Hwang-ching islands	226, 227, 229, 230, 231
Hwangkwa islet	100
Hwang-shan hill	204
—— bay	204
Hwang-shih-kang	213
Hwayuen-chin	210
Iago, San, fort	19, 20
Ia-sha island	205
Ibayat island	274, 276, 277, 278
Ibugos island	274, 276
Ichey island	309
Idsu cape	340, 342
—— peninsula	339
Ilary point	384
Ile de la Selle	260
Image point	289, 290
Imperial port	403
Inago, mount	313
Incog islands	140, 141
Ingersoll patches	305, 344
—— rocks	328
Ingoda river	407
Inflexible reef	291
Inside island	15, 77, 136
Insular point	186
Iona, St., island	418
Irada mount	275, 278
Isaac island	301
Island head	225, 230
Islet point	397
Ismennai bay, point, and rock	373, 374
Isthmus island	136, 165
Iturup island	366
Ivana, port of	275
Iwoga sima	327, 336
Iwo sima	324, 356
Iye sima	308

	Page
Jansen rock	162
Japan islands	266, 333–358
——, climate	334
—— sea	266, 376–400
——, current	10, 335, 336, 341, 346, 376
——, winds	334, 376
Jar point	145
Jardine point	58
Jardines, Los, islands	320
Jin island	65
Joachim bank	86
—— bay	251
—— harbour	252
John Peak island	162
Jokako peak	92
—— point	11, 92, 93
Jona, St., island	418
Jones cove	67
Jonquière bay	405
Joss House bay	30
—— islet	282, 285
Jow rock	176
Julie island	327
Julo island	324
Junk channel	160
—— creek	51, 58
—— harbour	302
—— head	108
—— island	58, 123, 146
—— passage	55, 291
—— tides	290
—— Sail rock	119
Justina shoal	106
Just-in-the-Way islet	12, 169, 184, 185, 186
Kabe sima	353
Kada	338
Kagosima	336
Kai-chu fu	239
Kaikong island	18, 19
Kai-yik-kwan	61
Kakisaki	340
Kalcewan river	292
Kamchatka, S.E. coast	370–375
Kamida creek	361
Kamino sima	356
Kamisaki, cape	343, 344
Kanagawa	345

INDEX.

	Page
Kanlan point	174
Ka-o islet	20
Kao-shan island	227, 229, 231, 234
Katona sima	325
Kau-li-tau-shan	93
Kaulun bay	33
—— peninsula	33
—— village	68
Kawatsu bay	343
Ke-chau	212
Kee-ow island	21
Kei island	302
Kellet bank	41
—— island	34
Ke-lung	289–291
—— harbour	6, 289–291
—— island	289
Kemong harbour	146
Kem-suc islet	79
Kendrick island	310
Kerama islands	302, 304
Kera sima	302
Kerr island	116
Keshen point	52, 53
Ke-sin-she island	77
Ke-tau	12, 212, 213
Ketau point	160, 168, 215
—— shore	160
Ketoy island	368
Ketsu island	173
—— pass	173
Keui island	65
Keun shan	204, 205
Keu-shan island	175, 178
Kharim-kotan island	369
Khing-khan Ulla mountain	407
Kiabtsz	82
Kiangshan hills	118, 121
Kiang-yin	204, 205
Kibats bay	356
Kiddisol island	171, 173
Kieu-hien	208
Kien-tyau	152
Kieshi-wei	80
Kikai-sima	325
Kiku-saki	378
King-chu fu	239
King William isle	328
Kin-ho island	176
Kin-men island	150
Kino channel	333, 338, 376

	Page
Kinpai bluff	133
—— pass	132
—— point	132, 133
Kin-shan bluff	214
Kinsiang point and village	79, 83
Kintang channel	184
—— island	6, 170, 172, 173, 184, 186, 188
Kin-yu islet	79
Ki-seu island	100, 103
—— pagoda	102, 105
Kite bank	188
Kiu-hien	208
Kiu Kiang	211, 216
Kiusiu island	350–357
Kiu T'oan beacon tower 192, 193, 196, 198
Klaproth point	393
Klokatcheff, cape	415
Knob island	116
—— point	223, 224, 396
—— reef	68
—— rock	96
Ko channel	185
Kohebi sima	327
Ko-ho island	14, 15, 18, 19
—— point	20
Koisakoi, cape	406
Kok-he-mung harbour	282, 284, 285
Kok-si-kon port	282, 283–285
Ko-kún-to group	259, 260
Komagadaki peak	361
Kolokozev point	390
Korean Archipelago	231, 236, 250, 258, 259, 262
Korea, east coast	379–391
—— gulf	382
—— strait	376, 378
——, west and south coasts	250–265
Kosiki islands	355
Ko sima	363
Kotian island	256
Kouching point	169
Kowloon bay	33
—— peninsula	33
—— village	68
Kowlui head	133
Kozu sima	329, 343
Krafto island	412
Krusenstern channel	378
Kudaka island	309
Kuï island	308

INDEX. 443

	Page
Ku-kien-san island	297
Kuktúi river	419
Kulangseu island	101, 102, 103
———, channel west of	103, 104
Kume sima	302, 304
Kumi head	303
——— island	296
Kum-kwoh-shek	61
Kunashir island	366
Kung-kung island	222, 223, 224
Kung-kung-shan islands	221–224
Kuper island	58
Kuprianof island	384
Kurama island	304
Ku-ree-mah island	297
Kuril islands	266, 365–370
——— strait	370
Kuro sima	328
Kuro siwo or Japan stream	10, 335, 336, 341, 346
Kusa-saki island	354
Kusau mount	108
——— pagoda	108, 110
Ku-shan hill	205
——— point	203, 204, 205
Kutsino sima	327
Kutsnose rocks	357
Kwa-fau rock	171
Kwan island	174, 175, 179
Kwang-li	60, 61
Kwang-si	61, 62
Kwei channel	174
Kweiling	61
Kwei-tau rock	29, 30
Kweshan islands	154, 155, 156
Kwimun channel	174
Kwing bay	118, 122
——— island	118
Kwi-si-hill	175
——— island	174, 179
Kwo-kan islet	171
Kwokeu	160
Kyau-chu	217
Kyoko island	309
Kypong islands	28, 41, 44, 45
Kysi lake	405
Ladrone islands	6, 14, 44, 45, 46, 311–320
Lafsami island	21, 27, 28, 43, 44
Lai-chau bank	232

	Page
Laï-chau-feu	231, 232
Lakeah island	165
Lakeati island	165
Lambay island	280
Lam-hong-ho bay	293
Lamma channels	31, 32, 41
——— island	25, 31, 32, 33, 41, 43
Lamon rocks	90
Lamock islands	83, 90, 103
Lam point	115, 116
Launtia island	97, 98, 99
Lam-yit channel	115
——— island	11, 114, 115, 116, 117, 122
Lang-kiang-ki	209
Lang-shan crossing	203, 204
——— pagoda	203
Lanjett islets	155, 158
Lankeet flat	39, 40
——— island	39, 40, 47, 48, 49, 59
——— road	39
——— spit	40
Lan-sew bay	165, 177, 181, 183, 184, 195
——— island	174, 175, 179, 184, 195
Lan-shan island	174
Lantao channel	27, 28, 43, 44, 45
——— island	27, 28, 32, 35–37, 43, 44, 47
Laoush rock	162, 184
Lapelin rocks	327
Lapérouse strait	376, 409–412
Lark bay	15
Larne islet	11, 135, 136
——— rocks	135, 328
Larva rocks	136
Latea island	171
Lau-mu ho	244, 245
Lava islet	270
Lazaref cape	400, 406
——— port	379, 382, 385–387
Lea-ming island	151, 152, 153
Ledge island	170
Lee rock	212, 213, 215
Leeehin point	125, 129
Leenhwa-yang	164
Leeo-lu bay	11, 105
——— head	103, 105, 106
——— hill	106
Lema channel	35, 42, 45, 70
——— islands	9, 25, 29–31, 32, 33, 41, 42, 45, 70
Lesseps, cape	404
Le-tau island	218

INDEX.

	Page
Leuconna island	177, 178, 181, 182
Leu-cung island	218, 219, 220, 221
Lexington reef	- 305
Liau-tung gulf	230, 231, 237–246
——— province	222, 247
Liang-kiau	- 281
——— bay	- 279, 280, 281
Liak cape	- 408
Liancourt rocks	- 379
Liau ho	- 241
Liau-tie-shan head	- 225, 246, 247
——— channel	- 226
Lights, Avatcha bay	- 372
———, Fisher island, Pescadores	- 127
Light vessel, Yang-tse kiang	- 192
Linden point	- 390
Lindsay island	255, 256, 257, 258, 311
Lindsey port	- 353
Lingting island	25, 26, 31, 32, 41, 43, 44
Lin island	171, 172
Linschoten islands	326–328
Lintin bar	- 38, 39, 40, 48, 49, 50
——— island	21, 37, 38, 39, 45, 46, 47, 48, 50
——— peak	- 21, 38, 47, 48
——— sand	- 47, 49
——— spit	- 38, 44
——— south sand	- 38, 49
Lishan bay	- 140
Li-tsin ho	- 232
Litsitah point	- 126
Little Black island	- 228
——— Botel-Tobago sima	277, 279
——— Ladrone island	-14, 21, 22, 44
——— Orphan islet	210, 211
——— pass	- 140
——— Quemoy island	- 105
——— Sau-pwan island	- 144
Liu-kiu islands	301–309
Liungnib island	- 22, 24, 26, 44
Liu-sia kwang	- 243
Liwan island	163, 177
Lloyd, port	- 322
Lo-chau island	32, 34, 35
Lockyung river	- 109
Loka island	- 165
Lokaup group	- 71
——— island	- 71, 72, 75
Lokea island	- 163, 176, 177
Loktaou islands	- 255
Loney bluff	- 263

	Page
Long harbour	67, 68
——— island	- 227
Look-out hill	- 187
Lopatka, cape	370, 416
Losef point	- 388
Losing island	- 130, 133, 134
——— spit	- 133
Los Jardines islands	- 320
Lot-sin bay	146, 147
Lot's wife rock	- 328
Louisa, cape	- 350
Louis, port	392, 394
Lourmel port	257, 258
Loutz rock	- 114
——— shoal	- 114
Low cape	- 402
——— Chikok island	148, 149
——— island	- 73
——— islet	359, 364
——— point	- 224, 225, 398
Löwcuörn, cape	- 413
Löwenstern, cape	- 415
Lu-chu islands	301–309
———, directions from Hong Kong	- 303
———, winds	- 321
Luhwang, cape	- 158, 159, 162
——— island	12, 156, 158, 159, 161, 162, 179, 183
Luké point	- 400
Luminan reefs	- 314
Lutai bay	- 232
Luzon, north coast of	- 1, 2, 3, 7, 8, 268–271
———, west	- 68
Lyemun pass	33, 41, 63
Lynx rock	110, 111
Lyons river	- 397
Lyra island	- 261
Ma-aou point	- 174
Mabag island	- 271
Mabatui point	- 276
Mabudis island	- 274, 277, 278
Macao	6, 7, 14, 16, 17, 19, 20, 21, 46, 55, 58
——— fort	- 17
——— harbour	18, 19, 20, 270
——— island	20, 21
——— road	15, 17, 19, 20, 27, 43, 44, 45, 46, 47

INDEX. 445

	Page
Macarira island	17, 18, 19, 20
Macclesfield bank	- - 40, 268
——— island	167, 169, 170, 171, 184
Mace point -	- - 76, 77
Macedonian mound -	- - 290
Mackau island	- - 260, 261
Madreporic hills	- - - 312
Magicienne bay	- - - 317
Magnetic head	- - - 380
Mah-chau island	- - 49, 50
Mahiloue hills	- - - 312
Mahon island	- - - 262
Mah-wan island	- - - 37
Makanruru island	- - - 368
Makanrushi island -	- - 369
Makung -	- - 125, 126
——— citadel	- - 126, 127
——— harbour	- 6, 125, 126, 127, 129
Mal abrigo islets	- - - 321
Malay peninsula	- - - 7, 8
Malmiusk island	- - - 419
Ma-lo-chau islets	- - - 20
Malte Brun point	- - - 393
Man-mi-chau islets -	- - 28
Mañañion bay	- - - 275
Mandarin bluff	- - - 345
———, Great, point	- 253, 258
Mandarins cap	- - - 5, 6
Mangs islands	- - 319, 320
Manila -	- - 270, 275
Man-san island	- - - 21
Maoutze island	- - 176, 177
Margaret islands	- - - 321
Maria, cape -	- - - 416
Mariana or Ladrones islands	311–320
———————, winds -	331
Mariner reef	- 158, 180, 181
Marizo harbour	- - - 312
Marjoribanks harbour	- 254–258
Marlin Spike peak -	- - 114
Marolles islet	- - - 253
Marryat island	- - - 263
Marshall islands	- - - 320
Mas-kong island	- - - 31
Mason island	- - - 202
Masou peninsula	- - 286, 290
——— valley	- - - 286
Matheson point	- - - 34
———, port	- - 111, 112
Mathilde group	- - 256, 257
Matna island	- - - 368

	Page
Matsou island	- 134, 135, 137, 138
Matsumae, cape	- - - 362
Matu sima	- - - 379
Ma-urh point	- - - 85
Mauzac islet	- - - 256
May port -	- - - 396
Meac-sima group	- 354, 355, 357
Mecoura island	- - - 329
Medusa creek	- - - 157
Meiaco-sima group -	- 296–301
———————, directions	- 298
Meichen island	- - - 113
——— sound	- 11, 112, 113, 114
——— village	- - - 113
Meih-ting island	- - 175, 176
Meih-yun island	- - - 175
Mei-shan group	- - - 225
——— island	- - - 160
Melros point	- - - 301
Melville channel	- 169, 170, 171, 184
——— port	- - - 308
——— rock	- - 170, 172
Memoyah -	- - - 359
Menalai rocks	- - - 379
Mendoza island	- - 74, 75, 76
Menschikof, cape	- - - 406
Merope bay -	- - - 290
——— shoals	- - 97, 98
Mesan group	- 155, 156, 158
Me sima -	- - - 354
Miako -	- - - 337
Mian-tsin-liang point	- - 383
Miau-tau group	225, 226, 227, 228, 234
——— island	- 226, 228, 230, 231
——— strait	- - 225, 230
Michaud islet	- - - 383
Michael Seymour port	- - 397
Mid-channel reef	- - - 110
Middle Dog island -	121, 122, 131
Middle Ground 132, 133, 169, 170, 188, 190,	
	198, 199
Middle group	- - - 72
——— islet	- 91, 93, 94, 157
——— rock	- - 74, 75
——— rocks	- - 71, 75
Midway islands	- - - 174
Miles island	- - - 180
Millers Thumb	- - - 162
Min-ka -	- - - 288
Min river	- 9, 10, 11, 121, 130–134, 290
Mingan -	- - - 133

446 INDEX.

	Page
Mingan pass	133, 134
—— rock	394
Mino sima	350
Mirs bay	5, 64, 66–71
—— point	66, 67, 69, 70
Misana islet	341
Misima Mada	333
Mississippi bay	345
—————— rocks	355
Mitsuna island	297, 300, 301
Mitsuse rocks	357
Mochang-shi islet	229
Modeste island	260, 261
Mong-chau island	15
Montreal island	261
Monneron island	409
Monsoons in China sea	2, 3
Montagu island	151, 154
Montanha island	14, 15, 16, 18, 20, 22, 46
—————— peak	15
Montgomery group	307
Mordvinof bay	413
Morgan point	15
Morrison islands	183
—————— rocks	328
Morton point	207
Moto fort	17
Motoe islets	36
Motomi, mount	327
Motubn	307
Mouchez island	383
Moules, Anse aux	392
Mound peak	113, 223, 224
Mo-un islet	174, 175
Monse rock	155
Mud islet	149
Muravief point	385, 387, 408
Murray sound	259, 261
Musa bay	271, 272, 273
Muséef point	388
Mushroom rock	140, 151
Musir islands	369
Musoir rock	393, 394
Musquito islands	265
Mussel cove	392
—— point	360
Nab rock	168
Naeane point	340

	Page
Nacosi	307
Nadicjda cape	362
—— rocks	355
—— strait	368
Nagasaki harbour	334, 356–358
Nagiba, cape	417
Naguh	307
Naka sima	326
Nakazuni cove	307
Nakhimof peninsula	385, 388
Nambu, cape	346, 359, 364
—— peninsula	359
Namki islands	12, 142, 143
—— peak	142
——, port	142, 143
Namoa island	5, 9, 10, 84, 85, 87, 88, 89, 90, 91
—— peak	89, 91
——————, knolls off west end	87
Namoh village	69
Nam-pan	146
Nam-quan	141
—————— bay	137, 141
—————— harbour	141, 142
Namshan point	51
Nam-ye-kok point	16, 17
Nangaou bay	88, 89, 91
Nanho island	188, 189, 190
Nanking	191, 205–209, 212, 214, 216
—— reach	209
Nantai Wúshan pagoda	98, 99, 102, 103
Napha-kiang road	302–306
Napoléon gulf	396
—————— road	392, 379, 393
Narrow island	58, 72, 73
Natchijen mountains	307
Nau-tau-mun island	27
N.E. island	165
Nee sima	330
Needle rocks	25, 250
Neegata port	348
Nemesis rock	187
Nertchinsk	407
Net island	70, 71
Neva river	414
Nevelskoi, cape	408
New port	321
Nezumi island	356
Ngau river	146
Ng-chu	61
Niaow island	144, 145

INDEX. 447

	Page		Page
Niegata port	- 438	N.W. Outlier patch -	- 128
Nikolaevsk -	406, 408	Nyew-tew island	151, 153
Nikolski island	384, 487		
Nimrod point	- 157		
——— rock	142, 226, 229, 230	Oar channel	303, 306
——— sound	152, 154, 157	——— reef	- 303
Nine-feet patch	65, 131	Obman bay -	- 417
——— reef	- 123	Obree island	- 351
Nine islands	20, 38	Observation island -	- 383
Ninepin group	64, 65, 66, 70, 165	Observatory island 219, 220, 221, 264, 404
——— island	64, 150	——— point	- 285
——— rock 64, 65, 66, 69, 112, 113, 150, 395		Ockseu islands	- 114
		Oda -	- 348
Ning-hai -	241, 242	Oeste rock -	- 90
Ning-hau river	- 152	Oga sima peninsula -	- 347
Ning island	- 179	Ogle island -	- 326
Ning-po 9, 157, 158, 172, 176, 177, 181, 185, 186, 241		Ohosaka bay	- 337
		Oho saki -	- 358
——— river	157, 172	Oho sima -	325, 330, 342
Nin-le-been	- 138	Ohotake island	- 297
Nipon island, east coast	333–346	Oïe-haï-oïe harbour -	- 218
———, west coast	346–350	Oity lake -	- 337
Niu-chwang port	- 237, 240, 241	Okebets, cape	- 366
Niu-kung bay	- 127	Okhota river	- 419
Niupi-shan -	- 156	Okhotsk -	- 419
Nob rock -	- 164	———, sea of	417–420
Nobby reef -	- 73	Oki islands	- 349
Nomo, cape -	- 356	Okinawa sima	- 301
North Bashi rocks -	- 277	Okosiri island	- 363
——— bay -	173, 416	Olga bay -	- 397
——— breakers	- 130	Oliphant island	211, 216
——— channel	- 406	Olivutsa rocks	- 379
——— Cone rock -	- 73	Omae saki -	- 342
——— East island	- 165	Omega bank	- 110
——— Foreland islet	- 149	Omuru rock	- 327
——— Gau rock	- 67	Onanga sima	- 328
——— island -	128, 188, 274, 277, 278	One-foot rock	- 64
——— Ninepin rock	64, 65	Onekotan island	- 369
——— rock	- 64, 222, 223	Onnodake, mount	- 303
——— White rock -	- 24	Onon river -	- 407
——— Yit islet	- 117	Onting, port	- 308
Norton rock	- 121	Oon-sah -	- 307
Nose islet -	- 150	Opasnost rock	- 409
——— point -	- 264	Opium point	- 88
Nossyab, cape	- 409	Orange island	- 274
Nossyam, cape	- 366	Organ island	- 129
Notch island	- 97	Oroté point	- 313
Notches islets	- 159	Osborn reach	- 208
Noto, cape -	- 349	Ota rock -	110, 111
Notoro, cape	409, 410	Otméloi, cape	- 415
Nut island -	- 29	Otsisi, cape	- 405

INDEX.

	Page
Ousha island	162, 163, 184, 195
Outer knoll	131
—— Min reef	130
—— passage	40
Ow-chau islets	51
Owick bay	92
—— point	92
Oyster island	405
—— islet and rock	107
Pa-chau island	124
Pa-chung-san	297
Padaran, cape	2, 6, 7, 8
Pago harbour	312
Pagon island	319
Pagoda bay	89
—— hill	86, 87
—— island	74, 88, 91, 93, 94
—— islet	108
—— range	84
—— rock	133
Pa-ho reach	213
Pai rock	161
Paicpouc bay	312
Pak-kong-ho bay	293
Pakington reach	214
Pakleak island	22, 23, 24, 44
Paksa point	286
Pak-tang island	16
Pak-tsim island	28, 29, 30
Palabi island	269, 270
Palawan, coast of	1, 3, 7, 8
—— passage	270
Pallas bay	403
—— rocks	354, 355
Palm island	289
Pangpeto reef	143
Panuetan islet	272
Papenberg island	356
Paramushir island	369
Paracel islands	7
Parece Vela reef	311
Parker island	157
—— islands	182, 355
Parkyns rock	10, 84
Parry isles	321
Parseval group	257
Pass island	119, 133
—— islands	94
Passage island	22, 109, 110, 111

	Page
Passage islands	116
Pastel rock	135
Pas-yew island	185, 187, 188
Pata point	268, 269
Patahecock islet	154, 155
Patera island	17, 19
Patience bay and cape	413
Pating island	170
Patua point	392
Patung island	36, 37, 43
Pauk Piah rock	74, 76
Paukshao bay	80
—— point	79, 80
Paushan point	197, 198
—— town	198
Peak island	163
—— islet	187, 259, 260
—— rock	70, 71
Peaked rock	29
—— bay	175, 178
Pe-chi li gulf	12, 13, 225, 230, 231–236, 241, 262
——, south coast	231
Pedra Arecka	20
Pedra-mea rock	20
Pedro Blanco rock	76
Peel island	321, 322
Pehoe island	127, 128, 129
Peh-tang ho	232, 246
Pei ho	230, 232, 233–236, 237, 247
Peking	242, 243
Pe-kyau point	137
Pelées islands	393, 395
Pelican rock	164
Pélissier, cape	382
Pellion island	256
Penaud island	257
Penetration Pass	147
Peng-chau island	69
Pennell point	181
Périer point	384
Perry island	344
Pescadores islands	6, 7, 122–129, 283, 284
Peschurof, cape	382
Peshan island	146, 147
—— islet	179
Pe-ta-oa bay	286
Peter, St., island	328
Pe-ting island	123
Petit Thouars, cape	383, 389
Peton point and village	285

	Page
Petropaulski harbour	371–375
Pheasant point	198, 199
Philippine islands	9
Pic de Langle	409
Pico channel	366
Pih-hu shan	214
Pih-ki-shan islands	12, 143, 144
Pih-Kuen islands	121
Pih-lon island	163, 167
Phi pass	141
Pih-quan harbour	6, 141, 142
—— peak	141
Pih-seang islands	139, 140
Pih-sha island	165
Pih-sin chau	206
Pihting islet	164
Pilier rock	392
Pillar rock	119, 164
Pillars islet	72
——, the	71, 72, 207, 208
Pine Cone island	150
Ping point	114
—— rock	114
Ping-fong island	141
Pinghai	114
—— bay	113, 114
—— harbour	381
Pingyang point	142
Pinnacle island	295, 303, 326
—— point	373, 405
—— reef	373
Pinnacles group	265
Pio Quinto islet	273
—————— port	273
Pique bay	400
Pirate bay	183
Pisai island	110, 111
Pitew point	116
Plat island	90
Ploughman group	156
Plover cove	68
—— point	203
Plymouth rocks	344
Pocking-han island	21
Pogobi, cape	400
Pointed rock	323
Ponafidin island	328
Ponghou archipelago	122–129
—————— harbour	126, 127
—————— island	122, 123, 125, 126, 129
Pong-li	280, 281

[C.]

	Page
Popof mount	382
Port island	67, 68
Portsmouth breakers	330
Porzic point	393
Posiette harbour	379, 392, 393
Positions, table of	422–430
Potoe island	14, 15, 21, 22, 44, 46
Pou-no islet	179
Poun-tin island	30
Pou-ti islet	179
Povorotnoï, cape	370, 372
Powhattan bay	345
—— reef	327
Poyang lake	211
Pratas island and reef	266–268
Pratt rock	51, 52
Prevost peak	368
Prince Imperial archipelago	253
—— island	329
—— Jérôme gulf	252, 253
Prongé, cape	406, 407, 408
Providence reef	298
Puffin island	161
Pu ho	243, 244
Pulo Aor	8, 41
—— Obi	6
—— Sapata	1, 2, 3, 40
Pumice-stone bay	21
Putoy island	22, 30, 32, 42, 43
Putu island	164, 165, 176, 177, 183, 195
Pwanche island	167
Pwan-peen island	143, 144
Pyramid point	11, 111, 112
—— rock	65, 112
Quadra island	92
Quang-ta islands	146
Quang-tung province	55, 62, 91
Quantao shoal	133
Quarry island	152
Quar-see-kau bay	291
Quelpart island	262, 264
Quemoy bank	105
——, island	6, 101, 105, 106, 107, 108
——, point, pagodas on	106
—— spit	103, 105
Quoin island	227, 230
—— point	404
Raffles island	182, 183
Rag islands	138

F F

450 INDEX.

	Page
Ragged island	- - - 129
—— point	- - 135, 137, 138
Raikoké island	- ‥ - - 369
Rakovya harbour	- - 371, 374
—— shoal	- - 371, 374
Raleigh rock	- 24, 44, 296, 304
Rankin point	- - - - 308
Rasa island -	- - - 310
Rashau island	- - - - 368
Ratmanof, cape	- - - 414
Rebuntsiriboi island	- - - 368
Red bay -	- 7, 11, 96, 97, 98
—— Cliff island	- - - - 395
—— cliffs -	- ‥ - - 359
—— islet -	- - - - 72
—— point -	‐ - - . 36
Redang islands	- - - - 41
Redfield rocks	- - - - 329
Red rock -	- - - - 120
—— Yit island	- - - - 117
Reef island -	- - 117, 123, 134
—— islands, 78, 79, 123, 150, 302, 304,	
	305
—— islet -	- - - - 72
Rees islands	- - 11, 94, 95
—— pass -	- - 94, 96, 98
—— rock -	- 94, 131, 132, 134
Refun-siri -	- - - - 409
Regents Sword	- - - - 225
Retribution rocks	- ‐ - - 355
Richards island	- - - - 351
Richardson island	- - - - 262
Rickord, cape	- - - - 367
Rijutan islet	- - - - 271
Rimnik, cape	- - - - 414
Riposet, mount	- - - - 276
Ri-siri -	- - - - 409
Ritidian point	- - - - 314
River islet -	- - - - 395
Robben island	- - - - 414
Roberts pass	- 160, 161, 183, 184
Roche Percée	- - - - 260
—— Poncié islands	- - - 355
Rock islands	- - - 340, 342
Rocky harbour	- - 64, 65, 66, 69
—— head point -	- - - 84
—— point	- - 84, 242, 243
Rogues point	- - - 112, 113
Romanzov bay	- - - - 410
Rondo islet -	- - - - 173
Rosario island	- - - - 323

	Page
Roselskoï volcano	- - - 372
Ross head -	‥ - - - 75
—— island -	- - - - 15
Rota island -	- - - - 314
Rouge, cape	- - - - 388
Round hill -	- - - 79, 80
—— island, 31, 68, 72, 73, 129, 131, 132,	
	170, 219, 221
Roussin island	- - 256, 257
Roundabout island -	168, 170, 184
Rover group	‐ - - - 124
—— island	- - - - 210
Rover Knob cliff	- ‐ - - 124
Rowan islands	- - - - 114
Ruff Rock -	- - - - 90
Rugged islands	- - - - 116
—— islands 180, 181, 183, 190, 191	
—— rock	- - - - 29
—— point	- - - - 294
Sable island	- - 128, 129
Sabatan island	- 274, 275, 276
Saddle group	- - 181, 182
—— hill -	- - - - 232
—— island 113, 154, 196, 239, 241, 260	
—— islands 181, 182, 183, 192, 194,	
	195, 196, 197
—— mountain	- - - 360
—— peak	- - - - 91
Saddle-shape mountain	- - 382
Sado island -	- - - - 348
Sagami, cape	- - - - 343
Saghalin channel	- -. - 406
—— gulf	- - - - 405
—— island	- - 412–417
—— , east coast	- - 413
—— , north coast	- _ 416
—— , south coast	- - 411
—— , west coast	- - 401
Saghalinula	- - - - 407
Sah-lo-wung bay and village	- 36
Sai-nam -	- - - - 60
Sai-wan passage	- - - - 59
—— bank	- - - - 59
Sakara island	- - - - 336
Sakura point	- - - - 343
Salmon cove	- - - - 411
Saloupa river	- - - - 313
Salutation bay	- - - - 417

	Page		Page
Samarang island	- - - 326	Saratoga spit - - 344, 345
—— patch	- - - 270	Sarigaun island - - - 318
—— rock	- - - 263	Sarpan island - - - 314
Samasana island	- - - 294	Sasagota bay - - - 358
Sam-chau inlet	- - - 75	Sau-o-bay - - - - 292
Samcock island	- - 19, 26	—— reef - - - - 293
Samoun group	- - 25, 44	Sawa-umi bay - - - 348
Sam-pan-chau islet -	38, 40, 48, 49, 51,	Saw-chau island - - 36, 37, 38
	52, 53, 56	Saw-shan island - - - 174
Sampson peak	- - 247, 248, 249	Saw-shee hill - - - 54
Sam-sah bay	- ,, - 11	Scattered Yits islets - - 117
—— inlet -	- - 137, 138	Schlippenback, cape - - 389
Samun islets	- ,, 70	Schwartz, cape - - - 389
San Alessandro island	- - 320	Scout rock - - - 112, 133, 134
San Augustino island	- - 320	Scrag point - - - 106, 107, 108
San Carlos - -	- 275, 276	Sea Cat rock - - - 134
San Domingo bay -	- 275, 276	Sea Dog rock - - - 134
San Luis D'Apra, port	- 312, 313	Sea of Japan - - - 376
San Pio Quinto port	- - 273	—— Okhotsk - - - 417
San Vicente port -	- 269, 270	S.E. passage - - 162, 183
Sanchesan island -	- - 152	Seao-Seao island - - - 153
San-chau island	- 14, 15, 17, 46	Seao-tan island - - 100, 101
Sand island	- - 117, 128	Seaon-ken islamd - - - 167
—— patch rock -	- - 73	Seaou-yew island - - - 187
—— peak - -	- 121, 131, 132	Scatoi bank - 109, 110, 111
—— point -	- 131, 132	—— island - 109, 110, 111
Sandon rock	- - 326	Seau-sha island - - - 205
Sandy island	- - 299, 304	Sebastian Lobos islands - - 320
—— point	- - 395, 396	Second bar - - - 54, 57
Sang-kau bay	- - - 218	——, creek - - - 54
Sang-tau island	- - - 232	——, hills - - - 59
San-mun bay	151, 152, 153, 155, 157	—— pagoda - - - 54, 57
—— island	- - - 150	Sedlovaya mountain - - 382
San-pwan pass	- - - 144	Sedogawa river - - - 337
—— islands	- - 144, 145	Senhora de Penhos church - 20, 21
San shan islet	- - 185, 206	Senhouse island - - - 182
—— islands -	- - 248, 249	Seniavin, cape - - ◄ 413
San Shan tau island	- 247, 248	Sentinel island - - - 378
—— channel	- - 248	Sentry island - - - 117
San-shi islet	- - - 147	—— rock - - 223, 224
San-shui junction -	- - 60	Seoluk islands - - - 146
Sanson island	- - - 232	Séoul - - - 253, 387
Santa Crux fort	- - 314	Seou-ping-tao-bay - - - 247
—— Rosa shoal -	- - 313	Seriphos rock - - - 327
Saracen head	- - 282, 283	Serrated peak - - 132, 133
—— rock	- - - 264	Seshan islands - - 189, 190, 191
Sarah Galley channel	158, 162, 163, 176,	Sesostris rock - - - 187
	184, 195	Setei island - - - - 307
Sarah island	- 169, 170, 184	Seven Sisters group - - 189
—— Lucy rock	- - - 80	—— Stars islets - - 61, 141
Saraki, cape	- 360, 362, 364	Sewshan islet - - - 174

F F 2

	Page
Seymour bay	297
—— reach	211, 212
Seypan island	316–318
Shaaon harbour	166, 173, 174
Shackoff point	374
Shag island	161
—— rock	82, 83
Shah bay	308
Sha-ho island	151
Shallow bay	91, 92, 242
Sha-lul-tien banks	233, 234, 236, 245
—————— island	233, 235
Sha-mo island	226, 230
Shang-hai	4, 9, 191, 194, 199, 200, 201
Shang rock	148
Shang-ta island	148, 149
Shantar, Great, island	418
Shantarski islands	417
Shantau	86
—— estuary	84, 85
Shan Tung peninsula	218, 222, 236, 240
—————— promontory	217, 218, 220, 231, 247
—————— province	219, 248
Shao King	60, 61, 62
Sharp island	24, 65
—— peak	131, 132
—— peak	64, 66, 131
—— island	134
—— point	132, 134
—— Point bluff	132
—— Shoulder	131, 132
Shaweishan island	192, 195, 196
Shei-luh channel	177
Sheipu	152, 153
—— harbour	153, 154
—— road	152
Shekpywan harbour	31
Shelter bay	64, 399
—— island	64, 65
—— port	64, 65
Sheppey island	174
Shestoi strait	369
Shetung islet	147
Shendi hill	303
Shiash-kotan island	369
Shih-wuy-aon	212
Shilka river	407
Shingan island	120
Shingshimún pass	35
Ship point	228, 230

	Page
Shirinki island	369
Shoal bay	77, 88, 89
—— gulf	251, 252, 253, 254–258
—— island	72
—— point	242, 264, 415
Show islands	180
Shroud islet	144
Shumshu island	370
Shun reef	96
Shun-tak branch	60
Shwin-gan river	142, 146
Siah-kia-kau	211
Siam, gulf of	1, 2, 7, 8, 40, 41
Siau-Chu-shan, or Little Orphan island,	210–230
Siau-hi-shan island	227, 228, 230
Siau head	127
Siau Wuhu bay	397
Siayan island	274, 277, 278
Si-chi-tau channel	59
Side Saddle islets	181
Si-ki islet	79, 80
Si kiang or West River	17, 58–62
Sikini sima	329
Sikok island	333
Silver island	172, 184, 205, 206
Simago group	326
Simoda harbour	334, 340–342
—— road	340
Simplicia Wreck rocks	95
Simusir island	368
Sin island	153
Singapore	1
—————— strait	3, 7
Single island	70, 71
—— peak	287
—— rock	76
—— tree	202, 203
Sing-lo-san island	160
Sinkong point	171, 173
Sinta rock	90
Sir James Hall group	254
Sirahama bay	340
Sirakami, cape	362, 364
Siretoko, cape	411
Siriya saki	359
Sirofama, cape	346
Sisdro point	390, 391
Sisters islets	93
Sisuro peninsula	390
—— village	390

INDEX. 453

	Page
Si-ting island	75, 76, 77
Siwokubi, cape	360, 362
Six Feet rock	132
Skead islet	95, 180, 181
Sloping point	224
Slut island	119, 120
Small bar	54, 57
Smith island	328
Snares islets	369
Society bay	237
So-co hill	285
Soimonof, cape	413
Soko islands	28, 36, 43
Solitary rock	140
Songari river	407
Song-men	147
———— point	147
Sonson bay	275
Sorrel rock	113, 114
Sossan Hagno	315
Sossan Haya	315
Sotonohirase rocks	357
Soudan island	147, 148
Sour islet	108
South bank	131
———— bay	11, 89
———— cape of Formosa	277, 279
———— channel	304
———— chop-house	54, 58
———— Chukea island	163
———— Dog island	121
———— east island	95, 222, 223
———— Gau island	66, 69
———— isle	328, 405
———— Merope shoal	103
———— Ninepin rock	64, 76
———— reef	118
———— rock	299
———— shore bank	193
———— White rock	24
———— Yit island	11, 114, 115
Southampton rock	341
Southern head	322
Soya, cape	409, 410
Spanberg island	366
Spek strait	353
Spenberg mount	413
Sphinx rock	153
Spider island	136, 137, 138
Spire islet	96, 183
Spit point	225, 228

	Page
Spiteful island	133
———— rock	133
Spithead anchorage	168
Squall islands	148, 149
Square island	92, 185
———— rock	112, 322
———— point	350
———— rocks	137
———— Stone islet	168
Squat rock	85, 86
St. Andrew island	159
St. Clair island	328
St. François Xavier island	326
St. George island	152
St. Jona island	419
St. John island	7, 45, 46
St. Peter island	328
Sta. Rosa mount	276
St. Vladimir bay	398
Stanitski point	373
Stapleton island	321
Starboard Jack rock	155, 156
Starling island	204
Staten island	366
Station island	117, 118, 119
Staunton island	218
Steep island	63, 64, 292
———— rock	138
Steeple island	123
Steward rock	172, 173
Stewarts house	5, 88
Stewart, port	396
Stickup rock	223
Stolovaya mount	389
Stone-cutters island	37, 41
Stragglers islets	147
Strawstrack islets	140
Su-chau creek	200
Suco island	307
Suffren bay	402
———— cape	402
Sugar Loaf island	73, 85, 86, 147, 308
Sui-chan island	49
Sul rock	90
Sulphur islands	320, 324
———— reach	58
Sungseu bay	104
Sunharom	316
Sun-kong island	32, 33, 34, 35
Supply rock	341
Surface currents	419

INDEX.

	Page
Susaki, bay	346
—— cape	346
—— village	340
Susquehanna bay	345
Suwa sima	326
Suwo Nada	333, 376
—— sea	333
Swatau	86
Sybille bay	399
—— head	403
Sylock island	18, 19
—— islet	26
Symplegades island	355
Table chain	252, 257
—— hill	305
—— island	65, 125
—— islet	64, 65
—— mountain	255, 256
Tablet island	125, 126
Tabu sima	347
Ta-chen island	176
Ta-ching point	168
Ta Chü-shan island	227, 229
Tae islands	140, 141, 142
Tae-pan point	99, 103
—— shoal	100
Tae-pih islands	143
Tae-shan channel	178, 179
—— island	175, 177, 179, 180, 181, 195
Tae-tan island	99, 100, 101, 103, 105
Taewang rock	16
Tafou island	151
—— peak	151
Tagau point	96
Tago bay	339
Taheen rock	110, 111
Ta Ili-shan island	225, 227, 228, 230
Tahkut island	109
Tai-chau	149
—— bay	149
—— islands	148
—— river	149
Tai-cho ho	243
Tai-lung channel	59
—— Hill pagoda	59
—— rocks	59
Tai-ping	208
—— pagoda	207

	Page
Taï pin-san island	297, 298
Tai-ta-mi channel	25, 29, 30, 44
—— island	30
Tai-tzu-chi islet	209
Taï-wan fu	283, 284, 285
Taï-wan island	279-296
Tajo river	269
Ta-kan island	163, 177
Taka sima	354
Ta-kau	281
Ta-kau-kon, port of	281, 282
Take sima	327
Takeu island	169, 170, 171
Tak-hien	62
Takin island	229, 230, 231
Taking island	309
Tako sima	379
Ta-lien-hwan bay	247, 248, 249
Taluk island	146, 147
Talung island	187, 188, 189
Tamanu	301
Ta-maou island	167
Tamchau hills	59
—— passage	59
Tam-kan island	29, 30, 45
Tam-sui harbour	286, 288
—— mountains	285
Tam-tu island	5, 63, 64, 70
Tan point	118
—— rocks	118
Tanega sima	327
Tang-fau island	163
Tang rocks	16
Tang-tu	207
Tanto-shan island	150
Tanue bay	142
—— point	142
Tao Kwang mount	389
Tao-sao islet	100
Taotau	169
Taou-hwa island	162, 163
Ta-ou island	146
Taou-sau-mun channel	161
Ta-outse harbour	6, 172, 185
—— island	172
Taow-pung island	147
Taping island	172, 185
—— point	185
Tarara islet	297, 300, 301
Tareinski harbour	371
Tarofofo harbour	312

INDEX. 455

	Page
Tarofofo river	312
Tartary, coast of	266
———, gulf of	400–406
———, strait of	400, 406
Ta-san ho	234
Tas-de-Foin islet	251, 256
Ta-shih-tau harbour	218
Tathong channel	42, 63
——— point	63
——— rock	63
Tatoi island	109, 110, 111
Tatsang island	127
Tatsupi saki	358
Tau-tau point	180
Tau-tew point	153
Tavano, port	367
Tawu island	167
Taya islands	5, 41
Ta-yew	187
Tayung island	183
Tea island	167, 168, 169, 171, 184
Tebakh, cape	407
Teen islet	163
Tegnell	18
Teih-mei-heen or Black ink city	217
Teih-kiang	208
Teijo island	167
Tein-tung	185
Tei-tzuchi islet	211
Tei-yat-kok	17
Temple bay	230
——— island	227, 228
——— point	130, 131, 132, 225, 230
——— rock	132
Ten-fathom hole	623
Teng-chau	224
——— fu	225, 227, 231, 233
——— head	224, 225
——— point	230
Ternate rock	91
Tessara islands	120, 121
Teukcham port and village	286
Teyih point	70, 71
Thalia bank	107, 108
Theodolite point	206
Thornton island	175
Thistle island	259
Three Chimney bluff	90
——————— point	11
——— Fathoms patch	65
——— Feet patch	65, 66

	Page
Three Gates, or Samoun group	25
——— island	124, 129
——— trees	193
——— Hills island	328
Throat gates	37
Thunder head	93, 94
Tiara mount	414
Ti-a-usu island	295
Tide cove	8, 69
——— point	85
Tien-pak island	41
Tien-tsin	234
——— ho	233–236
Tiger island	48, 52, 53, 54, 56, 57, 59
Tigers claw	53
——— Head entrance	50
——— Tail rock	187
Tikmenef point	381
Ting-hai	166, 169, 170, 171
——— bay	11, 136, 137, 138
——— point	137
——— harbour	6, 12, 165, 171, 177, 183, 185, 186
Tingtae bay	11, 98
Tinian island	315
Tinker rock	155, 156
Tintao	133
Tinwan island	152
Toa-sik-tau harbour	218
Tok sima	324
Tokara sima	326
To-ki island	226, 227, 228, 229, 230, 231, 234
Tollatock river	281
Tolo channel	68
——— harbour	68
Tomb point	53, 56
Tongbu	109, 111, 112
Tong-ho island	22, 23, 44
Tong King, Gulf of	1, 2, 3, 6, 7, 40
Tong-ku harbour	37, 38
———, island	37, 49
Tong-lae	84
——— point	83, 84
Tongmi point	79, 80, 81
Tongsang basin	94
——— harbour	5, 9, 93, 94, 96, 98, 290
Tong-sha island	121, 122, 131, 134
Tongting islet	164, 165, 195
Tongue shoal	133
Tongyung	93, 94

	Page
Tonin, cape	413
Toriwi saki	359
To sima	330
Tortoise head	29
——— rock	128
Totomo-siri island	409
Tougouéne point	313
Toumon bay	314
Touron, cape	41
Towan island	146, 147
Tower Hill channel	168, 469, 184
——— island	12, 167, 168, 169, 184, 186
Tower rock	119
Towling flat	53, 56
——— island	56
Town island	66, 139
Treaty point	343
Treble islands	157
Tree-a-top island	73, 159
Tree island	49, 50
Triangle head	131
——— peak	264
Triangle-yit island	117
Triangles islands	185
Trifanof point	386
Trio islets	63, 64
——— rocks	135, 328
Triple island	73, 75, 152
Tripoint island	183
Trite island	118
Tronson point	403
Trumball island	167, 168, 170, 171
Trunk point	163
Truro shoal	268
Tsae island	175, 176
Tsai-shih-ki	207
Tsang-chau island	73, 74, 75
Tsang islets	142
Tsau-hia island	206
Tsau-liang-hai harbour	380
Tseenshan	190
Tseigh islands	143, 144
Tse le island	185, 187, 188, 190
Tsiang island	124
Tsiech point	78
Tsieching	78
Tsie-kie river	186
Tsien-tang estuary	188
Tsih-sing group	139
Tsih-tze island	172

	Page
Tst mi wan island	28, 29
Tsincoe island	74, 75
Tsing-hing river	390
Tsingluy Tau	170
Tsing-seu island	99, 100, 103
Tsiuka, cape	361, 362, 364
Tsugar cape	358, 364
——— strait	358–364, 370
———, directions from Yedo	346
Tsuki point	362
Tsukurase islands	355
Tsung-ming island	191, 202
Tsus sima	378
——— sound	378
Tullo island	403
Tumen river	390, 391
Tung islet	170
Tung kiang	165
Tungao road and village	82, 83
Tungchuh island	149, 150, 151
Tung-chung bay and village	36
Tung-ju	157
Tung-ki islet	79, 80
Tunk-ku	234, 235
Tung liu	210
——— reach	210
Tung lung island	63
Tungmum island	153
Tungpwan island	143, 144
Tungsha banks	192, 193, 194, 196, 197
——— island	135
Tung-ting island	76, 77
Tung-ying island	11, 136
Tuni-ang group	70, 71
——— island	69, 70, 71, 72, 75
Turnabout island	118, 121, 122
Turret island	143
Turtle rock	81
Tu sima	304
Tutomi gulf	338, 339
Twins islets	72, 73, 151
Twkaschi island	302
Ty-cock-tau fort	52
——— island	39, 40, 50, 56
Ty-fu island	53
Tygosan island	168, 185, 186
Ty-ho island	35–37
——— village	35
Tylo island	24, 26, 44
Tylock islet	26
Tylong head	34, 35, 63

INDEX. 457

	Page
Ty-lou island	14
Typa anchorage	14, 17, 18, 19, 20
—— island	18, 19, 20
Typhoou harbours	5, 6
Typhoons	3–5, 21, 271
————, Bonin islands	332
————, in China Sea	3–5, 45
————, Meiaco sima group	301, 304
Typung	72
—— bay	69, 75
—— harbour	71, 72, 73, 75
Ty-sami inlet	5, 78
—— mound	76, 78
Tytam bay	34, 35, 42
—— harbour	34, 35
—— head	34
—— village	34, 35
Tzee islands	302
Tzko-kang	213
U'cona rocks	340, 341
Udsi sima	355
Umata bay	313
—— river	313
Ung lo hill	285
Ung-shan islands	151
Unkofsky bay	381
Uraccas islands	320
Uragu channel	343
Urh Tao island	218
Urmston bay	37, 38
Urup island	367
Urusof point	390
Ushishir islands	368
U sima	363
Usúri river	407
Utoma island	330
Vachofski peninsula	398
Van Diemen strait	336
Vandalia bluff	340
Vangan pagoda	117
—— point	117
—— inlet	116, 117
Vashon, cape	380
Vele-Rete rocks	277, 279
Vernon channel	158, 161, 162, 183, 184
—— island	161, 162
—— point	161
Victoria bay	33, 248, 249, 397
—— peak	33

	Page
Victoria town	33
Video island	165, 177, 178, 195
Village bay	121, 221, 223
Villeuchinski mount	372
Vincennes bay	325
Vincente island	269, 275
Virginie mount	395
—— bay	385
Virst, cape	415
Vishnevski island	385
Vixen spit	133
Vladimir, St., bay	398
Vneshni, cape	418
Volcano bay	365
—— islands	321, 326, 327
—— islands,	178, 179, 180, 183, 190, 191, 320
Vries island	330, 342
—— strait	367
Vuyloy shoal	282
Wade island	207
Wac-wu channel	169
—— island	167, 169
Wag-lan island	33, 34, 35
Wai-ian-do island	256
Walker bay	321
Walvis islands	366
Wan-chu	142, 145
—— island	145
—— river	144, 145
Wangchi island	153
Wanki bay	137
Wan-tong islands	49, 50, 52, 53, 56, 57
—— rock	52
—— tower	57
Ward reach	212
Warning rocks	120
Washington reach	213
Watchful sand	62
Wateo island	171, 172
Water island	26, 27
—— islands	15, 17
Watson island	58, 102, 104
Waywoda rock	379
Webster island	345
Weï-haï-weï	219, 220, 222
—— barbour	218, 220, 221, 222, 223
West Blonde channel	201, 202
—— Entry point	248
—— hill	10

[C.]

G G

458 INDEX.

	Page
West peak	106
—— point	284
—— river	17, 58–62
—— Stork islet	189
—— Tiger hill	214
—— Water island	26
—— White stone	81
Whale-back hill	281
Whale rocks	76, 392
Whampoa channel	17, 18, 54, 55, 58
—— pagoda	54, 57, 58
—— island	54, 55, 58
Whang-hai or Yellow sea	217–225
Whang head	164, 175, 176
Whang ho or Yellow river	217
Whelps islets	155
White bluff	133
—— Cliff bay	395
—— Cliffs island	390
—— Dog islands	11, 121, 122, 130, 131, 134, 135
—— fort	133
—— head	68
—— island	116, 117, 121
—— rock	37, 38, 39, 41, 49, 50, 89, 108, 221
—— rocks	24, 107, 109, 111
Wild Boar reach	208, 209
Wilder bay	396
—— passage	59
Wilson island	351
Winchester mount	395
—— point	351
Winds, Bonin islands	331
——, Gulf of Liau-tung	236
——, Tartary	401
——, Japan islands	334
——, Kuril islands	370
——, Lu-chu islands	331
——, Sea of Japan	376
——, Okhotsk	420
——, Shanghai	200
Wodawara bay	343
Woga creek	132, 134
—— island	131, 132, 134
—— point	132
Wokeu islands	114
Wolf bay	164
Wolverine rock	132
Won-chu-chau island	37, 41
Wong-mou island	22

	Page
Wood hill	303, 305
—— point	306
Woodcock rock	59
Wo sima	354
Wonfou island	121, 130, 132
Wou-hou creek	179
Wu-an island	99
Wuchang fu	214
Wu-chang-hien	213
Wu-chu fu	61, 62
Wu-chu	61, 62
Wuhiutsun	212
Wu-hu	207, 208
—— reach	207
Wukido island	324
Wung Cum island	14
Wu-seu island	99
Wusung	191, 193, 194, 195, 198, 199, 216
—— bar above	194
—— river	191, 194, 198, 201, 202
—— spit	197
Wusung to Han-kau	201–216
Wyllie rocks	272
Yablonoi Krebit mountain	407
Yakai island	302
Yakimu, cape	304
Yakuno sima	327
Y'Ami island	274, 277
Yangho	243
Yang-ki	213
Yanglo	214
—— point	171
Yangsi islet	164
Yang-tse cape	191, 192
Yang-tse kiang	9, 12, 177, 181, 190, 191–216, 217
Yat-moun channel	29, 30, 31
Ye-chau channel	30
—— island	29, 30
Yedo bay	343–345
—— gulf, islands off	328–332
Yeirabu sima	324
Yellow river	217
—— sea	217–225, 231, 246–250
—— stone	80, 81
Yeng-rock	90
Yen-tai	221
Yen-tse-ki	206
Yerabu sima	327

	Page		Page
Yer-ra-bu island	- 297	Yori sima	- 324
Yesan, cape	- 360	Young, cape	- 380
Yetomo, cape	- 365	Younoi head	- 133
Yew islands	185, 187	Yuet-shing-reach	- 62
Yey-van bay	- 147	Yung-hing bay	382, 384
Yezo island	333, 365, 410	Yung islet	- 163
—— strait	- 366	—— river	172, 185–188, 190, 195
Yih bluff	- 63	Yungning islet	- 108
Yih-pan island	- 125	—— point	- 108
Yin-gar island	- 177	Yun-kai-tau rock	- 60
Yits	114, 117	—— village	59, 60
Yki island	351, 352	Yutsi sima	- 348
Ykima island	298, 299	Yuyao branch	- 186
Y-ki-mah island	- 297		
Ykitsk island	352, 353		
Ynarahan	- 312	Zamami island	- 302
—— bay	- 312	Zam-chau	- 18
Yoko sima	- 326	Zealandia breakers	- 318
Yokuhama bay	343, 345	Zelandia fort	281–285

www.ingramcontent.com/pod-product-compliance
Lightning Source LLC
Chambersburg PA
CBHW022112300426
44117CB00007B/679